59th Meeting of the Society for Machinery Failure Prevention Technology (MFPT 2005)

Essential Technologies for Successful Prognostics

Virginia Beach, Virginia, USA
18-21 April 2005

ISBN: 978-1-5108-8841-8

Printed from e-media with permission by:

Curran Associates, Inc.
57 Morehouse Lane
Red Hook, NY 12571

CURRAN ASSOCIATES INC.
proceedings.com

Some format issues inherent in the e-media version may also appear in this print version.

Printed by Curran Associates, Inc. (2019)

For permission requests, please contact Society for Machinery Failure Prevention Technology
at the address below.

MFPT Society
715 Twining Road
Suite 135
Dresher PA 19025

Phone: (215) 346-2870
Fax: (215) 352-6902

https://mfpt.org

Additional copies of this publication are available from:

Curran Associates, Inc.
57 Morehouse Lane
Red Hook, NY 12571 USA
Phone: 845-758-0400
Fax: 845-758-2633
Email: curran@proceedings.com
Web: www.proceedings.com

ESSENTIAL TECHNOLOGIES FOR SUCCESSFUL PROGNOSTICS

Proceedings of

The 59th Meeting

of the

SOCIETY FOR MACHINERY FAILURE PREVENTION TECHNOLOGY

VIRGINIA BEACH, VIRGINIA
April 18-21, 2005

Compiled by

HENRY C. and SALLIE C. PUSEY
and
W. R. HOBBS

Publication of the

Society for Machinery Failure Prevention Technology (MFPT)
(A Division of the Vibration Institute)

Special Notice
The U. S. Government retains a nonexclusive, royalty-free license to publish or reproduce, or allow others to publish or reproduce, the published forms of any papers in these proceedings authored by a government agency or a contractor to a government agency whenever such publication of reproduction is for U.S. government purposes.

TABLE OF CONTENTS

PREFACE

The 59th Meeting of the Society for Machinery Failure Prevention Technology (MFPT) was held April 18-21, 2005 in Virginia Beach, Virginia. MFPT sponsors include the Office of Naval Research, U.S. Army Research Laboratory, National Aeronautics and Space Administration and the Naval Surface Warfare Center. The cooperation of the Shock & Vibration Information Analysis Center (SAVIAC), the Society of Tribologists & Lubrication Engineers (STLE) and the Society of Experimental Mechanics (SEM) is acknowledged. Special thanks to **Mr. John V. Kelley** for organizing the session on Corrosion, to **Dr. Robert Bolton** for the Structural Health Management session and to **Dr. Fred Discenzo** for the special Session on Standards. **Mr. Paul L. Howard** of Paul L. Howard Enterprises was Technical Program Chairman and Chair of the Opening Plenary in which there were five outstanding Keynote Speakers.

The first three Keynote Speakers; **Mr. Paul L. Howard, Mr. William D. Marscher** and **Dr. Paul R. Muessig**; have published their papers, including contact information, in these proceedings. The other Plenary Keynoters were **Dr. Steven B. Chase** from the Federal Highway Administration who spoke on *Measuring Highway Bridge Performance*. We are grateful to Dr. Chase. This is his second invited presentation at our MFPT conferences. **Jack Kelley** from ARL and I both appreciate the Keynote Presentation of **Mr. Daniel J. Dunmire** from the Office of the Under Secretary of Defense who spoke on *DoD Corrosion Prevention and Control – Status and Update*

The rest of the conference consisted of four Technical Tracks with technical sessions as indicated in the Table of Contents and six embedded half-day tutorials. Sincere thanks for a fine job by the tutorial instructors. They were **John L. Frarey & Howard A Gaberson** on *Signal Analysis*, **David C. Zimmerman** on *Structural Health Management*, **James W. Taylor** on *Failure Analysis (FEMCA)*, **Gordon B. Hirschman & Thomas J. Walter** on *Wireless Systems*, **Alan K. Pride** on *Reliability Centered Maintenance* and **John E. Judd** on *Sensors Technology*.

There were twelve technical paper sessions, Panel Discussions on Tribology, Prognostics and Standards. If any sessions are missing some papers, no manuscript was available as these proceedings went to press. Those interested in unpublished papers should contact the author. I deeply appreciate the participation and support of all those on the Program.

I am grateful to all of our exhibitors. Their participation, both in the Exhibit Hall and in the New Products and Services Session contributes greatly to the success of the meeting. I thank my wife, Sallie, for her continuing outstanding management of all the conference details. Special thanks to our MFPT Society's Board of Directors and all the chairs, panelists and speakers who helped to make MFPT 59 another success story.

Henry C. Pusey
Executive Director
MFPT Society

Visit the MFPT Society Web Site

HTTP://www.mfpt.org

The Society's Board of Directors
Including Chairs of all Technical Committees
may be found there.

KEYNOTE PAPERS

Chair: Mr. Paul L. Howard
Paul L. Howard Enterprises

PROGNOSTIC TECHNOLOGY – NEW CHALLENGES

Paul L. Howard

Paul L. Howard Enterprises
149 Exeter Road, P.O. Box 362
Newmarket, NH 03857
plhoward@aol.com

Abstract: New Weapon System development programs have become highly reliant on the success of efforts to develop and integrate automated Diagnostic and Prognostic technology. These technologies are driving the implementation of new maintenance approaches and are central to achieving the reductions in manning and operating costs on these new systems. This has renewed the R&D focus on prognostic technology.

While prognostic solutions for some mechanical systems and electronic systems are emerging, they yet have to be verified and validated. Development of prognostics for structures, especially in the area of corrosion effects, presents its own set of challenges. There are new challenges in getting the right data to enable prognostics, training the maintainer to deal with failure prediction as opposed to failure detection, and applying the prognostic result in an operational scenario.

Emerging needs for thorough Verification and Validation (V&V) methods and independent "honest broker" technology evaluation must also be addressed to clear the way for the successful fielding of new systems with integrated prognostic capability.

This paper discusses the status of and current issues associated with prognostic technologies, sets some notional definitions, frameworks, and metrics for measuring future development progress and outlines some of the results of recent related workshops.

Key Words: Diagnostics; frameworks; metrics; prognostics; validation; verification.

Historical Background and the Emerging Need for Diagnostics and Prognostics: Why is there such a pressing interest in Diagnostics and Prognostics now? For decades the defense sector diagnostic technology market has been largely a "push" market with a few successes in niche corners of the market. The chief barrier has seemingly been the formidable "business case - cost benefit analysis", or colloquially "earning your way onto the platform". Oddly, a maintainer doing machinery diagnostics manually costs about $35/hour to add to the force, but seems to only produce a cost savings of $15/hour when he is removed from the force by automating that diagnostics role. Even if the personnel portion of the business case could be made, there is still the issue of actual equipment maintenance cost savings. While the annual cost of maintaining a given type of equipment is well documented in the DoD cost accounting records, the savings or cost avoidance (usually an estimated %) that a given diagnostic system could claim is usually argumentative. The issue of cost avoidance of lost missions afforded by diagnostics systems is also problematic to define. While the operational cost of performing missions can be captured, the cost penalty of an aborted mission seems to be hard to define. And then there is the (often well deserved) reputation that diagnostics enjoys. Predominating here are high false alarm rates and the dilemma of setting realistic limit values on diagnostic system readings that indicate the need for maintenance action.

So, what has caused the renaissance and largely changed the diagnostics market to a "pull" market? For one thing, in the present Defense Acquisition climate, new weapon systems only become economically acceptable if accompanied by promises of lower O&M costs, predominantly achieved by fewer operating and support personnel. This establishes the need for automation of a significant number of previously manual functions, many of them maintenance associated. Factor in the increasingly complex technology of new weapon systems, the new extended deployment models (such as Sea Basing), the higher operations tempo planned as well as the longer repair parts pipelines, and the need for automated diagnostics with early warning (prognostics) becomes clear.

Dimensions of the Challenge: Not only does the need become clear, but the diagnostic/prognostic system characteristics required to address these new challenges also become clear. In short, the diagnostics/prognostics technology suppliers need to solve the false alarm issue, provide higher reliability systems, and configure the data output to provide actionable and timely maintenance information to the part time maintainer and/or their distance support function. On the other side of the coin, the user community has obligations also. The development of a successful diagnostic/prognostic technology base will require data and a way to verify design performance. Specifically, the following types of data:

- Alpha test data sets and environmentally tagged field data sets to develop the detection and analysis algorithms to fit the needs of emerging equipment,
- Beta test data sets to validate the detection and analysis algorithms,
- Terms of reference and metrics for prognostics,
- A technology vetting and performance verification and validation capability/facility operated in an "honest broker", non-competitive environment.

Another dimension of the challenge is the scope of the diagnostics/prognostics applications present. The diagnostics community faces a fundamental focus change in the near future. In the past, diagnostics has largely been synonymous with fault detection in machinery and mechanical structures. The sensor industry has primarily developed around this focus and the premise that mechanical faults are the predominant source of maintenance cost and unanticipated failures are the main cause of downtime. The exponential increase in electronics content of weapon systems and the introduction of new structural materials that don't have SN curves analogous to those for aluminum have drastically altered the diagnostic and especially the prognostic requirements landscape. The diagnostic/prognostic technology base needed to address the new weapon system and support the emerging maintenance and logistics paradigm with its broader scope isn't there. The data sets needed to develop it, the data sets needed to prove it, the field operating data needed to confirm, and the capability to V&V it are not in place.

This is the challenge the diagnostics/prognostics community - developer and user - must address and solve for the new weapon systems to be effectively supported in the emerging warfighting environment.

One way to visualize the challenge might be the Diagnostic and Prognostic Technology Base Maturity Matrix shown in Figure 1 below which outlines the key needs areas and provides an estimate of the current state of maturity as noted by:

- "AP" (Technology Currently Available and Proven Effective),
- "A" (Technology Currently Available, But V&V Not Completed),
- "I" (Technology In process, but not completely ready for V&V), and
- "NO" (**NO** Significant Technology Development Effort In Place For This).

Note:
D = Estimated Diagnostic Technology Status
P = Estimated Prognostic Technology Status

Diagnostic/Prognostic Technology For:	AP	A	I	NO
Basic Machinery (motors, pumps, generators, etc.)	D		P	
Complex Machinery (Helicopter Gearboxes, etc.)		D	P	
Metal Structures	D		P	
Composite Structures			D&P	
Electronic Power Supplies (Low Power)		D	P	
Avionics and Controls Electronics	D		P	
Medium Power Electronics (Radar, etc.)		D		P
High Power Electronics (Electric Propulsion, etc)				D&P

Fig. 1 - Diagnostic and Prognostic Technology Base Maturity Matrix

Note that the "Ps" are all to the right of the "Ds" in this matrix. This is because the prognostic technology development follows the diagnostic and prognostic community axiom:

<u>Good Prognostics depends on Good Diagnostics</u>

Perhaps an even more fundamental challenge here is the lack of common definitions, terms of reference and metrics for "prognostics" itself. The term is currently used to describe everything from extrapolation of trend plots to statistical data models and physics based predictive models. The term has often been confused with data trending and equally poorly defined as a statistical probability of an occurrence. Therefore as part of a recent workshop on electronics prognostics, the following definition and terms of reference were developed to assure that the workshop presentations at least had a common frame of reference. They are offered here as a suggested starting point for an MFPT Society generated set of standard prognostic terms of reference to serve the prognostic technology developer and user communities.

A. Definitions:

Prognostics is a "forecast of future performance and / or condition".

Prognostic Engineering is therefore the discipline, body of knowledge, procedures, computation programs, and other resources, which supports Prognostics development and implementation.

Prognostic Technology is then the body of underlying science and mathematics, which supports Prognostic Engineering.

Prognostic Technique is a specific approach to Prognostics that may include development of specific features, feature vectors, models, statistical sets, etc. to be implemented in hardware / software and applied to a given component, subsystem, system, etc. These techniques may be broadly grouped in four classes:

- Those based upon statistical treatment of data or information.
- Those based upon a model of the application and the responses of the model to introduced faults or degradation.
- Those based upon direct sensor measurement, feature and feature vector development, and analysis and prediction based upon these parameters.
- A combination of any or all of the preceding three.

- The term "prognostics" may be applied at any level such as component, SRU/SRA, LRU/WRA, subsystem, system, etc.
- Prognostic Technology may include (at least) data based, statistical, and model based algorithms and analytical methods individually or in combination to develop a "forecast of future performance or condition". There are some "Terms of Reference" described below that may apply to and improve the understanding of these classes of methods.
- The data / information used to implement Prognostics may come from all relevant sources ranging from component raw material data to component fabrication and test data, through to system acceptance and fielded system maintenance historical data.
- Life Assessment and / or Remaining Useful Life may also be included in "prognostics" under the term "condition" in the definition. Thus Prognostics may include estimating the remaining useful life of a component, subsystem, etc.

- Inherent in the definition is the term "forecast". The parameters that may apply to better understanding the context of this term are provided in the "Terms of Reference" section below.

B. Terms of Reference:

The following is a short list of some of the terms of reference useful in setting the context of the previous definitions.

1.) Prognostic Accuracy or Confidence Level – the accuracy in terms of difference between the future forecast of performance or condition and the actual future value achieved expressed as +/- an amount or as a percentage of the forecast. It may also be applied to the accuracy of the predicted time to failure, time to a given performance degradation point or percentage, remaining useful life, etc.
2.) Prognostic Horizon – the maximum time or related parameter (such as number of missions, etc.) for which a given Prognostic Technique will achieve a set accuracy or confidence level. For example, technique "A" may achieve a 90% prognostic accuracy with a horizon of 200 operating hours, or Prognostic Technique "B" may achieve a 75% prognostic accuracy with a prognostic horizon of 3 missions.
3.) Prognostic Data – raw or processed sensor outputs and/or visual observations used directly or indirectly as a basis for prognostics.
4.) Prognostic Information – the information derived from data and / or other sources used directly or indirectly as a basis for prognostics.
5.) Prognostic Features – measures (or Figures of Merit) derived from data or information that are used by a given prognostic technique.
6.) Prognostic Feature Vector – a specific combination of features that are used to generate a partial or total output for a specific prognostic technique.
7.) Prognostic Feature Vector Set – those combined feature vectors that are used to generate a partial or total output for a specific prognostic technique.
8.) Prognostic Metrics – those measures of performance of a prognostic technique or system that characterize the performance and predictive reliability of that technique or system for a specific application. These metrics may include:

    - Demonstrated versus design prognostic accuracy / confidence level.
    - Demonstrated versus design prognostic horizon.
    - Demonstrated reliability of the prognostic system versus the system it monitors.
    - Applicability or robustness of the prognostic technique or system – how many other applications can the technique be applied to with commensurate accuracy, reliability and horizon attributes.

And then there's the issue of getting the data needed to enable prognostics when getting sensors in the right location for diagnostics to work is a continuing battle. Probably the main contributor to poor performance of some of today's diagnostic systems is the low fault signal to background noise ratio that has resulted from sensors being located for the convenience of the system designer rather than for the quality of data acquired to detect and identify faults. The resulting complexity of analysis algorithms and the unreliable identification of faults has led to high false alarm rates that are compensated by raising the fault detection severity thresholds. The end result is first the loss of user confidence due to the high false alarm rate and the subsequent disabling the monitoring system. Achieving a prognostic capability with this kind of diagnostic foundation is not likely.

Still another challenge will follow success in the development of prognostic technology. After the challenges of getting the technology working reliably in the operational (field) environment, and getting the output in the form that enables actionable maintenance and mission planning, there is the challenge of acceptance of prognostics by the maintainer and distance support personnel. As one maintainer put it, " you can't reliably tell me now exactly what's broken, but you want me to believe that you can tell me what's *about to break*" Prognostics clearly has a longer road to success than the technology developer may think.

Some Emerging Pieces of the Solution: Over the past year or two there has been considerable activity in the prognostics community. New Weapon Platforms such as the JSF have developed system support processes that depend upon successful development of prognostic capabilities for the major subsystems. The innovation needed to achieve this is to a large extent coming from Small Businesses through the SBIR process. This effort continues to produce innovative additions to the electronic as well as mechanical prognostics technology base.

National attention has been focused on prognostics with the sponsorship of workshops and conferences by organizations such as the National Defense Industrial Association's Systems Engineering Division. They have produced not only a workshop focused on Electronics Prognostics, but a follow on task to define and describe the Electronics R&D Program Needs and Development Roadmap. A Task Group has been formed and is now is now in the process of developing the products of this task for the various classes of electronics shown in the Diagnostic and Prognostic Technology Base Maturity Matrix.

With the success of prognostic technology development efforts in the electronic, machinery and mechanical structural areas comes the challenge of demonstrating the true capabilities of the resulting technologies as well as providing the data sets to develop the next generation of improvements in this technology. There is an emerging need for an "honest broker" based center of excellence to address this challenge. Several months ago Sandia National Laboratories initiated such a program and has now held the first Advisory Board meeting on how such a V&V capability might be developed.

Finally, there is a long road ahead to establish the solid diagnostic and prognostic technology base needed to support the future maintainer and warfighter. The MFPT Society has, in my view, a clear leadership role in front of it and the right talent pool in its membership to contribute to the growth of the diagnostic and prognostic technology base and to the success of the new weapon systems it supports. The time for MFPT to weigh in on these National Challenges is now.

HOW VIBRATION LEVELS ARE RELATED TO TRIBOLOGICAL FAILURE IN ROTATING MACHINERY

By William D. Marscher, P.E.
Mechanical Solutions, Inc.
1719 Route 10 East Suite 305
Parsippany NJ 07054
www.mechsol.com

Abstract:
Vibration is a key machinery parameter that experts use to perform diagnosis and prognosis for a wide variety of potential mechanical or fluid dynamic problems. Excessive vibration in and of itself can cause problems involving rubbing wear, brinnelled bearings, and chipped or cracked mechanical seals. The interpretation of specific vibration readings to determine if they are damaging, or if damage has already occurred, has been performed for many decades, but is (surprisingly) still a developing science. This paper will explain the scientific basis for the relationship of vibration to tribological problems, illustrated by an example calculation. It will provide insight concerning telltale signs in vibration signals and how to interpret them quantitatively.

Key Words: Vibration, Machinery Prognosis, Machinery Diagnostics, Fatigue, Wear, Tribology, Rotordynamics

Introduction:
Injury to rotating and reciprocating machinery due to excessive rotor vibration takes the form of wear or fatigue to the internal components, such as bearings, seals, and shafting. In general, wear rate and fatigue tendency for a specific machine is proportional to the oscillating displacement associated with vibration. However, testing by various sources over the last 70 years has indicated that for *classes* of machines (e.g. centrifugal pumps), operating at perhaps a range of different speeds and loads, vibration velocity rather than displacement, correlates best with likelihood of wear (tribological) or fatigue (metallurgical) failure. The reason for this is not made clear in most references concerning acceptable vibration levels and machinery vibration acceptance standards. Early references on this subject were Rathbone 1939 (insurance company data), Blake 1964 (the so-called "Haystack Chart"), Baxter and Bernhard 1967 (the so-called "IRD Chart"), and Hancock 1974 (petrochemical plant experience with vibration level vs. failure rate for a large population of industrial pumps). A typical result of these studies is the chart that originated from the paper of Baxter and Bernhard, which has become copied so many times by various organizations through the years that it is now public domain, and is reproduced for reference in Figure 1.

VIBRATION SEVERITY CHART

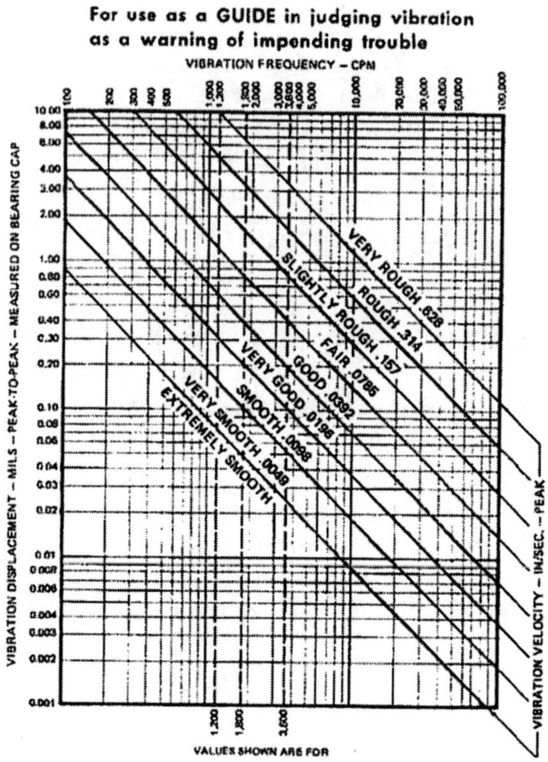

Figure 1: Vibration Severity Chart Derived from Baxter & Bernhard 1967

Each of the classical references concluded that for large groups of machinery of various sizes and speeds, vibration velocity (i.e. the first differential in time of the vibration displacement) correlated better than displacement with machinery deterioration rate and current state of health.

However, for a given machine that vibrates primarily at a constant running speed, as the great majority do (because variable speed machines are still the exception), and when residual imbalance is the main forcing function (as it indeed is for most in-service machinery), it was accepted that either displacement or velocity level could be used for trending purposes. This is because velocity was merely displacement times speed times a units constant, as shown in Fig. 2.

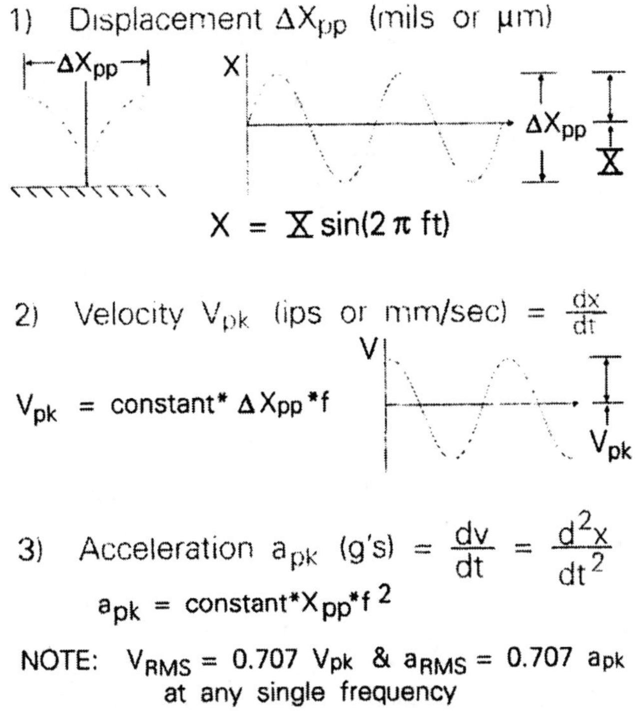

Figure 2: Relationship of Vibration Displacement, Velocity, and Acceleration

It is a relevant fact, as will be seen, that nearly all of the data which makes up the database of the classical references, was actually vibration displacement versus running speed, since vibration "reed vibrometers" were used to capture the unfiltered (i.e. "all pass", not filtered to consist of a single specific frequency) vibration displacement, and since the assumption was made that the exciting frequency was running speed (FFT spectrum analyzers were not yet available). Comprehensive studies such as those prior to 1975, but using modern spectrum analysis equipment, have been contemplated from time-to-time, but to the author's knowledge have not been accomplished. One of the purposes of the MIMOSA taskforce has been to accumulate such a database from a wide variety of operating machines and plant environments, organized according to various types, where all data would be from existing plant operating data, where the conditions were well enough documented for "apples and apples" data correlations. This has been a laudable effort, but is still a work-in-progress, and getting frequency (other than running speed) documentation from operating plant archived data is a considerable challenge. As discussed by Marscher 1987, and as will be illustrated below, this situation has led to a possible misinterpretation of the data from the classical references.

Example of the Relationship of Vibration to Internal Forces & Displacements:
The following example illustrates how vibration displacement can lead to directly to wear, and how it also can lead indirectly to wear or directly to fatigue through forces at the bearings and bearing supports. For simple fluid-element-between-bearings machines, rotordynamic analysis can be simplified without significant loss of accuracy, and simply supported beam calculations can be used to determine natural frequencies and mode shapes. If the impeller mass is M, the mass of the shaft is not large compared to the impeller, and the shaft length and moment of inertia ($= \pi D^4/64$) are L and I, respectively, g_c is the gravitational units constant, and E is Young's Modulus, then the following relationships can be derived from handbook beam formulas. If the vibration displacement of the impeller relative to the bearing rotational centerline is δ, then the unbalance force is:

$$F_{ub} = M\delta\omega^2 / g_c \qquad (1)$$

Half of this force is transmitted to each bearing, and as the force increases, bearing wear would be expected to increase (typically with roughly the cube of the force). In a journal bearing as opposed to a rolling element bearing, a certain critical force level would need to be passed to penetrate the lubricant film and lead to any significant wear at all. Based on such issues, a relationship between vibration δ at the impeller, and wear and/ or fatigue rates can be established, for the specific machine in question. Unfortunately, the impeller in generally inaccessible to a measurement of δ, but through the bearing housing and pedestal support stiffness k, an indirect relationship can still be established since the displacement x on the bearing housings (easily accessible) is related to F_{ub} by the simple expression (if no natural frequencies are nearby)

$$x = F_{ub}/k \qquad (2)$$

Alternately, for a given level of force F_{ex} independent of impeller motion (for example, from the fluid surrounding the impeller instead of imbalance) the vibration displacement expected at the impeller can be expressed, using beam theory again, as:

$$\delta = (F_{ex} * L^3)/(48EI) \qquad (3)$$

Once again, vibration of the impeller can be related to exciting force, and through inversion of equation 2, bearing housing vibration levels can be used to infer the level of F_{ex}, and then this knowledge of F_{ex} can be used (in principle) to predict bearing life and/ or fatigue of the bearing housing supports, for example.

The most common form of deterioration in rotating machinery, however, is wear (tribology), not fatigue, although both certainly can and do occur. In rotating machinery, typically wear that requires machinery refurbishment is at close running clearances (e.g. pump "wear ring" or compressor "laby ring" annular seals, or across the fluid films of mechanical seal faces) . Figure 3 illustrates the potential wearing zones of a typical multistage centrifugal pump or compressor, and although the fluid elements look somewhat different, the same situation is true for an axial gas or steam turbine.

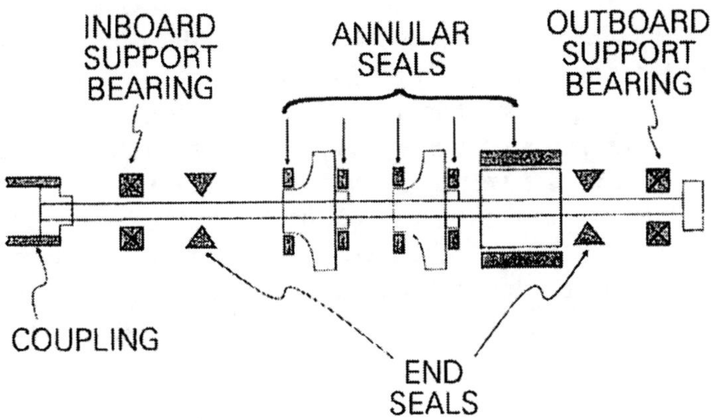

Figure 3: Illustration of Components Subject to Rubbing Wear in a Rotating Machine

Physical Reason for the Universality of Vibration Velocity:
As can be seen from the equations (1) through (3), vibration damage mechanisms seem to correlate most naturally with vibration displacement, not vibration velocity. However, as shown in Figure 2, displacement may be transformed into velocity by multiplying it by the running speed, and therefore for a given machine at a fixed speed, a doubling of displacement, for example, results in a doubling in velocity. In this manner, either displacement or velocity can be related to the forces and/ or close clearance utilization at potential troublespots within the machine. The classical references chose velocity rather than displacement to accomplish this, because their data of allowable vibration displacement versus running speed for a wide variety of machinery indicated that *allowable vibration displacement was inversely proportional to running speed.* On a displacement vs. speed plot, a line whose slope represents this fact is a line of constant vibration velocity, since displacement divided by frequency (of which speed is a type) is velocity. Therefore, nearly all modern specifications (particularly those intended to relate to a wide variety of machinery, or many different sizes and speeds of machines within a given type) use vibration velocity limits as the basis for their allowable specifications.

This begs the question of why the field test results indicate (on average) the inverse velocity vs. speed relationship. There may be several reasons for this, and in particular people who have considered this question often use heuristic arguments that this is because rate of oscillation (and therefore accumulated fatigue cycles, for example) are proportional to velocity, or because higher velocity for a given displacement reflects proportionately higher momentum or higher kinetic energy (in the latter case the square of the velocity term in conveniently neglected). However, there is a more straightforward reason why, on average, vibration velocity was found by the classical references to best correlate across machinery types and applications.

Consider the question of how wear rate at close clearance areas such as those of Fig. 3 correlates with increasing vibration displacement. Obviously, this depends on radial load, and therefore eccentricity, and it depends on the relative vibration of the part being measured versus the opposing part that probably is not being directly measured. However, most importantly it depends upon *what the initial clearance is*. If an attempt is made to draw a plot of wear rate versus vibration level, there would be many factors involved in terms of the shape of the curve of the absolute wear rate at any given vibration level, but the uncertainty would end when the vibration displacement reached the available diametral clearance. In such an event, in general the only option is for wear rate to assymtotically increase, because any vibration greater than the clearance quickly creates a larger clearance.

This seems like a trivial conclusion, but it is not when it is combined with a plot of typical clearance versus running speed. For a given class of machines, for various reasons (stress in compressors and turbines, and cavitation in centrifugal pumps), fluid component diameters are limited as a function of speed, e.g. the "tip speed" of impeller wheels cannot exceed a certain limit. For industrial compressors, for example, peak stress in the impellers is proportional to material density times tip speed squared, and in a system of consistent units the proportionality constant is about 1/3 for typical well-engineered impeller designs. The resulting stress level cannot be greater than the yield (or in the limit, the ultimate) strength of the material. Instead of the resulting impeller diameter allowable at a given speed becoming merely an upper limit, on average it is very similar for the majority of compressors running at that speed, in order to get optimum performance out of the compressor. Specifically, by making tip speed as high as its limit, the compressor can be smaller and lighter weight for a given output, and therefore it is more cost competitive than its competition that is not optimized in this fashion. While not all compressors would run at approximately the same tip speed (and therefore wheel diameter) for a given running speed, the majority will. Therefore, typically, compressor impeller diameter will be inversely proportional to running speed.

A second point that is key is that, in order to maintain efficiency, the running clearance must reduce as the wheel diameter reduces. Otherwise, leakage through, for example, the labyrinth eye seals will result in greater and greater efficiency decrement as the impeller diameter (and therefore its flowrate at a given pressure ratio) decreases. In spite of significant differences between pumps, compressors, and turbines, it is interesting that their running clearances for a given size impeller or bladed disk are all of the same order of magnitude for a given disk diameter. An exception is that, at very low speeds, the manufacturing tolerance vs. cost and ease-of-assembly benefits that clearance proportional to diameter provides disappear. For a 150 rpm centrifuge, it is relatively easy to close clearances tighter as a fraction of diameter than on a 2 inch diameter high speed cryogenic expander. For this reason, below a given speed, the expected increase of clearance in proportion to speed decrease diminishes, and eventually disappears.

For these reasons, a plot of clearance versus speed for a wide spectrum of industrial machinery is expected to approximately look like the bottom of Fig. 4. On the other hand, a plot of wear rate versus vibration should look something like the top of Fig. 4, where clearance provides a "stone wall" that long term allowable vibration level cannot cross without extraordinary wear rates.

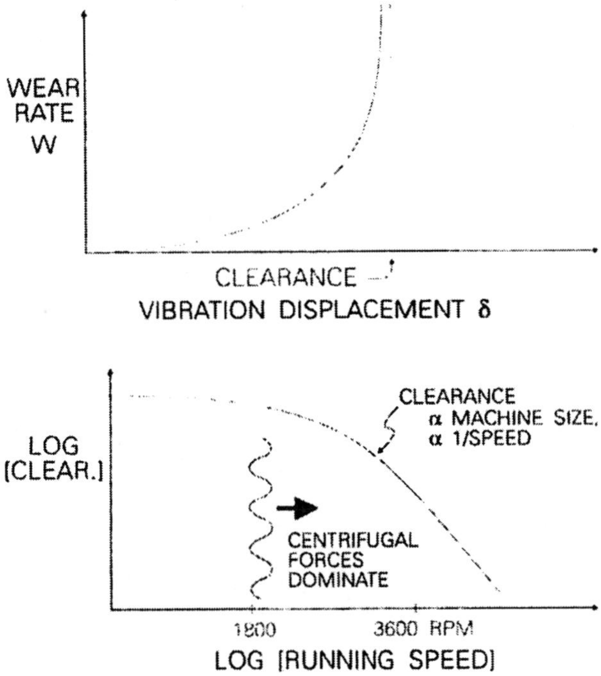

Figure 4: Dependence of Top): Wear vs. Vibration; (Bottom): Clearance vs. Speed

Combining the information in the two plots of Fig. 4 results in an expected dependence of wear on vibration that looks like Fig. 5. It is interesting that Fig. 5 is close in shape to the "haystack" charts of Rathbone 1939 and Blake 1964. This curve explains the shape of these plots, based on actual data versus running speed, as opposed to the simplification of strict constant velocity criteria, as was proposed by Baxter & Bernhard 1967.

However, the proposed relationship of allowable vibration level being inversely proportional to running speed (particularly at speeds above about 1000 rpm) is based on the fact that the data plotted by the classical references in fact reflects vibration displacement versus running speed, *not necessarily vibration versus frequency*. This raises the unsettling question of whether the same acceptable vibration level applies to vibrations being forced to occur at, say, blade passing frequency, as the acceptance level for vibration at 1x running speed. This has been a subject of considerable controversy. Some machinery OEM's suggest experience-based criteria which state that allowable vibration displacement for frequency harmonics above running speed varies inversely proportional not to running speed, but rather to the square root of running speed.

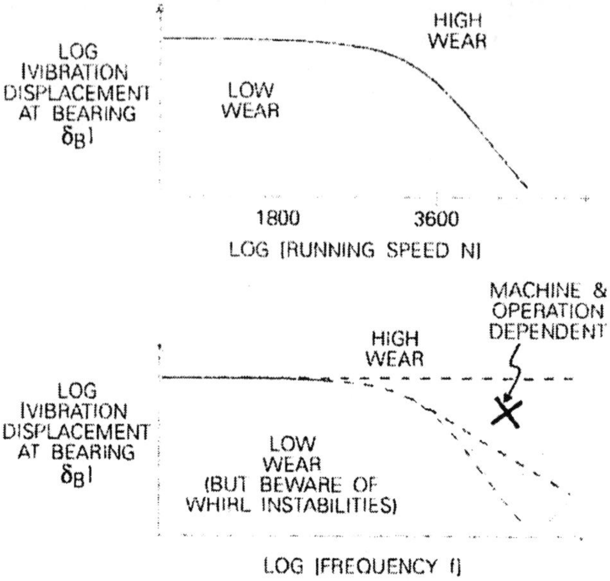

Figure 5: Expected Dependence of Wear on Vibration at Various Running Speeds & Frequencies

Re-interpretation of strict velocity acceptance criteria has been occurring in recent years, after high frequency FFT analyzers became inexpensive enough to come into common use. Taken in the strict context of modern vibration specifications, a very high frequency structure-borne noise could result in automatic rejection of a machine because of a vibration of less than a millionth of an inch, because the vibration's high frequency resulted in a higher than typical velocity value of an otherwise acceptable (from a tribological and fatigue standpoint) vibration level. There is common agreement among machinery diagnosticians that unusually high vibration velocities and/ or accelerations always merit investigation and may represent some problem that is being observed indirectly from deep inside the machine. For example, rolling element bearing or gear tooth contact faults often provide their first strong indication of problems in the high frequency range. However, often such high frequency (i.e. much higher than vane or blade passing frequency) anomalies are found to represent some sort of acoustic issue harmless to the machine in question, such as cavitation in a downstream discharge valve in a centrifugal pump.

This raises the issue of accidental rejection of good equipment, versus the more commonly considered issue of unwanted acceptance of faulty equipment. Both situations have an economic cost. Althought the OEM's bear the cost of unmerited rejection during the warranty period, such costs absorbed by all OEM's across the industry will eventually lead to increased equipment prices, and therefore be passed on to the End User. Therefore, it is important that vibration specifications not simply be *tight*, but rather be *economically appropriate*. Taking centrifugal pumps again as an example, the author's experience is plotted approximately in Fig. 6. Since most specifications in recent years have gradually decreased to roughly the neighborhood of 0.15 inches per second velocity, peak, (ips-pk), as can be seen in Fig. 6 this leads to the rejection of a large proportion of pumps that would actually run well many years. In fact, until recently most plants ran equipment successfully at vibration levels higher than 0.15 ips-pk). This is verified by the results of Hancock 1974, for example, who discovered that the relationship of vibration to maintenance problems in centrifugal pumps lost statistical significance when vibration levels were less than 0.25 ips-pk, with maintenance problems at lower vibration levels being dependent almost entirely on other factors. However, because of the cost of the few pumps that do fail unexpectedly at 0.15 ips-pk, the specification writers believe that "setting the bar" at this level merits the high likelihood of rejection of good pumps.

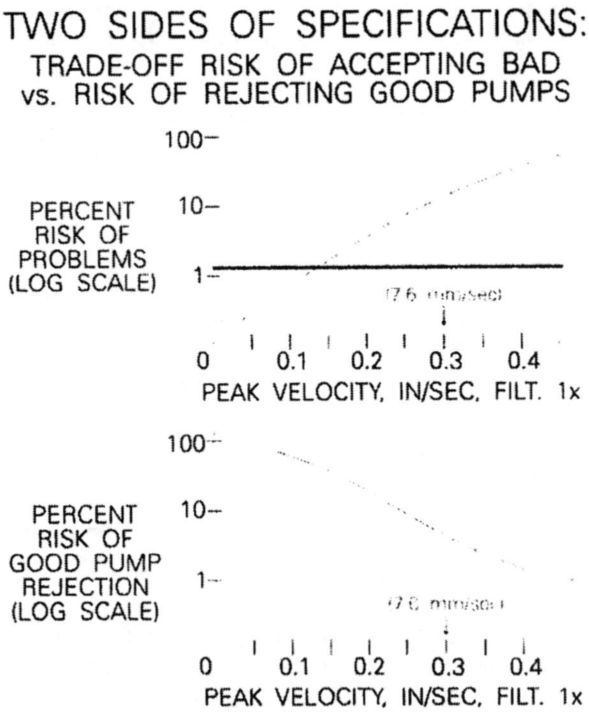

Figure 6: Author's Impression of Prognosis Risk Factors for Various Vibration Levels

The purpose of Fig. 6 is not to accurately represent all centrifugal pump vibration prognosis characteristics, but to provide an appropriate "ballpark" impression of the level of risk of good pump rejection vs. bad pump acceptance. Both sides of the issue have an economic cost, but only the latter has been considered in most cases.

Closure:

The goal of specifications should be to protect all parties - manufacturers, consultants, contractors, and owners - by clearly defining and applying appropriate vibration specification levels. In this context, injury to pumps due to excessive rotor vibration takes the form of wear or fatigue damage to the machine's internal components, such as bearings, annular seals, and mechanical seals. Vibration velocity has been found to correlate relatively universally with these issues if vibration occurs predominantly at a frequency equal to running speed. For other frequencies, the relationship is more complex, but vibration velocity remains a good place to start, in terms of assessing potential anomalous behavior of the system. However, it is important to not be overly strict in application of blanket acceptance levels at frequencies far in excess of vane or blade passing frequencies, and to furthermore not set specification acceptance levels too "tight" in general. There is an economic cost of good machine rejection that is just as real to the bottom line as the cost of acceptance or continued operation of a machine that in fact is in serious trouble. More research is needed to clarify both sides of the issue, particularly with regard to appropriate acceptance levels of very high frequencies.

REFERENCES:

BAXTER, R.L., BERNHARD, D.L., "Vibration Tolerances for Industry", ASME 67-PET-14, 1967

BLAKE, M. "New Vibration Standards for Maintenance", Hydrocarbon Processing and Petroleum Refining, January 1964

HANCOCK, W.P. "How to Control Pump Vibration", Hydrocarbon Processing, March 1974

MARSCHER, W., "The Relationship Between Pump Rotor System Tribology and Appropriate Vibration Specifications for Centrifugal Pumps", Proc IMechE 3rd European Congress on Fluid Machinery for the Oil and Petrochemical Industries, The Hague, Netherlands, May 1987

MARSCHER, W. D., "Accurate Predictive Maintenance for Rotating Machinery," Proc. ASME Rotating Machinery Conference & Exposition (RO-CON '93), Somerset, NJ, November 10-12, 1993.

MARSCHER, W.D., "Wear and Erosion of Pumps", Chapter in Vol. 18, ASM Metals Handbook, c. 1993.

MARSCHER, W., Vibration Chapter and Predictive Maintenance Chapter, STLE Tribology & Lubrication Handbook, STLE Park Ridge IL, 1997

MARSCHER, W., "Bearing Application and Lubrication Chapter", Modern Marine Engineer's Manual 3rd Edition, Cornell Maritime Press, Centreville MD, 1999

MARSCHER, W., "Avoiding Failures in Centrifugal Pumps", Tutorial in Proc. TAMU International Pump Symposium, Houston TX, Feb. 2002

RATHBONE, T. "Vibration Tolerance", Power Plant Engineering, v. 43, November 1939

DETECTION AND DEFEAT OF IMPROVISED EXPLOSIVE DEVICES: A NEW CHALLENGE FOR THE SOCIETY

Dr. Paul R. Muessig
Office of Naval Research, 800 North Quincy Street, Arlington, Virginia 22217
703.432.1307; paul.muessig@usmc.mil

Abstract: Improvised Explosive Devices (IEDs), more commonly known as "roadside bombs", constitute a large and growing threat in the global war on terrorism (GWOT). They are not only the single largest cause of coalition casualties in Iraq, they have also been used to terrorize civilian populations around the globe and influence public opinion in democratic societies. Both the Department of Defense (DoD) and the Department of Homeland Security (DHS) are keenly interested in finding ways to defeat IEDs across a spectrum of activities that includes prediction and detection of IED emplacements, prevention of IED detonation, neutralization and disposal of detected IEDs, and protection against the effects of IED explosions. This paper provides an unclassified overview of the IED threat, as well as DoD's efforts to mitigate it, including a description of representative approaches and technologies. The paper concludes with a summary of unresolved technical issues, and challenges the Society to consider non-traditional applications of its expertise and technology to develop creative solutions to this most pernicious threat to national and international security.

Key Words: Improvised explosive devices; Detection; Technology challenges; Innovation.

Figure 1: IED Detonation

Introduction: The threat posed by IEDs to national and international security is large and growing. IEDs are not a new tactical concept; they have been used frequently by the weak against the strong to assassinate leaders, destabilize governments, and foment insurrection by the imposition of a regime of terror and mayhem. Europe, for example, has extensive experience with the use of IEDs as terror devices, having suffered through attacks during the 1970's and 80's from the Irish Republican Army (IRA), the Bader Meinhof Gang and the Red Brigade. More recently, IEDs were used during the run-up to the Afghan elections to discourage turnout and delegitimize the results. In Spain, IEDs were effectively used to influence election results and change Spanish policy on Iraq, while in Saudi Arabia IEDs are routinely used to destabilize the government of an unpopular ruling class. In Iraq, IEDs are the single largest source of coalition casualties, with obvious implications for continued public support of

Figure 2: IED Detonation Effects

Figure 3: Typical IED Blast Crater

the war among coalition partners. These small, inexpensive, easily hidden and lethal devices have political and military consequences far out of proportion to their cost, and they may well determine the course and outcome of the GWOT.

Development of an effective counter-IED strategy requires an understanding of common IED technologies and implementation approaches. Analysis of historical data on IED attacks reveals patterns that can be used to develop a portfolio of effective counter-IED technologies and tactics, techniques and procedures (TTPs).

The Nature of the IED Threat

Figure 4: IED Source Material

Construction: IEDs can be constructed of metal, wood, plastic or any combination thereof, making it very difficult to detect them with a single technology. Almost any explosive will do for an IED, and we have seen combinations of almost every type imaginable, from aircraft bombs to mortar shells to anti-aircraft missiles, etc. The most common explosive used for IEDs in Iraq is conventional ordnance left over from the previous regime. Artillery shells of the 152mm, 130mm and 120mm size are the most common, usually fused with blasting caps and plastique. Because the explosive material is encased in a metal shell, detection at any reasonable standoff distance (say 100m) is very difficult. Conventional mines placed on top of each other are another common construction technique.

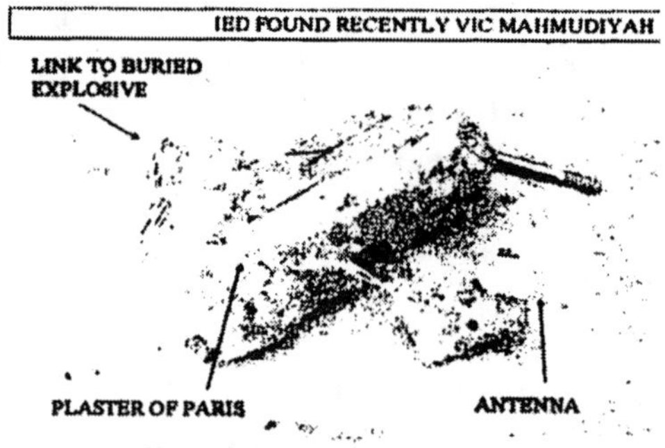

Figure 5: Well-camouflaged IED

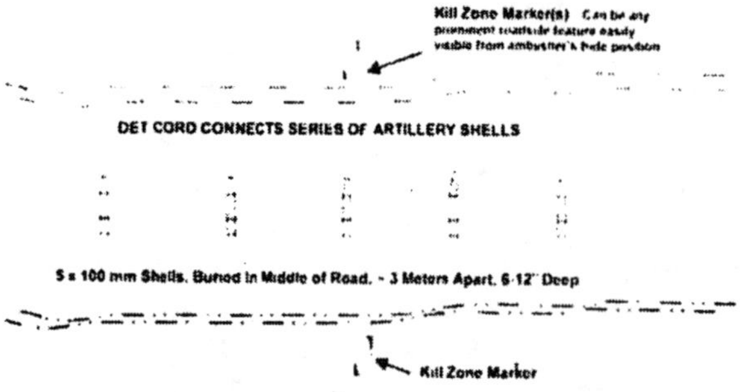

Figure 6: Typical IED Attack Layout

Emplacement: Most IEDs are surface laid. They are commonly concealed with roadside garbage, canvas sacks, or even dead animal carcasses. They are sometimes encased in Plaster of Paris or concrete, or hidden behind roadside guardrails, posters or pictures. More sophisticated attempts at concealment involve road repair or trash crews leaving a pile of dirt or trash along a route and then emplacing an IED there days later. The environment in Iraq is conducive to this type of emplacement, since rubbish piles and trash of all types are prevalent along roadways in populated areas. IED emplacements have also included placing explosives within concrete curbs and underneath asphalt roadways in order to conceal their presence. We have also seen IEDs placed underneath overpasses, on light poles, and in trees in an attempt to attack armored vehicle crewmen positioned in open hatches atop vehicles. IEDs are also placed in abandoned vehicles or in operable vehicles manned with suicide drivers. In some cases, vehicle-borne IEDs with suicide drivers have been rigged with remote control backup devices to prevent the driver from changing his mind.

Location: The majority of IED attacks occur in urban or suburban, vice rural, areas. IED attacks are targeted along frequently used routes, and we have seen terrorists use the exact same spot or hole that was used for a previous IED to plant another IED. Operational data have shown a high number of IED attacks in areas close to known ammunition storage areas or mosques. It is clear that the terrorists would prefer not to travel any distance with their devices, and are utilizing coalition reluctance to enter mosques to their advantage. Most IEDs are emplaced during the hours of darkness, and initiated during the day when visibility is good.

Figure 7: Cell Phone IED Initiator

Figure 8: Toy Truck IED Initiator

Initiation: Between 70 and 80 percent of all IEDs encountered are triggered by some kind of remote control device. A remote device offers the obvious advantage of allowing standoff distance between the terrorist and the explosive. This allows him to conduct the attack from a concealed position, enhancing the opportunity to escape the area after carrying out the attack. The majority of these IED remote triggers have been low-power, cheap, easily available devices such as those used to open car doors, garage doors, etc. Recent attacks have used long-range cordless telephones as initiators. These provide higher power and longer-range capability for the attacker. Additionally, they are ubiquitous and easy to use. As the commercial cellular telephone infrastructure is established in both the OIF and OEF theaters, the use of these devices as IED initiators is likely to increase. Some IEDs have utilized a length of wire from the fuse to the remote, enabling a longer standoff range in order to circumvent jamming. In some cases an amplifier has been used to ensure that the detonation signal travels through the length of the wire. There have also been IEDs triggered by simple pressure plates. On occasion these will be combined with a remote initiator. When this occurs, the terrorist will observe a target approaching and activate the device through a remote, so that it is now armed and ready once pressure is applied. This allows a measure of control, defeats jamming and prevents "wasting" an IED on a small or uninteresting target.

Figure 9: Key Fob IED Initiators

Attacks: Most IED attacks occur against mounted coalition troops moving along a road in either a convoy or a patrol. The device explodes and the terrorist escapes. On occasion we will see an IED with multiple explosives rigged together to create a larger killing zone. We have also seen dummy IEDs set up in order to attract first responders (e.g., medical or explosive ordnance disposal [EOD] personnel) who are then attacked with a second, more cleverly concealed IED when they approach the vicinity of the dummy to treat casualties or defuse it. Coordinated IED attacks are also common, and consist of an IED initiation followed by direct (e.g., small arms) and/or indirect fire (e.g., mortar) engagements. In all cases there is an unwillingness to close with coalition forces, and coordinated engagements frequently break off quickly. Vehicle-borne IED attacks are most commonly conducted in populated target areas against Iraqis cooperating with the government or against coalition troop convoys.

Counter-IED Approaches and Technologies

The Services have identified five tenets of IED defeat, arranged along a rough time continuum as shown in figure 10:

- Prediction: Solutions that disrupt enemy IED operations prior to the emplacement of an IED. This includes the ability to forecast specific enemy operations related to the leadership, funding, recruitment and training of insurgents, and the construction and emplacement of IEDs.

- Detection: Solutions that locate and identify emplaced IEDs, IED manufacturing facilities and sources of supply, IED handlers and carriers, and the leadership and funding network that supports them.

- Prevention: Solutions that render remotely controlled IEDs temporarily inoperative. The most common approach to date has been the use of electronic countermeasures (ECM) to disable or disrupt the devices used to arm and detonate IEDs.

- Neutralization: Solutions that render IEDs permanently inoperative. Typical means include pre-detonation, controlled detonation, and render-safe (defusing) and disposal operations.

- Mitigation: Solutions that contribute to the physical protection of personnel and equipment from the blast and fragmentation effects associated with an IED detonation.

The Services are looking for technologies and associated TTPs that span the spectrum of IED defeat described above. To date, much has been accomplished along the right side of this spectrum (near the "boom"), but accomplishments dwindle as one moves left along the continuum, partly because the early phases of IED defeat are less amenable to simple technology solutions. Predicting attacks and emplacement sites has proven particularly difficult, as the enemy has a decentralized organizational structure that is only loosely controlled. Consequently, patterns are hard to recognize and may not exist at all. Moreover, supplies of explosive material are plentiful and accessible, and the components required for IED initiation are cheap, and easy to obtain and operate. As a result, current efforts in this area focus heavily on non-technological approaches such as human intelligence (HUMINT), information operations (IO) and psychological operations (PSYOP) to penetrate and exploit IED construction and emplacement networks. One technological approach that may have some merit is persistent surveillance, in which an aerostat with high resolution EO/IR cameras "stares" down at a fixed area on the ground, typically several square kilometers in size, digitally recording everything it sees. After an IED event occurs, "backtracking" software is used to analyze imagery before and after the event to reconstruct and correlate the paths of people in the area, some of whom are insurgents taking up observation posts prior to the event and fleeing to safe havens after the event.

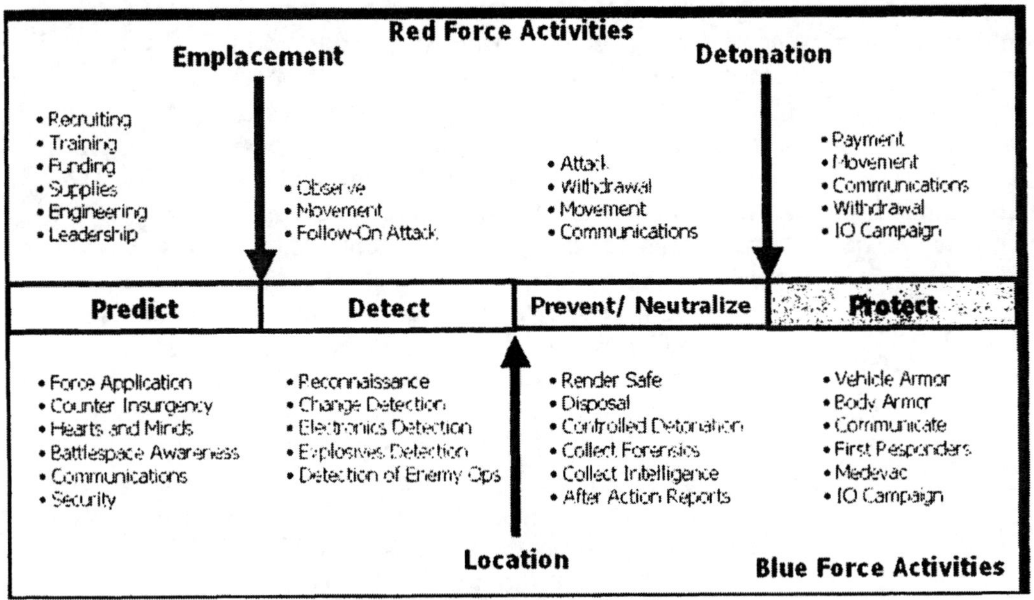

Figure 10: IED Defeat Continuum

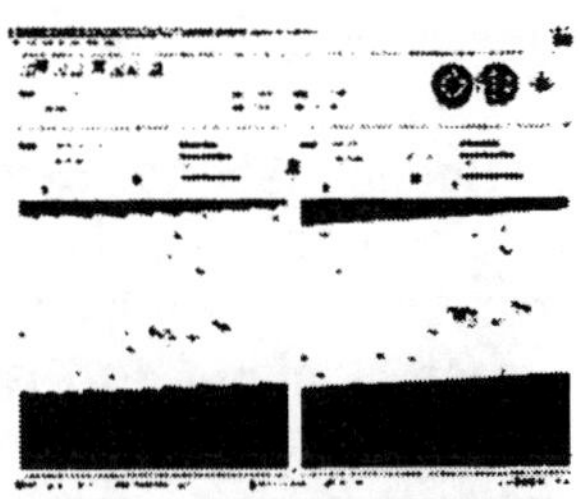

Figure 11: Change Detection

As mentioned above, detection of IEDs is hampered by the wide variety of their construction materials and emplacement characteristics. Short-range (less than 10m) detection efforts have focused on chemical vapor "sniffers" of various types (e.g., ion mobility spectrometers, photoluminescence, etc.) that can detect the characteristic vapors given off by explosives, and a variety of imaging techniques (X-Ray, infrared (IR), active and passive millimeter wave, etc.) that can detect explosives concealed in vehicles or on terrorists. The drawback to these approaches is that their short detection range limits their applicability to vehicle or personnel inspection applications, e.g., at entry control points. Longer-range (out to about 100m) detection efforts have focused on various types of ground penetrating radars (GPRs), and a variety of imaging techniques (video, electro-optic (EO), infrared (IR), hyperspectral, etc.). Of these latter, change detection (figure 2) has generated significant interest. In change detection, digital images of the same location taken a two different times are correlated in space and then processed pixel by pixel to identify changes that could indicate the location of an IED emplaced during the time interval between the images. The major drawback of this approach is the potential for high false alarm rates in areas characterized by high human and vehicular traffic, as well as random clutter such as refuse. Image processing and automatic target recognition software is being developed to minimize these effects while maintaining the ability see and declare relevant changes, but advances in signal processing technology are clearly needed to make the technique more useful.

Prevention technologies focus on jamming the initiation mechanisms of remotely controlled IEDs with RF energy along a continuum stretching from low frequency (a few 10's of MHz) up to about 6 GHz (figure 3). While technical specifications and descriptions of these systems are classified, the technical challenges associated with putting enough power into all of the frequency bands across so wide a spectrum within a small, mobile and easily concealable package should be obvious.

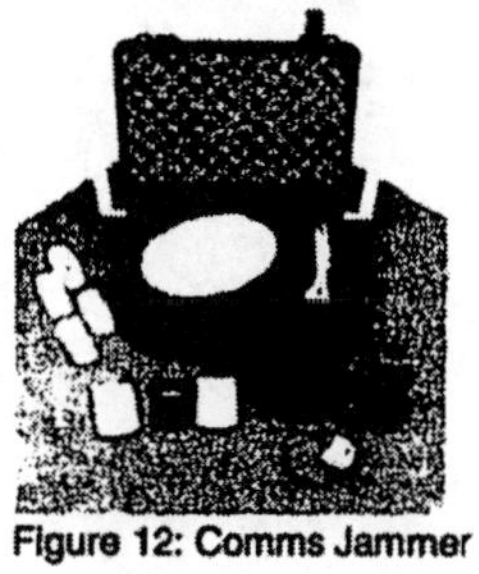
Figure 12: Comms Jammer

Figure 13: Explosive Ordnance Disposal (EOD) Robot

Neutralization technologies focus on remote detonation and explosive ordnance disposal approaches. Typical approaches span the spectrum from the mundane (e.g., blast rings and tents) to the innovative (robots that perform various EOD functions remotely (figure 4), remote detonation via high power RF energy) to the speculative (deployable armored blankets with explosive netting to allow destruction in place while minimizing blast and fragmentation damage). Key challenges in this area include improved reliability, availability and maintainability of these complex machines; improved robotic functionality, especially autonomous operations; and low power, highly focused pre-detonation approaches.

Protection technologies focus on various types of blast resistant body and vehicular armor (figure 5). Typical examples include ballistic glass for windshields and windows, ballistic armor blankets and panels for retrofit of existing vehicles. Key technical challenges in this area include the need for lightweight and flexible materials that can protect personnel and equipment from the effects of blast and fragmentation at short distances while not impeding basic combat operations.

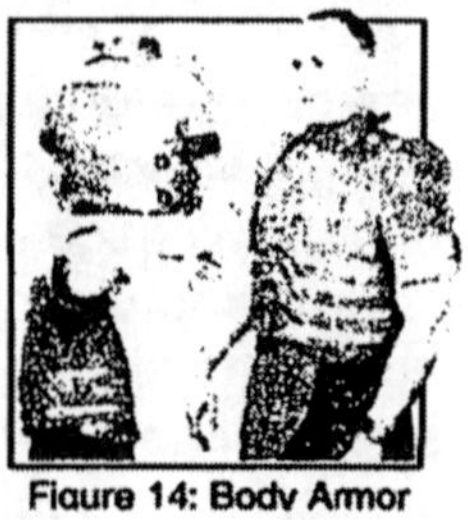
Fiaure 14: Bodv Armor

Last summer, Deputy Secretary of Defense Paul Wolfowitz established a Joint Integrated Process Team (JIPT) for IED Defeat to coordinate and focus DoD and Service efforts to find and field technologies to defeat IEDs. This IPT is divided into sub-IPTs that focus on rapid identification, evaluation and fielding of promising counter-IED technologies in each of the IED defeat areas described above. Other key organizations involved in the development and fielding of counter-IED technologies for both military and homeland security applications are the Technical Support Working Group (sponsored by the Department of State (DOS) and managed by the Assistant Secretary of Defense for Special Operations and Low Intensity Conflict, ASD(SO/LIC)); the Department of Homeland Security Advanced Research Projects Agency (DHSARPA); the Combating Terrorism Technology Task Force (CTTTF) sponsored by the Undersecretary of Defense for Acquisition, Technology and Logistics (USD(AT&L)). The military Services also have a rapid response fund to field counter-terrorism technologies in theaters across the globe – namely, the Combating Terrorism Readiness Initiatives Fund (CbTRIF), sponsored by the Chairman, Joint Chiefs of Staff (CJCS). The IED threat has also caused numerous agencies to reconsider and tailor their primary missions to respond to this threat. The Defense Threat Reduction Agency (DTRA) is one such example, which used to focus on reducing threats to US national security in the area of weapons of mass destruction (WMD), but has now sponsored numerous efforts to defeat IEDs as well. These and many other efforts, both large and small, are focusing the combined energies of some of America's best and brightest minds and resources on developing innovative solutions to a pervasive, cheap, low-tech threat with enormous ramifications for national and international security.

Challenges for the Society:

Which brings us to the Society. How can a professional society devoted to "gaining a better understanding of the process of mechanical failure" contribute to an effort that seems so far out of its purview? I would argue that many of the technologies and analytical approaches of interest to some of the Society's subgroups and interest areas are ripe for innovative application to IED defeat. For example, might not some of the approaches developed in Sensors Technology or Non-Destructive Evaluation and Testing be applied to detection of IEDs? Might not Root Cause Failure Analysis and Failure Modes and Mechanisms research and technology be applied to known IED construction methods and techniques to identify the best way to make them fail? While I would certainly defer to the judgment of experts in this field with regard to the applicability of MFPT to IED defeat, I would also encourage everyone in this area of research to look for fresh ways to apply existing technology and expertise. The threat to national security is real, and the stakes couldn't be higher.

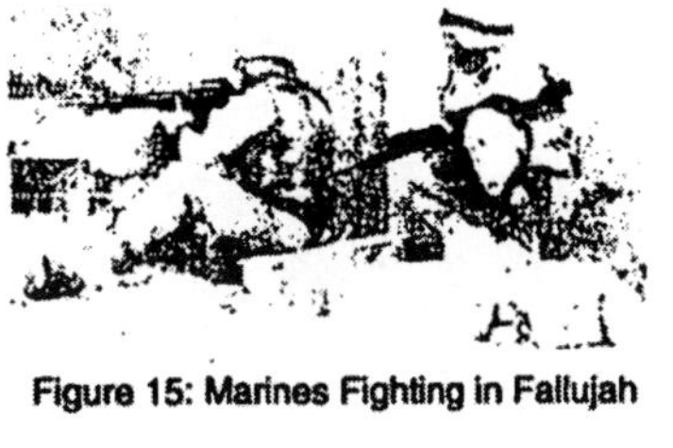

Figure 15: Marines Fighting in Fallujah

Acknowledgements: The author extends his most grateful appreciation to his co-workers on the USMC IED Working Group at the Marine Corps Warfighting Laboratory, for their deep and abiding friendship, for their dedication to saving the lives of Marines, and for their assistance with the preparation and review of this article: Mr. Jean T. Malone, Maj Joseph F. Monaghan, Maj John E. Moore, Major Bruce G. Paterson, Mr. Peter Pham, and SSgt Brian J. Rhoney.

PROGNOSTICS 1

Chair: Dr. C. James Li
Rensselaer Polytechnic Institute

AUTONOMOUS IMPACT DAMAGE DETECTION AND ISOLATION PREDICTION FOR AEROSPACE STRUCTURES

Michael J. Roemer, Jianhua Ge, Alex Liberson
Impact Technologies, LLC
200 Canal View Blvd., Rochester, NY 14623
Tel: (585) 424-1990
mike.roemer@impact-tek.com

G. P. Tandon, R. Y. Kim
University of Dayton Research Institute Dayton, OH 45433

Abstract: This paper presents a practical yet innovative impact damage identification and prognosis approach for aerospace structures that uses an optimized suite of reliable COTS sensors coupled with advanced damage detection and modeling algorithms. The presented methodology utilizes a monitoring approach based on acceleration measurements that are analyzed using advanced signal processing and dispersive wave theory models that capture frequency and orientation dependent wave propagation effects. The acceleration measurements and associated processing modules are used to provide immediate detection and isolation estimates, while an energy amplitude feature allows for assessments of damage severity after the impact. By embedding wave theory model results with the adaptive signal processing algorithms, a more accurate understanding of the time-frequency behavior of the dispersive waves produced at impact is gained. Damage localization is performed based on the comparison between the predicted and measured wave group velocities, with a genetic algorithm used to optimize the parameters of a triangulation procedure. This combination of model and feature-based algorithms allows the system to make use of limited, but readily available accelerometer data. This procedure also minimizes the learning and modeling difficulties associated with other techniques that are based solely on models or measurements. A few selected demonstrations are presented that illustrate the impact location prediction capabilities in realistic carbon fiber reinforced polymer (CFRP) composite panels.

Key Words: Damage Detection; Diagnostics; Prognostics; Structural Health Monitoring

Introduction: Applications for structural composites used in the aerospace industry often undergo high-stress loading profiles and operational impacts that can produce various types of damage, including fiber breakage, matrix cracking, delamination and interface de-bonding. The result of such damage can have detrimental effects on the overall strength of a vehicle in terms of its ability to complete its mission or compromise the safety of the crew. The need for practical, reliable and non-intrusive technologies for continuously assessing and monitoring the health state of complex aerospace structures is required in order to improve vehicle availability and reduce the safety risks and total ownership cost through elimination of costly inspection processes.

In addition, current structural fatigue prediction and damage detection methods have not kept pace with the increasing performance requirements and the use of new materials, nor with the need to use aircraft beyond their designed service lives. Next generation structural health monitoring systems (SHM) need to be capable of operating at both the local or global levels, and provide a means to significantly reduce life cycle costs of aerospace structures by providing accurate real-time assessments of structural damage. These considerations highlight the pressing need for innovative signal analysis and damage classification techniques based on practical measurements such as accelerometers embedded in CFRP structures.

A review of the literature [1-6] indicates that there continues to be a need for reliable strategies in structural health monitoring (SHM) that are robust enough to be of practical use on aerospace systems. A few of the drawbacks of current SHM methods include:

- Current structural monitoring implementations are typically accomplished with a limited number of sensors, dispersed over a relatively large area of a structure. These sensing systems can only provide the spatial resolution to detect fairly significant damage and can not locate damage accurately. Recent advances in sensing and on-board processing technologies are making it more feasible and affordable to collect larger amounts of data for long-term continuous monitoring of in-service structures.

- Current off-line, non-destructive inspection systems require data analysis to be manually performed by experienced engineers, with automation of this data analysis remaining largely unsolved. For continuous, real-time monitoring, it will be critical to reduce unnecessary interfaces by end-users and to automate the data analysis process throughout a distributed health monitoring system architecture.

- Current damage detection techniques have been successfully applied to scaled models or specimens tested in controlled environments, the performance of these techniques in real operational environments needs to be further developed and validated. Varying environmental and operational conditions produce changes in the system's dynamic response that can be easily mistaken for damage. Hence a need still exists for damage detection algorithms that will minimize false positive and negative indications of damage.

The technology components necessary to develop a more reliable impact damage assessment and prediction system should include proven, air-worthy measurement technologies and an associated set of validated, real-time analysis techniques that can provide a comprehensive and updated correlation with the health state of the structure. Hence, establishing quantitative links between specific damage types, size and location based on measured changes from highly sensitive signal processing and model-based technologies such as wavelet transforms and dispersive wave theory models is the recommended approach presented herein. Figure 2-1 shows a simple representation of a combined model and feature based procedure for performing damage detection in an aerospace structure. The major elements of this damage prediction process include: data pre-processing and feature extraction, model-based wave arrival time prediction, error pattern classification, data/information fusion and impact predictions. Where the core impact analysis exists in the classification of the differences between the predicted wave group velocities and those measured with the wavelet-processed features.

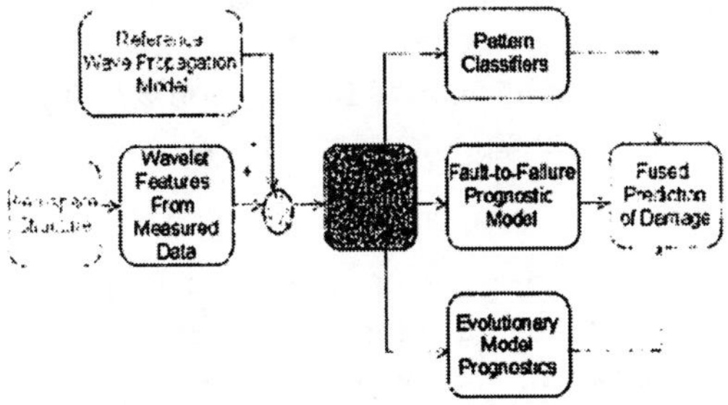

Figure 1: Model-Based and Signal Processing-Based Analysis for Damage Detection in Aerospace Structures

The project team of Impact Technologies and the University of Dayton Research Institute has developed a series of structural monitoring test beds with the objectives of identifying (a) damage initiation, (b) damage location and (c) severity of damage in composite structures. Figure 2 shows the implementation stages for the damage detection and localization process discussed herein that uses both data-driven and model-based approaches. Specifically, a robust monitoring concept will be presented based on an accelerometer-based sensing system using a series of flat and curved composite panel test beds that was able to provide comprehensive insight into the structure's observed health state. Two different, but collaborative approaches have been developed and assessed in terms of their respective abilities to detect, localize and assess severities of damage events. The first approach described is based on advanced signal processing techniques that utilize wavelet packets, high-order moment statistics, neural networks, and a triangulation procedure. This is also referred to as "feature-based" approach. The second approach is referred to as "model-based" and utilizes either a Finite Element (FE) model or dispersive wave theory equations of motion to evaluate and extract similar impact damage characteristics. A brief description of each of these approaches and associated results from a subset of tests are presented in the following sections.

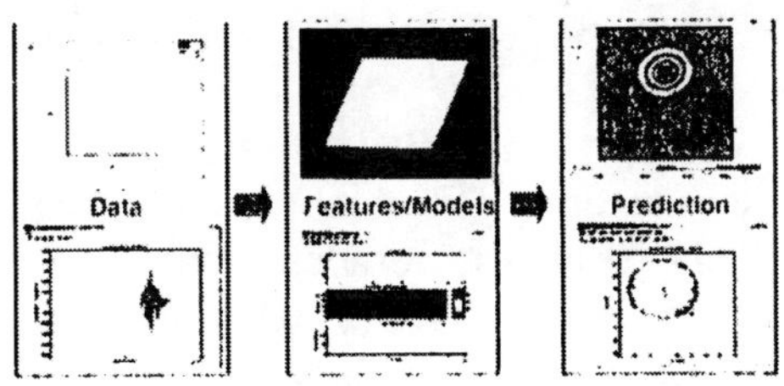

Figure 2: Damage Identification and Localization Stages

Feature-Based (Signal Processing) Approach: Complex structures, such as those utilized in aerospace applications, often undergo damage profiles that require non-stationary analysis techniques in order to be capable of detecting them [1, 2]. It is of critical interest to monitor these dynamic signals from inherent structural energy changes and vibration that reflect both the non-stationary components due to the surrounding excitation and those changes due to the damage itself. Traditional Fourier series analysis models are often inappropriate when used alone, since they require stationary data and do not account for time-dependent considerations. Multi-scale, time-frequency analysis techniques such as wavelets can be applied to this problem to provide a robust framework for efficient analysis of such non-stationary processes [3, 4, 5].

Wavelet transforms have been studied since the fifties by mathematicians, but it has only been in the last ten years that they have made huge advancements in the engineering and signal processing community [6]. The wavelet transform has emerged in many areas of signal processing, such as image processing, data compression, sub-band coding, communications and speech processing. The wavelet transform is favored over the Fourier transform techniques by speech researchers because of the non-periodic segments in a speech signal and the flexibility of the transform. These features make it attractive for structural response analysis as well.

The wavelet transform is particularly effective for representing various aspects of non-stationary signals such as trends, discontinuities, and repeated patterns that have both time and frequency components in them. Wavelet analysis has become one of the most powerful methods in data analysis for damage detection [3, 4, 8]. The advantage of algorithms based on the wavelet transform is that they are able to reveal important temporal and spectral patterns in a signal. These patterns may constitute the signatures of the dynamic events in question and may offer the key to more effective and reliable interpretation of the sensor signals. Wavelets have a built-in "spatial adaptivity" that allows efficient estimation of functions with discontinuities in derivatives, sharp spikes, and discontinuities in the function itself. The power of wavelets is that they capture local scale activity in the data. Transient-detection algorithms based on these adaptive signal-processing concepts can be used to detect, automatically, transient loads so as to provide an accurate statistical characterization of the extreme load history experienced by a structure [11]. A sudden structural stiffness loss may result in a stiffness discontinuity or singularity in acceleration response data, and can be detected by the wavelet approaches due to sensitivity of wavelets to singularities in a vibration signal.

Wavelet Basics: The wavelet transform is an operation that transforms a signal by integrating it with a kernel function. The kernel function or mother wavelet is a function that will be scaled and delayed as it is correlated with the signal of interest during the integration process. The wavelet transform of f(t) with respect to the mother wavelet g(t) is defined as

$$W_g f(s,\tau) = \frac{1}{\sqrt{|s|}} \int_{-\infty}^{\infty} f(t) g\left(\frac{t-\tau}{s}\right) dt \qquad (1)$$

where s is the scale and τ is the delay of the mother wavelet. In the wavelet transform, the kernel function, g(t), is delayed by an amount τ and scaled by an amount s. The scaled mother wavelet is energy normalized by the $1/\sqrt{s}$ term. A function must be admissible for it to qualify as a mother wavelet. A function, g(t), is admissible if

$$c_g = \int_{-\infty}^{\infty} \frac{G(\omega)}{|\omega|} d\omega < \infty \tag{2}$$

where $G(\omega)$ is the Fourier transform of g(t). The constant c_g is the admissibility constant of the function g(t), and the requirement that it is finite allows for inversion of the wavelet transform. An example of a signal that is not admissible is a sine wave, because the signal has infinite energy. To make a function like this admissible, a Gaussian function can be multiplied by the non-admissible function. A Gaussian weighted tone, which is called a Morlet mother wavelet, has the following form.

$$g(t) = e^{-\alpha t^2} e^{j\omega t} \tag{3}$$

The signal used for the mother wavelet can be user defined and is very flexible. Basically any signal may be used as long as the energy of the signal is finite. This means that a section of the processing signal that represents some significant event may be used. Figure 3 is a block diagram of the implemented wavelet processing procedure presented herein. Utilizing the Morlet mother wavelet as shown above, the signal transient behavior is extracted and retained.

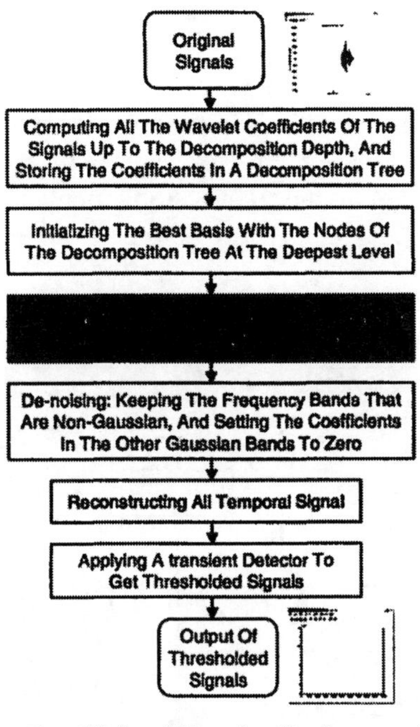

Figure 3: Transient Detection Using Wavelet Packets and Higher-Order-Statistics

Figure 4 shows the results of the process described in Figure 3, starting with raw data from an accelerometer taken from the test bed discussed in a later section. The data sample acquisition rate was 1.0e-6 samples per second, with the start time of 0 and an end time of 9.549e-3 seconds. Figure 4(b) shows the results using the Morlet wavelet packet decomposition, and Figure 4(c) is the final output of the thresholded signal using an adaptive energy detector.

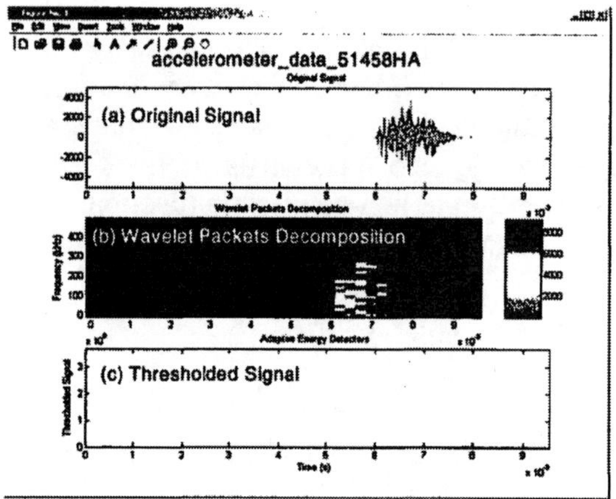

Figure 4: Original Signal and Wavelet Processing Results for Example Accelerometer

Based on the wavelet transform analysis described above, the next step is to extract specific features that can be used for assessing the impact damage characteristics. The features of interest in determining impact location and severity are related to the time of arrival of the propagating dispersive wave and magnitude of energy received at the accelerometer locations, respectively. Figure 5 shows the entire process associated with the feature-based approach, where the block diagram on the left is specifically used in the training stage and the diagram on the right is implemented during the evaluation stage.

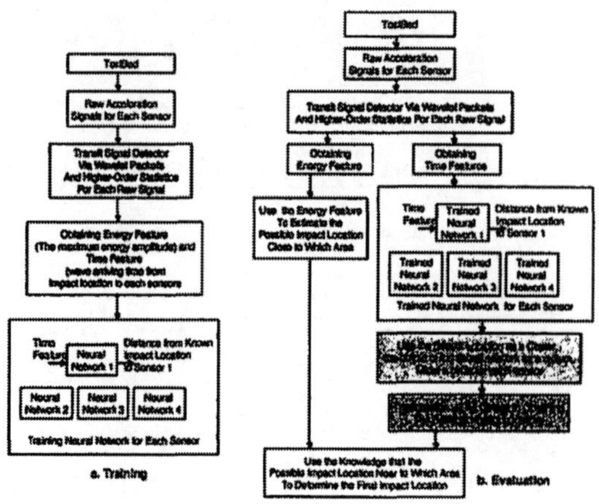

Figure 5: The Feature-Based Approach for Impact Damage Detection and Localization

Based on the output from the wavelet transform, thresholds are set on the magnitude of the output energy detector in order to calculate the time of arrival of the dispersive wave group velocity. In addition to this time-based feature, an integration of the overall signal magnitude represents an energy feature that is used to locate the approximate area of the impact. Next, a combination of these features is used to perform the final localization with a modified triangulation process. Figure 6 is an example of the features extracted from four accelerometers located at the corners of a 24-inch square plate used in the evaluation process described in a later section.

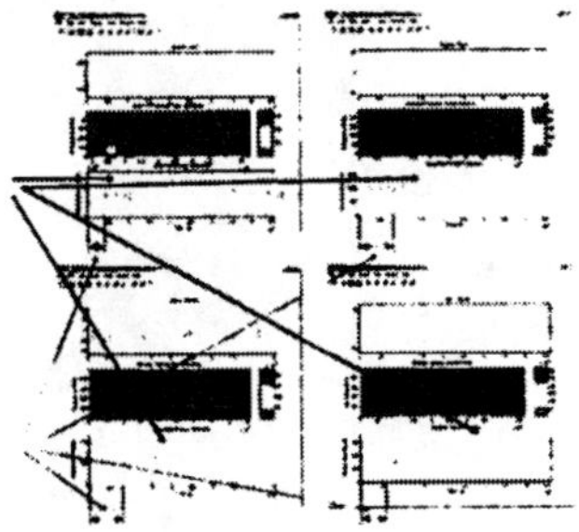

Figure 6: An Example of Feature Extraction Using Wavelet Packets and Statistics

After performing several impact tests, where the features described above are extracted with known information about the exact locations of the impact, a series of files are logged for training purposes. The energy features are then calculated from the maximum amplitude experienced from the Kurtosis of the output from the wavelet transform. This maximum amplitude feature is then compared with those experienced by all accelerometers to determine the general area of the applied impact. Next, time of arrival features are calculated from each accelerometer based on when the adaptive energy feature exceeds a particular threshold. Based on the number of known cases used in the training file, a simple back-propagation neural network was developed that relates the actual distance of the true impact location to each of the accelerometers. Figure 7 is an illustration of this process as related to the energy and time features.

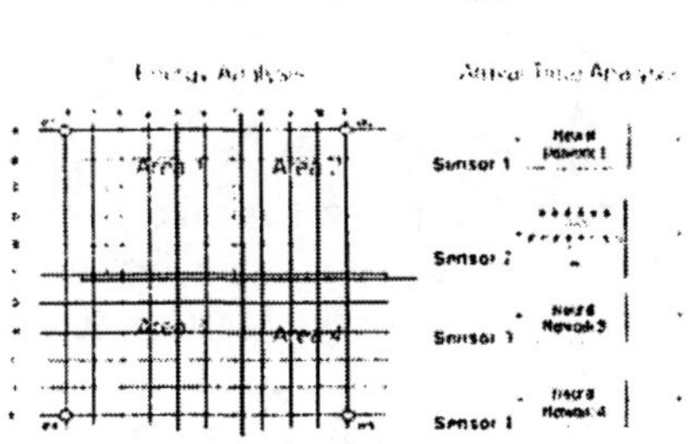

Figure 7: Energy and Time-Based Features Used in Training Algorithms

A neural network is a data-driven modeling approach that begins its training process with initial weights and constants associated with a selected set of basis functions. Hence, the approach taken in this paper was to train multiple neural networks with initial weights and basis function constants provided by a Monte Carlo simulation procedure. Therefore, the output of each of these trained neural networks with different initial conditions would provide a distance prediction based on the input group velocity arrival time extracted from threshold crossing. An example of the outputs of these networks for a given example is shown in Figure 8. Finally, a prediction on the location of the impact is made based on the intersection of each of the distances predicted from the accelerometers. These predicted distances are shown as concentric circles, with each of the accelerometers representing the centers.

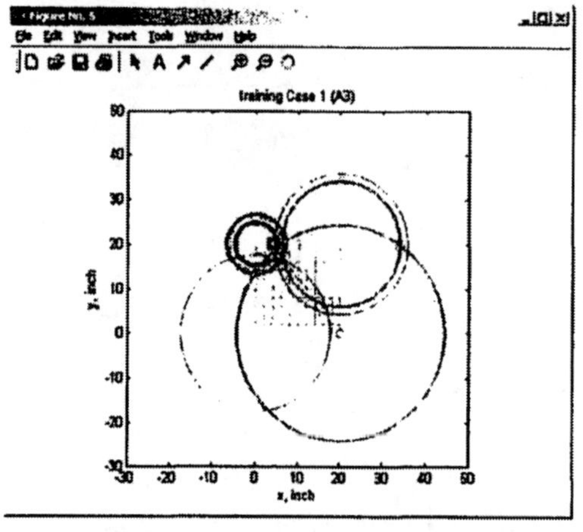

Figure 8: Example Outputs from Trained Neural Network for Distance Prediction

Model-Based (Finite Element and Dispersive Wave Theory) Approach

Method 1: Finite Element Transient Simulation: The first model-based method described is based on a finite element model transient simulation. In this process, the measured sensor outputs are compared with the predicted responses from the model to determine the impact location. If two responses are not similar in terms of a proximity norm, the algorithm revises the model run to characterize the response from a different impact location. The process is reiterated until the model response and the registered response of the structure match with a required accuracy.

Due to uncertainties associated with boundary conditions, material properties, damping characteristics and shape of the applied impulse, the theoretical response will not have an absolute match with the sensor output. Hence, a good feature for comparison was found to be the time of wave propagation from impact location to the sensor position, as was used in the feature-based approach already discussed. As a result, the numerical algorithm implemented was based on a minimization of the norm identifying the proximity of FE model and measured results in terms of the arrival time of wave propagation, as shown below.

$$J = \min_{x,y} \sum_{\omega} \sum_{s} \left\| T_{s,recorded}(\omega) - T_{s,FE}(\omega, x, y) \right\| \quad (4)$$

where $T_{s,recorded}$ – is the time of propagation as a function of frequency, as extracted from the raw sensor data; $T_{s,FE}$ – is the time of propagation as a function of frequency, as extracted from the FE simulation; x,y – unknown coordinates of the impact. Propagating flexural waves are naturally dispersive, whose velocities are a function of frequency ω. The function J is built by summation of the arrival time norms over all sensors, s=1,2,.... and over all frequencies ω picked for the data analysis.

Figure 9 illustrates the transient response of an elastic plate composed of 16 orthotropic plies, subject to the impact by a rectangular impulse load with duration time of 0.001 seconds. Figure 10 illustrates the 2-D wavelet transformation from the measured signal in the time – frequency domain, with corresponding contour lines. For each scale on this plot, the distance to the maximum peak indicates the arrival time of the wave group velocity. Note: that the arrival time of the group velocity from the simulation in Figure 9 is approximately 2.5 milliseconds and that matches the peak of the wavelet transform output as measured on the vertical time scale axis of Figure 10.

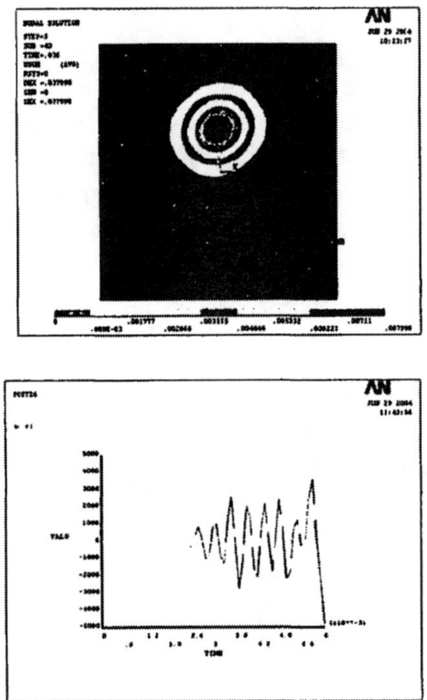

Figure 9: FE-Based Transient Response under Impulse Load with Duration 0.001 sec

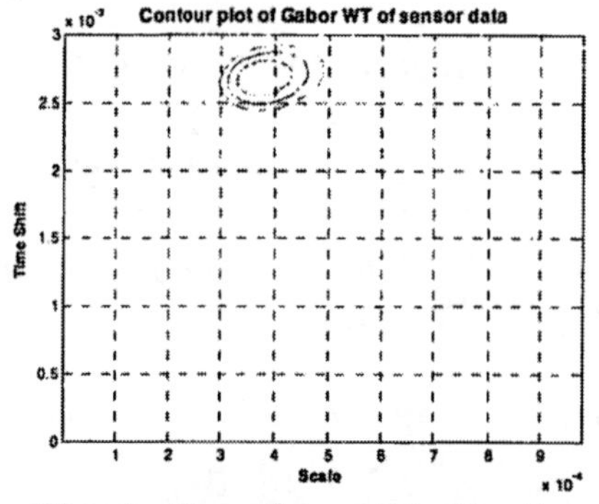

Figure 10: Wavelet Transform of the Predicted Response

The main drawback from using this approach is that practical implementation of this model-based approach using the FE model dynamic transient solution are very CPU intensive. Hence, an additional method that directly utilizes the equations of motion for dispersive waves was also developed and is presented next.

Method 2: Dispersive Wave Equations of Motion in Laminated Composites: Impact detection localization is an inverse problem that can also be addressed based upon theoretical principles of dispersive wave propagation. Based on the Mindlin–Reissner shell theory, the relationships between the in-plane stress vector $N=(N_x, N_y, N_{xy})^T$, bending moments $M=(M_x, M_y, M_{xy})^T$, in-plane strains $\varepsilon=(\varepsilon_x, \varepsilon_y, \varepsilon_{xy})^T$, and curvature vector $\kappa=(\kappa_x, \kappa_y, \kappa_{xy})^T$ can be found using the equation below.

$$\begin{Bmatrix} N \\ M \end{Bmatrix} = \begin{bmatrix} A & B \\ B & D \end{bmatrix} \begin{Bmatrix} \varepsilon \\ \kappa \end{Bmatrix} \tag{5}$$

where, the sub-matrix A represents the extensional stiffness, B – bending extension stiffness, and D – bending stiffness are calculated by the summation of the stiffness matrices Q_k which represent all of the elementary plies across the thickness (Jones,1975).

$$A=\sum_{k=1}^{N} Q_k (Z_k - Z_{k-1});$$

$$B=\tfrac{1}{2}\sum_{k=1}^{N} Q_k ({Z^2}_k - {Z^2}_{k-1});$$

$$D=\tfrac{1}{3}\sum_{k=1}^{N} Q_k ({Z^3}_k - {Z^3}_{k-1})$$

Z_k – the transverse coordinate of the k-th layer.

In the case of harmonic wave propagation, in-plane displacements, u, v, normal to plane displacements w and normal angle rotations ψ_x, ψ_y take the form

$$(u, v, w, \psi_x, \psi_y) = (U, V, W, \Psi_x, \Psi_y) e^{i(k_x x + k_y y - \omega t)} \tag{6}$$

Making the substitution into the equation of motion yields the following generalized Hermitian eigenvalue problem

$$\left\{ \begin{bmatrix} K_{11} & K_{12} & K_{13} & K_{14} & K_{15} \\ K_{12} & K_{22} & K_{23} & K_{24} & K_{25} \\ K_{13} & K_{23} & K_{33} & K_{34} & K_{35} \\ K_{14} & K_{24} & -K_{34} & K_{44} & K_{45} \\ K_{15} & K_{25} & -K_{35} & K_{45} & K_{55} \end{bmatrix} - \omega^2 \begin{bmatrix} \rho h & 0 & 0 & 0 & 0 \\ 0 & \rho h & 0 & 0 & 0 \\ 0 & 0 & \rho h & 0 & 0 \\ 0 & 0 & 0 & \rho h^3/12 & 0 \\ 0 & 0 & 0 & 0 & \rho h^3/12 \end{bmatrix} \right\} \begin{Bmatrix} U \\ V \\ W \\ \Psi_x \\ \Psi_y \end{Bmatrix} = \begin{Bmatrix} 0 \\ 0 \\ 0 \\ 0 \\ 0 \end{Bmatrix} \tag{7}$$

where ρ is the mass density, h is the thickness of the plate, K – stiffness matrix of laminate, and ω - frequency of the wave. Once ω as a function of k_x and k_y is known, the phase and group velocities are determined using the relations;

$C_p=\omega/k$; $C_g=d\omega/dk$;
$k=\mathrm{sqrt}(k_x^2+k_y^2)$;
$\theta = \arctan(k_y / k_x)$

The objective is to extract the group velocity Cg as the function of frequency ω and direction of propagation θ. Based on these equations, the dimensionless group velocities, corresponding to different angles as a function of frequencies, are plotted in Figure 11.

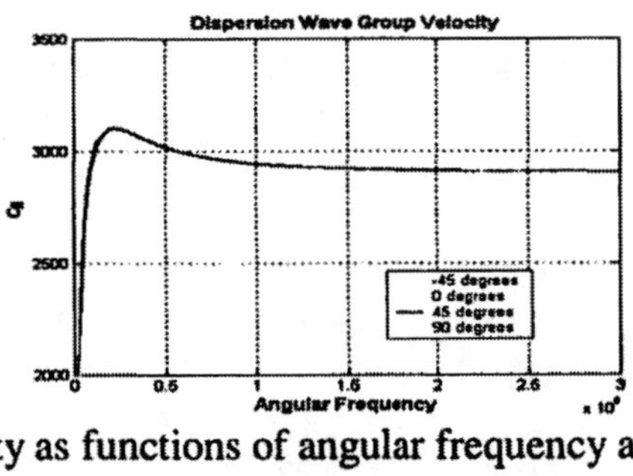

Figure 11: Group velocity as functions of angular frequency and direction of propagation

Once the group velocity is known as a function of frequency and angular of propagation, and the wave arrival time to each sensor is extracted from the corresponding sensor transient record, one can determine the coordinates of the location based on the following algorithm.

The relation between time of arrival t and unknown coordinates x, y of the impact is given by the simple mechanics based nonlinear expression below.

$$t_i(\omega) = t_{delay}(\omega) + \frac{\sqrt{(x - x_{sensor,i})^2 + (y - y_{sensor,i})^2}}{V_g(\omega, \theta)} \tag{8}$$

where $t_i(\omega)$ – time of propagation of the dispersive wave, corresponding to the angular frequency ω, from the point of impact to the i-th sensor; x,y – the unknown coordinates of the impact; x_{sensor}, y_{sensor} – the coordinates of the current sensor; $V_g(\omega,\theta)$ – group velocity as a function of frequency and direction of propagation, given by dispersive wave theory; $t_{delay}(\omega)$ – delay time, associated with the unknown local 3D and non-linear effects, as well as local damage phenomena. Three unknown variables are introduced into the equation above: two coordinates x,y and a delay time. For the number of measurements, this equation is satisfied in terms of the best fit, based on the least square approach.

Experimental Test Bed and Results: A flat, quasi-isotropic composite plate was used as a test bed for validating the proposed impact damage detection and isolation approaches. The objective was to assess the developed methodologies using acceleration-based sensors with the purpose of identifying (a) damage initiation and location in a composite structure. The combined data-driven and model-based approaches utilize analysis of wave characteristics (frequency, amplitude, velocity, type and damping characteristics) and source location using triangulation methods (in conjunction with time of arrival information and energy amplitude of signal received).

The material investigated is IM7/5250-4, quasi-isotropic $[0/45/90/-45]_{2s}$ laminate having the following elastic properties (x-y is in-plane while z is the thickness direction): $E_x=E_y=9.15$ Msi, $E_z=1.5$ Msi , $G_{xy}=3.49$ Msi, $G_{xz}=0.5$ Msi, $\square_{xy}=0.32$, $\square_{xz} = 0.45$

The composite plate was simply supported (on wood panel) on two edges, as shown in Figure 12(a). Four accelerometers were mounted in the locations CH1, CH2, CH3, and CH4, as shown in Figure 2-12(b). The plate was then tapped with a hand held hammer at several different chosen locations, and measurements were made at all four accelerometers. The data set was investigated to relate the basic signal characteristics to laminate properties in absence of damage and provides the training set for developing the methodology for source location.

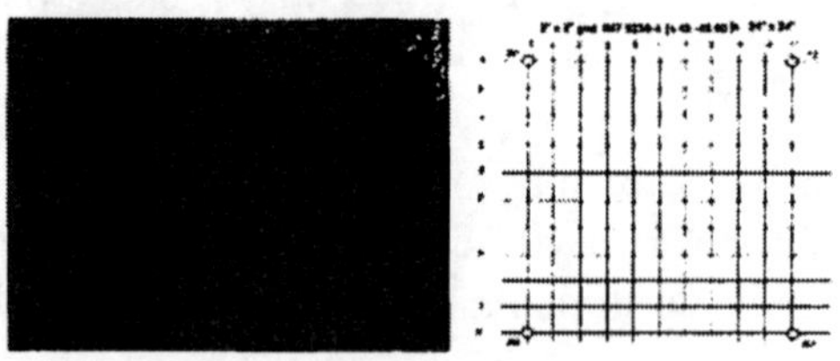

Figure 12: Composite Plate Showing Support Conditions, and Location of Accelerometers

A total of 12 different locations were selected on the composite plate for tapping measurements. The identities of tapping locations for 5 measurements were kept secret while that of the remaining 7 were disclosed for training the algorithms. Figure 13 is a comparison of the predictions using the feature-based intelligent approach and model-based FE approach with the truth data for the 5 blind test cases.

Using the data-driven (feature-based) approach, an exact prediction was obtained for one of the tests and was within 1 grid spacing for the remainder tests. For the model-based approach, the prediction was within 1 grid space for one test, within 2 grid spaces for three of the tests and within 3 grid spaces for the final test. As seen in this figure, both approaches are reasonably good in predicting the location of the impact using the measurements from the accelerometer sensors.

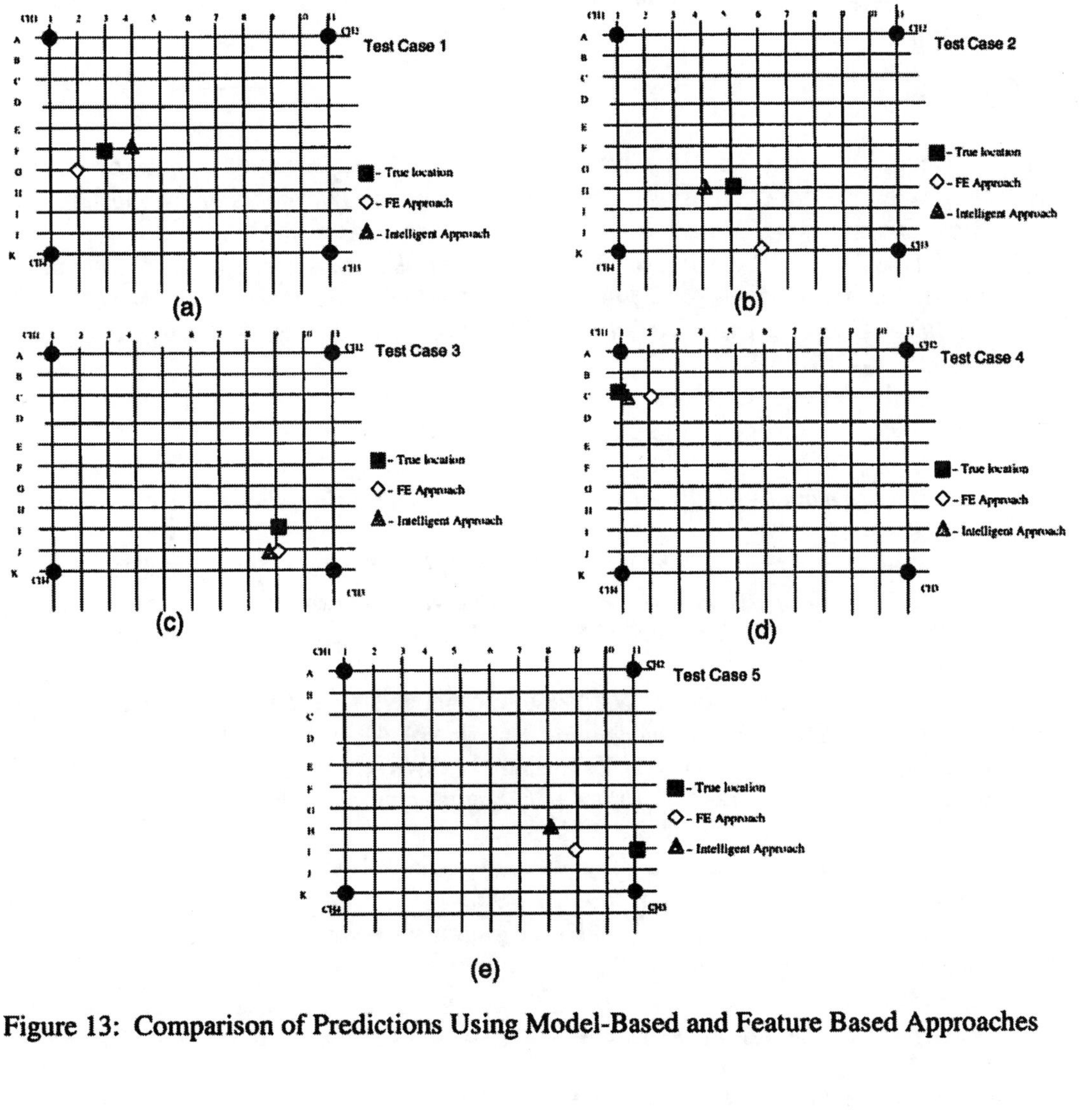

Figure 13: Comparison of Predictions Using Model-Based and Feature Based Approaches

Conclusions: Impact damage, which may occur during service or be maintenance induced, is a leading cost driver for composite structures used in various aerospace vehicle applications. Impact Technologies, in collaboration with the University of Dayton Research Institute (UDRI), have developed both feature-based and model-based impact damage detection and localization algorithms for composite structures. Specifically, this paper presented: 1) A robust and low-cost sensing approach using only COTS accelerometers to provide comprehensive insight into the structure's health state; 2) Advanced signal analysis and feature extraction techniques that included time-frequency based wavelet transforms and higher order statistics; 3) An intelligent approach for impact damage localization using triangulation and neural network techniques; 4) Two different model-based impact damage detection techniques; and 5) Experimental validation of the proposed algorithms using a composite plate test bed. Although these results are considered preliminary, our research team believes that the robust nature of the sensors and algorithms will allow for future scalability to larger and more complex structures.

References:

[1] X. Deng, Q. Wang, and V. Giurgiutiu, "Structural health monitoring using active sensors and wavelet transforms," *SPIE's 6th Annual International Symposium on Smart Structures and Materials*, March, 1999, Newport Bench, CA.

[2] S. W. Doebling, C. R. Farrar, M. B. Prime, and D. W. Shevitz, *Damage Identification and Health Monitoring of Structural and Mechanical Systems from Changes in Their Vibration Characteristics: a Literature Review*, Los Alamos National Laboratory, Report LA-13070-MS.

[3] S. R. Hall and T. J. Conquest, "The total data integrity initiative – structural health monitoring, the next generation," *Proceedings of the USAF ASIP*, 1999.

[4] R. Ikegami, E. Haugse, A. Trego, L. Rogers, and J. Maly, *Structural Technology and Analysis Program (STAP) Delivery Order 004: Durability Patch*, AFRL-VA-WP-TR-2001-3037, 2001.

[5] M. Jansen, *Noise Reduction by Wavelet Thresholding*, Springer Verlag, 2001.

[6] S. S. Kessler, *Piezoelectric-based In-situ Damage Detection of Composite Materials for Structural Health Monitoring Systems*, Massachusetts Institute of Technology, Ph.D. Thesis, 2002.

[7] H. Krim, and J. C. Pesquet, "Multiresolution analysis of a class of nonstationary processes," *IEEE Trans. on Information Theory*, vol. 41, no. 4, pp. 1010-1020, 1995.

[8] Masuda, M. Noori, A. Sone, and Y. Hashimoto, "Wavelet-based health monitoring of randomly excited structures," *15th ASCE Engineering Mechanica Conference*, June, 2002, Columbia University, New York, NY.

[9] M. Nielsen, "Highly nonstationary wavelet packets," *Applied and Computational Harmonic Analysis*, vol. 12, pp. 209-229, 2002.

[10] D. B. Percival and A. T. Walden, *Wavelet Methods for Time Series Analysis*, Cambridge University Press, 2000.

[11] P. Ravier, P.-O. Amblard, "Wavelet packets and de-noising based on higher-order-statistics for transient detection," *Signal Processing*, vol. 81, no.9, pp 1909-1926, August 2001.

[12] B. W. Suter, *Multirate and Wavelet Signal Processing*, Academic Press, San Diego, CA, 1998.

TURBINE ENGINE HOT SECTION PROGNOSTICS

Harrison H. Chin

Applied Concept Research, Inc.
23 Pine Street
Bedford, MA 01730
Harrison.Chin@AppliedConceptResearch.com

Abstract: The design life of engine hot section components is typically shorter than their cold section counterparts, due to the fact that they have to operate under elevated temperatures and stresses. Previously unseen failure modes are being discovered due to the increasing demand for modern engines to operate at higher temperatures and stresses, coupled with the use of new materials for the engines. These unanticipated failure modes can keep engines from achieving their expected design life and may even result in loss of engines. This paper describes an integrated prognostic approach designed to monitor hot section component degradation under elevated temperature, pressure, and corrosion, as well as to infer their useful remaining life. This approach combines both safe life and damage tolerance concepts, and takes into account of all the common hot section failure mechanisms such as creep, oxidation, low and high cycle fatigue, and corrosion. Successful integration of this approach into modern Prognostic and Health Management (PHM) systems can reduce the cycle cost for both the military and the commercial gas turbine engines while maintaining the equivalent safety.

Key Words: Creep; Damage tolerance; Engine hot section; High cycle fatigue; Low cycle fatigue; Oxidation; Prognostics; Safe life

Introduction: Traditional engine health maintenance and life management relies mostly on Time Based Maintenance (TBM), which is done by tracking the operating time and cycles of the engine. Maintenance for a component is performed based on a fixed operating hour/cycle schedule, and/or using exceedance data from basic instruments and algorithms. These maintenance and retirement schedules are typically derived from material properties and fatigue data, based on the assumption of worst-case usage. To help ensure safety, the life limits of flight critical components represent only a fraction of the probable usage that can accumulate before failure indications arise. As such, it is generally referred to as the Safe-Life (SL) approach. The SL approach does not account for the component that fails early or would have lasted for many times the safe life limit.

To minimize the possibility of early component failures, diagnostics are employed for flight critical components. Diagnostics in TBM consists of visual inspections and analysis of operator reports, sometimes aided by data on temperatures, vibrations, and gas path properties. Empirical limits are set for individual measurements as to determine whether a fault is present. Component health determination is often based upon circumstances as determined by flight line or operational personnel. This type of health management is sometimes referred to as the Damage-Tolerance (DT) approach. It is now well understood that TBM-driven lifing of engines leads to both costly unnecessary retirements and maintenance, as well as continued operation of engines that have suffered degradation that may endanger crew and mission performance.

The ability to properly diagnose engine hot section components that are failing due to unknown causes, or prematurely depends on new sensor technologies that can characterize and localize the component that is damaged or degrading. Examples of such sensor technologies include optical combustion constituent characterization, optical pyrometry to determine and map the temperature of blades surfaces, dynamic pressure sensing of combustion stability and inter-blade pressure profiles, bearing vibration diagnostics, electrostatic debris monitoring, and fuel nozzle thermal/acoustic signatures.

One such new sensor technology is an optical system developed by Meggitt/Vibro-Meter that provides enhanced hot section diagnostic and prognostic capability. This system detects the photonic emissions of the combustion flame, and filters the signal to classify and quantify the constituents of the flame. It is believed that this signature can be associated with hot spots, fuel contamination, uneven flame patterns (pattern factor), and upstream (nozzle and compressor) problems that lead to hot section degradation.

Diagnostic systems can be further enhanced by the incorporation of prognostics, which is the enabler for Prognostic Health Maintenance (PHM) by accurately relating the detected fault to the remaining useful life of a component. This requires the development of a reliable prognostic strategy that integrates the history of diagnostics in conjunction with the operational/structural usage, and the physics of fault progression.

The Useful Life Remaining (ULR) concept, which is illustrated in Figure 1, is designed to provide a prognostic life management system that can maximize the useful service life of engine hot sections while providing a greater level of safety than the current TBM lifing and inspection approach does. This goal is achieved by combining the continuous analysis of state/usage parameters and advanced diagnostics capabilities.

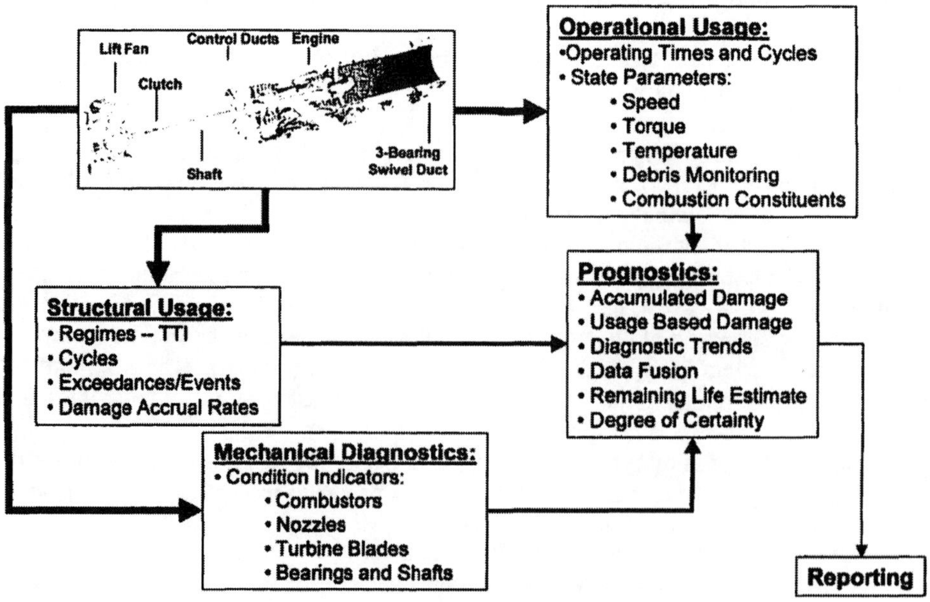

Figure 1. Overview of the ULR prognostic concept.

Hot Section Failure Mechanisms: The engine hot section is generally referred to where the components are subject to high temperatures, such as in the combustor and turbine. These components include: nozzles, combustor liners, transition pieces, 1st and 2nd stage vanes and blades of the gas turbine. Historically turbine blades are the primary contributor to the maintenance cost of gas turbine engines [1, 5, 11].

In general, hot section life is known to be critically dependent upon the following factors:

- Maintaining temperatures within specified ranges
- Maintaining uniform spatial distribution of temperatures at the turbine inlet
- Minimizing incidents of overspeeding of rotating components
- Avoiding/managing corrosive fuel/air mixtures
- Hot start intensity and frequency

Due to elevated temperatures and operational stresses, the design life of a hot section component is only about half as long as that of a cold section component. As engines are being operated at higher temperatures and stresses, and with the use of new materials, previously unseen failure modes are being discovered. These unanticipated failure modes can keep engine components from achieving their expected design life [9, 13, 14]. Common failure mechanisms, which can arise from normal operations or improper management of the engine, include:

- Low Cycle Fatigue (LCF)
- High Cycle Fatigue (HCF)
- Creep/Rupture
- Oxidation
- Corrosion
- Foreign Object Damage (FOD)

The first four mechanisms are typically related to the engine design, while the last two can be attributed to the operating environment.

<u>Low Cycle Fatigue (LCF):</u> Engine hot section components develop so called Thermo-Mechanical Fatigue (TMF), which is a unique type of fatigue because the material is simultaneously subjected to fluctuating loads and temperatures. Isothermal life prediction techniques are often not applicable to TMF because different damage mechanisms can arise under extreme temperature conditions. This type of fatigue failure can be furthered classified into In-Phase (IP) TMF that occurs when the maximum strain and peak cycle temperature coincide, or Out-of-Phase (OP) TMF where the maximum strain and lowest cycle temperature coincide. One example of OP TMF cycles is the TMF damage occurred on the leading edge of a gas turbine blade from repeated turbine starts and stops [4, 9].

<u>High Cycle Fatigue (HCF):</u> HCF represents metal fatigue that results from cracking or fracture phenomenon at stress levels much lower than stresses associated with steady loading. HCF can result from a combination of steady stress, vibratory stress, and material imperfections that initiate the formation of a small microscopic crack. Since today's advanced turbomachinery blading is designed to have high steady stress levels, HCF occurs because of high mean stress - low amplitude vibratory loading of the airfoils. The most common unsteady aerodynamic flow-induced vibrations are wakes generated by an upstream airfoil row. A reduction in the relative velocity in the wake causes a decrease in the absolute velocity as well as an incidence increase to the downstream stator vanes, which then leads to fluctuating forces on the downstream airfoils [4, 9, 13].

HCF resulting in the loss of gas turbine engine blades or disks is currently the leading surprise failure mode. This is because of the design of high thrust-to-weight ratio engines, accomplished by increasing the mass flow and utilizing fewer parts.

Oxidation: As a result of the extreme temperatures, turbine blades are quite susceptible to high-temperature oxidation. Oxidation problems occur due to the formation of brittle oxide surface layer on the turbine blades, which can lead to premature fracture damage in fatigue. Turbine blades rely on cooling air delivered from upstream in the engine to cool the airfoils and to provide a film of cooler air to protect the airfoil from high temperature. If the continue flow of cooling air is compromised or if the external temperature environment changes, the blades can fail before their design life [3, 6].

Creep: Creep is permanent material deformation under stress. A certain amount of thermal activation energy is required for creep to occur. Under high temperature conditions, materials are weakened thus the creep process accelerates. Creep can initiate small cracks in components, which can then grow in size and eventually lead to component failures. These failures can be catastrophic, especially when the failed components are compressor or turbine rotors. This crack growth and component failure process is known as stress rupture. The evolution of creep damage can be represented by a creep curve consisting of three stages: primary (transient), secondary (steady-state) and tertiary (unstable) which then leads to rupture [2, 3, 4].

Creep rupture occurs primarily due to microcraking and void growth along grain boundary, and is associated with boundary sliding. In single crystal blades where there is no grain boundary, failure is associated with void growth and the subsequent microcracking of voids. The time to rupture (t_r) is often found to increase linearly as the steady-state creep rate ($\dot{\varepsilon}_s$) decreases, based on the Monkman-Grant relationship:

$$t_r = M\frac{1}{\dot{\varepsilon}_s}$$

where M is a material dependent variable, ranging from 0.1-10. As $\dot{\varepsilon}_s$ increases, the creep curve gets more vertical and therefore the time to rupture would decrease.

Creep strains in the low temperature range (less than 30% of the material's melting point) are usually very low and as such creep rarely leads to failure. For intermediate temperatures (between 40% to 90% of the material's melting point), the steady-state creep rate can be shown to be a power-law function of stress (σ) and temperature (T):

$$\dot{\varepsilon}_s = AB\sigma^a e^{-(Q_C/RT)}$$

where A is a constant, B and a are material-dependent constants, R is the universal gas constant (≈ 8.314 JK-1mol-1) and Q_C is the material-dependent activation energy for creep to occur.

Figure 2 shows example of five creep curves for several different temperatures and stresses. These curves are derived using the equation above with S-590 alloy properties.

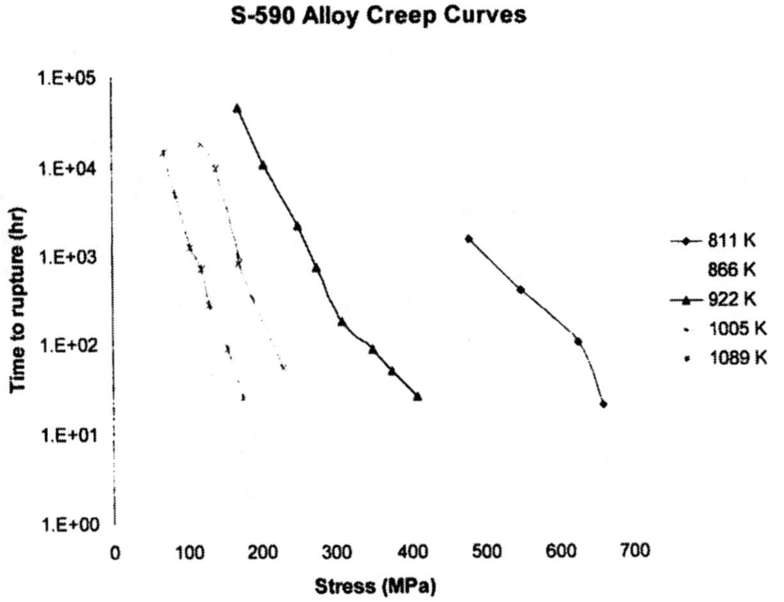

Figure 2. Example of creep curves under various temperature and stress conditions. Shorter time to rupture occurs under higher temperature and stress.

After unloading, all materials demonstrating creep also display creep recovery, i.e. some of the strain will gradually disappear. For metals however, the recoverable strain is usually quite small (≤ 1%) and can be ignored.

Generally speaking the performance of current industrial gas turbines is limited by the temperature and strength capabilities of the materials used. Since industrial gas turbines are designed to run continuously for extended periods of time, the primary failure modes for the blades are HCF and creep. As such, lifing research has been concentrated on developing methods for describing the effect of these phenomena on blades. On the other hand, jet engines are cycled between start up and maximum thrust for each flight (about 10 flights a day for commercial aircraft). Therefore LCF is the dominant failure mechanism, though some unexpected failures can be attributed to HCF.

Several damage control approaches have been proposed in the past to mitigate the hot section failure modes, and some have been integrated into the Full Authority Digital Electronic Control (FADEC) module for modern engines. One such example is the Life-Extending Control (LEC) approach originated from rocket engines, which was designed to control LCF/HCF and creep by optimizing the engine fuel flow rate [13, 14].

ULR Prognostic Model: As mentioned earlier, lifing models can generally be classified into two categories: total life and crack growth. Total life models, such as the Palmgren-Miner model, only calculate the time to failure and do not consider the way failure is reached. These models are representatives of the SL approach, aiming to retire a component before a crack initiates. To ensure safety, a substantial amount of conservatism is built into a total life model since it is an open-loop approach [7, 8, 9, 13, 14].

In contrast crack growth models accept the presence of material defects and aims to monitor crack growth. Because of this, they are also known as the DT approach. In this approach components are removed before the crack becomes unstable. Local analysis, using stress histories at specific locations on the component (i.e., locations of crack initiation or high stress points) is performed based on crack growth models. As such they are only applicable to certain locations. Recent sensor advancements enable some level of monitoring of crack initiations and their subsequent growth. However the effectiveness of this approach is not fully understood due to the uncertainty associated with the data. It is further complicated by the fact that most direct measurements are not possible due to the extreme environment in the engine hot section.

Temperature is the key factor in hot section component life. Heat transfer models are usually used to model the distribution of component temperatures. The metal temperatures are then used in conjunction with the pressure and centrifugal loads, as well as the material stress-rupture properties to determine the hours of life used or remaining, typically through finite element analysis. These models however, are difficult to implement for automated real-time analysis.

The hot section component ULR prognostic model incorporates both SL and DT approaches, as shown in Figure 3. In this model, the parameter data such as temperatures, pressures, speeds, torques, load factors, vibrations, and stress waves from various sources are utilized to compute the time in each operating condition for each flight. This usage data, combined with OEM supplied damage rates, are then used to compute the projected time to retirement for each component.

The acquired data is also used by the crack monitor to detect any presence of cracks. If no crack is detected, the projected remaining life of a component is reported based on the SL estimate. Once a crack is detected and confirmed, the crack growth estimator will use the data to determine the rate of crack propagation and to report the time to failure, based on various mission spectra.

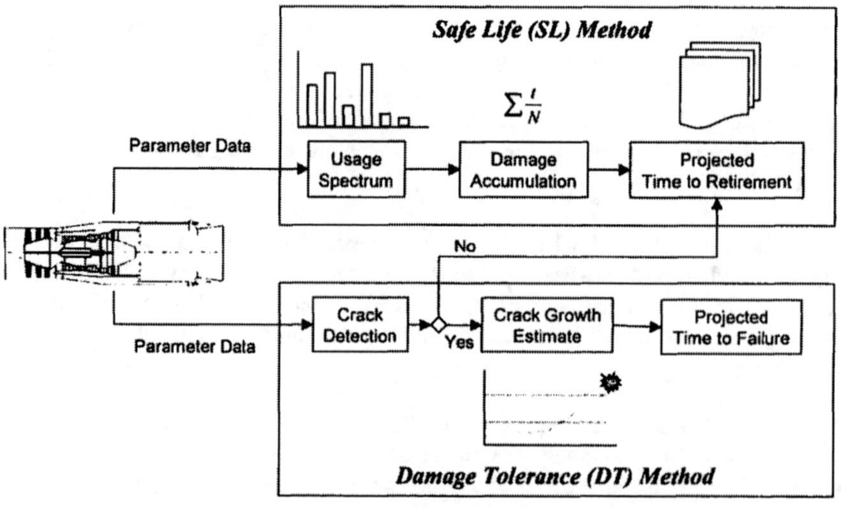

Figure 3. The ULR prognostic model includes both safe life (SL) and damage tolerance (DT) methods.

Damage accumulation in the ULR model for engine hot section components incorporates the following algorithms:

For LCF/HCF, damage accumulation is done by Miner's rule:

$$D_f = \sum \frac{n_i}{N_i}, \qquad i = 1,\ldots m$$

where D_f is the accumulated damage, n_i is the number of cycles at a given load, N_i is the number of cycles to failure, and m is the total number of load conditions. End of life is implied when D_f reaches one.

Temperature dependent creep can be accumulated in a similar manner as described below:

$$D_c = \sum \frac{t_i}{t_{ri}}, \qquad i = 1,\ldots m$$

where D_c is the accumulated creep damage, t_i is the time at a given strain and temperature, t_{ri} is the time to rupture, and m is the total number of strain/temperature combination.

Turbine blade oxidation damage accumulation is calculated as:

$$D_o = \sum D_o^l, \qquad l = 1,2,3$$

$$D_o^l = \sum \frac{t_i}{t_{oi}}, \qquad i = 1,\ldots m$$

where D_c is the accumulated oxidation damage, for a blade consisting of three layers of different materials it is the sum of damages from each layer, t_i is the time at a given temperature, t_{oi} is the time to failure, and m is the total number of thrust/temperature combination.

Since components in the hot section are subject to the combination of the above three failure modes, they are tracked by the ULR model. At any given time, one of the three will be the dominant failure mode (i.e., the highest accumulated damage) and the final component life is projected based on the damage from this dominant mode.

Crack detection and tracking involve detecting crack initiations, diagnosing the location and severity of a detected crack, and predicting the crack growth rate. Typically this is done by collecting and analyzing data from one or more sensors. For example abnormal stress waves, vibrations, and/or uneven temperature distributions may indicate a cracked blade.

The ULR prognostic model provides a data fusion algorithm that combines various indicators from multiple sensor sources to produce a component health indicator. The prognostic model provides the ability to determine if a trend is developing in the health indicator for a given component. Furthermore, it includes the ability to determine how quickly that indicator is approaching (or might cross) a preset limit, based on projected usage.

Flame Contaminant Detection (FCD): Corrosive contaminants such as sodium in the combustor can weaken parts and thus reduce life. An optical system developed by Meggitt/Vibro-

Meter has the ability to detect the photonic emissions of the combustion flame and filters the signal to classify and quantify the constituents of the combustion flame [10]. This signature can be associated with hot spots, fuel contamination, uneven flame patterns (pattern factor), and upstream (nozzle and compressor) problems that lead to hot section degradation.

Extensive data analysis showed that the FCD is capable of detecting sodium contaminants for the entire gas turbine load range, and has the potential for quantifying the amount of contaminants in the combustor. Figure 4, which represents one of the results obtained in this study, shows the amounts of sodium contaminant predicted based on the FCD data. The result indicates that the predicted sodium amounts match the actual amounts injected to the combustor relatively well.

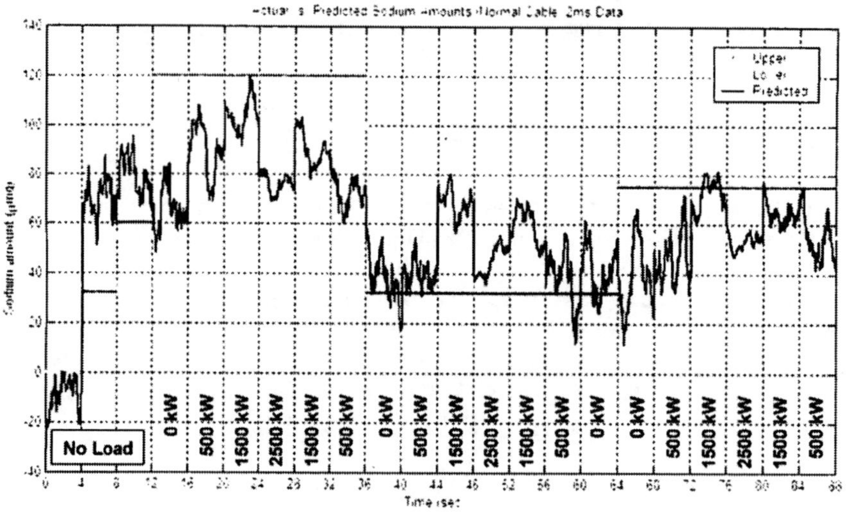

Figure 4. Actual vs. predicted sodium contaminant amounts.

Summary: The ULR prognostic methodology combines both SL and DT approaches to achieve complete and accurate prognostics. The SL approach utilizes damage accumulation to determine the safe-life limit for a component, while the DT approach relies on using sensor data to detect and monitor cracks.

Damage accumulation in the ULR methodology is performed through the use of aircraft/engine parameter data mapped to a set of component load conditions. Component load data, damage rates, and their failure modes are then combined to derive damages after each flight. Damages due to LCF/HCF, creep, and oxidation are monitored and accumulated.

Crack detection and monitoring is done by the fusion of sensors designed to identify the presence of abnormal stress waves, vibrations, pattern factors, discharges, in conjunction with a physical model of the engine hot section. When no crack is present, the ULR prognostic model projects the component life based on damage accumulation. Once a crack is detected, the ULR model switches to crack tracking prognostics for useful life remaining estimation. This functionality provides a safeguard to failures that might occur before the safe-life limit, and addresses the safety and reliability concerns among aviation community.

Investigation and testing of the flame contaminant detector showed that it is capable of detecting sodium contaminant for the entire gas turbine load range. In addition it shows the potential for quantifying the amount of contaminant in the combustor. This capability is especially important for establishing the relationship between hot section degradation and various contaminants in the engine. Data from the frame detector can also be used to infer temperatures, which is a crucial parameter for hot section component lifing.

Acknowledgments: The work presented in this paper was funded by the US Navy SBIR Program (Contract # N68335-03-C-0192). The author would like to thank Meggitt/Vibro-Meter Inc. for providing the flame contaminant data.

References:

[1] K. Hunecke, Jet Engines – Fundamentals of Theory, Design and Operation, 6th Impression, Biddles LTD, 2003.
[2] D. P. Shepherd, A. Wisbey, G. F. Harrison, T. J. Ward and B. Vermeulen, "Cyclic Operation of Aero Gas Turbines – Materials and Component Life Implications," Science Review, Materials at High Temperatures, 18 (4), pp. 231 –239, 2001.
[3] D. Zhu and R. A. Miller, "Thermophysical and Thermomechanical Properties of Thermal Barrier Coating Systems," NASA/TM—2000-210237, 2000.
[4] S. Fleeter, C. Zhou, E. Houstis, and J. Rice, "Fatigue Life Prediction of Turbomachine Blading," School of Mechanical Engineering and Department of Computer Sciences, Purdue University West Lafayette, Indiana 47907.
[5] D. S. Aldridge, "Component and System Life Distribution Prediction Using Weibull and Monte Carlo Analysis with Reliability Demonstration Implications for an Electronic Diesel Fuel Injector," 2003 SAE World Congress, Detroit, Michigan, March 3-6, 2003, Paper # 2003-01-1363, 2003.
[6] V. SidWell and D. Darmofal, "Probabilistic Analysis of A Turbine Cooling Air Supply System: The Effect on Airfoil Oxidation Life," Proceedings of ASME Turbo Expo 2003 Power for Land, Sea, and Air June 16-19, 2003, Atlanta, Georgia, USA, Paper # GT2003-38119, 2003.
[7] S. I. Zamrik, "Life Prediction of Advanced Materials For Gas Turbine Applications," DOD-FETC/SCEC Contract # AGTSR-93-01-SR012D.
[8] T. Tinga, W.B. de Wolf, W.P.J. Visser and S. Woldendorp, "Integrated lifing analysis of a film-cooled turbine," National Aerospace Laboratory NLR-TP-2001-402, 2001.
[9] M. Kemppainen, J. Scheibel and R. Viswanathan, "Combustion Turbine Hot Section Life Management," CT Hot Section Life Management, OMMI (Vol. 1, Issue 2) August 2002.
[10] Flame Contaminant Detector LBES Test & Evaluation Plan For Allison 501K GTE NAVSEA – Philadelphia, Contract # N65540-03-M-0253.
[11] O. Younossi, M. V. Arena, R. M. Moore, M. Lorell, J. Mason, and J. C. Graser, Military Jet Engine Acquisition – Technology Basics and Cost-Estimating Methodology, RAND, 2002.
[12] H. Weber, D. Hartmann, O. Faber, and H. J. Niemann, "Process Analysis and Reliability Estimation," 8th ASCE Specialty Conference on Probabilistic Mechanics and Structural Reliability, 2000.
[13] L. D. Jaw, D. N. Wu, and D. J. Bryg, "Tracking and Control of Gas Turbine Engine Component Damage/Life," NASA Center for Aerospace Information Report #20030020435.
[14] M. A. McGaw, "Life Prediction Technologies for Aeronautical Propulsion Systems," NASA Center for Aerospace Information Report #92N22520.

PHYSICS-BASED PROGNOSTICS FOR OPTIMIZING PLANT OPERATION

Don B. Jarrell and Leonard J. Bond[a]

Pacific Northwest National Laboratory[b]
PO BOX 999, K5-26
Richland, WA 99352
Tel: (509) 372-4096; (509) 375-4486

Abstract: Scientists at the Pacific Northwest National Laboratory (PNNL) have examined the necessity for optimization of energy plant operation using "DSOM®[c]" – Decision Support Operation and Maintenance and this has been deployed at several sites. This approach has been expanded to include a prognostics component and tested on a pilot scale service water system, modeled on the design employed in a nuclear power plant. A key element in plant optimization is understanding and controlling the aging process of safety-specific nuclear plant components. This paper reports the development and demonstration of a physics-based approach to prognostic analysis that combines distributed computing, RF data links, the measurement of aging precursor metrics and their correlation with degradation rate and projected machine failure.[d]

Keywords: DSOM®; Prognostics

Introduction:

Component failures run the gamut from mildly annoying, (e.g., a leak in a compressed air line), to disastrous, as in the stuck open pressurizer relief valve at Three-Mile Island. In all such events a price is to be paid in terms of the economic penalty paid in lost revenue resulting from component replacement (downtime, maintenance hours and parts), a loss in production reliability (schedule impingement), or the public perception of the safety of the process itself.

There is a compelling need to provide new and improved measurement science and technology, together with appropriate methodologies, for the analysis, monitoring, protection and management of the nation's critical infrastructure. Many critical systems face threats as a result of random acts, including terrorism, and threats that result from degradation caused by the effects of aging.

---

[a] From February 1, 2005, Director, Center for Advanced Energy Studies, Idaho National Laboratory.
[b] Operated for the U.S. Department of Energy by Battelle Memorial Institute under Contract DE-AC05-76RLO 1830.
[c] DSOM® is a registered trademark of Battelle Memorial Institute.
[d] This work was performed with the support of the U.S. Department of Energy's (DOE's) Nuclear Energy Research Initiative (NERI) and PNNL's Laboratory Directed Research and Development (LDRD) funding.

Much of the critical civil infrastructure in the US is rapidly approaching, or in some cases has already passed, its original design life. The total replacement cost for all aging systems that are found in the national critical infrastructure – including the electrical power system, oil and gas production and storage, transportation and water supply – is enormous and far beyond reasonable levels of expenditure for "on-time" replacement [1].

Producing an improved methodology for optimizing asset lifetime productivity was a direct result of plant operations and maintenance (O&M) experience being applied in a research environment to develop and quantify the fundamental communications and diagnostics relationships between the necessary O&M functional elements of an operating process plant [2].

Diagnostics: understanding the problem and its root cause.

The DSOM® system was designed to diagnose plant problems and provide the operations, maintenance, engineering, training and administrative (termed OMETA) agents necessary to allow the formulation of each of these five plant perspectives in a software model. The DSOM® program aims to improve the communication linkages necessary to facilitate the information flow that each element needs to cope with process plant operational problems.

Prognostics: understanding component stressors and the rate of process and material degradation that will result in a future failure.

Stressors are defined as those physical conditions that would cause a specific degradation mechanism to be active such as *low inlet pressure* causing fluid cavitation in a pump or *high temperature* causing a rapid drop in electrical insulation resistivity. The DSOM® physics-based approach to prognostics is really a conjunction and codification of several well accepted and related facets of the O&M world [2].

A First Principals Approach to Prognostics:

Condition-based maintenance (CBM) stressor-based analysis is based on understanding which stressor characteristics provide an early indication for mapping subsequent damage resulting from a degradation mechanism. The resulting physical damage and the associated decrease in asset performance start with the application of a stressor to the component. The design engineer sets the desired stressor intensity level so the degradation in the physical state of the component happens slowly enough for the equipment to last for a specified design life. In general, when the design limit of a stressor is exceeded, during operation, the component life expectancy starts to shorten to less than the projected design duration. Conversely, careful control of operational parameters can extend the component life beyond that normally expected for the design failure point.

The premise of the prognostic methodology is that, by not trending a performance metric per se, but by focusing on trending the stressor characteristics, a relationship can be derived that will allow a much more accurate projection of the remaining useful life. Figure 1 shows the expected result in narrowing the uncertainty by keying on the stressor itself.

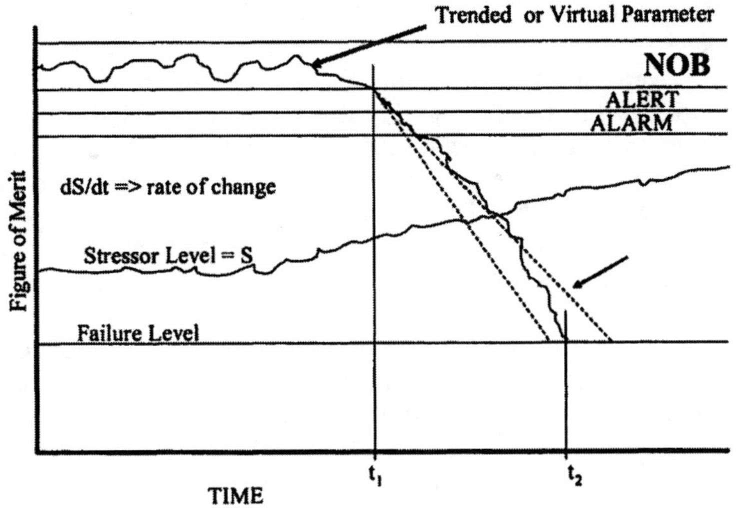

Fig. 1. Stressor Measurement Effect on Prediction Uncertainty

In the trend analysis example above, the slope of the trended parameter is thought to give a measure of the degradation rate of the component performance. This is assumed to be a function of the rate of decline in the physical characteristics of the equipment as well. Experience from measurements has shown this assumption to be true if one accounts for the nonlinearity between physical attributes and their effects on performance [2].

In the stressor-based analysis, the instantaneous degradation rate can be correlated to the stressor intensity by a functional relationship. This can then be taken one step further, to:

$d^2P/dt^2 = dS/dt$, which is nothing more than stressor trend or slope (1)

So by following the slope of the stressor intensity, we have a precursive measure of the rate of change in the performance degradation. Thus, the stressor slope can be used to predict and to refine the path of the performance vector.

The rate of change in the stressor slope gives an even more sensitive precursive dimension for narrowing the uncertainty of the predicted performance path. This stressor gradient (termed the root precursive indicator) is the most sensitive time-linked correlation from stressors through to failure.

$d^2S/dt^2 \Rightarrow$ Root Precursive Indicator (2)

If a measure of this root indicator can be accurately determined, each level in the derivative chain can be integrated to provide an accurate physical description of the future condition and performance of the component. It also follows that as diagnosticians, our quest is a complete mathematical description of the stressor's derivative chain – its level, slope and gradient.

Formulation of Stressor-Degradation Correlations:

A set of experiments were performed at PNNL to provide a proof of principal demonstration for this methodology. A single-stage centrifugal pump was chosen as the active component that would bear the greatest impact on common production facilities. By examination of failure and root cause histories, the two most predominant mechanisms resulting in centrifugal pump failure are found to be cavitation and vibration. Careful consideration was given to the instrumentation required to quantify appropriate stressor intensities and examine the physical effects of degradation [2].

The ultimate goal is to allow a first principles approach to a prognostic algorithm that will accurately predict the reduction of impeller and bearing residual life as a result of cavitation and misalignment of the driver from its driven component.

Pump Cavitation Experimental Results:

The initial goal for the cavitation test series was to characterize the operational data as well as the spatial and spectral nature of the cavitation produced in a single stage centrifugal pump. To this end, highly accurate operational instrumentation was used to measure the motor current; suction pressure and temperature; and the discharge pressure, temperature and flow. Specialized acoustic sensors were then installed in the test pump. These sensors were placed in direct contact with the pumped fluid to provide a clear view of the acoustic energy impacting the wall of the volute.

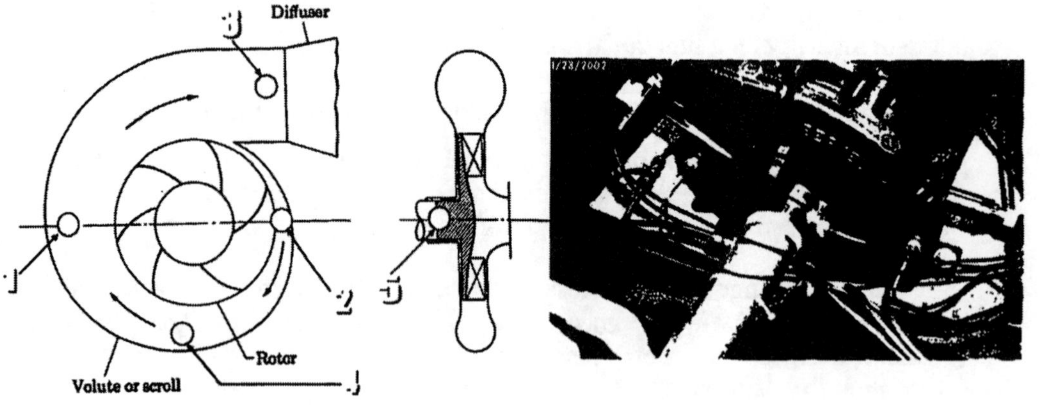

Fig. 2. Initial Acoustic Sensor Placement.

Acoustic emission (AE) techniques [3] were used in an attempt to detect incipient cavitation and to quantify the spectral and spatial (source location) intensity of the degradation mechanism they present. The objective was to not only show that AE can be used to detect the vapor implosion acoustic signals, but to discriminate cavitation intensity from signals generated by the mechanical and fluid noise in the pumping system.

To accomplish this end, an array of Pinducer VP-1093 acoustic emission sensors were placed in the pump suction and volute (pressure recovery) sections to map acoustic impingement intensity. Special plugs were drilled and the sensors were epoxied into the plug without metal-to-metal contact. The depth of sensor extension was gauged to place the active tip at the exact flow stream interface (flush with the pump casing interior wall). These transducers are miniature piezoelectric crystals specified to have a frequency response of 0 (DC) to ~1.2 MHz. Five separate pre-amplifiers and amplifiers were connected to a LeCroy LT374M digital oscilloscope. Each transducer was sampled from 0 to 400 MHz, and a filter used to remove system noise. The resulting spectra were then compared and correlated to the physical degradation observed at each location during the course of an extended cavitation run. Severe electromagnetic interference was experienced. Comparison tests were then performed on a process pump that was not driven by a variable frequency drive (VFD), and a clear cavitation signature was obtained. Subsequent trials on our test pump were performed following removal of the VFD, and spectral signatures were obtained for both direct fluid contact and contact with the exterior surface of the pump volute. Highest values of signal-to-noise ratio were obtained using a non-intrusive probe located near the suction of the pump, as shown in Figure 3.

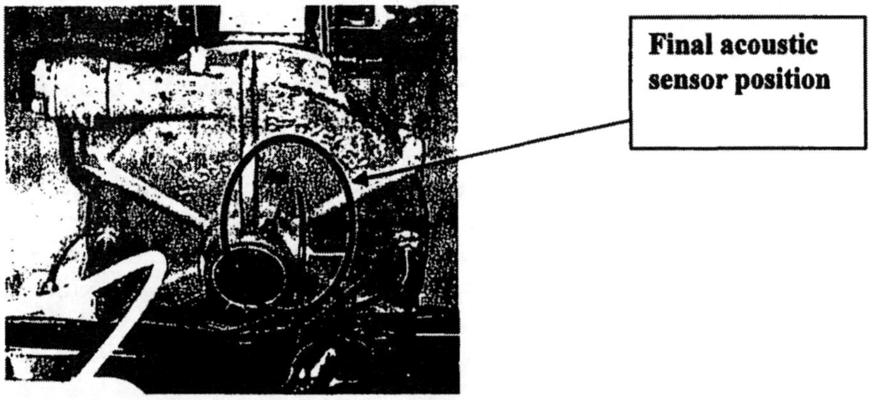

Fig. 3. Surface Mount Position of Acoustic Probe.

Cavitation Analysis:

Once the measurement hurdles were overcome, it remained to quantify the cavitation stressor intensity and to perform a long duration run so that the physical degradation effects could be determined. Before degradation testing could commence, a series of tests was performed to characterize the operational and experimental acoustic measurement system response to cavitation and to develop a methodology for quantifying stressor intensity. Initially the primary circulation pump and associated valving were configured to produce the design net positive suction head plus 10 psi. Subsequent runs performed by reducing the suction pressure in 2-psi increments, produced a clear acoustic cavitation pattern in the 30 to 55 kHz range.

To make intuitive sense from this data, the baseline (no cavitation) curve was used to normalize, the data sequences to produce a positive acoustic value in association with increasing cavitation. Additionally, following normalization, the 30 to 55 kHz interval was integrated for each run to capture the overall nature of the incident acoustic energy. The quantification of this incident acoustic bandwidth as a function of the pump suction pressure produced the curve characteristic shown in Figure 4.

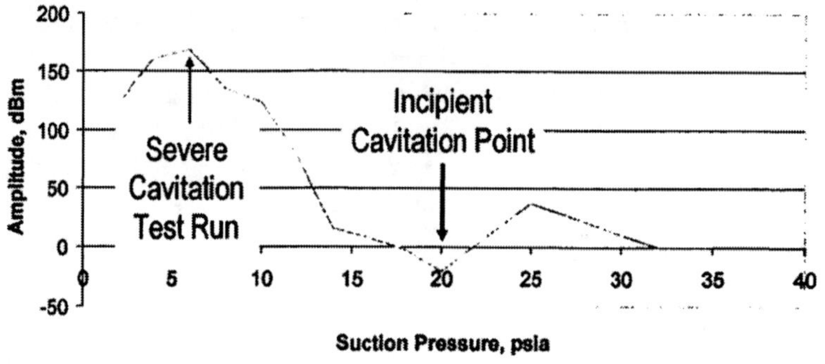

Fig. 4. Cavitation Intensity as a Function of Suction Pressure.

Several points should be noted from this curve. The inflection point at 20 psia appears to be a clear indication of incipient cavitation. According to the pump manufacturer, the NPSH requirement was 18 psia, indicating either an incorrect pump characterization or perhaps a willingness by the designer to allow the pump to operate in a mild cavitation condition.

The fact that the cavitation inception point is a minimum acoustic signal would be expected and that by going to the left (decreasing suction pressure) one would expect increasing acoustic energy because of the increase in cavitation. The increase in going to the right (increasing suction pressure) has logically been explained by other researchers [4] as a higher acoustic transmission of inherent pump noise caused by the "hardening" of the now single phase fluid in the pump casing. Basically, this is caused by a lack of attenuation from the vapor phase.

A maximum acoustic energy signal is seen at about 6 psia, below which it was observed that very low pressures produce a slug flow phenomena that again reduces the acoustic coupling to the casing and reduces the impacting energy level. This correlates well with slug flow observations in the Lexan suction pipe of our test rig at these pressures.

A continuous cavitation run was initiated and continued for 4 weeks, with the exception of a 4 hour power outage. The test pump was then secured, drained and disassembled to obtain wear readings relative to the baseline. With the exception of the wear ring clearances, very little metal removal was observed. The impeller to volute gap (wear ring) did indicate a 10-mil increase in clearance.

As a result of the extensive troubleshooting and test apparatus modifications that were necessitated by the unanticipated difficulty in establishing a satisfactory acoustic signal, only 4-weeks remained following the acoustic baseline characterizations. Without performing further experiments, only a simple linear correlation can be derived from the available two point wear data set. When combined with the acoustic intensity measurement, this gave us a "zeroth order" approximation to a deterministic correlation that relates suction differential pressure to incipient cavitation (the primary stressor) and ultimately the degradation rate of the pump.

The data from the slope of the logarithmic increase of acoustic signal with increasing pressure drop from cavitation inception was used to produce a cavitation-damage correlation. Making several assumptions about the validity of our logarithmic intensity scale and of its relationship to metal removal rate, we derived an equation of the form

$$MRR = K[10\exp(13.9x(PSID_{NPSH}))] \quad (3)$$

where MRR is the metal removal rate (in mils)

$PSID_{NPSH}$ is the differential pressure between the operating point and the pump NPSH limit

K is a material and geometric constant dependent on the specific pump

The coefficient 13.9 is the slope of the (logarithmic) acoustic intensity line from Figure 4 in db/psid.

It is suggested by the literature [2] that the metal removal rate is not, in fact, linear with time given constant cavitation intensity, but follows an asymptotic exponential approach to a limiting degradation rate, as shown in Figure 5.

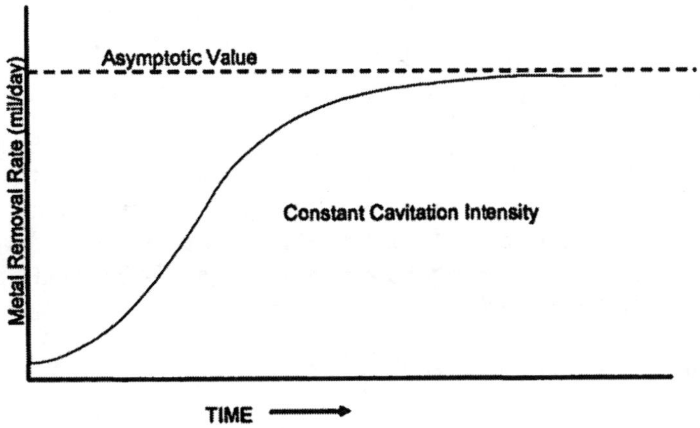

Fig.5. Suggested metal removal rate at constant cavitation intensity.

This variable rate is explained by observing the cavitation shock front initially as it impinges on a relatively smooth metal surface, and it is difficult for even a high velocity (shock) wave to create sufficient impulse to remove metal particles from the surface. As time goes on, the gradual pitting of the surface causes a roughening effect that allows better "purchase" for shock fronts to cause removal by more direct impact on the metal fragments. Once a fully roughened surface is created, the fragment removal rate approaches a constant value.

When combined with the available acoustic intensity measurements, this gives us an approximation to a deterministic correlation relating suction pressure differential to the actual degradation rate of the pump casing. Making several assumptions about the validity of the logarithmic intensity scale and the linearity of its relationship to metal removal rate, we can come up with a final equation of the form -

$$MRR = \underbrace{K10^{a\Delta P+b}}_{\text{Pump Design, Pump Material, Cavitation Intensity}}\ \underbrace{e^{(\Delta P \div D)t}}_{\text{Damage Factor}} \quad (4)$$

Pump Design
Pump Material
Cavitation Intensity

Damage Factor

The constants K, a, b, and D are dependent on pump-and case-specific inputs as noted above.

In conclusion, this cavitation experimental test run has:

- Identified the "active" cavitation frequency range
- Provided location guidance for cavitation acoustic measurement
- Developed a unique method for identification of incipient cavitation
- Developed a definitive method for quantifying cavitation intensity
- Established a metal removal rate as a function of cavitation intensity.

This final bullet represents the achievement of the ultimate goal of this project in that it gives an enabling demonstration of prediction through a first principles correlation between the stressor level and the degradation rate. When integrated over the component lifetime, it will provide a tool for the determination of component residual life based on the environment in which it operates.

A Field Application of the Methodology:

Following completion of the laboratory cavitation development work, an opportunity presented itself to attempt to extend the theory to an industrial application. A large hydro-electric dam reported severe pump cavitation with evident surface vortices in their auxiliary service water (ASW) pumps. Their engineering staff requested the laboratory team apply the damage prognostic technique to define the intensity of the cavitation, the expected damage rate and the expected residual life of the pumps.

Machinery and Measurements Description:

The three ASW pumps are of a low-head axial flow line-shaft design. These large capacity pumps find uses in many applications such as irrigation and flood control where their rugged design and reliability are a requisite. The pump's 7 foot diameter "propeller" is driven by a 450-Hp 6-pole induction motor (1100) through a single reduction gear reducer (8.5:1) via a 46-foot drive shaft. The oxygenated water from the discharge of these pumps is directed to a location near the entrance to the dam's fish ladder to act as an attractant to enhance salmon migration up the ladders.

The vane axial or propeller pump is the most sensitive pump to hydraulic instabilities caused by flow conditions at its inlet [5]. Following replacement of the ASW pumps to increase in their capacity by 50%, unacceptable hydraulic instabilities in the pump sump region were observed and later verified by hydraulic modeling. Because modifications to the existing sumps would be expensive and the result would still be outside the recommended Hydraulic Institute design criteria, pump prognostic expertise from the Pacific Northwest National Laboratory was sought in an attempt to quantify the life factor reduction that would result from operation under the existing conditions.

Following the procedure demonstrated in the laboratory, tri-axial vibration and acoustic emission signals were obtained by epoxying the transducers directly to the pump casing. The amplified signal was then fed to the high speed oscilloscope, where a fast Fourier transform (FFT) was produced from the time serial data. A ten-fold sampling of the FFT was used to present a representative trace of the acoustic spectrum. The acoustic traces for all three AWS pumps have been compiled and are presented in Figure 6. A clear cavitation "bulge" is seen in the acoustic signature for all three pumps.

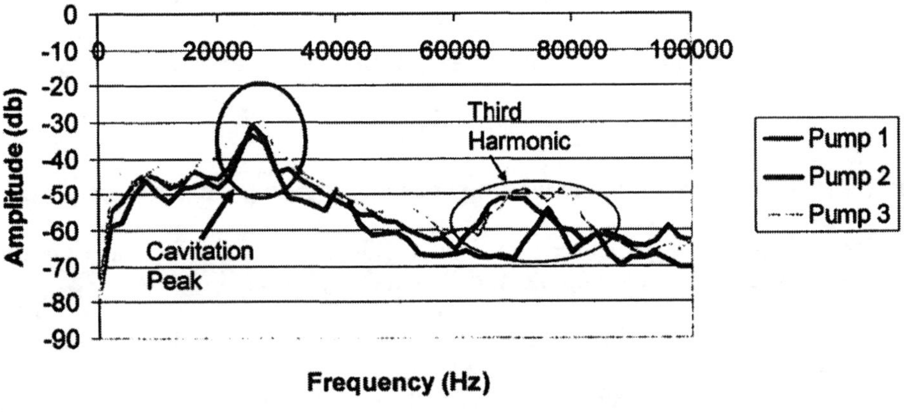

Fig. 6. Cavitation Spectral Indication.

To accurately quantify the cavitation intensity would require a reference non-cavitation state measurement. Lacking that, an estimate based on laboratory operational experience was used to evaluate the energy potential found in the cavitation peak. A comparison of integrated energy and the height of the peak values will then allow an extrapolation from the laboratory degradation values [6] to what might be expected for wear rates in the field. A complicating factor is the large distance that separates the primary cavitation site (the propeller tip) and the acoustic sensor. The vertical flow does act to sweep the collapsing bubbles toward our sensors. The result of all these caveats is that an uncertainty estimate of ± 50% must be placed on the resulting damage rate.

It is expected, based on the above discussion and knowledge of the propeller materials (ASTM A242 steel), that the cavitation wear rate under the conditions found at the time of testing will be limited to approximately 5 ±3 mils (thousandths of an inch) of metal per year. This is a fairly mild cavitation situation and by itself would lead to pump failure in approximately 78 years, thus it does not significantly impair the life expectancy of the pump. Further flow impediment caused by debris collection on the sump trash rack could, however, change this condition dramatically leading to much higher rates of degradation.

Conclusions:

The physics-based prognostic methodology has been demonstrated both in the laboratory and at a dam. The economic impact of the approach is clearly seen in the dam study.

The existence of cavitation and sump-wall vortices has been confirmed in the dam AWS pumps. Degradation in the impeller from cavitation is not expected to be excessive nor lead to dramatic shortening of pump life. It is probable, however, that the pump output is somewhat less than the expected flow because of the reduced efficiency of the impeller. Quantification of the flow reduction would not be possible without additional measurements. Additionally, increases in vibration severity induced by further degradation of the hydraulic conditions could be expected to cause early pump failures. The savings identified from this small (~50 K) predictive analysis were dramatic in that a $1.2 million dollar sump modification for these pumps was avoided.

References:

[1] Bond, L.J. 1999. "Predictive Engineering for Aging Infrastructure." *SPIE* 3588, pp. 2-13.
[2] Jarrell, D.B., D.R. Sisk, and L.J. Bond. 2004. "Prognostics and Condition-Based Maintenance: A New Approach to Precursive Metrics." *Nuclear Technology*, Vol. 145, pp. 275-286.
[3] Holroyd, T.J. 2000. The Acoustic Emission & Ultrasonic Handbook. Coxmoor Publishing Company, Oxford, England.
[4] Neill, G.D., R.L. Reuben, P.M. Sandford, E.R. Brown, and J.A. Steel. 1997. "Detection of Incipient Cavitation in Pumps Using Acoustic Emission." In *Proc. Instn. Mech. Engrs*, Vol. 211, Part E, pp. 267-277.
[5] Younes, M.A., S.M. Abdel-Rahman, M.A. Helal. "Dynamic Response and Instability of Propeller Pumps." National Water Researcher Center, Mechanical & Electrical Research Institute, Delta Barrage, Egypt.
[6] Bond, L.J., et al. 2003 "On-Line Intelligent Self-Diagnostic Monitoring System for Next Generation Nuclear Power Plants." Final Project Report, PNNL-14304. Available through NTIS on http://www.ntis.gov?ordering.htm.

A SELF-ASSESSMENT APPROACH TO CONFIDENCE MODELING IN DIAGNOSTICS AND PROGNOSTICS

Eric Wemhoff* and Asif Khalak
BAE Systems
6 New England Executive Park
Burlington, MA 01803, USA
*eric.wemhoff@baesystems.com, asif.khalak@baesystems.com

Abstract: We discuss the problem of modeling and tracking uncertainty when making estimates and predictions. This is important in the course of doing diagnostics or prognostics, due to the high levels of uncertainty typically seen. Further, we consider the problem of detecting the situation that explicit or implicit assumptions used by the estimator are not satisfied or have become invalid. In such cases, conclusions are potentially inaccurate and decisions based on them unwarranted. One way to test assumptions is to perform explicit checks. A more interesting implicit approach that becomes available when information about prediction uncertainty is available, is to compare predictions with their realizations for consistency. When a self-monitoring system is in place, this allows detection of problems with assumptions, and switching from a maintenance practice that depends on possibly errant conclusions to a more conservative "safety" maintenance procedure. We demonstrate uncertainty tracking and implicit assumption checking with a generic damage progression model.

Key Words: Bayesian filtering; fault detection; optimal nonlinear filtering; prognostics and health management; sequential state estimation.

PROGNOSTIC POTENTIAL OF THE ACOUSTIC EMISSION TECHNIQUE FOR SPUR GEARS

C K Tan[1], P Irving[2] and D Mba[3]

[1,2] School of Industrial and manufacturing Science and [3] School of Engineering,
Cranfield University, Bedfordshire. MK43 0AL, UK
E-mail : d.mba@cranfield.ac.uk

Abstract: Research in prognosis of gear life using the AE technique is relatively new in condition monitoring of rotating machinery. This paper describes an investigation undertaken on fatigue tests of gears in which natural pitting was allowed to occur. Acoustic Emission parameters were monitored throughout the test period in an attempt to relate natural life degradation to AE activity. Furthermore, vibration was monitored for comparative purposes. It is concluded that the AE technique is capable of detecting the onset on pitting much earlier than vibration analysis. In addition, opportunities for prognosis with the AE technique are very encouraging.

Keywords: Acoustic Emission; condition monitoring; gear pitting; machine prognosis; vibration analysis

Introduction

Acoustic emission (AE) was originally developed for non-destructive testing of static structures [1], however, over recent years its application has been extended to health monitoring of rotating machines and bearings [2-7]. It offers the advantage of earlier defect detection for gearboxes in comparison to vibration analysis [8-10]. However, on seeded faults in gearboxes, this is not without difficulties [11].

The use of vibration analysis for gear fault diagnosis and monitoring has been widely investigated and its application in industry is well established [12-14]. This is particularly reflected in the aviation industry where the helicopter engine, drive trains and rotor systems are fitted with vibration sensors for component health monitoring. In this paper, the authors present results from an experimental programme that observed the relationship between AE and vibration with natural progressive pitting in a pair of spur gears.

Acoustics Emission (AE)

AE is defined as the transient elastic waves generated due to a rapid release of strain energy caused by structural alteration in/on a solid material under mechanical or thermal stresses. Primary sources of AE are crack initiation, crack propagation, plastic deformation and friction. AE was originally developed as a method of Non-Destructive Testing (NDT) where it was used to monitor crack initiation, propagation and location. Attempts to apply this technique to condition monitoring started in the late 1960's [15].

Understanding the source of AE activity at the gear mesh is a fundamental pre-requisite if this technique is to be successfully employed for gear diagnostics and prognostics. Toutountzakis [11] highlighted limitations in the current understanding of the source mechanism of AE during gear meshing. Tan et al [16] ascertained the AE source mechanism through a series of experimental programmes. These experimental programmes consisted of isothermal tests on undamaged gears to explore the effects of rotational speed and applied torque on AE levels. From the isothermal test results, it was observed that variation of the applied torque had a negligible effect on the AE r.m.s levels, similar to the negligible effect of load on film thickness under Elastohydrodynamic Lubrication (EHL) of non-conforming mating surfaces. It was noted that the variation in rotation speed had a more pronounced effect on AE r.m.s levels relative to the load. Tan et al [16] concluded that the source of AE during gear mesh was asperity contact under rolling and sliding of the meshing gear teeth surfaces.

Vibration characteristics and gear damage

Yesilyurt et al [17] utilised the vibration analysis technique for damage detection and assessment, and stated that gear tooth damage would cause a reduction in tooth stiffness and the extent of this damage could be monitored. Drosjack et al [18] presented an experimental and theoretical study on the effect of simulated pits located on the pitch-line using vibration technique. Drosjack concluded that pitting on gear teeth surfaces altered the vibration characteristics via the change in stiffness of the gear tooth. In summary, the presence of damage such as pitting, either localised or distributed, will alter the stiffness of the gear due to modification of the Hertzian contact zone. In addition, an impulsive reaction between gears that have lost the original involute profile will change the vibration levels from the gears.

Experimental set-up

The test-rig employed for this experimental work consisted of two identical oil-bath lubricated gearboxes, connected in a back-to-back arrangement. The gear set employed was made of 045M15 steel (without any heat treatment) which had a measured hardness of 137 Hv30. The gears (49 and 65 teeth) had a module of 3 mm, a pressure angle of 20°, and a surface roughness (Ra) of between 2-3 μm. A simple mechanism that permitted a pair of coupling flanges to be rotated relative to each other, and locked in position, was employed to apply torque to the gears.

The AE sensor used for this experiment was a broadband type with relatively flat response in the region between 100 KHz to 1MHz. The sensor was placed on the pinion with 49 teeth. The cable connecting the sensor placed on the pinion with the pre-amplifier was fed into the shaft and connected to a slip ring ('IDM' PH-12). This arrangement allowed the AE sensor to be placed as close as possible to the gear teeth. The sensor was held in place with strong adhesive superglue. The output signal from the AE sensors was pre-amplified at 20dB. An accelerometer was fitted onto the bearing casing of the pinion wheel to record vibration data. The accelerometer used for vibration measurement in this experiment was a resonant type sensor with a frequency response between 10 Hz and 8000 Hz. The accelerometer was mounted on the base of the bearing casing connecting to the pinion shaft. The accelerometer was connected to a charge amplifier, and the signal output from the pre-amplifier was fed to a commercial data acquisition card. Under typical gearbox operation, an anti-wear lubricant is usually employed to prevent or slow down wear on the gear teeth. In order to initiate surface pitting in a relatively short time frame, lubricant oil without anti-wear properties was employed for the accelerated gear fatigue tests; SAE 20W-50. Also, to accelerate the pitting process the face width of the pinion employed was half that of the wheel.

Experimental procedures

The fatigue gear tests were performed at a rotational speed of 745 rpm and applied torque of 220Nm. At regular intervals (ranging from 15 to 55 hours), visual inspection of gear surface damage was undertaken. Continuous AE r.m.s values were calculated in real time by the analogue-to-digital converter (ADC) controlling software. This software employed a hardware accelerator to perform calculations in real time for a programmable time interval set by the user, 10ms in this instance and a sampling interval of 90ms was employed. Anti-aliasing filters were also employed prior to the ADC. Raw vibration waveforms, sampled at 8192 Hz, were recorded for a period of 1 second at intervals of 30 minutes. Vibration r.m.s values were calculated over the recorded duration (1-second). During the inspection interval, gear teeth surfaces on both the pinion and gear were visually inspected for pitting or other abnormalities such as scoring and scuffing. The largest pitted area on any single tooth was recorded. The authors set the failure, or test termination, criterion at 50% pitted area of the gear tooth surface area. The visual inspections were performed by two separate inspectors independently and repeated for consistency. This inspection error was determined to be ± 5% of pitted area.

Results

Figure 1 shows percentage of the gear surface pitted area plotted against the test operating time.

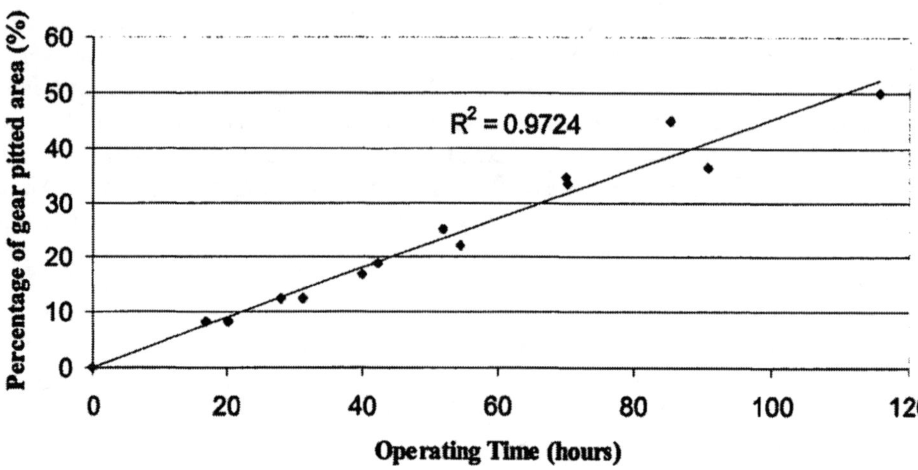

Figure 1 **Pitting rates of the test gear under 220Nm at 745 rpm**

All original data from are presented in figures 2 and 3. From figure 2, it can be seen that despite the application of the same torque, both tests started at different normalised AE r.m.s values. However, in one of the tests the AE r.m.s initially decreased, whereas in the other it increased from the start. After approximately 15 hours the AE levels in both tests increased at very similar rates (gradients) but different absolute AE r.m.s values. Figure 3 shows a plot of AE r.m.s versus the percent gear pitted area illustrating a linear relationship between the two for both tests.

Figure 4 shows normalised vibration r.m.s values against time for the 220Nm test. It demonstrates that the application of the same torque produced similar normalised vibration r.m.s values until 60 hours when the tests departed from each other. Also it was observed that there was an initial increase of vibration level from 0 to between 10 and 15 hours, thereafter the vibration level remained relatively constant until 60 hours. Figure 5 shows the original vibration r.m.s values plotted against percentage of pitted area. Following the run-in period vibration levels remained constant until 30% pitted area, after which levels rose steadily.

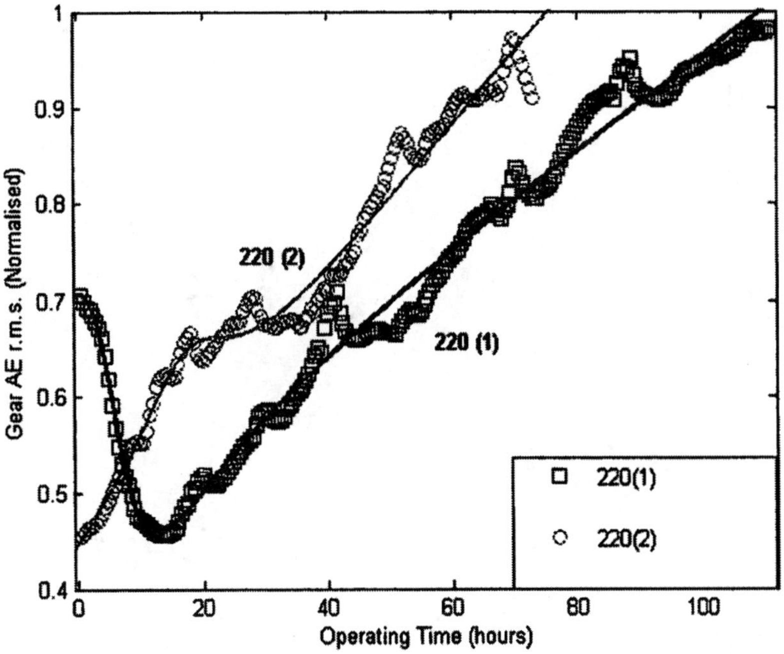

Figure 2 Normalised AE r.m.s against operating time at 220 Nm; 745 rpm

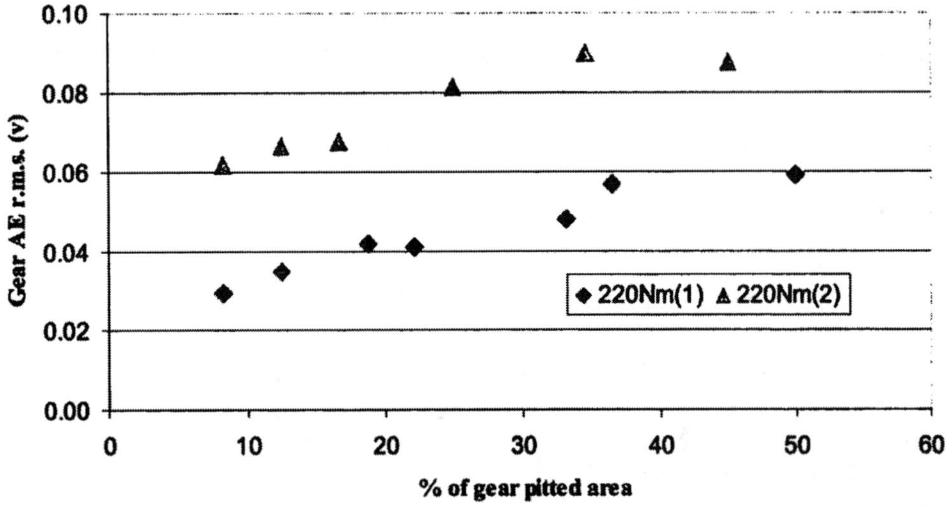

Figure 3 AE r.m.s against % pitted area; 220 Nm, 745 rpm

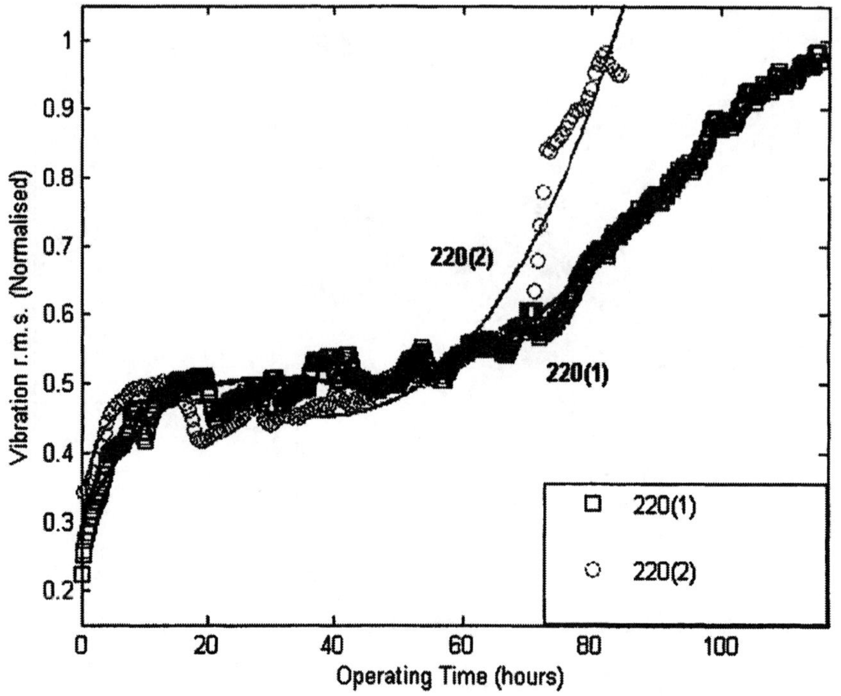

Figure 4 **Normalised vibration r.m.s against operating time at 220 Nm; 745 rpm**

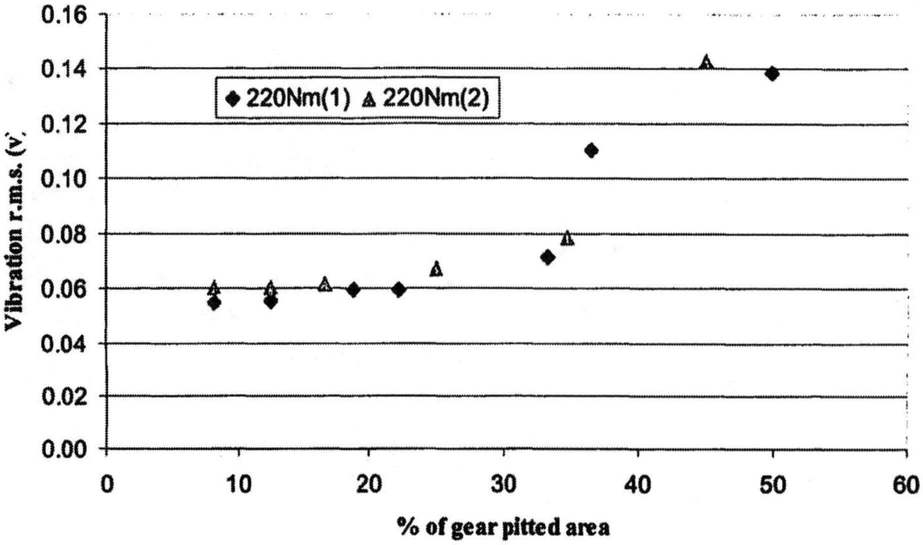

Figure 5 **Vibration r.m.s against % pitted area; 220 Nm, 745 rpm**

Figure 6 shows the progression of gear tooth pitting from 13% to 50% of gear pitted area, under the test condition of 73 Nm and 745 rpm.

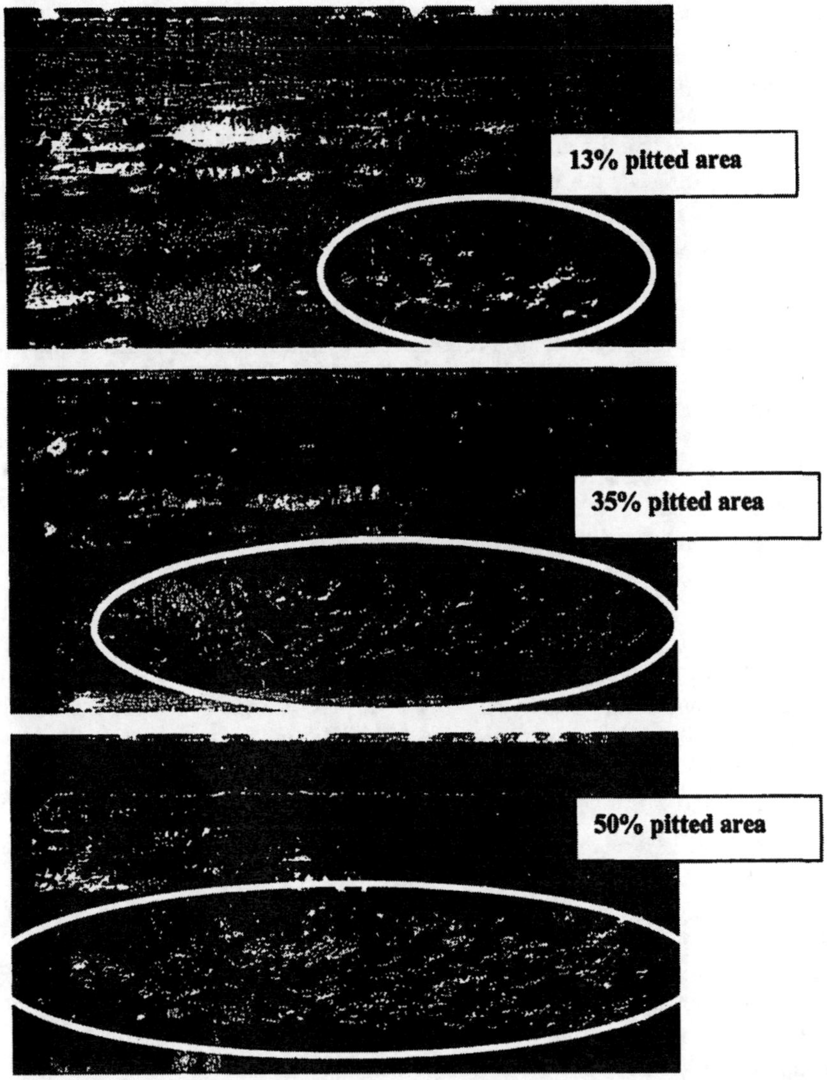

Figure 6 **Gear pitted area at various operating times**

Discussions

In relating AE activity to pitting rates cognisance of the effects of surface roughness, lubrication regime, friction and the slide-to-rolling ratio of the meshing gears must be considered. Xiao et al [19] noted that friction coefficient of mating surfaces increased with increasing average surface roughness. Diei [20] proposed a relationship between AE r.m.s and the rate of frictional energy dissipation from sliding contact whilst Dornfeld [21] et al have shown the high sensitivity of AE to sliding speed and applied load. Suh [22] defined asperity deformation as the main determinant of friction in metal to metal contact and Tan et al [16] concluded that the source of AE during the gear mesh was attributed to asperity contact. Based on the observations of AE activity and pitting progression during this investigation, and conclusions of the various researchers detailed above, the authors postulate that AE levels will increase with increasing gear pitted area. A consequence of the increase in pitted area is an increase of surface roughness and friction, leading to an increase in AE levels. This deduction was confirmed by the observations from the experimental results presented which show AE levels increasing with operating time / gear pitted area.

Clearly, there existed an initial period during which the gear teeth surface smoothened out, oil sump temperatures increased and dynamic stabilisation of the rotating systems (such as bearing, alignments etc) took place. Because of the complexity involved during this process, it was deemed inappropriate to relate any of the monitoring indicators to this period; 0 to 15hrs. However, after this initial period defined as wear-in, the monitoring indicators behaved differently with pit progressions. As discussed earlier, AE r.m.s level exhibits a linear relationship with running time after this initial period, which was not necessarily the case for vibration observations.

The AE technique

Throughout the duration of the tests it is believed that there are two processes affecting the generation of AE. Firstly, the wear-in process which causes a smoothing of surface roughness with a consequent decrease in AE levels. The second involves the increased surface roughness due to pitting progression/development which will increase surface roughness and AE levels. At the beginning of the tests AE levels will also be influenced by the oil film temperature and dynamic characteristics of the test-rig arrangement. Increasing oil temperature will lead to a reduction of oil film thickness; this encourages more asperity contacts resulting in increased AE levels. On the other hand, the smoothing of the gear teeth surfaces due to the wear-in process will reduce surface roughness which will result in lower AE levels. Furthermore, the authors postulate that the factors that determine the onset value of the AE level are the initial surface roughness of the gear teeth surfaces, assembly of the gear components and bearings, and, the initial oil temperature.

The two test results at 220Nm, see figure 2, exhibited different trends at the start of the tests; '220(1)' showed decreasing AE levels up to 15 hours operational time, whereas AE level for '220(2)' increased from the start of the test. It is postulated that the difference is due to the balancing process between increasing oil temperature and reducing surface roughness of the gear teeth, which have opposing effects on AE levels. The linear relationship between AE levels and pit growth rate was encouraging and emphasised the sensitivity of the AE technique.

The Vibration technique

Figure 5 shows the plot for vibration r.m.s against gearbox operating time under an applied torque of 220Nm. Vibration r.m.s increased from 0 to between 10 to 15 hours, which was indicative of increasing oil temperature and decreasing oil film thickness. As oil film thickness reduced, the damping effect of the oil film between the meshing gear teeth surfaces will reduce resulting in increasing vibration levels. A plateau was observed for the vibration r.m.s between 15 to 55 hours of the running time, even though gear surface pitted area increased to 30%, see figure 6. This showed that vibration technique was unable to monitor the pit grow process until the pit development was advanced. Hence at this point, it can be concluded that AE technique has an advantage over vibration technique in terms of pit growth monitoring. In summary, the vibration response increased when minimum criteria of 30% pitted area was reached which is attributed to alterations in the stiffness of the gear due to modification of the Hertzian contact zone.

Conclusions

From the results presented it was clearly evident that the AE monitoring indicator could be linearly correlated to the gearbox pitting rates for all torque conditions, with detection of onset of pitting as early as 8% of the pitted area. This offered much earlier diagnosis than vibration analysis where only after approximately 30% of pitted area did this technique offer capability for defect identification. This near linear relationship between AE and pit progression offers great potential, and opportunities, for prognostics in rotating machinery.

References

1. Miller, R.K. and McIntire, P. Acoustic Emission Testing, Volume 5, 2nd ed., Non-destructive Testing Handbook. Ed., American Society for Non-destructive Testing, pp 275-310, 1987.
2. Toutountzakis, T. and Mba, D. Observation of Acoustic Emission Activity During Gear Defect Diagnosis. NDT and E International. **Vol. 36, No.7**, pp 471-477. 2003.
3. Mba, D. and Bannister, R.H. Condition monitoring of low-speed rotating machinery using stress waves: Part1 and Part 2. Proc Inst Mech Engrs. **Vol 213, Part E**, pp 153-185. 1999.
4. Morhain, A, Mba, D, **Bearing** defect diagnosis and acoustic emission, **Journal of Engineering Tribology, I Mech E**, Vol 217, No. 4, Part J, p 257-272, 2003. ISSN 1350-6501.
5. L. D. Hall and D. Mba, Acoustic emissions diagnosis of rotor-stator rubs using the KS statistic, **Mechanical Systems and Signal Processing**, Volume 18, Issue 4, Pages *849-868, 2004.*
6. Mba, D. Applicability of acoustic emissions to monitoring the mechanical integrity of bolted structures in low speed rotating machinery: case study. NDT and E International. **Vol 35, No. 5**, pp. 293-300. 2002.

7. Mba, D., Cooke, A., Roby, D. and Hewitt, G. Opportunities offered by acoustic emission for shaft-seal rubbing in power generation turbines; a case study sponsored by the British Institute of NDT. International Conference on Condition Monitoring. Oxford, UK, 2-4 July 2003. pp 280-286, ISBN 1901892174. 2003.
8. Tandon, N. and Mata, S. Detection of Defects in Gears by Acoustic Emission Measurements, Journal of Acoustic Emission. **Vol. 17**, Issue 1-2, 23-27, 1999.
9. Singh, A., Houser, D. R., and Vijayakar, S. Early Detection of Gear Pitting, Power Transmission and Gearing Conference, ASME. **DE-Vol. 88**, 673-678, 1996.
10. Singh, A., Houser, D. R., and Vijayakar, S. Detecting Gear Tooth Breakage Using Acoustic Emission: A Feasibility and Sensor Placement Study, Journal of Mechanical Design. **Vol. 121**, 587-593, 1999.
11. Toutountzakis, T., Tan, C.K. and Mba, D. Application of acoustic emission to seeded gear fault detection. NDT and E International. **Vol. 38, No. 1**, pp 27-36. 2005.
12. Gadd, P. and Mitchell, P.J. Condition monitoring of helicopter gearboxes using automatic vibration analysis techniques. AGARD CP 369. Gears and power transmission system for helicopter turboprops. 29/1 – 29/10. 1984.
13. Leblanc, J.F.A., Dube, J.R.F. and Devereux, B. Helicopter gearbox vibration analysis in the Canadian Forces-applications and lessons. 1st International Conference, Gearbox noise and vibration. IMechE, Cambridge, UK. C404/023, pp 173 -177. 1990.
14. Cameron, B.G. and Stuckey, M.J. A review of transmission vibration monitoring at Westland Helicopter Ltd. 20Th European Rotorcraft Forum. Paper 116. pp 116/1 – 116/16. 1994.
15. Holroyd, T.J. (2000). Acoustic emission and ultrasounics. 1st ed. Coxmoor Publishing Company. Oxford, UK.
16. Tan, C.K. and Mba, D.U. Identification of the Acoustic Emission source during a comparative study on diagnosis of a spur gearbox. Tribology International. *Article in press.* Ref: T1584, 2004.
17. Yesilyurt, I., Gu, F.S. and Ball, A.D. Gear tooth stiffness reduction measurement using modal analysis and its use in wear fault severity assessment of spur gears. NDT & E International. **Vol. 36, No. 5.** 357-372. 2003.
18. Drosjack, M.J. and Houser, D.R. An experimental and theoretical study of the effects of simulated pitch line pitting on the vibration of a geared system. ASME report 77-DET-123. 1977.
19. Xiao, L., Rosen, B.G., Amini, N. And Nilsson, P.H. A study on the effect of surface topography on rough friction in roller contact. Wear. **Vol 254 (11),** 1162-1169. 2003.
20. Diei, E.N. Investigation of the milling process using acoustic emission signal analysis, *PhD. Thesis*, Department of Mechanical engineering, University of California, Berkeley, CA. 1985.
21. Dornfeld, D. and Handy, C. Slip detection using acoustic emission signal analysis. IEEE. 1868-1875. 1987.
22. Suh, N.P. Tribophysics. Prentice-Hall. Englewood Cliffs. NJ. 1986.

PRESCRIPTION BASED MAINTENANCE MANAGEMENT SYSTEM

G. Scott Valentine
James E. Dzakowic
Impact Technologies, LLC
200 Canal View Blvd
Rochester, NY 14623
scott.valentine@impact-tek.com

Thomas Galie
NSWC
Carderock Division
Philadelphia Naval Business Center,
Philadelphia, PA, 19112-5083

John Scharschan
NSWC
5001 South Broad Street
Philadelphia, PA 19112

Abstract: In recent years, significant focus has been placed on the development and implementation of advanced prognostic and health management (PHM) technologies in military and industrial applications. The term PHM encompasses anomaly, diagnostic and prognostic algorithms as well as higher level reasoning algorithms for isolating root causes of faults/failures and directing optimal operational or maintenance actions. In such systems, two current deficiencies exist. First, for a variety of reasons, component and subsystem interactions in such systems are poorly realized. The issue manifests itself as multiple dependent "boxes" indicating faults with shotgun tests or valuable domain expertise required to de-conflict and reduce ambiguity groups. Secondly, complex systems still largely rely on expert rule-bases for reasoning which are notoriously difficult to maintain over a life cycle and are prone to logical conflicts. This paper begins to address these deficiencies by outlining a simulation-based process for automatically 1) realizing complex system interactions for optimal PHM system design and 2) building and maintaining model-based reasoning architectures where decisions and conclusions naturally precipitate out of a more manageable system model.

Keywords: Condition Monitoring; Open System Architecture; Optimal Maintenance Decisions; Prognostic and Health Management

Introduction: Prognostic and health management embraces a philosophy that works to maximize a piece of equipment's availability to perform a task or mission, while simultaneously minimizing its maintenance cost. In recent years many PHM packages have entered the market incorporating near real time automated monitoring, data analysis, diagnostic routines and more advanced, embedded algorithms driven prognostic applications predicting remaining useful life. The plethora of information derived throughout these multiple layers all garner inputs for decision support tools. These inputs give maintenance planners the knowledge to take appropriate actions. In light of the data now made available to maintainer's, the next natural progression of questions to be asked is "What action should be taken first when sensors may be indicating that several available maintenance actions could correct the failure at hand?" A reasoning engine that learns from its previous experiences and provides guidance on the most likely next course of action is a decision support tool that will compliment the PHM philosophy.

Prescription-Based Maintenance Management System (PBMS) is such a maintenance reasoner engine that directs the performance of maintenance tasks only when there is sufficient evidence of a need. Towards that, a system of monitoring, detection, diagnosis and prognosis will guide the users to a probable set of maintenance tasks to correct an occurring discrepancy. The next appropriate step will be to prescribe the most likely corrective action, "prescription," to remedy the situation. This process flow is displayed in Figure 1 below. An effective Prescription-Based Maintenance Management System will ultimately reduce the overall operation and support (O&S) costs of shipboard systems, through the efficient use of manpower, equipment and spare parts.

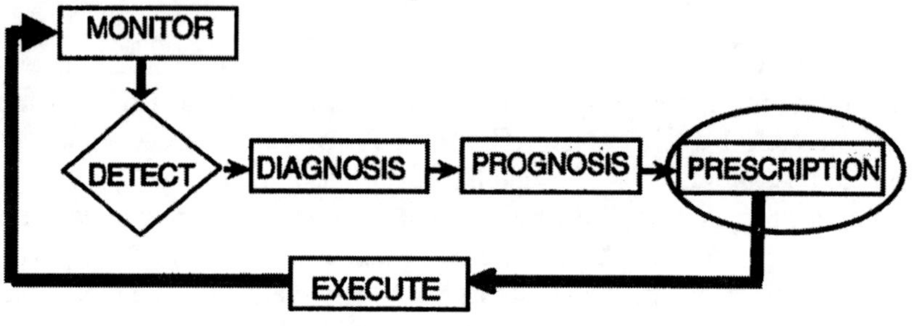

Figure 1: Prescription-Based Maintenance Concept

Overall System Architecture: As with any support tool emphasis must be placed on creating a system architecture that has scalable attributes and that easily communicates with other systems through an open architecture environment. The PBMS application interacts with existing Navy systems such as the Integrated Condition Assessment System (ICAS), existing diagnostic or prognostic agents, scheduling software, electronic instructions, and maintenance databases. Open system standards are used where interaction with other software modules is expected. Evolving Open Systems Architecture for Enterprise Application Integration (OSA-EAI) standards is used for undefined external interfaces where compatibility is required. Capabilities and provisions are being provided for future interfacing with a mission readiness assessment system. The PBMS external interfaces are outlined in Figure 2.

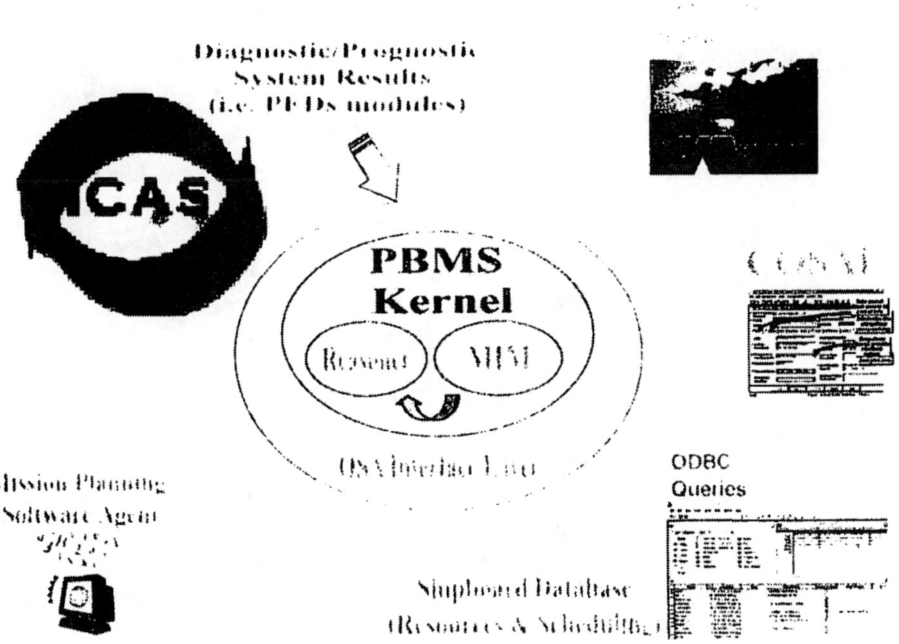

Figure 2: PBMS Data/Info Integration

During a SBIR Phase II program, Impact Technologies, LLC developed the PBMS shell in the Tcl/Tk programming language displayed with XML for web base capability. Tcl/Tk is an extremely versatile high-level language with a large pallet of Graphical User Interface (GUI) elements and Open System Architecture object links such as COM, DCOM, CORBA and XML. The XML modeling approach is used to model and analyze the reasoner-based system. The developer can use standard notation to model system components, behaviors, and users. This approach is widely accepted for distributed object-oriented systems. This architecture enables systems in dynamic link library (DLL) files to interface with the evaluation system through a standardized OSA protocol. An XML-based Maintenance Integrated Model (MIM) will be integrated into the core application program. The MIM incorporates a fault tree type database with the appropriate condition monitors and maintenance tasks in a graphical portal.

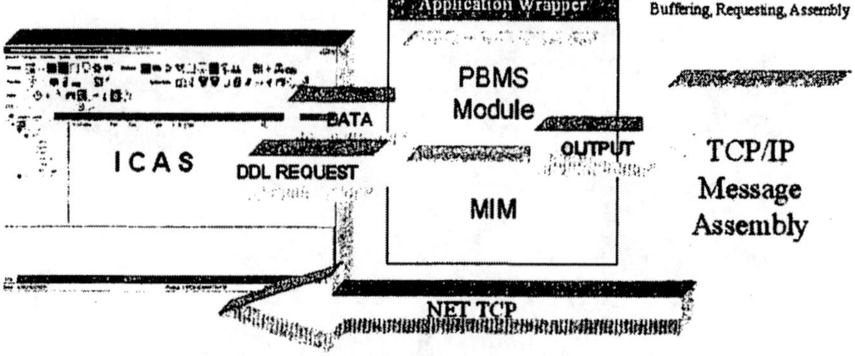

Figure 3: PBMS Integration with ICAS

As an integral part of the PBMS, information from diagnostic or prognostic agents is utilized to trigger appropriate maintenance actions. Machinery health assessments from diagnostic or prognostic agents are passed through ICAS to the PBMS as failure mode evidence sources for maintenance reasoning. Other classes of information and data such as operating mode are also being exchanged as required. The PBMS application wrapper is interfaced between the ICAS application and TCP/IP messaging service, Figure 3, where a data demand function pulls the appropriate data and information from ICAS to the PBMS application. The PBMS module itself directly interfaces with the MIM that has previously captured the knowledge associated with the CBM technologies and maintenance tasking. The overall system architecture envelops three distinct agents: Maintenance integrated model, PBMS Reasoner, and Reinforced Learning.

Maintenance Integrated Model (MIM): The Maintenance Integrated Model incorporates a fault tree type database with the appropriate condition monitors and maintenance tasks in a graphical portal. The majority of the information in the MIM should come from system Failure Modes, Effects, and Criticality Analysis (FMECA). The FMECA information that is required to build a complete MIM model is listed below.

- Functional Elements (Systems, Subsystems, Components) and hierarchy
- Relationships between functional elements (Mechanical, Electrical, Fluid)
- Failure Modes or Faults
    - Failure Rate (1/MTTF – related to Weibull Parameters)
    - Weibull Shape Parameter
    - Severity (1:Negligible, 2:Marginal, 3:Critical, 4:Catastrophic)
- Symptoms and Effects
- Relationships between Failure Modes and Effects (Fault Tree)
- Probability of Propagation between each element of the fault tree
- Monitors or Algorithms (Built in Test or Discrete, Diagnostics, Prognostics)
    - Probability of Detection for BIT or Diagnostic
    - Probability of False Alarm for BIT or Diagnostic
    - For Prognostics, use Accuracy
    - Algorithm Life Cycle Cost
- Sensors that are required for the Algorithms
    - Sensor failure rate

- Sensor Life Cycle Cost
- Relationships between Monitors and Symptoms/Effects or Failure Modes
- Maintenance Tasks that mitigate failure modes (Unit, Intermediate, Depot)
  - Hardware cost to perform task
  - Labor cost to perform task
  - Hours of downtime required to perform task
  - Unscheduled task penalty factor
  - Maintenance Task Effectiveness
- Connectivity between Maintenance Task and Failure Mode
-

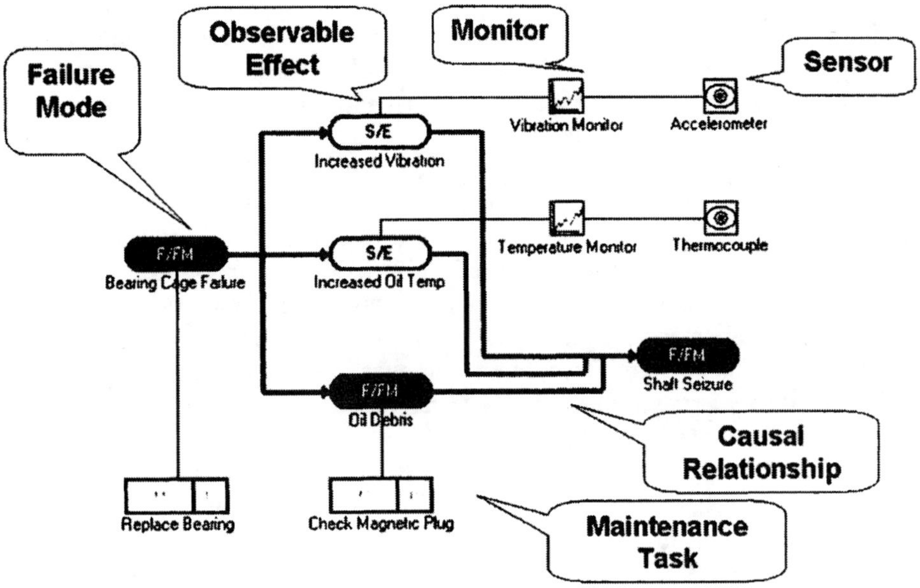

Figure 4: Example MIM with annotations

All MIM information is stored in specially formatted XML database files. As needed, any of the layers may access the file to extract pertinent information. The XML files contain two types of elements – components and connections. Components refer to systems, subsystems, system elements, faults and failure modes, effects, monitors, sensors, and maintenance tasks. Each component instance has a variety of associated attributes. Connections can refer to mechanical, electrical, and fluid links between system elements, or associations between various health management components (faults and failure modes, effects, monitors, sensors, and maintenance tasks). The structured XML approach allows standardized code to be used to read and interpret the data.

PBMS Reasoners: The reasoning algorithm being developed under this task represents the logical process through which decisions are made by the PBMS. In this case, a Reasoning algorithm must be able to identify root cause failure modes and corrective actions (maintenance tasks) based on all available evidence, information/knowledge sources and a practical set of restraints.

Most present day maintenance practices involve an experienced person to troubleshoot an observed problem with a system and decide upon the appropriate maintenance action. Currently used Interactive Electronic Tech Manuals (IETM) step the maintainers through a series of tests / procedures on the basis of convenience, cost or, at best, likelihood of success. The most modern IETM have "meta-tags" which act as handles for indexing to certain tasks. If something in the IETM is incorrect, it remains incorrect until a human changes it. The suggested maintenance tasking is static and most importantly, human interpretation is required to characterize and classify evidence. The automated reasoning algorithms and agent-based approaches demonstrated in this application can circumvent these issues of classification and characterization of sometimes incongruent data, to make a logical progression from failure modes to most appropriate maintenance tasks.

Evidence-Based Maintenance Reasoning: In the Phase II program, Impact Technologies is developing a knowledge-based fusion algorithm that is capable of performing failure mode and maintenance task ranking, as well as root cause isolation within a system containing a variety of information and knowledge sources already mentioned. While novel in its integration capabilities, the technique is based on fundamental mathematical principles and can significantly improve upon the performance of the results from the techniques previously introduced. A standard FMECA of the system provides the basic requirement of information for the proposed Impact Reasoner implementation.

The Impact Fusion Reasoner (IFR) will have two modes under which it receives/processes information to act upon: Initialization and Evidence Activation. Essentially, Reasoner information calculated in the initialization mode will be based on static model attributes such as evidence to failure mode dependencies, false alarm rates and real fault probabilities. During initialization, two steps will be required before reasoning can actually take place: Connectivity Matrix generation (CM) and Evidence Sub Graph (ESG) identification. The CM is generated by exploring the relationships between failure modes and monitors configured in the model. Figure 5 illustrates an example where Boolean numbers are used to generate the connectivity (note: the term "monitor" is equated with any piece of evidence). A value of "1" indicates that a piece of evidence is able to directly observe the effects of a specific failure mode. No direct observation corresponds to a value of "0".

Connectivity Matrix				
	Failure Mode 1	Failure Mode 2	Failure Mode 3	Failure Mode 4
Monitor 1	1	0	0	1
Monitor 2	0	0	1	0
Monitor 3	0	1	0	1

Figure 5: Connectivity Matrix

The Evidence Sub Graph (ESG) can be extracted from the Connectivity Matrix. An ESG contains generic monitors (i.e. evidence) that are directly or indirectly related to the same failure modes in the system. Two generic monitors observing a fault would be considered indirectly related through a failure mode while the relation between a monitor and an observable fault is direct.

The Reasoning Process Invoked Upon Monitor Activation: The Reasoning process is broken into three major steps: Time-based (temporal) considerations, Failure Mode Ranking and Failure Rate Adjustment. Each monitor could have time delays associated to its responsiveness to fault detection, as well as failure propagation paths with associated monitors that activate in a temporal sequence. As an example, consider the fact that an oil leak may only be identified after a prescribed amount of oil has been lost. Accounting for delays is an important feature that insures that all potential evidence that may be used to indict a root failure mode has been considered before a conclusion is reached (see Figure 6 as an example).

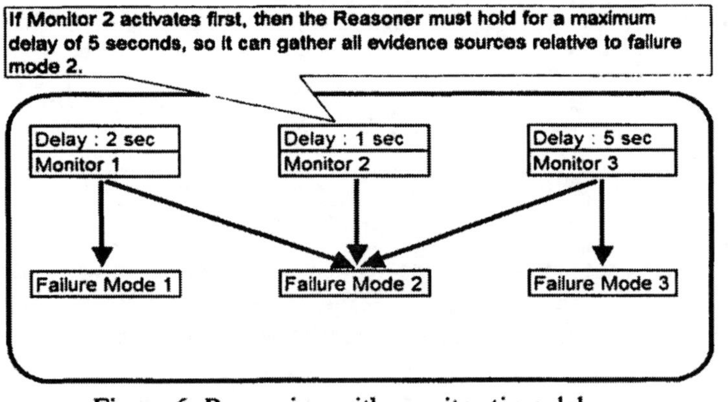

Figure 6: Reasoning with monitor time delays

Accounting for Positive and Negative Evidence: Positive and Negative Evidence, along with Real Fault Probabilities and False Alarm Rates need to directly contribute to the failure mode ranking process. Positive Evidence may be defined as evidence that indicts a failure mode consistent with the evidence sub-graph (ESG). An absence of evidence when it is expected may be defined as Negative Evidence. Figure 7 illustrates these concepts for three Monitors capable of observing a single Failure Mode in a generic example. Consider Failure Mode 1 to be a lube oil leak for a generic piece of machinery. One could characterize at least three pieces of evidence that, when combined, should indict this problem 1) loss of pressure 2) loss of oil level 3) observation of oil on the ground. If evidence 1 and 2 are present but not 3, there is imperfect evidence indicting the lube oil leak failure mode. It is entirely possible; however, that the oil may be evaporating, the leak wasn't found or that the pressure and level readings are inaccurate.

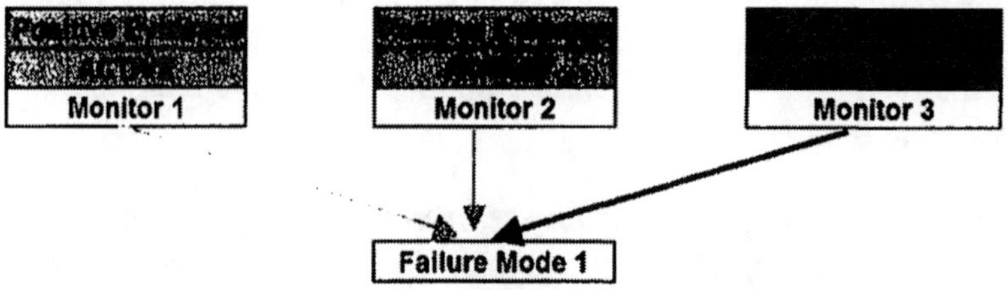

Figure 7: Positive and Negative Evidence Sources

Real Fault Probabilities and False Alarm Rates: The probability of a fault existing given that a monitor has indicted it is (P (FM/Monitor)) and corresponds to the Real Fault Probability (RFP). The False Alarm Rate (FAR) is related to the potential of a monitor activating given the absence of a fault (P (Monitor/~FM)). Equation 1 illustrates how the ranking of a given failure mode is determined through the consideration of Positive evidence, Negative evidence, FAR and RFP. This is a fundamental equation of the Impact Reasoner.

$$\tilde{Rank}_{FM(i)} = 1 + \frac{\sum_{j=1}^{PosE}(1 - FAR_j) - \sum_{k=1}^{NegE} RFP_k}{(PosE) + (NegE)} \quad (1)$$

The Positive Evidence contribution is reduced by 1 minus the potential of a monitor going active without the presence of a fault (False Alarm Rates). Negative Evidence contribution is reduced by 1 minus the potential of a fault existing but going undetected (1-Real Fault Probability). The numerator is normalized by the sum of the total potential of the system. The Rank ranges from 0 to 1, with the highest number being the best score. A second step is required to complete the ranking process, which addresses failure rates and can be used to break "ties" on ranking.

Accounting for failure rates: The initial ranking algorithm alone will not account for a-priori failure rates. For example, if the 2 failure modes have equal initial ranks, consideration of the historical failure rate of one relative to the other can act as the tie-breaker. The premise in the failure rate adjustment process is that, while failure rate statistics have no place in the initial ranking they can be appropriately mapped into a ranking adjustment of the potential set of root cause failure modes.

The ranking adjustment can be performed through a predefined look up chart, which converts the original rank to an adjusted rank based on failure rate values. The chart is a function of initialization variables, failure rates and original ranks. The initialization algorithm provides crucial parameters necessary in mapping the charts dynamic nature. Such a chart is illustrated in Figure 8. The values on the abscissa represent the original ranking values based upon Positive and Negative evidence. The ordinate includes two separate axes: a failure rate log-scale axis and a scale factor axis. The failure rate values are converted into a scale factor, which defines the y-intercept of a specific line. The log to linear scale conversion is based on the fact the failure rate are usually orders of magnitude apart (i.e. 1E-6 vs. 1E-5). The chart contains a mapping of two main regions separated by a variable ranking limit.

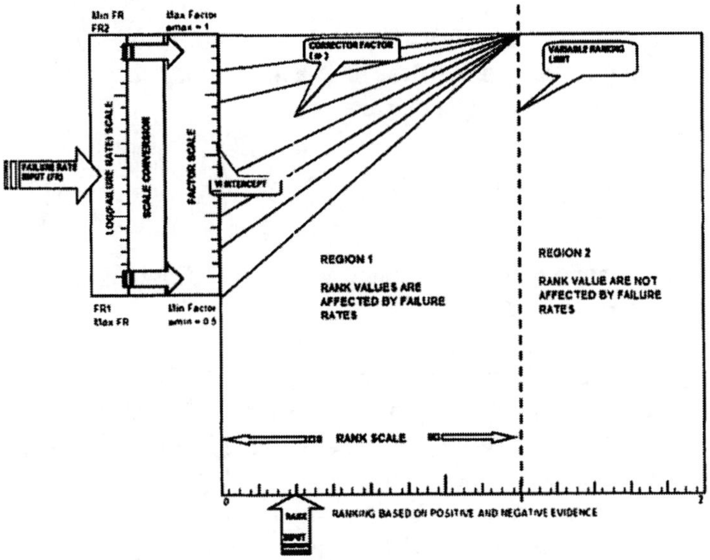

Figure 8: Failure Rate Adjustment Chart

The ranking limit can be adjusted along the abscissa changing the distribution of the set of lines and subsequently affecting the two regions. Failure rates will be irrelevant for any rank values that fall to the right of the limit. Any values to the left of the limit will be affected. The adjusted rank for a given failure mode is a calculated ratio of the original rank and a factor number provided by the chart as shown in Equation 2.

$$Adj\_Rank_{FM(i)} = \frac{Rank_{FM(i)}}{\omega_{FM(i)}} \tag{2}$$

The derivation of the correction factor requires the y-intercept and the slope of the line. Interpolating between two axes defined on the ordinate outputs a y-intercept measure on the factor scale axes.

$$Y_i = \left[\frac{(\omega_{\min} - 1)}{(\log(MaxFR) - \log(MinFR))} \cdot (\log(FR) - \log(MinFR)) \right] + \omega_{\max} \tag{3}$$

The slope of the line is calculated from Equation 4. The Rank Scale ranges between the origin and the ranking limit.

$$Slope = \frac{MaxFactor - Y_i}{Rankscale} \quad (4)$$

The corrector factor is calculated from Equation 5 and substituted into Equation 2. Using the original rank input, the slope of the line and its y-intercept the corrector factor is calculated and substituted into Equation 2.

$$\omega_{FM(i)} = (Rank_{FM(i)} \cdot Slope) + Y_i \quad (5)$$

The ranking adjustment process as stated will only allow failure mode rankings to improve based on failure rates.

Reinforced Learning: Q-Learning is among a series of Reinforcement Learning (RL) algorithms with a common objective aimed to successfully solve problems that require complex decision processing. Reinforcement Learning is a combination of fields such as dynamic programming and supervised learning, which lead to powerful intelligent machine learning systems. One of the biggest advantages of these techniques is the model-free aspect. The applicability of reinforcement learning algorithms is independent from the structure of the system taken into consideration.

Q-Learning has been integrated in the MIM environment to reinforce the maintenance task decision processes and address the evolvable aspect of the paradigm. Suppose a maintainer is presented with a problem or symptom which leads to the choice of multiple solutions that may not necessarily lead to an effective maintenance action. No matter which solution the maintainer decides to perform, the reinforcement learning system records these problem/solutions relationships and their outcome. During the course of time the RL system will ultimately lead the maintainer to choose the most effective solution to the given problem.

The following relationships formally introduce the reinforcement learning model.

- S : A discrete set of environment states
- A : A discrete set of agent actions
- R : A set of scalar reinforcement signals

An agent is connected to its environment through actions. For every interaction, an agent receives an indication of the current state, s, of the environment. Once the agent knows the current probable state, it chooses among a set of actions. The selected action 'a' will consequently change the state 's' of the environment. During the transition from the current state to the new state a scalar reinforcement signal 'r' is used to reward or penalize the action based on its outcome. After numerous interactions and transitions the agent will gain useful experience that will lead to choosing the optimal action 'a' for a given current state 's'. Once the system has learned over time, a policy can then be used to choose which actions should be performed when a state is encountered. The value of the state is defined as the sum of the reinforcements over a discrete number of interactions. The optimal policy can be calculated by indexing the mapping between state and actions that maximize the sum of the reinforcements.

The advantage of Q-learning is that initially, when the optimal state to action relationships are not entirely known, the agent may have the option of randomly selecting an action. Q-learning allows arbitrary experimentation and simultaneously preserves the current best estimate state values. The final results in Q-learning are not contaminated by experimental actions, which will only ultimately reward or penalize the state action pairs. In order to obtain meaningful values, it is necessary for the agent to try out each action in every state multiple times allowing the reinforcement updates to assign penalties or rewards based on the outcome.
The Q-learning equation is presented below:

$$Q(s_t, a_t) = (1-\alpha)Q(s_t, a_t) + \alpha(r(s_t, a_t) + \gamma \max Q(s_{t+1}, a\ )) \tag{6}$$

Where, Q is defined as the sum of all reinforcements for the state action pair (s, a) at a given time (t).

- α is the learning rate parameter [0,1]
- γ is the discount parameter [0,1]

The learning rate parameter α carries a significant importance in adjusting the algorithms' learning behavior. α controls the weighting on current and previous knowledge represented in the state-action pair values. As the learning rate parameter approaches 1, new experience outweighs previous experience. As it approaches 0, previous experience outweighs new experience. For the state – action pairs, representing task effectiveness of a maintenance action relative to a fault, α is set closer to 0, outweighing previous experience to new, which translates in assigning more importance to the maintenance task history performance.

The discount parameter γ controls the weighting on the new state action pair. For the current application, the parameter is fixed to a constant and provides scaling to the state-pair values, which does not affect the overall results of the Q-Learning algorithm.
The mapping between state and action pairs can be represented in an n x m matrix where

- n represents the number of possible states
- m represents the number of possible actions

[illegible]	Q(s1,a1)	Q(s1,a2)	Q(s1,a3)	Q(s1,a4)
[illegible]	Q(s2,a1)	Q(s2,a2)	Q(s2,a3)	Q(s2,a4)
[illegible]	Q(s3,a1)	Q(s3,a2)	Q(s3,a3)	Q(s3,a4)
	a1	a2	a3	a4

Figure 9: Q Learning Matrix

The Q-Matrix may be a representation of the Task Effectiveness of a particular maintenance action (a) directly connected to a failure mode (s). The outcome of the performed action leads to a state transition. The state transitions are defined in the following two cases and illustrated in Figure 10 below.

- Case 1 : Given a Failure Mode (S), if any particular selected action is successful, the state will transition from the current fault state to a healthy state.

- Case 2 : Given a Failure Mode (S), if any particular selected action is not successful, the state will transition from the current fault state back to its current fault state or to a different fault state. The transition to a different faulty state could be an indication that the failure mode was not clearly identified until after the maintenance action was performed.

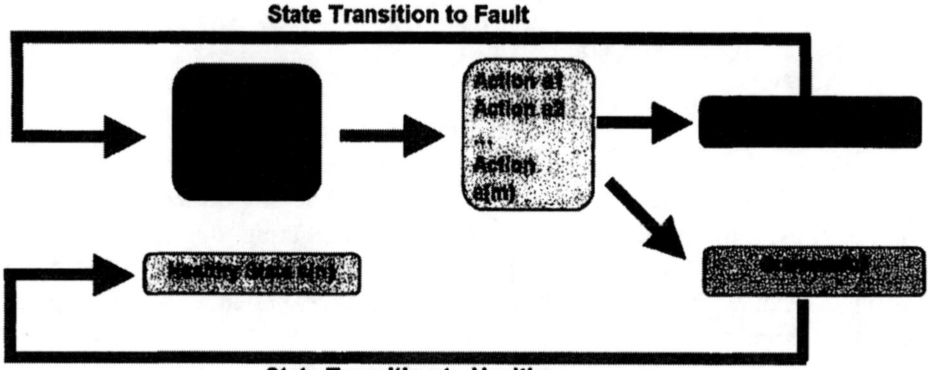

Figure 10: Decision Process Flow Chart

Implementation of the Application: The overall PBMS software package will be used to enhance existing CBM processes currently used for naval applications. The PBMS offers "plug and play" capability with software packages such as ICAS and enables the ability to intelligently recommend maintenance actions based on health management evidence, logistics, economics, and other factors. The design of the PBMS software architecture is flexible in that it may receive both archived and real-time data. Because it will be used to assess real-time information, the domain logic was designed to minimize response time. The reasoner also relies on close interaction with the MIM to perform its function.

The reasoner layer takes in evidence sources in the form of CBM monitor activations, processes the evidence in conjunction with a MIM, and outputs a list of probable failure modes and ranked maintenance task recommendations. The overall maintenance ranking can be modified to account for criteria weighting factors, availability, and scheduling. The current suite of ranking criteria includes: hardware and labor cost, availability or downtime considerations, safety, and criticality. Considerations such as common procedures, part/tool/labor availability, and upcoming scheduled tasks are factored into the domain logic.

The Human System Interface (HSI) provides a real-time link between the outputs of the PBMS (the system) and its user (the human). It is represented by a Graphical User Interface (GUI) that will present the information to the user in a manner that is easily understood as seen if Figure 11.

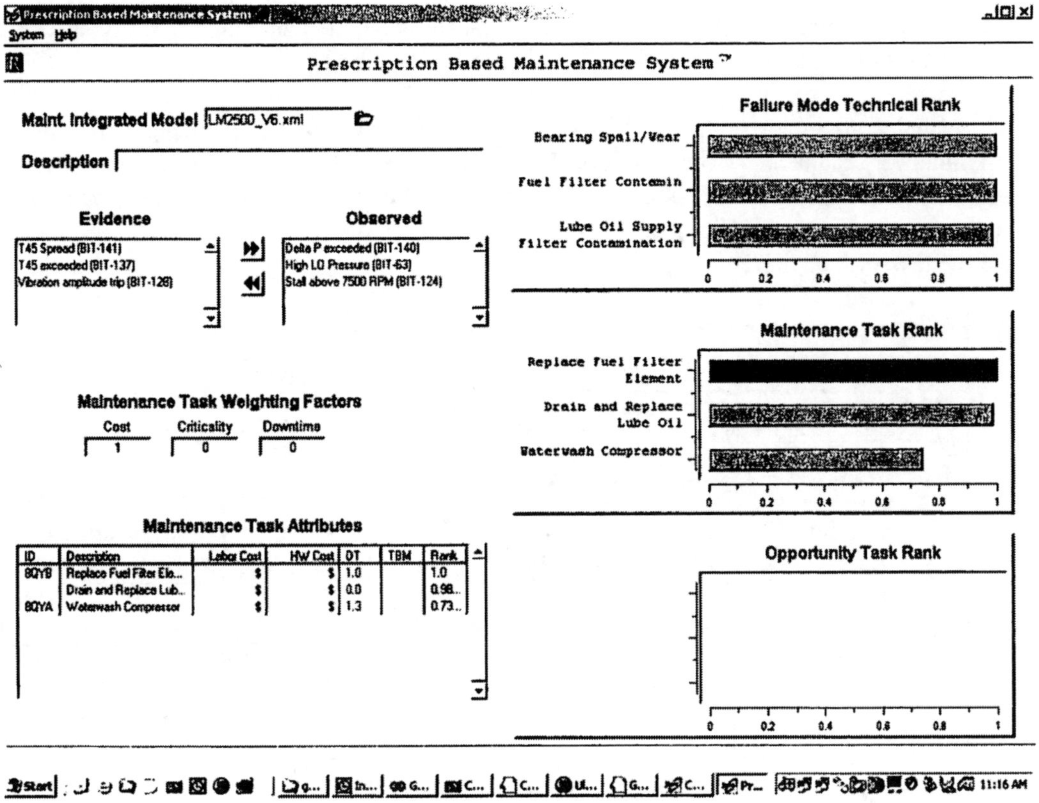

Figure 11: PBMS Graphical User interface

The HSI presents the maintenance technician with recommended maintenance actions and accepts maintainer feedback. Work order generation capabilities have been incorporated into the design based on the PMS Viewer format. Feedback provided by the user shall be used to update failure mode and maintenance task rankings, and if necessary update the MIM to make it more effective. Specifically, the PBMS reasoner must know which maintenance tasks were performed and whether or not they were successful.

The ship's information systems contain data that are updated in near real time and can be utilized by the PBMS. The Integrated Condition Assessment System (ICAS) contains event, fault and Built-in-Test (BIT) data. This evidential data is input to the PBMS and activates evidence when a fault condition is detected. Typically, the evidence trigger is a BIT or diagnostic, but several prognostic enhancements are available that will submit Remaining Useful Life (RUL) data as evidence to the PBMS.

The PBMS reasoner takes in the evidence provided by ICAS and, utilizing a subsystem's predefined Maintenance Integrated Model (MIM), determines the most likely failure modes and consequential maintenance tasks required to correct the fault. Once the maintenance tasks have been defined, the reasoner queries the Consolidated Onboard Ships Supply data base to determine whether or not the appropriate tools, parts, materials and personnel are available to complete the maintenance task. The reasoner flags the tasks that are missing resources so the human-in-the-loop can make a more informed decision for maintenance scheduling.

The final component of the PBMS is the activation of a learning mechanism through maintainer feedback. The feedback system is used to determine if a prescribed maintenance task was successful at correcting a fault condition. The link between failure modes and maintenance tasks is accordingly strengthen or weakened based on the success of the task or lack thereof.

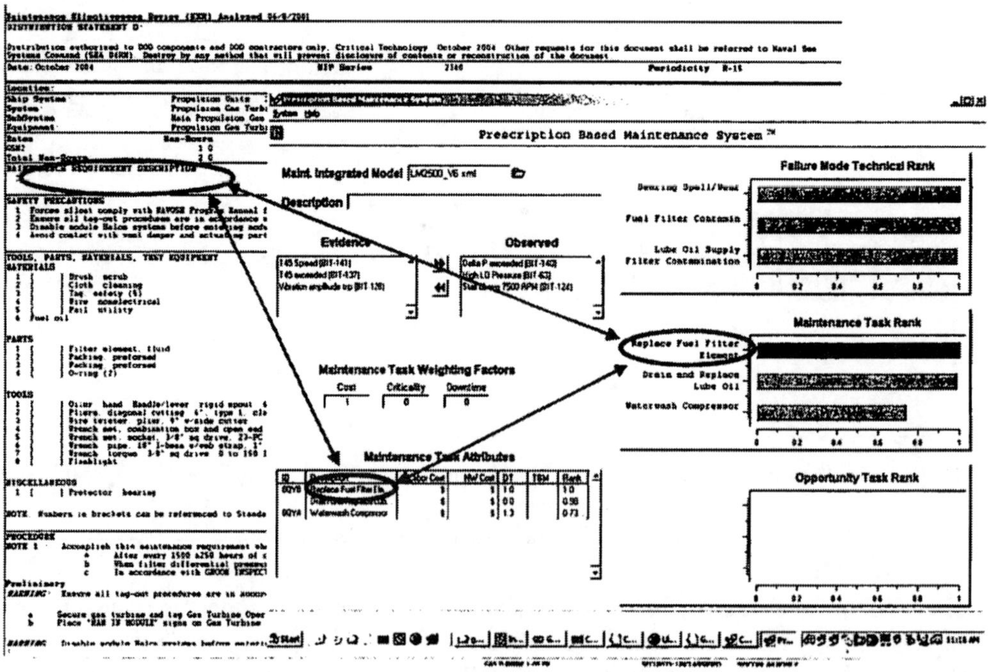

Figure 12: Connectivity between PBMS & PMS Viewer

Conclusions: A comprehensive approach has been presented to guide maintainers to the optimal maintenance actions as diagnosis through automated Condition-based maintenance systems. This integrated approach operates in open system architecture, using accepted OSA standards, to easily communicate with disparate data systems. The PBMS reasoners implement a variety of techniques such as Fusion reasoning Real fault probabilities, False alarm rates and advanced Reinforce learning. Finally, a human system interface concept was presented for illustrating how information from a complicated health management system could be presented to an end user.

References:

1. Bonarini, "Evolutionary Learning, Reinforcement Learning, and Fuzzy rules for Knowledge Acquisition in Agent Based Systems", Proceedings of IEEE, Vol. 89, No. 9, September 2001

2. Sakai, Ikeda, Iwata, "A New Criterion using Information gained for Action Selection Strategy in Reinforced Learning", IEEE Transaction on Neural Networks, Vol. 15, No. 4, July 2004

3. Kacprzynski, Palladino, "Self-Evolving Maintenance Knowledge Bases", SBIR Phase I Final report N68335-04-C-0181, November 2004

TRIBOLOGY

Chair: Mr. John M. Lucero
NASA Glenn Research Center

CLUTCH SYSTEM REMAINING USEFUL LIFE PREDICTION USING WEAR AND DYNAMIC MODELING EFFECTS

Carl Byington P.E., Matthew Watson, Sanket Amin

Impact Technologies, LLC
2571 Park Center Blvd.
State College, Pennsylvania 16801
United States
814-861-6273
Carl.Byington@impact-tek.com

Abstract: A model-based technique is presented for Remaining Useful Life (RUL) prediction of highly dynamic, high power dry clutch systems that combines physics-based simulation and wear prediction models. Primary load and engagement shear drivers (i.e., torque, speed, clutch surface temperature) are modeled using a first principle approach. An extension of Archard's law is presented in which life usage is predicted using multiple stochastic models to determine a wear coefficient for each applicable wear mechanism. The models consider the physical wear process, including debris particle and protective layer formation, using parameters such as surface roughness, particle size, and surface temperature. These stochastic variables are evaluated in a probabilistic framework using statistical methods, such as Monte Carlo Simulation and importance sampling, that consider both measurement and modeling uncertainty. Confidence interval prognostic results are provided to predict the RUL of the clutch throughout its limited life in near-real time.

Key Words: Abrasive wear; Adhesive wear; Clutches; Dynamic modeling; Life prediction methods; Prognostics and Health Management; Remaining Useful Life; Statistical analysis.

Introduction: The fundamental purpose of propulsion coupled drive train systems is to provide and direct mechanical power to lift and vectoring thrust elements. Gas turbine engines, gearboxes, rotor assemblies, fans and other rotating members are subject to finite design lives and susceptible to life limiting damage. Furthermore, the drive train and clutch systems of modern military systems are highly dynamic and carry a potential for high load densities, which further accelerates the life of these critical components. This paper describes an approach that was developed to predict the RUL of high power clutch systems used in such extreme applications.

Model-based prognostics methods have proven especially appropriate for fatigue-related failures in shafts, gear teeth, and ball bearings. These methods typically implement mathematical models that can be used to simulate steady state and transient performance under various commanded conditions, loads, and operational regimes. In the case of drive train systems, these models should be dynamic in order to simulate the transient interaction between speeds, loads, and power sources. These dynamic simulation models enable the identification of diagnostic indicators, which are necessary inputs for a prognostic health management (PHM) system.

Figure 1 depicts the general approach that has been developed to monitor and predict clutch plate wear, which has proven to be the dominant life driver of high power clutch systems. This model-based method incorporates physics-based models, stochastic variables, statistical sampling techniques, virtual temperature sensing, uncertainty characterization, and probabilistic wear forecasting for remaining useful life prediction. The process begins with a physical model of the clutch that is used to simulate the mechanical and thermal response of the system. A wear model is then implemented to estimate wear-induced material loss based on multiple, competing mechanisms that affect the life of the component. Statistical approaches and parameter trending approaches are then used to predict wear propagation and, ultimately, remaining useful life.

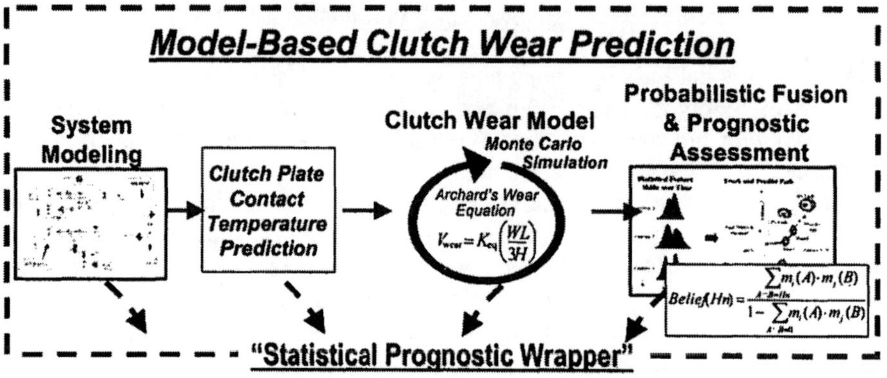

Figure 1 – Stochastic, modular wear-modeling architecture

Physical Clutch Modeling: As with any model-based approach, the physics-based model plays a central role in the overall PHM architecture. As such, a dynamic clutch model was developed using the Simulink modeling environment (Figure 2). The developed model can be used to (1) simulate system response (to generate fault progression data for validation of the developed approach) and (2) enable temperature prediction at the clutch plate interface.

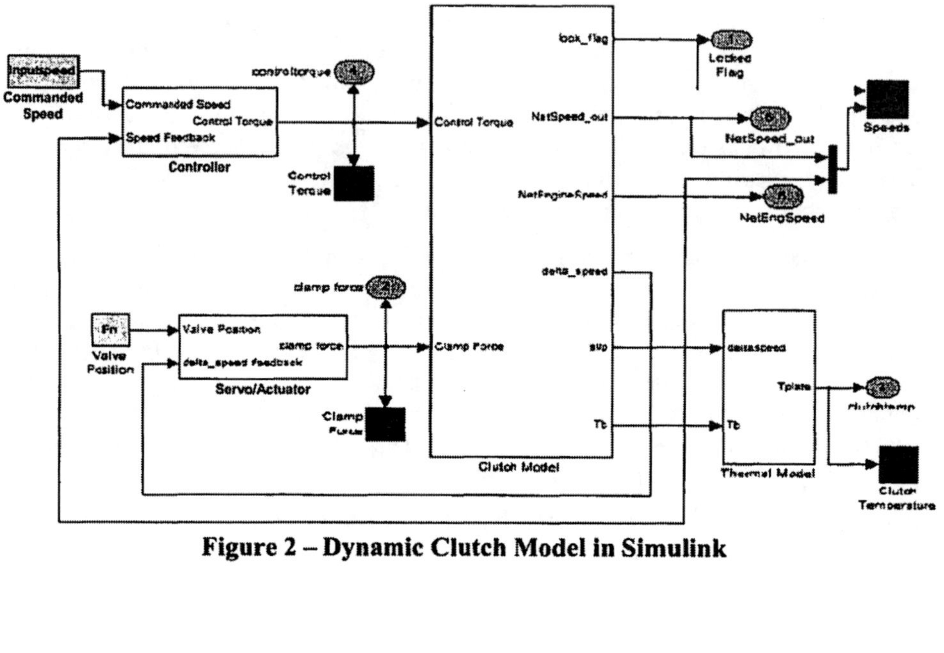

Figure 2 – Dynamic Clutch Model in Simulink

Dynamic Clutch Modeling: The developed clutch model is based on four distinct operating regimes that are typical of an actual clutch assembly (fully disengaged, engaging, fully engaged, and disengaging). These regimes were further classified into two operational modes: locked operation (fully engaged) and unlocked operation (fully disengaged, disengaging, and engaging). From a modeling standpoint, the transition from one mode to another is the most challenging. As such, consideration to both dynamic and static friction was needed to produce a high-fidelity model [1]. As part of the developed model, equations for frictional torque were used to determine both the operational mode of the system (static friction) and the amount of torque transmitted across the clutch (kinetic friction). For example, if the transmitted torque is less than the static friction of the plate, then the clutch will be locked. However, if the transmitted torque surpasses the static friction, the clutch will slip. The equations of motion of the system are described according to the two operational regimes. For the unlocked case, the clutch motion is described as:

$$\left[I_1\dot{\omega}_1 + b_1\omega_1\right] = T_{in} - T_{Transmit} \tag{1}$$

$$\left[I_2\dot{\omega}_2 + \omega_2 b_2\right] = T_{Transmit} \tag{2}$$

where I is inertia, $\dot{\omega}$ is rotational acceleration, b is damping coefficient, ω is the rotational speed, and T_{in} is the torque delivered to the system. In these equations, the driving plate is represented as Component 1 and the driven plate is represented as Component 2. For the locked condition, the two plates rotate together and their inertias are combined, resulting in the following equation of motion:

$$(I_1 + I_2)\dot{\omega} = T_{in} - (b_1 + b_2)\omega \tag{3}$$

This first principle's approach allows the primary load and engagement shear drivers to be modeled and enables accurate simulation of clutch operation.

Control System Modeling: In order to correctly simulate the dynamic operation of the clutch, a control system was also included in the model. This included a physical model of the actuator that is used to move and clamp the clutch plates, as well as some simple control loops to mimic operational commands to the system. The actuation system, modeled using a servo-cylinder and its accompanying hydraulics, was adapted from [2] and [3]. For this case, the hydraulics serves two purposes, depending on the state of the system. Initially, hydraulic pressure is used to displace the clutch plate. Clutch movement can be determined by balancing hydraulic pressure and resistance forces (in this case, force was lumped and modeled as a spring) using Newton's second law:

$$\ddot{x} = \frac{1}{m}\left[-F_s + A \cdot p_2\right] \tag{4}$$

Once contact is made between the plate surfaces, no further displacement can occur and changes in pressure directly relate to changes in normal force, creating the frictional torque needed to engage the system. Once again Newton's second law can be used:

$$F_N = -F_s + A \cdot p_2 \tag{5}$$

Virtual Temperature and Thermal Modeling: It is well known that wear is highly dependent on the interface temperature of the surfaces. As such, a critical component of this wear monitoring strategy is the accurate, reliable estimation of interfacial clutch temperatures. A technique for inferring, or "virtually sensing," temperature is therefore very important, as there are currently no practical means for taking this measurement. A physics-based temperature module was therefore included to enable reliable, autonomous temperature sensing. Data driven approaches, such as neural network and regression techniques, can also be implemented and fused with the physics-based virtual sensor for increased performance, but were not performed here.

Thermal modeling techniques are based on the physics of heat transfer and the dynamics of clutch operation. The two primary thermal mechanisms effecting the temperature of clutch plates are heat generation (caused by the slipping that occurs during engagement and disengagement) and forced convective heat transfer (from cooling air). A simple energy balance equation was used to calculate heat generation at the clutch plate surfaces during engagement and disengagement. This approach assumes that all energy losses in the system translate directly to heat generation, which in this case is a reasonable assumption. Convective heat transfer was modeled as forced convection flow across parallel plates for the fully disengaged state (Equation 6) and as forced convection over a cylinder in cross-flow (Equation 7) during the other three states of the system:

$$m_{plate} c_{p,plate} \frac{dT_{plate}}{dt} = Q_{stored} - \dot{m}_{air} c_{p,air} \left(T_{m,o} - T_{m,i}\right) \qquad (6)$$

$$\frac{dT_{plate}}{dt} = \frac{Q_{stored} - q_{cross}}{m_{plate} c_{p,plate}} \qquad (7)$$

This lumped parameter approach assumes that heat is dispersed uniformly across the clutch pack and does not consider heat conduction that would occur axially throughout the clutch plates or radially outward towards the cooled clutch plate's outer surfaces and nearby metal structures. This is a valid assumption for high performance clutch plates, which are made from very conductive materials, such as carbon-carbon composites, that effectively dissipate generated heat.

Multi-Mechanism Wear Model: Wear is generally defined as surface damage resulting from the removal of material during relative motion between a surface and another contacting body. Wear involves both physical and chemical interactions at the contact surface and is typically classified in two ways: by mode (i.e., sliding, fretting, etc.) and by mechanism (i.e., adhesive, abrasive, corrosive, thermal, etc.). Mode is a classification of the type of contact, while mechanism defines the process by which material is removed from the contact surface. It is important to note that these wear mechanisms are not mutually exclusive; two or more wear mechanisms frequently occur together. A difficulty often encountered when developing wear prediction models is that most of the models focus only on an individual wear mechanism, which is usually not good practice for practical applications. Therefore, it is imperative that the relevant wear mechanisms and their interdependencies are thoroughly understood and that any corresponding models are capable of accounting for each mechanism and interaction.

Although numerous models exist for calculating and predicting wear, one of the fundamental models for adhesive wear is Archard's Law [6], which postulates that wear is proportional to material hardness, contact pressure, and relative surface velocities. Specifically, Archard's Law is described as follows:

$$V_{Wear} = K\left(\frac{WL}{3H}\right) \tag{8}$$

where V_{Wear} is wear volume, H is material hardness, W is the contact load, L is the sliding distance, and K is the proportionality constant or wear coefficient. While Archard's Law has been adapted or expanded numerous times as researchers seek a greater understanding of sliding wear, it remains a standard in predicting wear under general sliding conditions. Often, adaptations of Archard's Law focus on experimental or theoretical derivations for the wear coefficient, K, which is considered the most critical term in Archard's equation. Although K was initially intended to represent the probability of adhesive wear, it has since been used to model abrasive and oxidative wear in various adaptations of Archard's sliding wear theory. One example of a more complex, physics-based wear model is Quinn's extension of Archard's model for oxidative wear, which includes a specific equation for determining the wear coefficient [9]. In addition, Hockenhull presents several other wear coefficient models for the sliding contact of metallic surfaces [10]. This flexibility allows the use of Archard's Law to generate separate models for each wear mechanism or a single model to represent all wear mechanisms. Regardless, these models illustrate the diversity of opinions on wear modeling and justify the implementation of a modeling approach that is modular and adaptable, allowing for improvements as better models become available.

In the case of drive train clutch systems, it is likely that adhesive wear will be the dominant wear mechanism. However, abrasive wear and oxidation could also play a substantial role, depending on the severity of interfacial temperatures [4], [5]. With these competing mechanisms in mind, a multi-mechanism wear modeling architecture was developed for drive train clutch systems (Figure 3). Two important characteristics were desired for the developed clutch wear modeling approach. First, a modular architecture that is flexible and facilitates updating or augmenting of the approach was desired. This flexibility allows consideration of multiple wear mechanisms. In the case of high performance clutch plates, for example, the flexibility should exist to not only model adhesion or abrasive wear but also oxidation, which is necessary due to the vulnerability of carbon-carbon composites to oxidize at high temperatures. A second important focus of the wear modeling approach is the use of statistical variables. Due to the inherently stochastic nature of wear, it is more appropriate to model certain parameters and material properties, such as hardness, surface roughness, and contact area, by using a distribution of values. Modeling wear in this manner provides the most robust capability for monitoring wear at clutch plate surfaces. Two important research sources for this effort were Stolarski's [7] probabilistic approach, which addresses the inherent uncertainty of wear prediction, and Stott's [8] protective layer model.

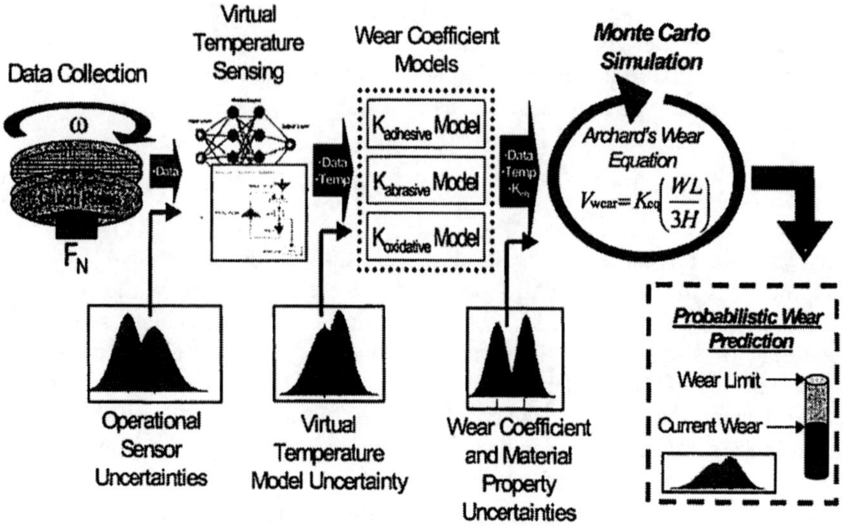

Figure 3 – Overall Wear Modeling Approach

Remaining Life Prediction: In the case of fault-to-failure prediction, the time (or number of engagements) remaining before the current health state progresses to functional failure is desired. The time to reach this condition is determined by tracking and projecting the path of wear predictions using a statistical trending method. One such method, double exponential smoothing (DXS), was employed here to effectively track and project wear. DXS employs an exponentially weighted averaging function that can forecast future values of a vector based on past observations. DXS is represented mathematically as:

$$S_T = a_s y_T + (1-a_s)S_{T-1} \qquad \hat{y}_{T+\tau}(T) = \left(2 + \frac{a_s\tau}{(1-a_s)}\right)S_T - \left(1 + \frac{a_s\tau}{1-a_s}\right)S^{[2]}{}_T$$
$$S^{[2]}{}_T = a_s S_T + (1-a_s)S^{[2]}{}_{T-1} \qquad (9)$$

where a_S is the smoothing constant and S_T and $S^{[2]}{}_T$ are the smoothing statistics. These past observations are then weighted using an exponential function. Other methods, such as Kalman filters, could also be applied.

Uncertainty Estimation: In any modeling effort, uncertainty plays an important role. In applications such as wear modeling, this role is even more critical as several physical parameters (i.e., surface roughness) must be determined by reasonable estimates or experimentally determined relationships. In this effort, uncertainty is considered in each step of the process and takes two primary forms: (1) measurement uncertainty, and (2) modeling uncertainty. Measurement uncertainty is dictated by the accuracy of the sensor and manifests itself as error propagation. Capturing this propagation is important to quantifying uncertainty in the final prediction. Modeling uncertainty appears as a residual difference between the validated model and the actual system. This is a consequence of a limited capability to exhaustively model all of the system's physical characteristics, as well as inherent system non-linearities and random system variations. These variations exist and can be accounted for by analyzing actual system data and quantifying the variability of monitored system variables.

Statistical Wrappers: Within the developed approach, uncertainty is accounted for using statistical wrappers and a probabilistic framework. Monte Carlo (MC) methods provided a practical means for reaching an accurate prediction within this framework. MC methods refer to statistical simulation techniques used to predict the behavior of an engineering system. The methods employ system modeling and random combinations of system parameters to perform this analysis. Unlike MC simulation, conventional simulation starts with system modeling involving discrete differential equations. These relationships are then solved algebraically to determine the unknown states of the system. In performing MC simulations, however, the system is assumed to be a stochastic process; therefore random number simulation is applicable. Because MC theory is dependent on random numbers and emphasizes the importance of randomness and chance, sampling is important. Common sampling techniques, such as variance reduction and importance sampling, can be used to improve the statistical efficiency of the simulator and decrease the number of samples needed to generate an estimate within a desired error tolerance. This is important for near real-time systems where processing time is a concern.

Wear Modeling Results: The operational data and virtual temperature predictions generated by the Simulink model were used to validate the model-based clutch fault progression module. Using this model, a number of fault progression data sets were generated by increasing the distance the clutch plates must be moved before engaging (simulating plate material loss). Parameter uncertainties were then estimated and coupled with simulation data for use within the developed probabilistic methodology. Operational sensor, wear coefficient, and material property uncertainties were estimated from past experience and literature. Virtual temperature model uncertainty was calculated from a sensitivity analysis using factorial Design of Experiment methodology, as discussed next.

Virtual Temperature Uncertainty: Uncertainty estimation is critical to the developed probabilistic approach. As such, techniques are needed to determine the uncertainty of model-determined values. As an alternative to real-time simulation techniques (i.e. Monte Carlo Simulation), a sensitivity analysis was implemented to determine virtual temperature model uncertainty. This approach reduces the amount of processing necessary for near real-time implementation by predetermining uncertainty values with experimentation. The basic idea of the approach is captured in the following equation:

$$\Delta T = \delta T / \delta P_1 * \Delta P_1 + \delta T / \delta P_2 * \Delta P_2 + \cdots + \delta T / \delta P_N * \Delta P_N \qquad (10)$$

where temperature (T) uncertainty is calculated using the uncertainty of each parameter (ΔP) and the effect (or sensitivity) that each critical model input has on temperature ($\delta T/\delta P$). Sensitivity parameters ($\delta T/\delta p_i$) were determined using experimentation, Design of Experiments methodology, and statistical analysis. An analysis was conducted after generating sufficient training data from the Simulink model to determine each primary effect and interaction. A regression estimate of the given data was then implemented and the sum of squares for each effect was calculated (similar to Yates [11]) as the regression progressed. In addition, factorial design was used to reduce the number of experiments (2^N) needed to fully explore the effects of the parameters. Five model parameters were selected for consideration in the analysis (coefficient of friction, specific heat, convective coefficient, ambient air temperature, and speed differential), with speed differential and coefficient of friction producing the greatest effect.

Stochastic Wear Results: For this initial effort, only the wear coefficient for adhesive wear was implemented. However, the algorithms were constructed in a modular fashion to allow for future updates to incorporate abrasive and oxidative wear coefficient models. Normal parameter distributions were assumed and probability density functions (PDFs) were created for each wear model input. An example of these parameter distributions can be seen in Figure 4. A Monte Carlo simulation was then used to solve Archard's wear equation, using 500 simulations, to produce an estimate of the lost wear volume from the surface of the plates. This process was repeated for each simulated engagement-disengagement cycle.

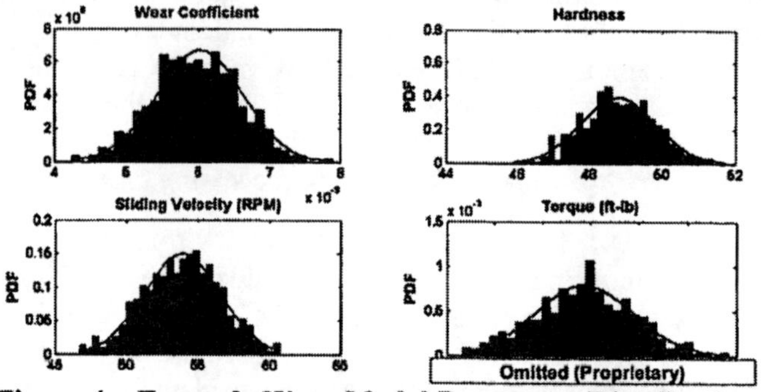

Figure 4 – Example Wear Model Parameter Distributions

Figure 5 (left plot) is an example of the results obtained for a series of engagements at various speeds. In the figure, confidence bounds are also included and appear as dashed lines around the mean prediction (solid line). These values are a result of the Monte Carlo simulation. An important characteristic of this approach is its ability to predict wear using the actual usage of the clutch. As seen in the figure, changes in the operation of the clutch result in different wear rates.

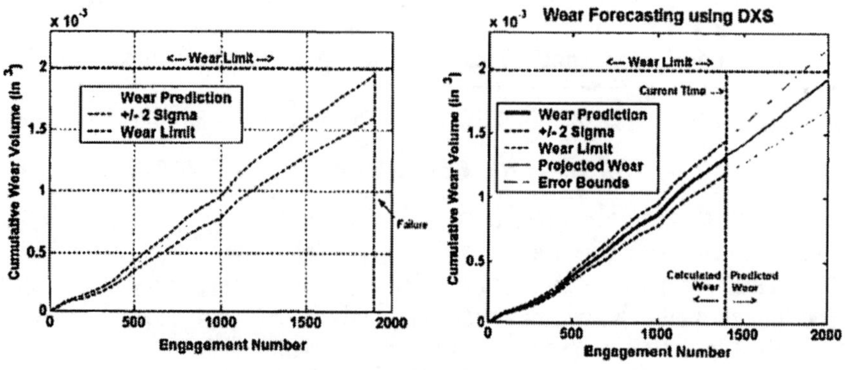

Figure 5 – Example wear prediction results (left plot) and forecast/RUL prediction (right plot) for engagements at various RPMs

An example of the RUL predictions for this data set can also be seen in Figure 5 (right plot). In the figure, a forecast is performed at ~1400 engagements (current time). The mean wear (solid line) and 2-sigma confidence bound (dashed line) represent the probabilistic wear predictions previously described. The lines to the right of the current time represent the predicted wear progression of the clutch using the DXS routine. The error bounds (dashed line) represent a fusion of uncertainty in the current health and past RUL error predictions. As seen, a conservative estimate of clutch wear (using the upper bound) results in failure at about 1800 engagements. Statistical methods, such as Monte Carlo simulation, could be used to generate a full distribution at each point, but this was not performed in the current effort.

Software Demonstration: A software demonstration of the model-based clutch fault progression module was also created to illustrate the operation of the module using the developed SIMULINK clutch model and wear prediction algorithms (Figure 6). As seen, the Graphical User Interface shows (1) raw operational data (friction torque and speed differential) and virtual temperature prediction, (2) statistical and uncertainty information used in the Monte Carlo Simulation, and (3) wear prediction and fault-to-failure prediction results. The ability to convert the predictions from 'number of engagements remaining' to a 'remaining useful operational time' is planned by allowing the user to enter the planned usage of the component, but is not active at this time.

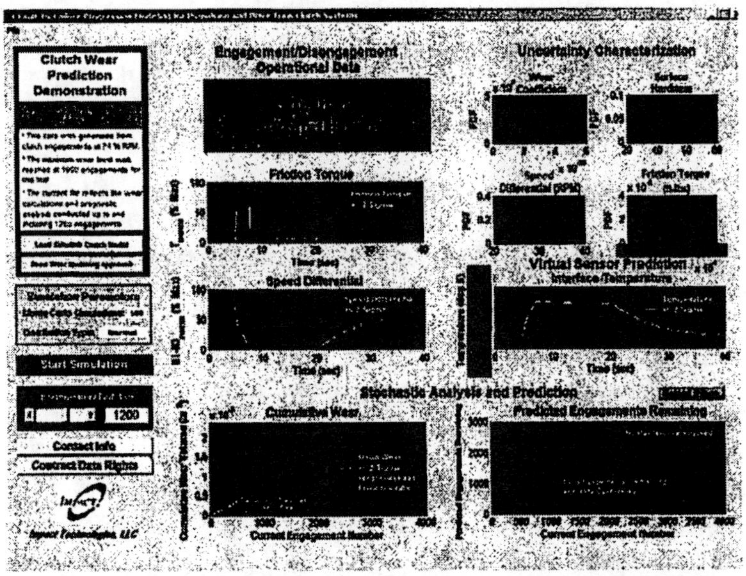

Figure 6 – Model-based Clutch Fault Progression Module Demonstration

Conclusions: An overall architecture was developed for Remaining Useful Life (RUL) prediction of highly dynamic, high-power dry clutch systems. This architecture integrates a physics-based model of the system with a stochastic, modular approach to wear modeling. Debris particle formation, protective layer development, surface roughness, particle size, and surface temperature are considered, as well as uncertainty due to both measurement error and modeling error. Within the developed approach, Monte Carlo methods replace the traditional deterministic analysis and are used to capture these stochastic variables within a probabilistic framework. Sampling methods, such as importance sampling, are also treated and confidence interval prognostic results are provided to predict the RUL of the clutch throughout its limited life in near-real time. The approach was validated using a number of fault progression data sets, which were generated from a model by increasing the distance the clutch plates must move before engaging (simulating plate material loss). The model demonstrated a good ability to follow the seeded progression of cumulative wear.

Future Work: As discussed, this stochastic, modular wear monitoring approach was initially validated using a generic model within Simulink. The next step in the validation process is to perform wear testing to produce actual failure progression data. A system-specific model will then be developed and executed with the collected data to validate its operation. In addition, the specific modes of heat transfer (instead of a lumped parameter) will be accounted for in the temperature prediction and fused with other prediction approaches, including neural network and regression techniques. Multiple wear mechanisms will also be modeled and combined using the methods described. Finally, additional prognostic approaches will be applied and approaches with reduced processing requirements will be considered (for on-board implementation).

Acknowledgements: This work significantly benefited from the support and technical consult of Michael Begin and Doug Gass of the Naval Air Warfare Center, Aircraft Division. The data support and consultation of Greg Colman from Rolls-Royce plc contributed to this work as well. Financial support for this work was provided by the NAVAIR SBIR program office through a Phase I Contract # N68335-03-C-0190 and is also gratefully acknowledged.

References:

[1] "Using Simulink & Stateflow in Automotive Applications." MathWorks, Inc. 1998.

[2] Sohl, G. and Bobrow, J. "Experiments and Simulations on the Nonlinear Control of a Hydraulic Servosystem." *IEEE Transactions on Control systems Technology*. V. 7, N.2., Mar 1999.

[3] Montanari, M., Ronchi, F., Rossi, C., Tilli, A., Tonielli, A. "Control and Performance Evaluation of a Clutch Servo System with Hydraulic Actuation," Control Engineering Practice, in press.

[4] Glaeser, W.A. "Friction and Wear." *IEEE Transactions on Parts, Hybrids, and Packaging*, Vol. PHP-7, No. 2, June 1971.

[5] Engel, Peter A. "Failure Models for Mechanical Wear Mode and Mechanisms." *IEEE Transactions on Reliability*, Vol. 42, No. 2, June 1993.

[6] Archard, J.F. "Contact Rubbing of Flat Surfaces." *Journal of Applied Physics*. V. 24 N. 1, pp 18-28.

[7] Stolarski, T. A. "A Probabilistic Approach to Wear Prediction." *J. Phys. D: Applied Physics*. 24 (1990) 1143 – 49.

[8] Stott, F.H. "High-Temperature Sliding Wear of Metals." *Tribology International* 35 (2002) 489-95

[9] Quinn, T.F.J. "The effect of 'Hot Spot' temperatures on the unlubricated wear of steels." ASLE. Trans. 10 (1967) 158-168.

[10] Hockenhull, Kopalinsky, and Oxley. "Mechanical Wear Models for Metallic Surfaces in Sliding Contact." *J. Phys. D: Applied Physics*. 25 (1992) A266-A272.

[11] Stat-Ease, Inc. Design-Expert Version 6.0 software documentation

PROGNOSTICS II
Expeditionary Fighting Vehicle Program

Chair: Mr. Vernon S. Fox
United Defense LP

VERIFICATION AND VALIDATION OF DIAGNOSTIC/PROGNOSTIC ALGORITHMS

Carl S. Byington, P.E.,
Michael J. Roemer, Ph.D.,
Patrick W. Kalgren

Impact Technologies, LLC
2571 Park Center Blvd., Suite 1,
State College, PA 16801
Carl.Byington@impact-tek.com

George Vachtsevanos, Ph.D.

Georgia Institute of Technology
School of Electrical and Computer Engineering
Atlanta, GA 30332-0250
gjv@ece.gatech.edu

Abstract: This paper describes metrics and strategies used to verify and validate diagnostic and prognostic algorithms. Metrics presented for fault detection, diagnosis and prognosis functional elements form the core performance assessment of a deployed PHM system. The authors discuss the use of simulation and real fault data, and the strategy employed to project fielded performance of the system. The effects of signal noise, measurement uncertainty, and threshold settings are also addressed. In addition, several tools are highlighted that aid in this testing and evaluation. The concept of 'performance metric growth' as more field data becomes available is also addressed in the work. Specific application of these techniques and tools to the Expeditionary Fighting Vehicle is offered.

Key Words: EFV; PHM Design; Statistical Analysis; Performance Metrics; Verification and Validation

Introduction: Verification and validation (V&V) techniques for Prognostic and Health Management (PHM) technologies serve to ensure that delivered capabilities meet system design requirements. System performance metrics provide the foundation for V&V and are useful for system evolutionary design. Few formal or universally accepted techniques and methodologies exist. In the following sections, we discuss strategies and metrics useful for PHM system V&V. Through the course of this discussion, three distinct performance measurement points emerge. Detection, diagnosis and prognosis form an upward chain of interdependencies that effect overall PHM system performance. Although good PHM system design techniques (i.e., incorporation of multiple strategies such as condition monitoring, usage monitoring, and physics of failure modeling) may mitigate the effects of these dependencies, absolute failures at lower levels result in serious consequences in following levels.

Detection: The EFV Drive-Train Prognostic System (DTPS) builds upon sound machinery condition monitoring technique. Utilizing a broadband, distributed data acquisition and signal processing system, the DTPS capability and overall effectiveness relies on high-fidelity fault detection. Assuming the presence, or absence, of a certain condition to be binary in nature, statistical signal detection techniques can form the basis for development of performance metrics.

The Neyman-Pearson theorem, illustrated in Figure 1, provides a solid foundation to begin a discussion of detection analysis [1, 2]. Bayes risk is another valid technique typically employed by pattern recognition systems. For convenience, we limit our discussion to the Neyman-Pearson theorem. Referring, once again to Figure 1, note that the density function on the left representing the value of a particular feature, given a healthy state, while the function described by the curve on the right represents the feature value from a damaged health state. The Probability of False Alarm P(FA) and the Probability of Detection P(D) are interdependent due to their measurement with respect to a particular threshold. If the threshold is raised to decrease the P(FA), a consequence occurs with a decrease P(D) and a increase in probability in missed detection. Therefore, once sensors are positioned, data is collected, and features are extracted, a choice must be made to either minimize the false alarm rate or maximize the probability of detection and then a threshold is set.

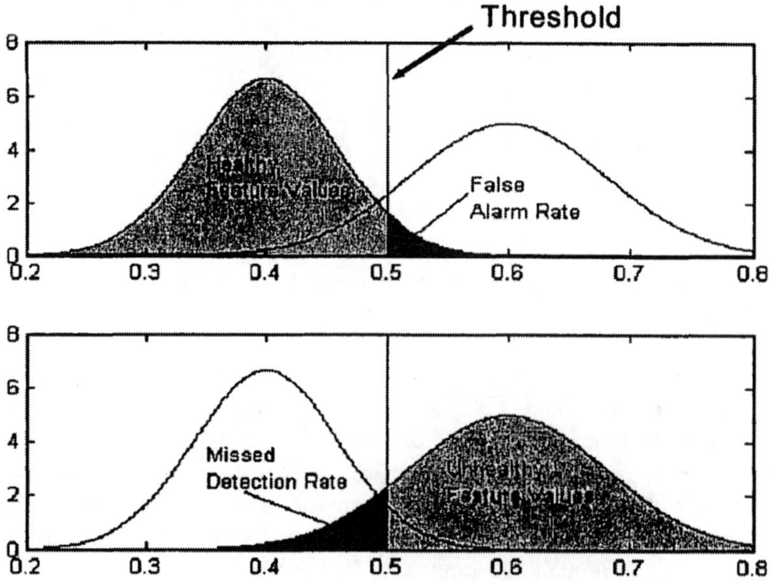

Figure 1 – Relationship of Statistical Detection Metrics

Assuming the computed feature is sensitive to a change brought about by a damage mechanism, the system designer's goal is to maximize that sensitivity, separating the mean values, while minimizing the variance to decrease the spread and resultant overlap of the distributions. For successful diagnostic systems, threshold setting is most effectively accomplished by balancing the P(FA) and P(D) using available data sets, simulations, and models as appropriate. Maximizing the separability and reducing the computed feature variances as much as possible is key to achieving good performance. Feature results can be treated in post-analysis and in near real-time processing as distributions.

For large sample sizes, an estimate of the fault detection performance can be computed directly using a decision matrix approach (Table I), which is based on a hypothesis testing methodology and represents the possible fault-detection combinations that may occur with an automated fault detection system. As seen in the figure, Column 2 of the matrix lists all the cases in which a fault is present in a monitored component. Column 3 consists of all cases in which no-fault occurs in the system. Row 2 represents an indicated fault by the detection logic and Row 3 represents when a fault is not indicated by the monitoring logic. Cells A, B, C, and D are populated by the results using the fault detection logic, and the performance metrics can by computed from these data.

Table I & II - Detection Metrics

DECISION MATRIX			
	Cases with a Fault (F1)	Cases with No Fault (F0)	Total
Fault Indicated (D1)	A: Number of Detected Faults	B: Number of False Alarms	A+B: Total Number of Alarms
Fault Not Indicated (D0)	C: Number of Missed Faults	D: Number of Correct Rejection	C+D: Total Number of Non Alarm
	A+C: Total Number of Faults	B+D: Number of Fault Free Cases	A+B+C+D: Total Number of Cases
	Detection Rate: A/(A+C)	False Alarm Rate: B/(B+D)	Accuracy: (A+D)/(A+B+C+D)

Metric	Equation
Probability of Detection	$P(D1/F1)=\frac{a}{a+c}$
Probability of False Alarm	$P(D1/F0)=\frac{b}{b+d}$
Accuracy	$P(D1/F1\&D0/F0)=\frac{a+d}{a+b+c+d}$
Receiver Operating Characteristic (ROC)	Plot of $P(D1/F1)$ vs $P(D1/F0)$

Table II lists the detailed metrics that are used for statistical detection performance analysis [2, 3, 4, 5]. Many well-developed metrics for analysis of detection exist, however a consolidated tool that incorporates raw data, algorithm interfaces, and metrics based analysis would prove valuable for development and long term support of specific applications. Due to the time dependent nature of prognostics, a method to predict a time-based metric (Mean Time Between False Alarms) may also be useful for future development. An example of the utility of these metrics, as applied to the EFV DTPS is given in [2].

Diagnosis: Perhaps better referred to as isolation, fault diagnosis is the next step in the formation of a PHM result. While the last section dealt with the detection of the existence of a fault, diagnosis refers to the identification of the failed, or, more importantly, failing component. Table III presents three useful metrics for analysis of diagnostic effectiveness.

Table III - Diagnostic Metrics

Diagnostic	Confusion Matrix	*Table of Classification Results*
	Probability of Isolation (FIR)	$FIR = \frac{A_T}{A_T + C_T} * 100$
	Kappa Coefficient	$kappa = \frac{N(observed) - N(expected)}{N(total) - N(expected)}$

The Confusion Matrix Table (see Table IV) shows the results of classifying data into several categories. The table shows actual defects (as headings across the top of the table) and how they were classified (as headings down the first column). The shaded diagonal represents the number of correct classifications for each fault in the column and subsequent numbers in the columns represent incorrect classifications.

Table IV - Confusion Matrix for Diagnostics

Confusion Matrix – Showing Number of Correct and Incorrect Classifications					
Classified as	**Collector Gear Crack Prop.**	**Quill Shaft Crack Prop.**	**Spiral Bevel Input Pinion Spalling**	**Input Pinion Bearing Corrosion**	**Total**
Collector Gear Crack Propagation	5	1	1	0	7
Quill Shaft Crack Propagation	0	6	0	1	7
Spiral Bevel Input Pinion Spalling	1	0	5	0	6
Input Pinion Bearing Corrosion	0	0	1	6	7
Total	6	7	7	7	27

The probability of isolation refers to percentage of all component failures that the monitor is able to unambiguously isolate The Kappa Coefficient represents how well an algorithm is able to correctly classify a fault (with a correction for chance agreement). The Kappa Coefficient is bounded by –1 and 1, with 0 representing chance agreement and negative numbers indicating that the algorithm performed worse than random guessing. Together, these metrics form a core analysis of diagnostic effectiveness.

Prognosis: Once a changing condition is detected and isolated, prognostics provide the final output of the DTPS. A prognostic vector typically indicates time to failure and an associated confidence in that prediction. However, the definition and implementation of prognostic performance metrics must consider multiple viewpoints and objectives. Two such viewpoints follow.

1. The maintainer's viewpoint: When should I perform maintenance? What is the most appropriate time (given such constraints as availability of spares, etc.) to maintain critical equipment?
2. The field commander's viewpoint: Are my assets available to perform the mission? What is the confidence level that critical assets will be available for the duration of the mission? Or, given a certain confidence level, what is the expected time to failure of a particular asset?

Answering the maintainer's question requires an estimate of the remaining useful lifetime of a failing component and assignment of uncertainty bounds to the trending curve that will provide him/her with the earliest and the latest (with increasing risk) time to perform maintenance and the associated risk factor when maintenance action is delayed.
Answering the commander's question will requires an estimate of the confidence level assigned to the completion of a mission considering mission duration, expected operating and environmental conditions, other contingencies, etc., or, given a desired confidence level, means to estimate the component's time to failure. Thus, the prognostic curve (the output of prognostic routines) must be modulated by a time-dependent distribution (statistical/Weibull or possibilistic/fuzzy) that will provide the uncertainty bounds. An appropriate integral of such a distribution at each point in time will result in the confidence level committed to completing the mission or, given a desired confidence level, the time to failure of a failing component.
Assume that a prognostic algorithm predicts the following progression or evolution of a fault with the associated uncertainty bounds: At current time t_o a fault has been detected and isolated and the prognostic routine is predicting a mean time-to-failure of T_{fm}, the earliest time to failure T_{fe}, and the latest time to failure of T_{fl}, as shown in Figure 2. The hazard line specifies the fault magnitude (dimension) at which the component ceases to be operational (failure).

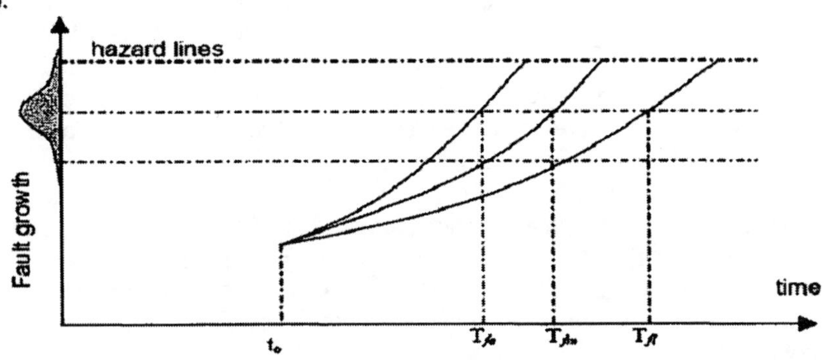

Figure 2 Confidence limits and uncertainty bounds

Let us further assume that failure data is available to superimpose a distribution or distributions, as shown pictorially in Figure 3. Strictly speaking, these "distributions" are either possibilistic (fuzzy) functions or probability density functions.

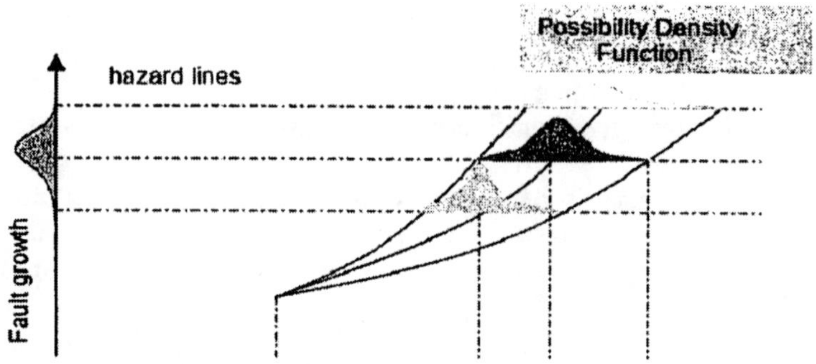

Figure 3 Possibility density functions for confidence bounds

Suppose further that the distribution at T_{fm} (crossing the hazard line) is as shown in Figure 4:

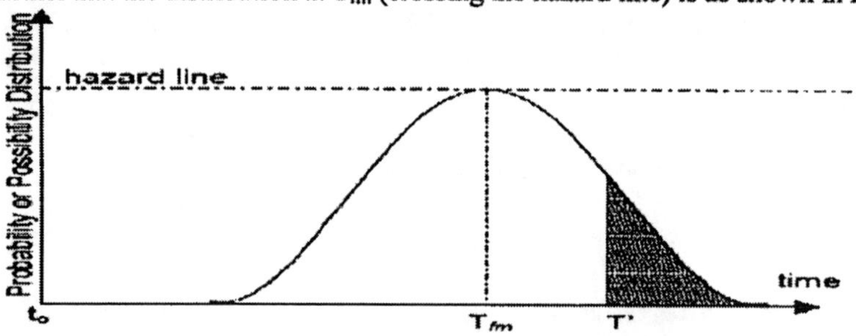

Figure 4 Distribution at the hazard line

As an example, assume that the pre-planned mission is estimated to require the availability of the asset under consideration for T' time. The integral under the distribution curve from T' to infinity will give us an estimate of the confidence level in terms of probability (or possibility), i.e. how confident we are that the asset will not fail before the mission is completed. Consider next the same distribution. Now, we specify a certain confidence level, say 95%, and would like to find the time T'' that the component will remain operational before a complete failure. Integrating the distribution curve from T'' to infinity and setting the result equal to the confidence limit (95% in our example), we solve for T'' thus arriving at the length of time (starting at to) that the asset will be available within the desired confidence limit, as shown in Figure 5. We view this procedure as a dynamic evolution of our estimates. That is, as more data becomes available and time marches on, new confidence limits are derived and the uncertainty bounds shrink through appropriate learning routines. The procedure outlined above may lead eventually to specification of performance metrics for prognostic systems. Such additional metrics may refer to the reliability of the confidence limits, risk assessment, safety specs, and others.

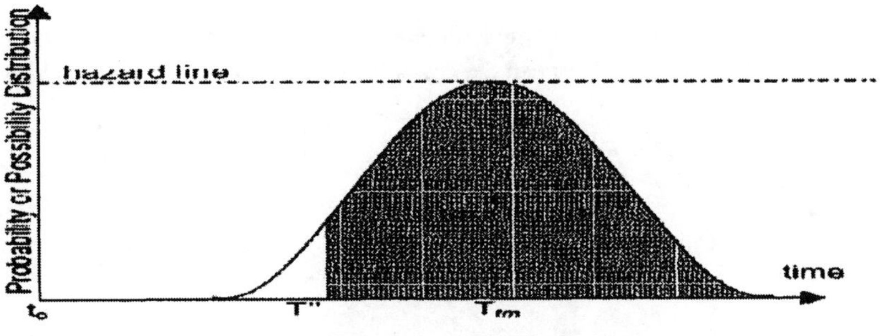

Figure 5 Distribution at hazard line

Similarity compares more than one predicted time series against the real series from a starting point to a certain future point in time. This measurement can be computed as

$$similarity(x,y) = \sum_{i=1}^{n} \left(1 - \frac{|x_i - y_i|}{\max_i - \min_i}\right) \quad (1)$$

where x_i and y_i are two i_{th} elements of two different time series, $\max_i$ and $\min_i$ are the maximum and minimum of all i_{th} elements.

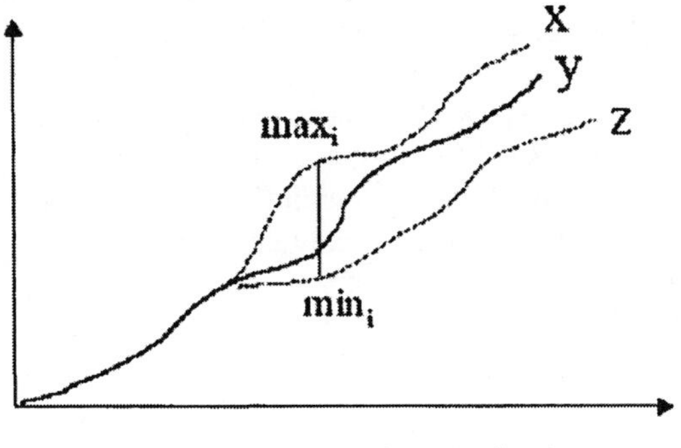

Figure 6 Determining similarity

Sensitivity measures how sensitive the prognostic algorithm is to input changes or external disturbances. It is defined as

$$SN = \frac{\sum_{i=1}^{N} \frac{\Delta_i^{Out}}{\Delta_i^{In}}}{N} \quad (2)$$

where Δ^{out} = distance measure of two successive outputs, and Δ^{in} = distance measure of two successive inputs.

The mean of the prediction-to-failure time is calculated as

$$E\{t_{pf}\} = \frac{1}{N}\sum_{i=1}^{N} t_{pf}(i) \quad (3)$$

where $t_{pf}(i)$ denotes the prediction-to-failure time for the i_{th} experiment and N is the number of experiments. The standard deviation of the prediction-to-failure time is calculated as

$$S\{t_{pf}\} = \left[\frac{1}{N}\sum_{i=1}^{N}(t_{pf}(i) - E\{t_{pf}\})^2\right]^{1/2} \quad (4)$$

The complexity metric refers to the computational aspects of the prognostic algorithm and suggests the expected relationship between the computation time and the time to failure.

$$complexity = E\left\{\frac{\text{computation time}}{\text{time to failure}}\right\} = E\left\{\frac{t_p - t_d}{t_{pf}}\right\} \quad \textbf{(5)}$$

where, t_p and t_d are the moments when the detection of the fault happens and when the remaining useful lifetime is predicted.

Importance of Ground Truth Choice in Metric Assessment: Ground truth is a representation of the actual condition of the system and serves as a baseline for comparison between a feature/algorithm result and the actual condition. Validation at all levels, detection, isolation, and prognosis, rely on ground truth data. Although the name itself implies an "absolutely true" representation, in most cases, ground truth is an engineering estimation of the actual condition, derived from a measurable parameter that closely matches the actual condition. The observable ground truth may be based on visible evidence, another feature or prediction result, or the damage severity curve. The ideal situation is for a feature/algorithm to possess perfect agreement with the ground truth in some measurable way. Metrics are used to assess and report the degree to which this agreement occurs. However, it is important to note that features and algorithms that score well against one ground truth could score lower against a second, different ground truth. Many factors may produce this effect. System overtraining is a large factor that can be controlled through careful advance planning. Acceptable control methods include robust system testing, use of multiple classifiers through knowledge fusion, and data reserve/hold-out techniques. The data hold-out technique, using a 70/30 development versus test data role is an effective preservation of ground truth integrity, given a substantive amount of data. Ground truth data for EFV will derive from test stand fault characterization efforts and from operational data.

Tools for Verification and Validation: As mentioned previously, robust tools for V&V can provide useful insights applicable to system development as well as ongoing validation and improvement. An ability to warehouse raw data, simulation models, and system configuration, along with an interface that permits easy exercise of developed detection, isolation, and prognostic algorithms would add significant value to many acquisition programs that seek to employ PHM technology. Such a tool could incorporate real data, simulation models, and noise augmentation techniques to analyze effectiveness and overall robustness of existing and emerging technology. Impact Technologies, in cooperation with the Georgia Institute of Technology, is developing a web-based software application that will provide JSF (F-35) system suppliers with a comprehensive set of PHM Verification and Validation (V&V) resources. The application includes; standards and definitions, V&V metrics for detection/diagnosis/prognosis, access to costly seeded fault data sets and example implementations, a collaborative user forum for the exchange of information, and an automated tool for impartially evaluating the performance and effectiveness of PHM technologies. This program is specifically focused on the development of the prototype software product to illustrate the feasibility of the techniques, methodologies, and approaches needed to verify and validate PHM capabilities. A team of JSF system suppliers, including

P&W, Northrop Grumman and Honeywell are being assembled to contribute, provide feedback and make recommendations to the product under development. The approach being pursued for assessing the overall PHM system accuracy is to quantify the associated uncertainties at each of the individual levels as illustrated in Figure 7, and build up to the system level through the PHM architecture.

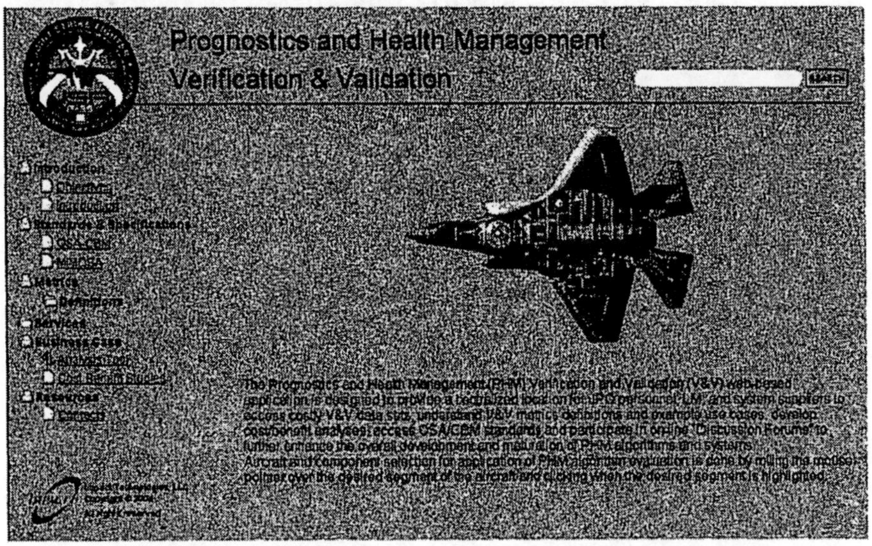

Figure 7 - JSF PHM V&V by Impact and Georgia Tech

Summary: This paper provided an overview and brief description of available metrics for analysis of detection, isolation, and prognostic algorithm effectiveness. Beginning with a solid foundation for monitoring and detection, a good PHM system deploys a wide array of techniques to achieve an ultimate result. Analysis of the effectiveness at each level provides not only the verification and validation required to measure overall performance, but provides legitimate feedback required for ongoing PHM system development and evolution. Techniques were discussed to mitigate effects of system specialization and algorithm overtraining. Prognostic measures presented encompass measures of effectiveness from two different viewpoints (maintainer/commander). Finally, the need for ongoing PHM metrics analysis and tool development was established. The Expeditionary Fighting Vehicle, and many other current acquisition programs, will benefit from a more in depth look at verification and validation tool development from performance and requirements development perspectives.

References:

1. Kay, Steven, "Fundamentals of Statistical Signal Processing," Volume II, Prentice Hall, ISBN 0-13-504135-X, 1998.
2. Kalgren, P., Byington, C., Amin, S., Kallappa, P., "Bearing Incipient Fault Detection: Technical Approach, Experience, and Issues," MFPT 2005.

3. Byington, C., Safa-Bakhsh, R., Watson, M. Kalgren, P., "Metrics Evaluation and Tool Development for Health and Usage Monitoring System Technology," AHS Forum 59, Phoenix, Arizona, May 6-8, 2003.

4. Orsagh R.F., Roemer M.J., et al "Development of Metrics for Mechanical Diagnostic Technique Qualification and Validation", COMADEM Conference, Houston TX, December 2000.

5. Roemer, M. Dzakowic, J., Orsagh, R., Byington, C., Vachtsevanos, G., "Validation and Verification of Prognostic and Health Management Technologies," IEEE Aerospace, 2005

6. I. Barlas, G. Zhang, N. Propes, and G. Vachtsevanos, "Confidence metrics and uncertainty management in prognosis," MARCON, Knoxville, May, 2003.

A PROPOSED DECISION SUPPORT AID FOR THE EXPEDITIONARY FIGHTING VEHICLE

David B. Goldstein[1], Darryl Busch[2], Carl Byington[3]

(1) Honeywell - DSES
699 Route 46 East, Teterboro, NJ 07608
David.Goldstein@Honeywell.Com

(2) Honeywell – HTC
Honeywell Technology Center, 3660 Technology Dr.
Minneapolis, MN 55418

(3) Impact Technologies
2571 Park Center Boulevard, Suite 1
State College, PA 16801

Abstract: A proposed decision aid for the EFV prognostics system will be discussed. The decision aid will provide mission-specific warnings and advice to the operator based on the diagnostic and prognostic outputs of the Drive Train Prognostic System (DTPS). Because prognostics are concerned with predicted Machinery State and future probable conditions, the benefits are unique from those available in a diagnostics system, which focuses on identifying a functional failure for the operator/maintainer

Executive Summary: The system detailed below, an add-on to the DTPS provided by Honeywell, will enable the vehicle operator/maintainer/mission planner/asset manager to effectively plan for mission specific use of EFV assets. Though primarily based on DTPS outputs, the proposed system can easily be adapted to provide decision aiding for other on-vehicle prognostics systems or even whole-vehicle decision reasoning.

The overall objective of this system add-on is to prevent mission critical failures during operation in and around the EFV. This system has three core functions:

- To provide accurate Go/No-Go advice to the soldier using diagnostic and prognostic outputs from the DTPS and other vehicle systems.
- Warn the crew before the mission of any impending failures and their impact on the mission.
- During missions, provide updates to the crew when the fault status changes in a way that is specifically important to the mission thereby avoiding nuisance warnings.

As a modification to the DTPS, the system will have the capability to refine the accuracy of DTPS prognostic predictions by tailoring them to the expected "loads" on the system. The proposed system will be capable of refining the accuracy of warnings and advice by projecting the prognostic predictions against the upcoming mission profile, assuming nominal loads. In addition, the system will be able to refine the accuracy of warnings and advice by determining which failures are critical and non-critical to the mission (ie. Consider "acceptable risk" when providing advice).

Beyond mission assistance, the proposed decision aid can be used as a useful embedded training tool allowing the crew to alter the mission and thus explore "what-if" scenarios. They can explore what types of missions under what loads with what durations *can* be completed.

Introduction: Effective military operations require accurate knowledge of an asset's status, which can be obtained by using various vehicle health aiding technologies including monitoring, prediction, and decision support. Implementation of these technologies on-vehicle provides high value benefits to the military warfighter.

For the asset user, simple component remaining useful life estimates are not enough. The vehicle operator needs to understand the remaining duration of healthy vehicle operation specific to the current (or future) mission. In addition, mission planners and fleet level planners need to understand and plan for mission success (as well as logistics support) based on the most realistic, mission specific, future vehicle state. Prognostic technologies coupled with the appropriate decision support tool will help increase the reliability and availability of assets while reducing the total cost of ownership.

To make the EFV prognostics system a more effective tool, additional efforts are required. First, when an incipient fault or other anomaly is identified which triggers a prognostic algorithm to modify its critical prediction horizon, it is imperative that this result be tied directly to the current (or future) vehicle use. Secondly, the system must have the ability to optimize the forecasts by adjusting the algorithms over time, using the same data source. In other words, the drive train prognostic system on the EFV requires a method to consistently derive parameters that reflect the operating conditions the drive train will be subject to from any present point in time to the desired prediction horizon. (i.e. the mission of the vehicle).

Description of the DTPS: The Expeditionary Fighting Vehicle (EFV) Drive Train Prognostic System (DTPS) program (awarded to Honeywell DSES – Teterboro by General Dynamics Amphibious Marine Systems) encompasses an effort to provide a strap-on diagnostic/prognostic proof of concept solution. The project scope includes the design, development, and demonstration of three DTPS systems to be mounted on two separate EFV's (two mounted, one spare). It is expected that the program will lead to the production of some 1013+ embedded DTPS systems for subsequent EFV's as they roll out of production.

The DTPS (shown in Figure 1) consists of a central small form-factor computer called the Subsystem Condition Monitoring Engine (SCME), a suite of Sensors with Embedded Processing (SEPs) and a communications and power distribution network. The SEPs (and associated transducers) and the SCME are the key physical elements of the DTPS. The SEPs are placed at strategic locations throughout the drive train elements and communicate information to the SCME. They monitor the drive train using sensors and perform several elements of the information processing including raw data pre-processing, data extraction, data forwarding and vehicle state determination. Each SEP consists of one or more sensors with its associated analog circuitry, a communication means (CANBus), and a small computer. The SCME is the primary prognostic activity center and central brain of the DTPS that produces diagnostic and prognostic results. The SCME coordinates the activities of the SEPs, and performs the higher-level processing of information, including detecting signs of faults in the features provided by the SEPs, isolating the faults, trending, and estimating remaining life (currently based on a set of nominal, pre-defined mission parameters). Included in the prediction processing is information from the vehicle data buses. The SCME possesses some data storage capability on board, but the primary data storage location is the vehicle data storage device called the Mass Memory Unit (MMU). The SEPs and SCME communicate via a dedicated communications bus. The bus also carries power from the SCMEs to the SEPs.[1] The output diagnostic classification and prognostic vectors are, at this point, passed to the vehicle.

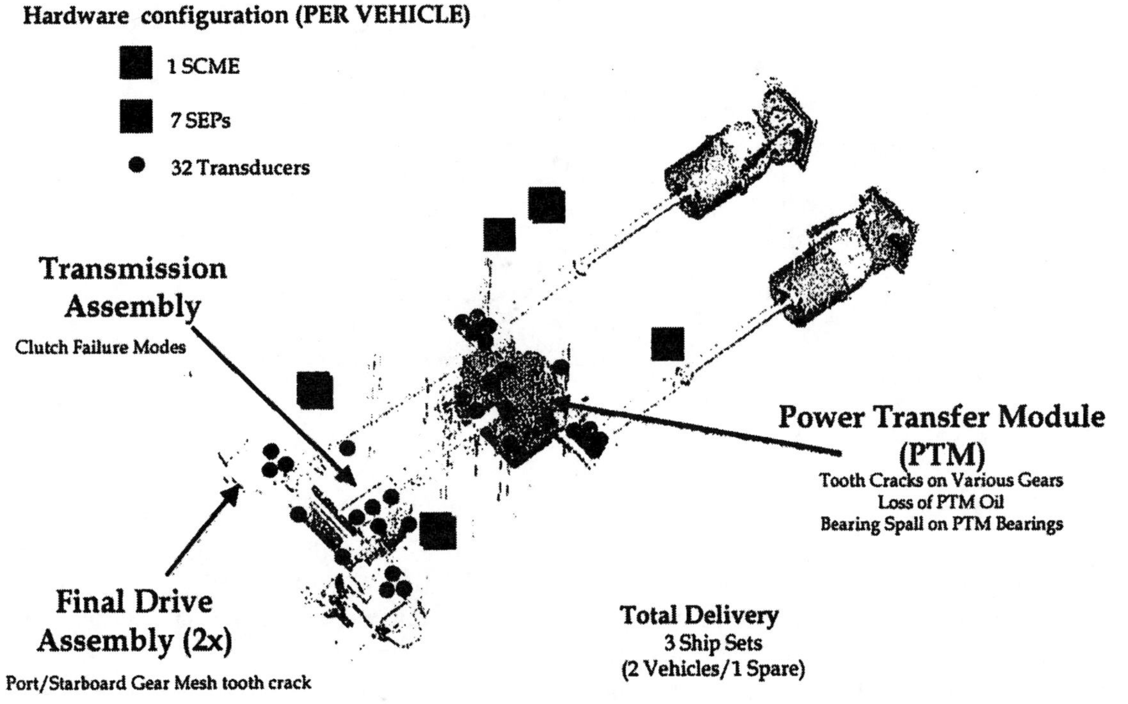

Figure 1 Current DTPS System

The output of the DTPS system is from the SCME and the system will provide detection, diagnostic and prognostic outputs. Depending on the failure mode, we will be able to provide different levels of output (e.g. severity, time to failure, confidence, etc.).

Current GAP: A detailed use case document has been developed by GDAMS[2] which clearly identifies the importance of providing the vehicle operator and crew with the most accurate decision aid possible:

(UD412.1.4 The Go No Go Recommendation), "A Go No Go recommendation is critical feedback to the operator and crew."

The OSA/CBM standard defines seven *Layers* that correspond to different stages of system maintenance:

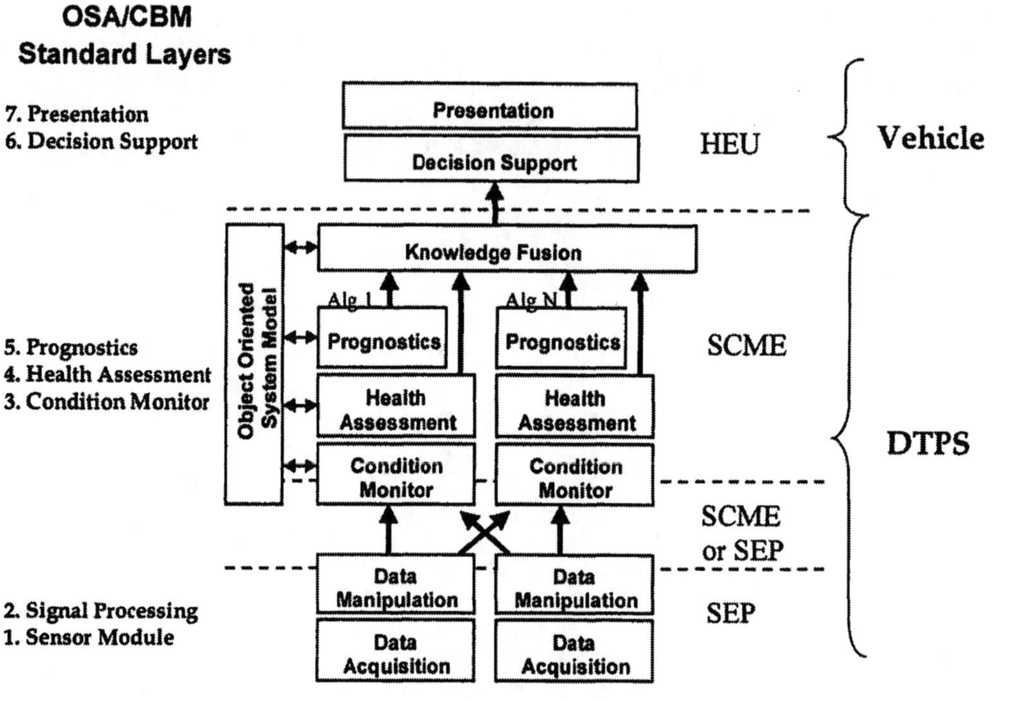

Figure 2 DTPS mapping to OSA/CBM

The current DTPS maps as follows (as shown in the preceding figure): The SEPs and onboard memory will correspond to the first and second OSA-CBM layers (Sensor Module and Signal Processing). The other subcomponent of the DTPS system, the SCME will, in general, embody layers 4 and 5, the health assessment and prognostics layers. Depending on system configuration, the third layer – condition monitor – will be embodied in either the SEP or the SCME. A significant gap exists between the current DTPS output and a meaningful, user-friendly decision aid. The system add-on will embody the final two layers of the OSA-CBM standard – the Decision Support and Presentation layers.

	DSF	BMO	TR	Theta
Algorithm 1:	0.3	0.3	0.3	0.1
Algorithm 2:	0.4	0.2	0.0	0.4
Algorithm 3:	0.0	0.0	0.9	0.1
DTPS System Output				
	0.13	0.1	0.75	0.02

Failure Modes:
Water jet drive shaft fatigue failure (DSF)
Bearing module overloading failure (BMO)
Tip rub failure (TR)
Other (Theta, Unknown)

Understanding the Problem - Man-Readability: One of the primary needs for the EFV Prognostics effort is a conversion of the machine-readable prognostic vectors into a man-readable form. The current DTPS outputs will take the form of a prognostic vector as detailed below.

As is easily seen, even in units related to the specific impending failure being reported (i.e. number of shifts of a clutch or revolutions remaining on a gear), the above detailed vector will require machine translation to become meaningful information for the user. At a minimum – a GUI/Display layer will be necessary. For those failures not easily man-associated with an impending failure (i.e. predicted number of remaining revolutions on a certain gear), some relation to the mission will be essential to providing clear vehicle health information.

Dynamic Outputs: The current DTPS output is not dynamic with the changing mission. The current vector output is based on a single, standard default mission. To provide optimal support for the vehicle, the prognostic system needs to have a continuously updated vector based on, not only the current Vehicle State, but also the expected future vehicle use.

Higher Level Decision Reasoning: The current DTPS has no higher-level decision reasoning above the drive train. This leaves a gap between the predicted future state of the limited number of monitored drive train components (based on a single, standard default mission) and the predicted future state of the vehicle. Without higher level decision aiding – tying the output vectors to the current (or future) use of the vehicle – the information usefulness is limited. In the field, the vehicle operator can make significant use out of a properly displayed vector identifying, for example, estimated remaining shifts of a clutch, but this information will quickly lose value as the current use and subsequently future use of the vehicle changes with the mission.

Mission Planning: Without knowledge of the future vehicle mission, the asset maintainer/mission planner has no means by which to interpret vector outputs and their relevance to current maintenance/planning actions. Higher-level decision reasoning is essential to maximizing a vehicles field time while minimizing its maintenance time.

Logistics Integration: Concurrent with the need for a higher-level decision aid in mission execution and mission planning, the prognostic outputs will not, in their current form, contain information sufficient to feed an integrated logistics system. A higher-level logistics aid, cleanly integrated with the EFV prognostics system will complete the system of support for the vehicle.

It is essential that the output of the EFV prognostic system provide a constantly updating vector, directly tied not only to past vehicle use and current Vehicle State but also to future vehicle use, future mission execution and vehicle maintenance

The Solution - Mission Analysis and Adaptation: The EFV Operational Mode Summary/Mission Profile (OMS/MP)[3] is derived from the operations of a Marine Expeditionary Force portrayed in a high-intensity conflict derived from a standard Marine Corps Scenario. The scenario depicts the first 28 hours and 50 minutes of an enabling force operation, beginning with launch from amphibious shipping, a high-speed surface assault from beyond the horizon, forcible entry into a defended beach, destruction of enemy occupation forces, and seizure of key objectives. The scenario ends with a mission change to defense of the area.

Monitoring machinery loads and conditions allows the enhanced regime recognition and awareness of component operational profiles. Due to the varying degrees of mission severity and randomness typically observed in military mission profiles, a combination of discrete and continuous analytical methods is proposed for mission regime recognition and usage monitoring. These methods will range from representing mission profile data with traditional statistical distributions, such as normal, exponential or Weibull, to more specialized regression and correlation techniques providing better representative missions for specific mission types. Rare/random vehicle operational events, such as waterjet FOD (foreign object damage), specific high G maneuvers, etc., which have a discrete occurrence in time, are simulated by discrete impulse functions, such as the Poisson impulse function. For a large time macro-scale, the typical mission environment will represent a continuous component, while the rare events represent a discrete component.

The combination of this continuous and discrete mission-level modeling paradigm is illustrated in Figure 3. In this figure, a continuous mission profile is simulated based on distribution statistics developed from mission data. Discrete rare/random events are then simulated in parallel and added to the continuous model output based on occurrence rates also identified from legacy data.

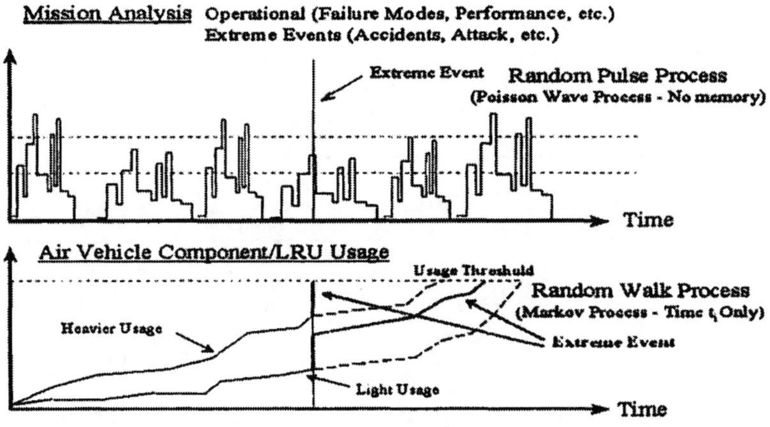

Figure 3 - Mission Loads Analysis and Usage Modeling

Providing a prognostics-based decision support system provides a continuous assessment of mission readiness and directly supports the adaptation of the mission during the EFV deployment.

Mission Regime Recognition and Interpretation: Based on the Team's past experience correlating mission profile statistics and aircraft life usage, the non-linear nature associated with many structural vehicle component damage mechanisms is dependent on both the inherent characteristics of the profiles and mission mix types. Significant component damage resulting from the large variability in the operating environment and severity of the missions directly affects the vehicle component lifetimes. Very often, component lives driven by fatigue failure modes are dominated by unique operational usage profiles or a few, rare, severe, randomly occurring events, including abnormal operating conditions, random damage occurrences, etc. The accuracy of this time-variant reliability aspect is the basis of the probabilistic mission simulation process.

Technical Approach for EFV: Figure 4 shows a block diagram of the system architecture, and its relationship with the DTPS and the PAM. Yellow symbols are components of the decision aid, blue symbols are existing components of the DTPS. The architecture breaks out the system into a set of specialized reasoners (yellow boxes) and their conclusions (yellow trapezoids), which are described in the following sections.

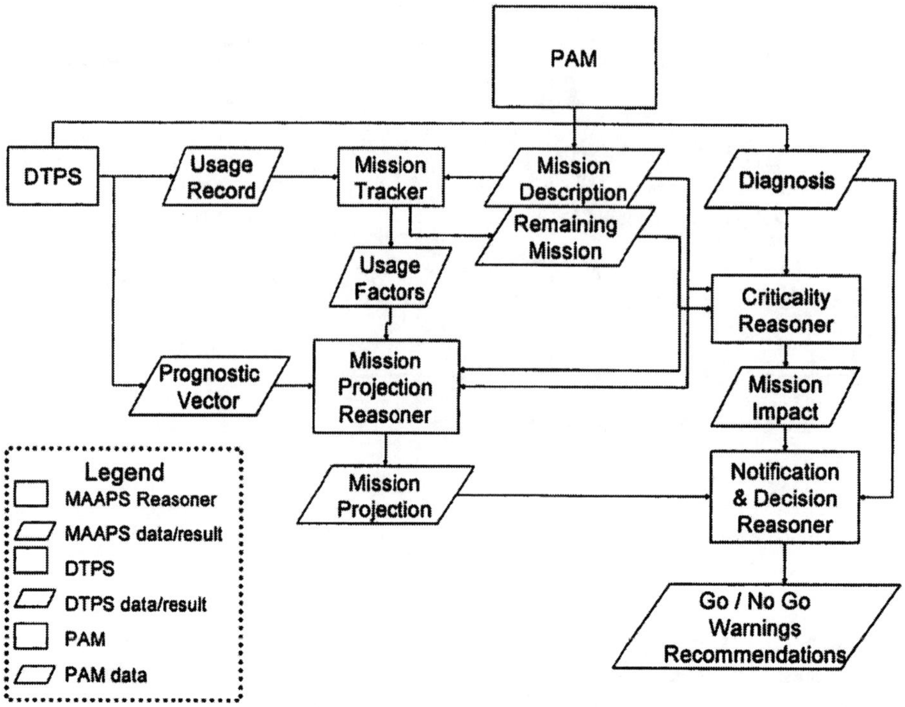

Figure 4 Decision Aid System Architecture

Mission Projection Reasoner: The Mission Projection Reasoner (MPR) inputs the profile(s) of the upcoming mission(s) and the prognostics outputs of the DTPS, and calculates the probability of any failures occurring during the mission(s). It does this by "projecting" the usage ordinate of DTPS prognostic vectors onto the description of usage for the upcoming mission(s), effectively converting usage units into missions.

Consider the following simple example. Suppose the mission description is:

Segment #	Duration	Segment Type	Sub Type
1	10min	transition mode,	
2	1.5 hr	water mode (planning)	SeaState=1; Temperature=Hot; Weather=Sunny;
3	10min	transition mode	
4	2.0 hr	Land mode (cross country)	Terrain=Desert; Temperature=Hot; Weather=Sunny;

Segment #	Duration	Segment Type	Sub Type
			Climate=Arid;

Also suppose DTPS is tracking an impending fault in one of the waterjet planetary gears, and produced the prognostic vector shown below. For reference, a prognostic vector describes the cumulative probability of failure as a function of some usage parameter. In this example, there is a 90% percent chance of failure somewhere in the next 14 hours of water mode operation. The first 8 hours are relatively safe, and the probability increases rapidly from there.

Hours of Water Operation	2	4	6	8	10	12	14
Failure Probability (%)	0.01	0.03	0.1	1	10	50	90

The MPR translates "Hours of Water Operation" into missions and segments of missions, and produces a new prognostic vector. The original vector and new vector are plotted in Figure 4. Inspection of the second subplot shows that the first five missions of this profile are relatively safe and the probability of failure ramps up quickly in later missions. Failure will likely occur in the 7th, 8th or 9th mission where the probability ramps from 3.5% to 87%.

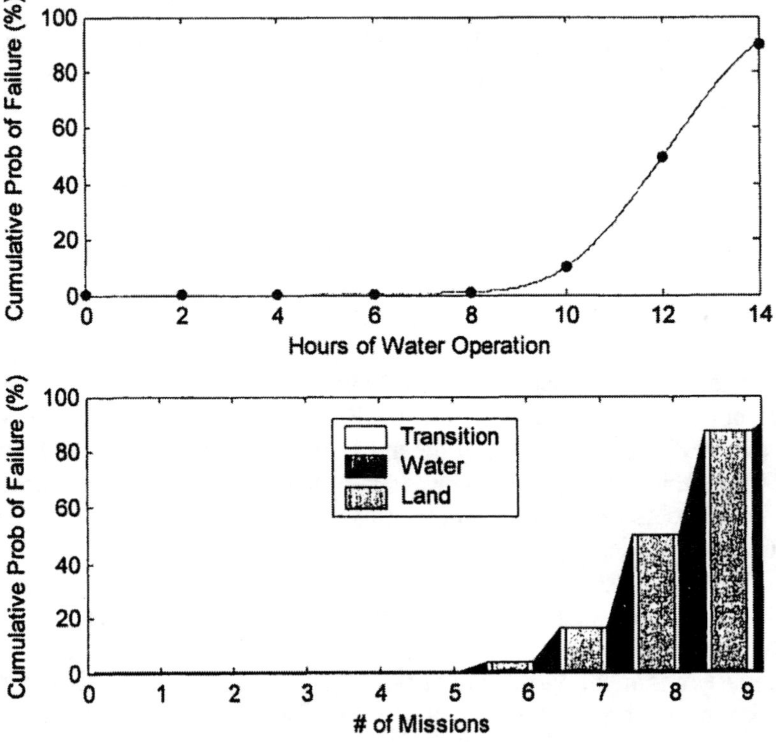

Figure 5 Prognostic vector (upper). Projection onto upcoming mission (lower).

The example above was relatively straightforward because the usage unit in the prognostic vector, hours of water mode operation, was readily projected onto the mission. In contrast, there are other failure modes where the projection requires assumptions, with associated uncertainty. An obvious example is the clutches, which will have prognostic vectors expressed in numbers of shifts rather than hours. No other unit has as much intrinsic meaning for remaining life. But translating shifts into missions requires assumptions with a high degree of uncertainty. For sake of argument, suppose experience shows there are typically 5 to 20 engagements of the C2 clutch per hour while in the "mode Land (cross country)". Then even a crisp prognostic vector that predicts failure in exactly 30 shifts would translate into a projection of failure in 1.5 to 6 missions.

There are at least several ways to address these potentially large uncertainties translating usage units into upcoming missions:

- Provide the soldier with additional information, like best and worst case, and let him interpret it. The drawback to this approach is that it could overwhelm the soldier with information.
- Provide an interface for the soldier to define the upcoming mission in more exact terms, so that the MPR can make more clear-cut assumptions. The drawbacks to this approach are that it forces the soldier to do more work, and do it accurately enough that it provides value.
- Develop a software module that can estimate the detailed attributes of the upcoming mission from other sources of information. For example, use map data and the planned path of the upcoming mission to estimate the number of starts, stops, shifts, turns etc. The drawback to this approach is that such a software module would be difficult to develop.
- Rely on recent missions to narrow the uncertainty, based on the assumption that conditions are similar. For example, if the cross country terrain was flat during yesterday's mission and so required few changes in speed and few shifts per kilometer, then bias the translation toward similar conditions in the upcoming mission.
- Accept that uncertainty is inherent in predicting the future.

The MPR operates slightly differently during missions versus outside of missions (e.g. before a mission). Before a mission, the MPR utilizes the latest prognostic vectors produced by the DTPS and the profile of the upcoming mission. Since the DTPS outputs are based on sensor data collected during operation, the DTPS outputs are basically static outside of operations, so the MPR would only change its projections if the soldier changes the mission profile.

During missions, the MPR will respond to changes in the DTPS outputs, changes in the mission profile, and estimates of how much remains of the current mission. This third input is particularly important during a mission when the DTPS is predicting that a failure will occur within a few hours. The MPR needs to know how much of the current mission remains in order to calculate the probability of failure from now to the end of the mission, and thus determine whether the mission can be completed. The source of this third input is discussed in the next section.

Mission Description: The Mission Description is a data construct provided to the system by the PAM. The "Perform Prognostics" use case document describes the various parameters that the soldier can adjust in order to describe the upcoming mission. The descriptiveness of the mission parameterization is probably sufficient for the MPR to use for projections before missions. One parameter that would be helpful is something that helps the MPR estimate the number of shifts and speed ranges that will be used. A step in the right direction is a parameter describing driving conditions such as: Cross-Country/Secondary Road/Primary Road/Urban.

However, the mission parameterization given in the "Perform Prognostics" use case is less helpful to the MPR during the mission. In particular, it is difficult for the system to deduce what remains in the current mission. There is one parameter, "Mission Hours to Go", that counts down the remaining hours in the mission, and that the PAM continuously updates. However this parameter is of questionable utility given that:

- The fidelity of the mission parameters is uncertain. They may be very accurate, very conservative, a WAG, or simply left at the defaults.
- The mission parameters can be adjusted during the mission by the soldier
- The parameters do not describe the sequence of the mission.

As an improvement, we suggest parameterizing the mission in terms of segments, as we did in the Mission Analysis section. Further improvements are provided by the Reasoner described in later sections.

Mission Tracker: The Mission Tracker (MR) operates during each mission and produces two outputs:

- The Remaining Mission: an estimate of what remains of the current mission, and how well the actual mission matches the Mission Description.
- Usage Factors: usage factors (e.g. shifts per hour) with narrower uncertainty than the default factors, and that are applicable during this mission and the next few missions.

It does this using the Mission Description data provided by the PAM and to usage data recorded by the DTPS describing the mission so far and recent missions. The usage data is utilized by the DTPS to produce prognostic vectors. For example, in order to trend a clutch fault, the DTPS must also count shifts in order to produce the usage axis of the trend. It is our expectation that the usage data recorded by the DTPS will be suitable for use by the MT. If not, the MT can record its own specialized usage data or the DTPS usage recording function can be extended.

The MT will produce Remaining Mission outputs like any of the following notional examples:

- Agreement of Mission Description with actual = excellent. Remaining Mission = 1hr of segment 4, through end of mission.
- Agreement of Mission Description with actual = none. Remaining Mission = Unknown.

The MT will produce and record Usage Factor outputs like the following notional example:

- This mission: C2 shifts=21 – 30 per hour (cross country). C2 shifts=2 – 5 per hour (primary road).
- Next missions: C2 shifts=15 – 35 per hour (cross country). C2 shifts=2 – 7 per hour (primary road).

The Usage Factors labeled "This mission" were obtained by giving extra weight to usage data obtained during the current mission, and are most applicable to the current mission. The Usage factors labeled "Next Missions" weight several recent missions and thus represent slightly longer-term trends. The Usage Factors produced by the MT are used by the MPR in lieu of the default parameters with the goal of reducing uncertainty in the projection. The MPR will revert back to the default usage factors if the Usage Factors produced by the MT are deemed inapplicable, as determined by calendar time and/or GPS location.

Criticality Reasoner: The implementation proposed here uses a criticality reasoner to apply mission specific criticality aids to the decision support tool output. This module will use knowledge of the vehicle and vehicle systems to identify the criticality of a pending failure. This module will allow the decision aiding system to provide a level of criticality of an impending fault. Not only with respect to the current system use, but, with inherent mission knowledge, with respect to future system use.

Notification & Decision Reasoner: The Notification and Decision Reasoner module will contain all of the higher level decision aiding and notification reasoning. The notification aid will take knowledge of the current and future vehicle mission and apply reasoning to the need for warnings. This capability will assist in avoiding nuisance warnings on systems not critical to the vehicles' current or future operation. The decision reasoner will merge current and future vehicle state information with current and future vehicle use information to provide Go/No-Go type decision aiding. This module can easily be used for in-use assets as well as for future mission planning.

Summary Recommendations and Next Steps: The current paper outlined a strategy for decision support development for the EFV program using a range of best technical approaches. The overall objective of such a development program is:

Develop a prognostic decision support system with the requisite analysis and reasoning algorithms to provide improved mission assurance and dynamic decision making on the EFV.

The suggested next steps include:

- Feasibility Assessment and Demonstration Prototype
A six-month analysis of methods to be applied and detailed problem definition will include characterization of likely sources and types of data available. Perform PC-based demonstration of the reasoning systems.
- EMD Prototype Demonstration
A two-year followon program to develop reasoners and demonstrate the functionality within DTPS use case analysis.
- Full EFV Decision Aid Development
Full development aligning with EFV schedule, spiral development, and technology refresh.

The Team believes this program represents a natural extension to the DTPS program and is consistent with the desires of GDAMS and the EFV Direct Report Program Manager. It also dovetails well with the motivation for full contractor logistics support through GATOR (Global Amphibious Total On-line Resource) Link and USMC plans for Autonomic Logistics.

References:

[1] Expeditionary Fighting Vehicle Use Case Specification: Perform Prognostics; UD-P3010017, Rev B, June 15, 2004. General Dynamics Amphibious Systems
[2] Ibid
[3] AAAV Operational Mode Summary-Mission Profile 2-24-97; United States Marine Corps Combat Development Command

HARDWARE AND SOFTWARE ARCHITECTURE FOR THE EXPEDITIONARY FIGHTING VEHICLE DRIVE TRAIN PROGNOSTICS SYSTEM

David B. Goldstein, William Cummmings, Kenneth Ferraro[1]
Carl Byington, Pat Kalgren[2]

(1) Honeywell
699 Route 46 East, Teterboro, NJ 07608
[illegible]

(2) Impact Technologies
2571 Park Center Boulevard, Suite 1
State College, PA 16801

Abstract: The Expeditionary Force Vehicle (EFV) Drive Train Prognostic System (DTPS) program (awarded to Honeywell DSES – Teterboro by General Dynamics Amphibious Marine Systems) encompasses an effort to provide a strap-on diagnostic/prognostic proof of concept solution. The project scope includes the design, development, and demonstration of the DTPS on the EFV. It is expected that the program will lead to the production of some 1013+ DTPS systems for subsequent EFV's as they roll out of production. Production systems will evolve from a strap-on solution to an integrated solution. Both the hardware and software architecture will be discussed.

Key Words: EFV; Prognostics; VHM; Vetronics; Drive Train, System Architecture

Overview: Because prognostics are concerned with predicted Machinery State and future probable conditions, the benefits are unique from those available in a diagnostics system.

The Expeditionary Force Vehicle (EFV) Drive Train Prognostic System (DTPS) program (awarded to Honeywell DSES – Teterboro by General Dynamics Amphibious Marine Systems) encompasses an effort to provide a strap-on diagnostic/prognostic proof of concept solution. The project scope includes the design, development, and demonstration of the DTPS on the EFV. It is expected that the program will lead to the production of some 1013+ DTPS systems for subsequent EFV's as they roll out of production. Production systems will evolve from a strap-on solution to an integrated solution.

System Architecture: The hierarchical design of the DTPS is provided in the figure below. The Drive Train Prognostic System (DTPS) consists of a central small form-factor computer called the Subsystem Condition Monitoring Engine (SCME), a suite of Sensors with Embedded Processing (SEPs) and a communications and power distribution network. The SEPs (and associated transducers) and the SCME are the key physical elements of the DTPS. The SEPs are placed at strategic locations throughout the drive train elements and communicate information to the SCME. They monitor the drive train using sensors and perform the majority of the signal processing in the DTPS, including preprocessing the gathered information and extracting salient features. Each SEP consists of one or more sensors, analog circuitry, and a small computer. The SCME is the

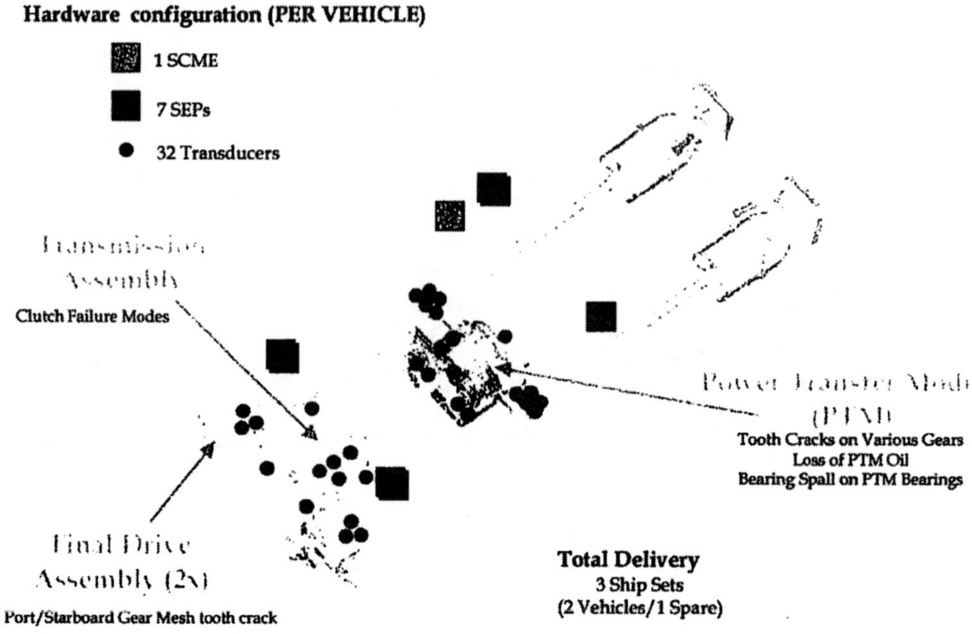

primary prognostic activity center and central brain of the DTPS that produces diagnostic and prognostic results. The SCME coordinates the activities of the SEPs, and performs the higher-level processing of information, including detecting signs of faults in the features provided by the SEPs, isolating the faults, trending, and estimating remaining life (currently based on a set of nominal, pre-defined mission parameters). Included in the prediction processing is information from the vehicle data buses. The SCME possesses some data storage capability on board, but the vehicle remains the primary data storage location. The output diagnostic classification and prognostic vectors are passed to the vehicle for decision aiding and presentation.

The functional architecture is shown below. Functionally, components are split between the SEP and the SCME. Processes are localized to optimize processing and data bus bandwidth.

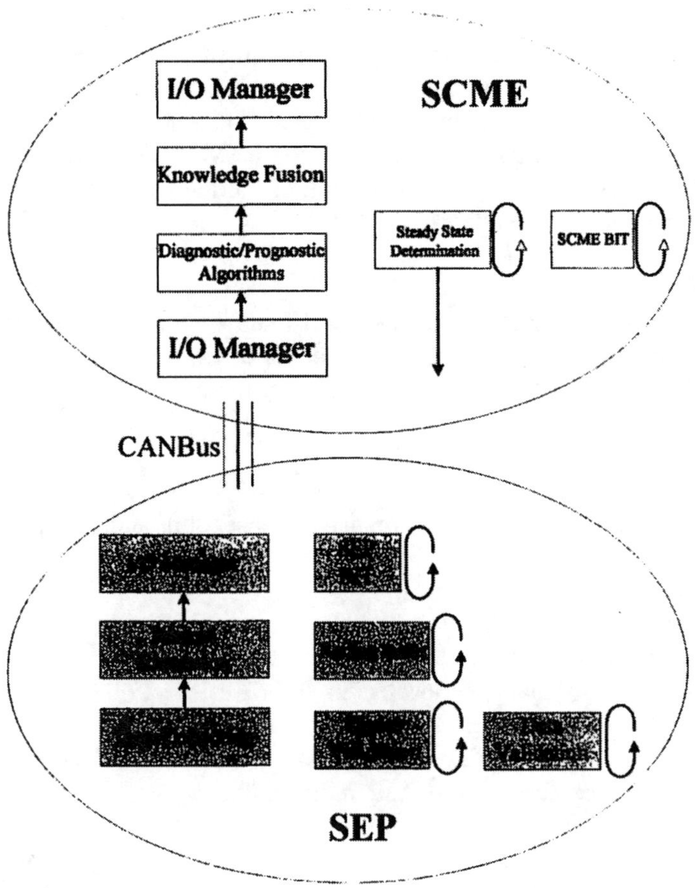

SEP Functional Components

Data Collection

The data collection module will be responsible for acquiring the raw data from the various transducers. This data will constantly be contained in a 10 – 15 second rolling buffer. Though not a software module, it is contained here for completeness (this will be taken care of via the SEP analog electronics).

Data Validation

All raw sensor data will be validated within the data validation module.

Sensor Validation

All sensors will be validated in the sensor validation module. A certain level of BIT-type circuitry will be built into the SEP electronics. This will be used on a periodic basis to verify the function of the various sensors in the system.

SEP BIT

Built In Test (BIT) will run on the SEP to meet program requirements. A combination of built-in circuitry and software exercising will be used to validate SEP operation. Three different types of BIT will be implemented; Startup-BIT, Periodic-BIT and Commanded-BIT. Startup-BIT will automatically run at system power-up, providing initial SEP health. This function will verify communication with all of the transducers and basic SEP operation. Periodic BIT will run on a predetermined cycle. Again, some basic level of SEP function will be verified including processor timeout, SEP/SCME communication and SEP/Transducer communication. Commanded BIT will run a deeper level of BIT than startup BIT or periodic BIT. When receiving a C-BIT command from the SCME, the SEP will run a deep self diagnostic, verifying SEP circuitry performance and isolating potential failures to the required level.

Feature Extraction

The Feature extraction process contains several feature extraction algorithms. Feature extraction algorithms are run on the raw data contained in the rolling buffer. The feature extraction module is only run when the SCME determines the vehicle to be in a Steady State.

SEP CANBus I/O Manager

As the name suggests, the I/O Manager manages the sending and receipt of data to/from the SEP via the CANBus.

SCME Functional Components

SCME CANBus I/O Manager

Like the communications manager located in the SEP, the I/O Manager manages the sending and receipt of data to/from the SCME via the CANBus (data communication with the SEPs), via the Ethernet (data communication with the HEU) or via the HEU RS-422 and HEU CANBuses for Vehicle Data monitoring.

Diagnostic/Prognostic Algorithms

The series of algorithms developed to diagnose and prognose drive train failures. Various algorithms will be run on the extracted features obtained from transducer raw data and other vehicle data. The algorithms will cover various mechanical failure modes of the EFV drive train including the following:

- Bearing failures
- Spur Gear Cracks
- Planetary Gear Cracks
- Clutch Failures
- Oil Loss

Knowledge Fusion Module

This module will take information from the vehicle and "fuse" it with the various outputs of the Diagnostic/Prognostic Algorithms creating a collective, intelligent output. One of the core components of the DTPS, the KF module will integrate information gathered from both on vehicle and off vehicle sources. In its initial manifestation, the only off vehicle information used will be relevant maintenance data provided by vehicle maintainers. In addition, the KF module will use data gathered from several monitored vehicle data busses (the Powertrain CANBus, the RS-422 Navigation-Auxiliary bus and the RS-422 Navigation-Primary Bus), prognostics algorithms results and additional vehicle state information provided within the MMU.

Steady State Determination Module
This module will monitor various vehicle conditions and data bus messages and continuously compare them to a series of steady state conditions. When one (or more) of these conditions is met, the vehicle will be determined to be in a Steady State. This determination will be broadcast to the SEPs for use in feature extraction.

SCME BIT
Built In Test (BIT) will run on the SCME to meet program requirements. A combination of built-in circuitry and software exercising will be used to validate SCME operation. Three different types of BIT will be implemented; Startup-BIT, Periodic-BIT and Commanded-BIT. Startup-BIT will automatically run at system power-up, providing initial SCME health. This function will verify communication with all of the necessary busses and basic SCME operation. Periodic BIT will run on a predetermined cycle. Again, some basic level of SCME function will be verified including processor timeout, SEP/SCME communication and SCME/HEU communication. Commanded BIT will run a deeper level of BIT than startup BIT or periodic BIT. When receiving a C-BIT command from the HEU, the SCME will run a deep self diagnostic, verifying SCME circuitry performance and isolating potential failures to the required level.

Main Functional Description
The functional flowchart is depicted in the figure below.

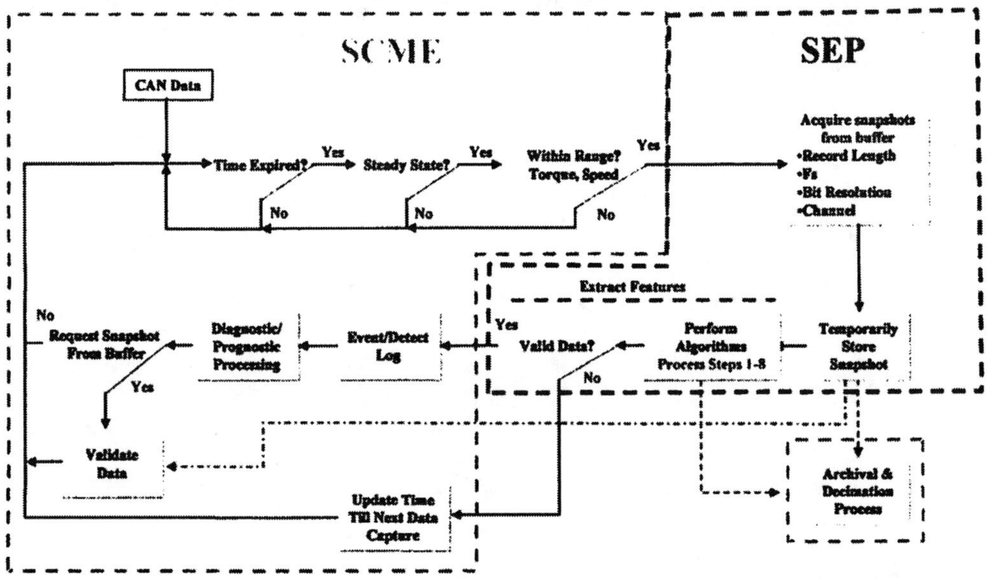

During nominal DTPS operation, the system is in a state of constant data collection. SEPs are continuously retrieving raw data from the various sensors. The SCME is continuously monitoring various MMU memory locations for vehicle data as well as monitoring three separate vehicle data busses for vehicle state information. This information is run through the Steady State Determination Module.

When a preset set of conditions are met, the module declares the vehicle to have been in a "steady-state" condition. A message is sent to the SEPs via the CANBus identifying the time at which the vehicle was identified to be in steady state.

The data collected from the various sensors at the specified time will then be sent through the feature extraction algorithms. During feature extraction the SEPs will run verification of the sensors and the raw data. With validated data, the SEPs will forward the extracted features to the SCME. This data will be fed into the various diagnostics and prognostics algorithms. The Knowledge Fusion module will use the outputs from these algorithms and vehicle state information collected from both the monitored data busses and select MMU memory locations to develop prognostics vector outputs.

Conclusions

Effective military operations require accurate knowledge of an asset's health, which can be obtained by using monitoring, prediction, and decision support technologies. Implementation of on-vehicle prognostics provides high value benefits to the military warfighter. Because prognostics are concerned with predicted machinery state and future probable conditions, the benefits are unique from those available in a diagnostics system, which focuses on identifying a functional failure for the operator/maintainer. Prognostic technologies coupled with the appropriate decision support will help increase the reliability and availability of engines and drive trains while reducing the total cost of ownership. The objective, enabled by automated condition assessments and prognostics, is to *enable the right asset on the right mission at the right time.*

UTILIZING PROGNOSTICS INFORMATION IN A DECISION SUPPORT SYSTEM

Arthur Brault
System Engineer
General Dynamics
Amphibious Systems Division
14041 Worth Avenue
Woodbridge, VA 22192
braulta@gdls.com

Abstract: Prognostics provides the information that can realize increased uptime, reduced maintenance costs, and improve the system lifetime. The prognostics information data set is one of the necessary contributors to optimize and achieve a fully robust decision support system. The goal is to integrate an information architecture that supplies data into a life cycle management system that supports program management, parts and services supply chain, an adaptive and responsive maintenance and repair organization, and the objective system platform. The Marine Corps' Expeditionary Fighting Vehicle (EFV) accounts for these requirements in the overall organizational framework for the planning, implementation, and execution of this integrated design approach. The information that feeds the decision support system originates from a number of sources. The resultant data from these sources are captured, stored, calculated, aggregated, and reported to the various organizational levels on a timely basis. This paper will describe the overall EFV decision support system, and how prognostics plays an integral part in contributing to reducing system downtime, assist in planning for periodic maintenance scheduling, and promoting system availability and readiness.

Key Words: Autonomic Logistics; Diagnostics; Decision Support; Electronic Logbook; Health Management; Prognostics;

Overview: The U.S. Marine Corps' Expeditionary Fighting Vehicle (EFV) is an armored, tracked amphibious combat vehicle that is capable of high speed maneuverability on land and water. There are two variants of the EFV, each providing complementing capabilities. The personnel variant has a crew of three and carries up to eighteen combat-ready Marines. The command variant vehicle provides real-time battlefield and datacommunications management and has a crew of six. The EFV program is currently in the system development and demonstration phase with General Dynamics Amphibious Systems (GDAMS) as the prime contractor/integrator.

The overall objective of prognostics on the EFV is to provide information to support better management of operational and maintenance resources. This will be accomplished using on-vehicle processing and data bus resources. The EFV prognostics system will collect, store and process prognostic system sensor and vehicle information. The resultant prognostics information will ultimately be displayed at one of the vehicle's crew stations if action by the operator is required, and will be made available for download to an off-vehicle processing device.

The prognostics information is but one piece of a decision support system puzzle. This information must be utilized with diagnostic, usage, mission profile, and sensor data to assess the vehicle's condition and maintenance requirements. This data can be used in a health management and autonomic logistics system to determine the vehicle's operational status, maintenance requirements for short and long term and predict the parts, tools, and labor requirements for vehicle life cycle support.

Vehicle Architecture: The EFV is made up of a number of electrical, hydraulic, and mechanical subsystems. These various subsystems all have sensors that monitor and report on applicable distance, time, quantity, temperature, data range and system parameter data. These data types are used for diagnostics, prognostics, and operational status information. This is the data collection which is utilized for the vehicle level decision support system.

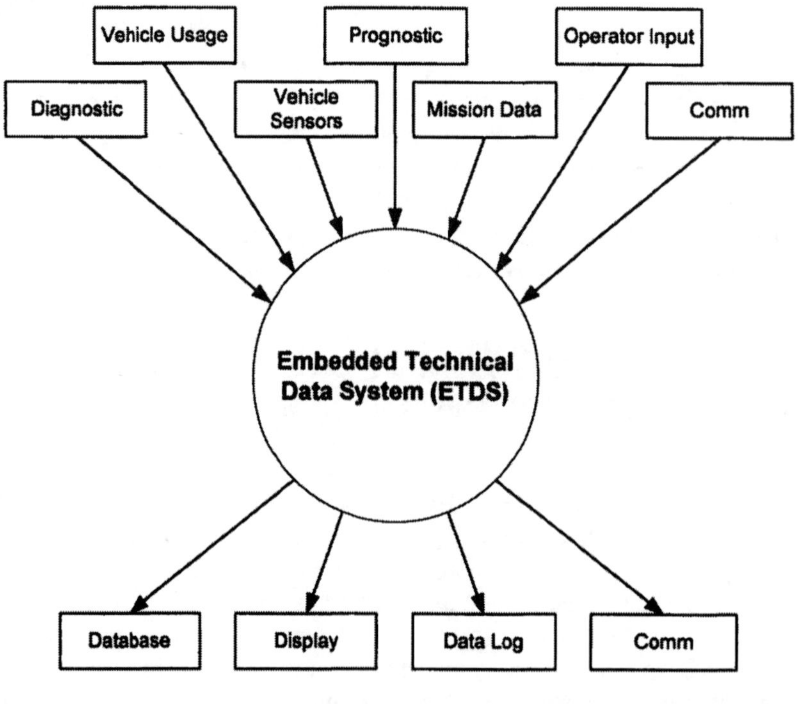

Figure 1: Embedded Technical Data System Overview

The Embedded Technical Data System (ETDS) is the software solution chosen to enable the implementation of a decision support system designed for the EVF. The components of the ETDS includes a user options menu that provides access to the Interactive Electronic Technical Manual (IETM), Diagnostics, Prognostics, Embedded Logistics System (ELS), Training System, Software Loader/Verifier (SLV), and Help functions. These components interrelate with each other as part of an integrated solution that resides on-board the vehicle for real-time analysis and reporting and off-board on a tethered ruggedized laptop to support maintenance and electronic logbook functions.

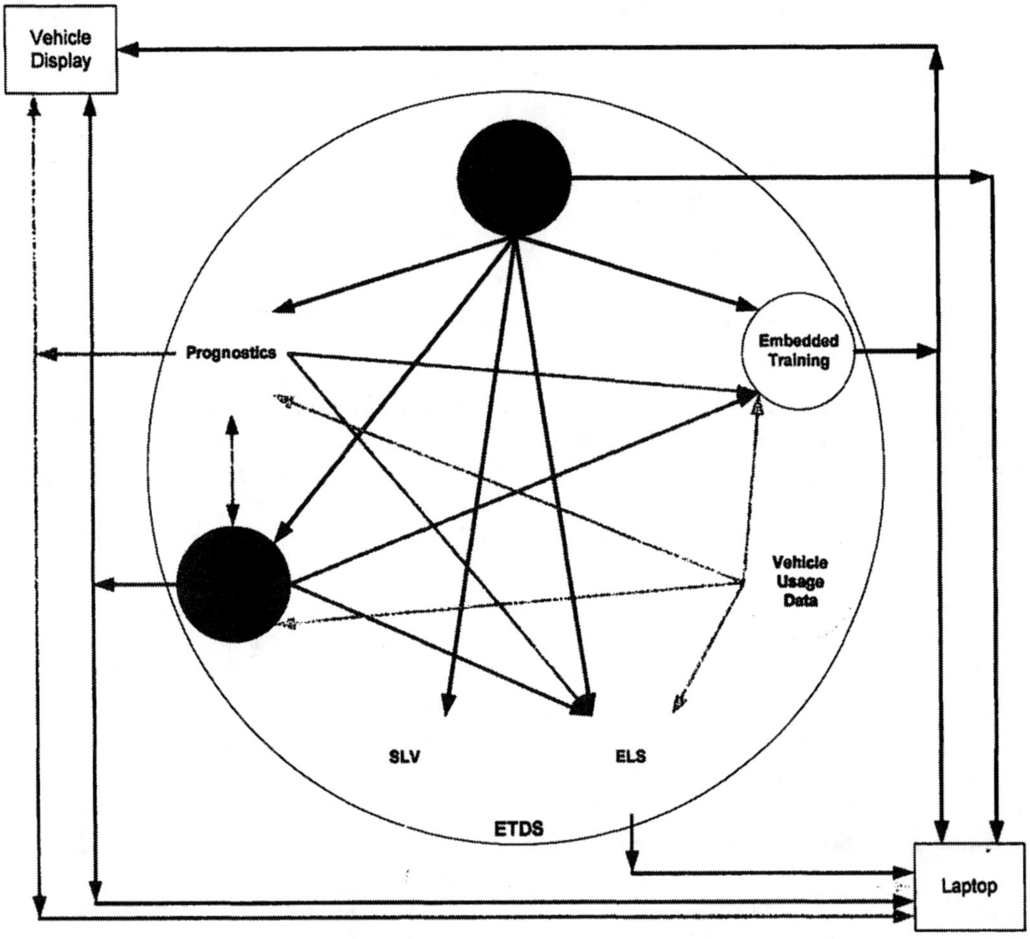

Figure 2: ETDS Module Architecture and Relationships

Prognostics: The prognostics system on the EFV is currently a prototype design. This design utilizes existing on-board sensors and data for the drive train, the engine, and the batteries. Each prognostic subsystem is designed to interface with the vehicle processing unit, storage memory, and displays as part of it's functionality. The aggregated data that the prognostic system produces is used to handshake with other modules in the ETDS.

Diagnostics utilizes prognostics state of health and remaining useful life information to correlate and bound ambiguity groups to determine if any potential fault containment, what the possible system overrides are, what is the most expedient system recovery, and how to expeditiously isolate, correct, and verify the fault correction to minimize vehicle system down time. Prognostics also augments usage data and provides useful life information to ELS for electronic logbook data and provides data that can be used for higher echelon analysis.

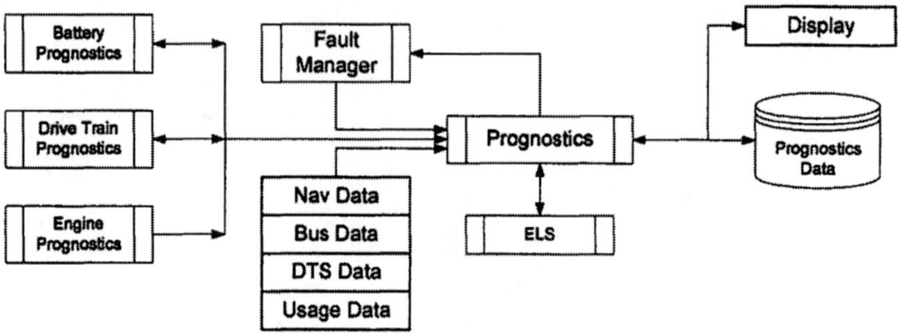

Figure 3: Prognostics Data Flow

The battery prognostics module provides state of health information for each of the eight batteries on the EFV, as well as aggregated information that can be used to inform the operator of remaining useful battery life. The drive train prognostics currently covers key gear and bearing information for the transmission and the Power Transfer Module (PTM). The PTM directs power to either the tracks for land operation or to the water jets. For the engine prognostics module, the voltage regulator/generator and the dust extraction fans are reporting such things as temperature and wear information through the engine's onboard processor. The prognostics system monitors existing data on the vehicle's data bus'. This data includes navigation data used for vehicle centric operational context, usage data which includes petroleum, oils, and other lubricants, and all data has an associated date-time stamp to enable the prognostics system to determine sequence of events as the vehicle operates in various types of conditions and mission profiles. Prognostic data is displayed to the operator on an as-needed basis. Data is stored for historical and electronic logbook purposes.

<u>Diagnostics:</u> The Fault Manager is the core of the diagnostics system for the diagnostics system for the EFV. This software module is responsible for monitoring the various vehicle subsystems by providing automatic, semi-automatic, and operator initiated built-in test capability, capturing the test results, and with the use of the IETM when applicable, providing fault containment and troubleshooting procedures for timely resolution. Prognostics utilizes the test results run by diagnostics in assisting with its determining state of health status for applicable subsystems. Diagnostic data is used for fault history log and electronic logbook entries in the ELS for maintenance and higher echelon reporting.

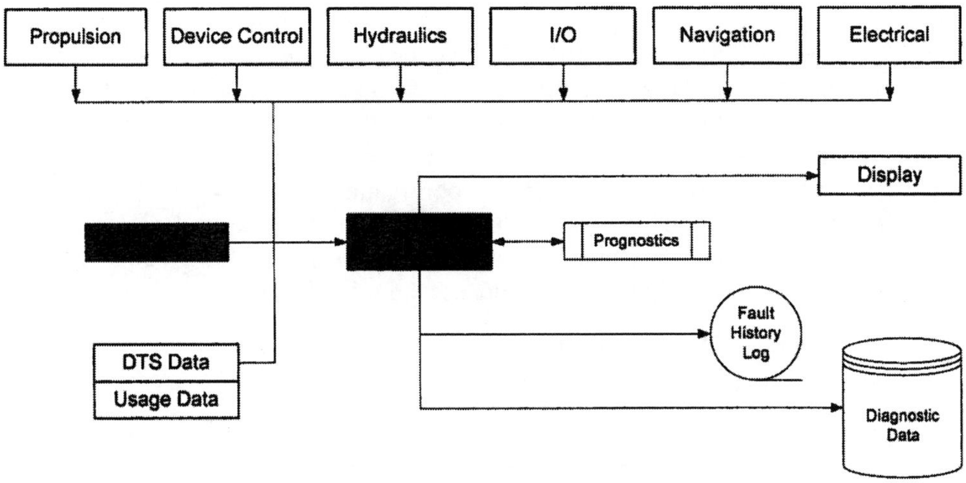

Figure 4: Diagnostic Data Flow

The diagnostic system receives information on the various vehicle data bus' from a series of sensors, resolvers, and switches. That information is used as input to run additional tests, through a series of automatic and on-demand BIT, to arrive at the lowest number of possible ambiguity groups and alert the operator of the fault event and possible corrective paths. Unresolved faults are stored in an operator retrievable fault log which will automatically link to applicable IETM troubleshooting tasks for resolution. Diagnostic data includes fault history recording as well as electronic logbook data used in ELS.

<u>IETM:</u> The EFV program is developing an IETM that provides all the maintenance, operational, troubleshooting procedures, and parts list information required for the operational crew and all levels of maintainers. The IETM provides procedural information that relates and interfaces with prognostics (explaining various operator screen displays), diagnostics (in the form of troubleshooting procedures), and assists in interfacing with electronic logbook data for ELS. As the IETM matures, the capability of commanding vehicle subsystem step functions will be valuable in automating test required for various troubleshooting and verification procedures.

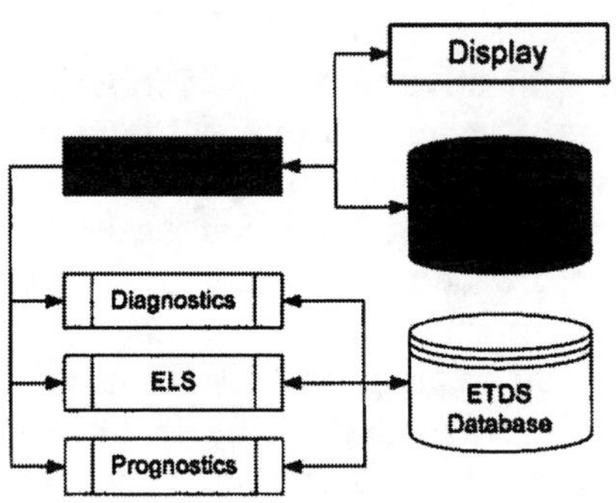

Figure 5: IETM Data Flow

ELS: The ELS EFV requirement provides for the design and implementation of an application that provides for the automation of the vehicle and weapon's logbook. This application is integrated with the vehicle software and interactive with operators and all level of maintainers through the use of the ruggedized laptop. ELS is to provide for the aggregation and reporting of vehicle usage, coordination of work order data for processing and transmission, and providing the information required by upper echelon embedded and contracted reporting and tracking organizations. ELAS logically links to other ETDS modules such as diagnostics, prognostics, and the IETM.

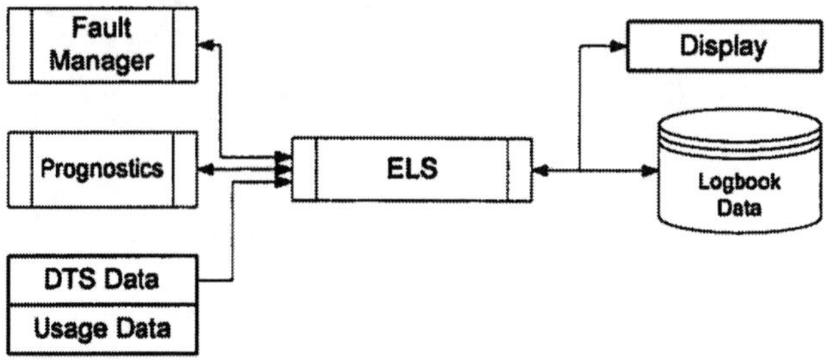

Figure 6: ELS Data Flow

The goal for ELS is to automate the traditional paper-based logbook function. This coincides with the IETM development effort which is to deliver a complete electronic version of the technical manual package. This paperless approach, while desirable, introduced the software development process into the technical manual production process. ELS and IETM are coupled closely together as sharing of parts information, usage data, and configuration data is necessary. Also coupled with ELS is the diagnostic and prognostic applications which also require usage and configuration data. The captured data elements are either provided by the vehicle or weapon system, input from operators and maintainers, or data received from external asset management systems. ELS also archives vehicle usage data. This data must also be aggregated to comply with the historical tracking logbook requirements. ELS also manages the checklists required by the operator and maintainer to perform for either operating or maintaining the vehicle.

Integrated Architecture: The collective modules within ETDS provides the data sources for the Marine Corps' command structure, current and future requirements for integrated autonomic logistics, conditioned based maintenance, and command and control systems. There are currently a few initiatives being developed for the Marine Corp that ETDS will interface with. These systems are designed to retrieve information from the vehicle platform, assist the command structure in utilizing the assets for maximum mission readiness, and streamline the parts requests and maintenance actions through higher tier analysis and feedback. Some of these initiatives include, Global Combat Support System, Global Command & Control Systems, and Autonomic Logistics. Tied in with the Marine Corps' architecture will be a GDAMS system that supports parts and service scheduling information as well as safety bulletins and system software release updates.

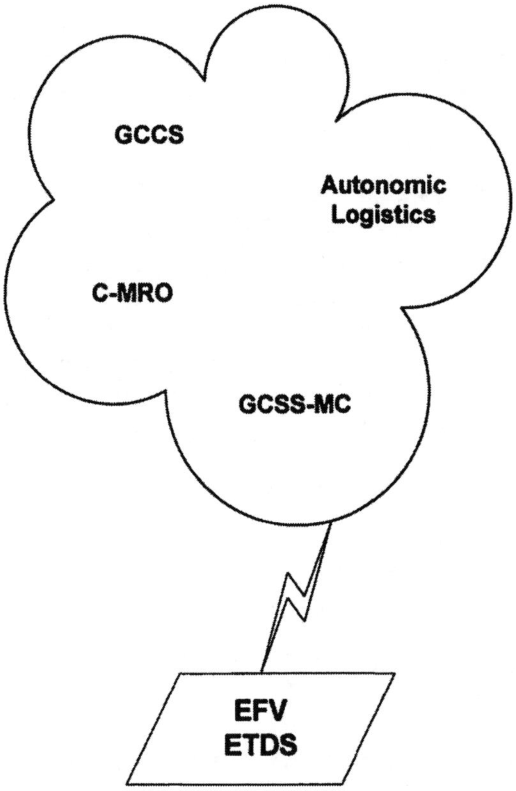

Figure 7: Decision Support Architecture

Global Combat Support System – Marine Corps (GCSS-MC) will be a Marine Corps wide tool that provides access to life cycle management, asset visibility, and decision support information. EFV platform data compliance with GCSS-MC will ensure that information can be shared not only among multiple logistics functions, but also with joint and coalition partners. GCSS-MC will support Marine Corps' operations by providing request mechanisms for products and services, and access to applicable shared data.

Global Command & Control Systems (GCCS) is a joint mandated Command and Control (C2) automated system providing command and control, communications, computers and intelligence (C4I) capabilities for Marine Corps commands participating in joint planning and execution. GCCS consists of common hardware, a common operating system, common software and C2 applications. The C2 applications or segments include joint segments, which are developed and maintained by the Defense Information Systems Agency.

Complex - Maintenance Repair & Overhaul (C-MRO) is GDAMS contributing piece of the life cycle support system supporting the EFV for the Marine Corps. C-MRO is closely tied into the GCSS and autonomic logistics system to provide spare parts, warranty support, field service support and the conduit for hardware and software upgrades. The C-MRO concept fits in with the Marine Corps' vision of an integrated logistics capability. This integrated approach provides for an enterprise wide, logistics centric focus which fulfills the demands for products and services when needed.

Summary: This is a high level overview of the EFV program's design goals for a decision support system. The development of this system is in the early stages and efforts are underway to build the core functionality. The ultimate goals are; to produce a condition based system which assess' the vehicle's operational status to determine short and long term maintenance and support requirements, utilize a robust prognostic and diagnostic system to efficiently realize the vehicle's maximum uptime with shortened maintenance cycles, and interface the vehicle's data structure with upper echelon data repositories for timely reporting and analysis to ensure mission readiness.

References:

Department of Defense, Office of Force Transformation "Integrated Dictionary For Sense And Respond Logistics Integrated Enterprise Domain Architecture," June 7, 2004

U. S. Marine Corps Logistics Command "Strategic Plan FY 2004 – FY 2009"
[illegible]

U. S. Marine Corps "Integrated Logistics Capability High Level Operational Architecture"
[illegible]Divisions/SOS/GCSS/GCSS_OA.pdf

James R. James "Modeling Of Information Dominance In Complex Systems: A System Partitioning And Hybrid Control Framework," *United States Military Academy, Department of Electrical Engineering and Computer Science,* Proceedings of the 36th Hawaii International Conference on System Sciences (HICSS'03)

U. S. Marine Corps "Marine Corps-Global Combat Support System/ Logistics Chain Management (GCSS-MC/LCM)"
[illegible]/sites/[illegible]/projects/GCSS-MC/GCSS_MAIN.asp

Defense Information Systems Agency "What is the Joint Global Command & Control Systems (GCCS-J)"
[illegible]

Blanchard, Benjamin S., Verma, Dinesh, and Peterson, Elmer L., "Maintainability: A Key To Effective Serviceability And Maintenance Management," John Wiley & Sons, Inc., 1995

Hvass, Paul Brian and Tesar, Dr. Delbert "Condition Based Maintenance For Intelligent Electromechanical Actuators," *Robotics Research Group, Mechanical Engineering Department, The University Of Texas At Austin*, June, 2004

SIGNAL ANALYSIS

Chair: Mr. Harry J. Decker
NASA Glenn Research Center

ZERO PHASE WAVELET TRANSFORM FOR DETECTING BEARING LOCALIZED DEFECTS

C. James Li

Department of Mechanical, Aerospace, and Nuclear Engineering
Rensselaer Polytechnic Institute
Troy, NY 12180
Lic3@rpi.edu
518-276-6192

Abstract: This study designed a wavelet in such a way that its Fourier transform has zero phase (hence the name zero-phase wavelet). Since the wavelet has zero phase, the resulted wavelet transform's phase part solely depends on the phase of the signal within the passing band of the wavelet used. With its ability to localize in time and frequency domains, the magnitude part reveals local amplitude modulation and the phase part reveals local phase modulations associated with the frequency band defined by the wavelet.

This study established the utility of the zero phase wavelet for bearing localized defect detection, examined the magnitude and phase maps of such wavelet transform of bearing signals to determine if it can detect periodic structural ringings as effectively as the prevailing envelope spectrum technique and if the phase part brings any additional diagnostic information.

It was found that wavelet transform is more effective than the envelope spectrum and the additional phase information complements the amplitude information and enhances the sensitivity for the detection.

Key Words: bearing; diagnostics; wavelet transform; zero-phase

Introduction: Condition monitoring has received considerable attention for many years because the majority of problems in rotating machines are caused by faulty bearings. The typical failure mode of rolling element bearings is localized defects, which occur when a sizable piece of material on the contact surface is dislodged during operation, mostly by fatigue cracking under cyclic contact stressing [1]. Thus, failure alarm for a rolling element bearing is often based on the detection of the onset of localized defects.

During the bearing operation, wide band impulses are generated when rollers pass over the defect at a frequency determined by shaft speed, bearing geometry, and defect location (inner-race, roller, or outer-race). The theoretical estimate of this frequency is termed "characteristic defect frequency" (Appendix A). Some of the vibration modes of the bearing and its supporting structure will be excited by the periodic impulses, and a distinct bearing signature will be generated by such ringings. The leading edge of each ringing usually comprises a very sharp rise that corresponds to the impact between a roller and the defect. The ringing then decays with an approximately exponential envelope as the energy is dissipated by internal damping (Fig. 1a). Therefore, the signature of a damaged bearing consists of exponentially decaying vibration ringing that occurs periodically at the characteristic defect frequency (Fig. 1b).

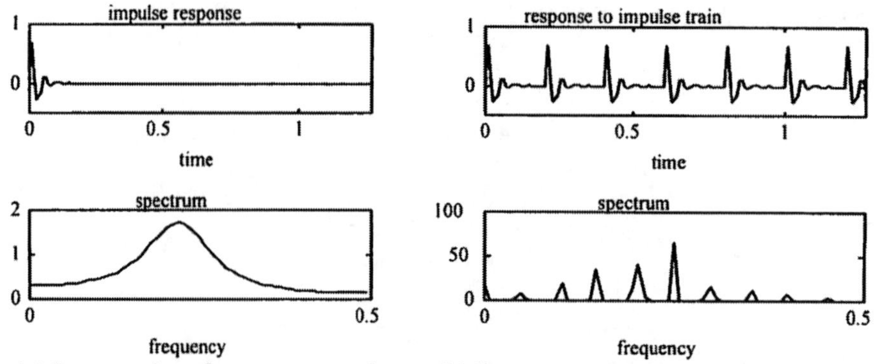

(a) Structure impulse response and its spectrum

(b) Structure response to impulse train and its spectrum

Fig. 1 Structure response to impulse input

The location dependent characteristic defect frequencies make it possible to detect the presence of a defect and to diagnose on what part of the bearing the defect is. The difficulty of localized defect detection lies in the fact that the energy of the signature of a defective bearing is spread across a wide frequency band and hence can be easily masked by noise. Its spectrum, as shown in Fig. 1b, consists of harmonics of the characteristic defect frequency, and the strongest one is in the neighborhood of the ringing mode frequency. Typically, the amplitude at the characteristic defect frequency (the first peak in the spectrum of Fig. 1b) is so small that this component may not be noticed at all in the presence of noise.

Both time and frequency domain methods have been developed to solve this problem. Almost all the existing technologies look for waveform changes due to the periodic ringings excited by the roller-defect impacts. Therefore, they loss their effectiveness when the ringings do not change waveform significantly. Time domain methods usually involve indices that are sensitive to impulsive oscillations, such as peak level, r.m.s. value [2], crest factor analysis [3], kurtosis analysis [4], and shock pulse counting [5]. These methods can yield satisfactory results as long as one selects a frequency band where the ringing modes due to the defects are dominant.

Alternatively, frequency domain techniques search for a train of ringings occurring at any of the characteristic defect frequencies. Approaches in this category include synchronized averaging [6], cepstrum analysis, sum and difference frequencies analysis [7-8], high frequency resonance technique (HFRT, or the so called envelope spectrum) [9-12], and short-time signal processing [13]. Both synchronized averaging and cepstrum analysis look at the total energy in the family of harmonics of a characteristic defect frequency. Naturally, they are more sensitive to the presence of a localized defect than is the spectral analysis, but are more likely to be influenced by noise that may be present at some of these harmonics. Of the remaining frequency domain methods, the most widely accepted is the HFRT, comprising bandpass filtering and enveloping. Like time domain techniques, all these frequency domain methods also require the availability and utility of a frequency band dominated by the ringings provided there is one. Because it is difficult to predict the bearing vibration modes and which one would be excited in the presence of roll-defect impacts, wavelet transform was proposed to facilitate the identification of the frequency band dominated by the ringings [14]. For more details about the state-of-the art, please refer to [15].

Since almost all the existing technologies look for waveform changes due to the periodic ringings excited by the roller-defect impacts, they are ineffective when the ringings do not change waveform significantly. A more comprehensive approach is needed. Thus, this study proposes the examination of both magnitude and the phase using a particular type of wavelet transform.

THE WAVELET TRANSFORM

Definition and Basic Properties: It is well known in the theory of functional analysis that an arbitrary square integrable function $x(t)$ can be decomposed into, or reconstructed from, shifted and dilated versions of another square integrable function $g(t)$ provided that $g(t)$, called analyzing wavelet, satisfies some additional conditions [16]. This decomposition and reconstruction pair is called wavelet transform (WT) in signal processing. Specifically, if $g(t)$ is square integrable

$$\int |g(t)|^2 \, dt < \infty. \tag{1}$$

and satisfies the admissibility condition

$$c_g = 2\pi \int \frac{|G(\omega)|^2}{|\omega|} d\omega < \infty. \tag{2}$$

where $G(\omega)$ is the Fourier transform of $g(t)$, then when $x(t)$, the function to be analyzed, satisfies

$$\int |x(t)|^2 \, dt < \infty \tag{3}$$

there exists the following analysis-synthesis pair:

$$W_x(a,b) = \int g^{*(a,b)}(t) x(t) dt \tag{4}$$

$$x(t) = \frac{1}{c_g} \int \int g^{(a,b)}(t) W_x(a,b) \frac{1}{a^2} \, da db \tag{5}$$

where $g^{(a,b)}(t)$ is the shifted and dilated wavelet, defined by

$$g^{(a,b)}(t) = \frac{1}{\sqrt{a}} g\left(\frac{t-b}{a}\right) \qquad (a > 0) \tag{6}$$

and the asterisk stands for the complex conjugate.

In practice, the admissibility condition (Eq. (2)) requires that $g(t)$ have zero mean value (no DC component):

$$\int g(t) = 0 \qquad or \qquad G(0) = 0 \tag{7}$$

Together with the finite energy condition (Eq. (1)), we know that the wavelet $g(t)$ must be oscillating and short-lasting.

Some mathematical manipulations can reveal the physical meaning of the wavelet transform. Let

$$h^{(a)}(t) = -\frac{1}{\sqrt{a}} g^{*}\left(-\frac{t}{a}\right) \tag{8}$$

then, Eq. (4) becomes

$$W_x(a,b) = \int h^{(a)}(t)x(b-t)dt \tag{9}$$

Clearly, for a given dialation a, $W_x(a,b)$, as a function of b, can be viewed as the output of a filter whose impulse response is $h^{(a)}(t)$ when the input is signal $x(t)$. When $G(\omega)$ is identically zero for all ω but $[\omega_{min}, \omega_{max}]$, the wavelet transform is just a noncausal band-pass filtering operation, with a passband $[\omega_{min}/a, \omega_{max}/a]$. This is also obvious in frequency domain when we express Eq. (4), using Fourier transforms, as:

$$W_x(a,b) = \sqrt{a} \int G^*(a\omega)X(\omega)\exp(jb\omega)d\omega \tag{10}$$

On the other hand, when b is fixed, if $g(t)$ is taken to be a function which is zero outside an interval around the origin, say $[-\tau, \tau]$, Eq. (4) says that the value of $W_x(a,b)$ depends on the values of $x(t)$ in the interval $[b-a\tau, b+a\tau]$. Smaller a makes the interval shorter and gives better localization in time domain. These properties make the wavelet transform a powerful tool for detecting and locating discontinuities or abrupt changes [17] and rapid transient of small amplitude. From the above analysis, we know that $g(t)$ should be of finite duration in both time and frequency domains.

Construction of the Basic Wavelet: For a particular application, there may be a wavelet better than the others. One interpretation of the WT, as defined in Eq. (4), is to consider it as the inner product of the signal with the complex conjugate of $g^{(a,b)}(t)$ defined in Eq. (6). Thus we can say that the WT measures the similarity between the signal and wavelets $g^{(a,b)}(t)$. For each wavelet, the wavelet analysis results in a series of wavelet coefficients which indicate how close the signal is to that particular wavelet. As we have stated above, the signature of a damaged bearing consists of periodic bursts of exponentially decaying sinusoidal vibrations. Therefore, it is natural to use a decaying sinusoidal function such as

$$\begin{aligned} g(t) &= e^{-\sigma t}\sin(2\pi f_c t) \quad & t \geq 0 \\ g(t) &= -g(-t), & t < 0 \end{aligned} \tag{11}$$

Fig. 2 shows an example of it and its magnitude frequency response which has the characteristic of a bandpass filter. f_c determines the center frequency of the passband and σ determines the bandwidth. To make a zero phase wavelet out of it, Fourier Transform is performed and the magnitude of the transform is inverse-transformed. Fig. 3 shows an example of such a zero phase wavelet.

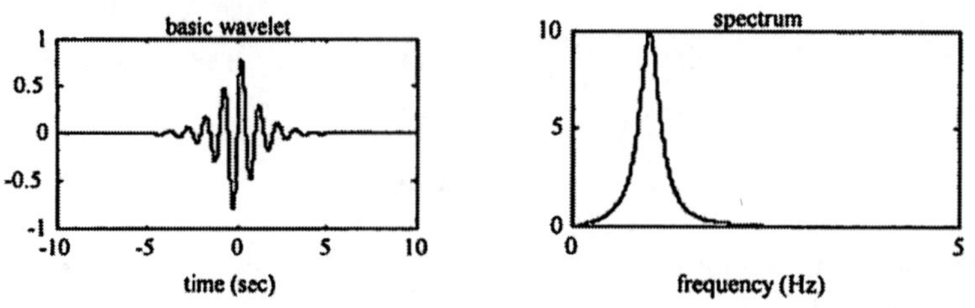

Fig. 2 The basic wavelet and its magnitude spectrum

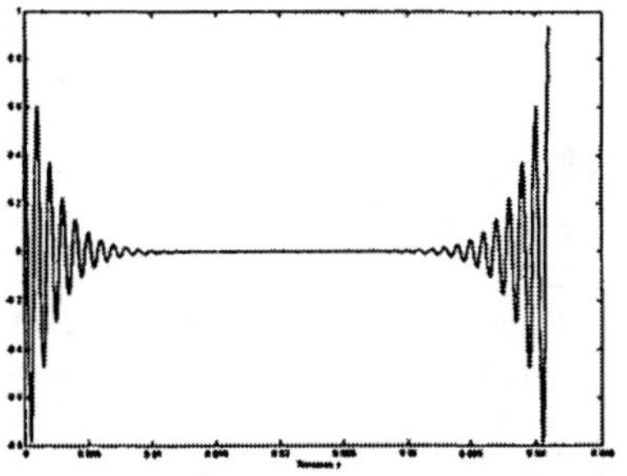

Fig. 3 Zero Phase Wavelet

Understand the zero phase wavelet transform: Fig. 4 shows a sine function of 500hz and magnitude and phase of its wavelet transform using a zero phase wavelet (f_c = 3,000hz and σ = 4,000). The magnitude part of the transform is similar to the output of a full-wave rectifier and it reflects the instantaneous magnitude of the signal waveform. The phase is either 0 or 180 degrees depending on the sign of the signal and the rising/falling edges of the phase plot highlight the zero crossing of the signal. (The vertical axis of the phase plot is in radians. So the phase switches between 0 and pi, i.e., 3.14rad, as the signal goes above and below the zero. Note that it does not work very well at either end of the signal.) If a transient event causes a local phase modulation, the zero crossing point will shift and that will be revealed by shifting in rising/falling edges. In other words, it is the spacing between neighboring edges that carries the information about the phase modulation.

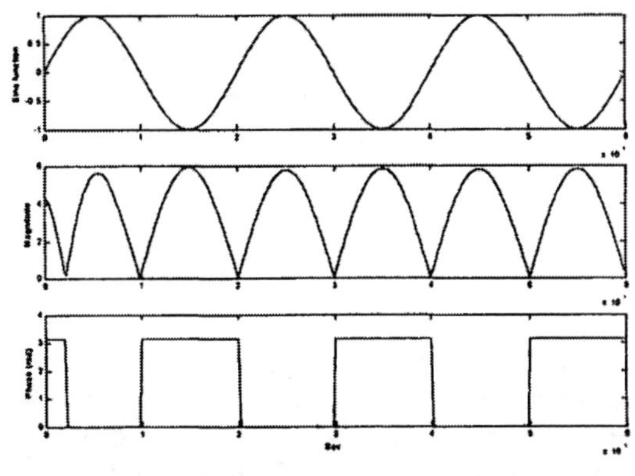

Fig. 4 Zero phase wavelet transform of sin 500hz

While it is easy to notice spacing irregularities due to phase modulation in a phase plot like the one shown in Fig. 4, that is not the case when a long segment of high frequency signal is considered. In such a case, many sign changes occur within a short period of time and therefore overwhelms the display resolution and human eye. To visualize the phase information carried by such phase plot, further processing is necessary. Additionally, it is desirable to quantify the phase variation more precisely.

To do so for a wavelet with a center frequency say f_c, a nominal interval is calculated as half of the period, i.e., $0.5/f_c$. Then, one will go through the phase of the wavelet transform and calculate a normalized deviation from the nominal spacing as

$$\frac{\text{time between neighboring edges - nominal interval}}{\text{nominal interval}} \qquad (12)$$

Once this variation is calculated, a linear smoothing is then employed to take out the high frequency noise.

BEARING DEFECT DETECTION

Defect Detection Scheme: Establishing the utility of WT has most to do with selecting appropriate f_c and σ. One of the most important things is to make sure the family of wavelets employed does cover the full spectrum so that a bearing resonance will not be missed. Additional tweaking helps but is not absolute necessary so long as the f_c and σ are not completely inappropriate. Such tweaking is usually guided by signal spectrum. Take the bearing vibration in Fig. 5 as an example. Its amplitude spectrum shows significant energy around 6k, 10k, 18k and 22k Hz. It was decided that the mother wavelet will have a center frequency of 3kHz and a σ of 4,000. Additionally, the center frequencies of baby wavelets will be 3kHz apart by choosing 'a' in Eq. 8 accordingly. So the six wavelets used have center frequencies at 3, 6, 9, 12, 15 and 18khz respectively. Fig. 6 shows the magnitude frequency responses of these wavelets and how they cover the full spectrum collectively.

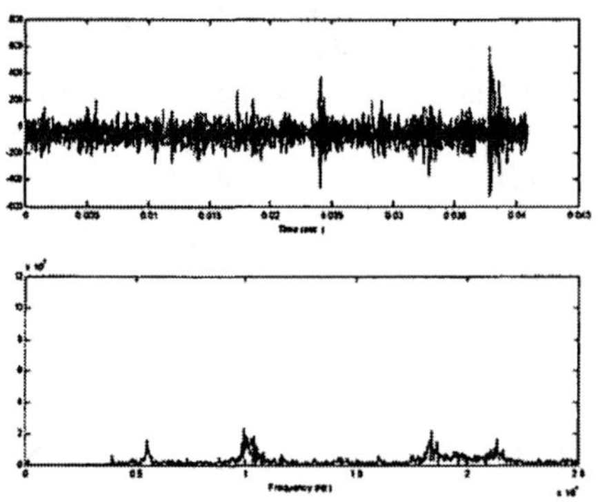

Fig. 5 Bearing vibration from a roller damaged bearing and its spectrum

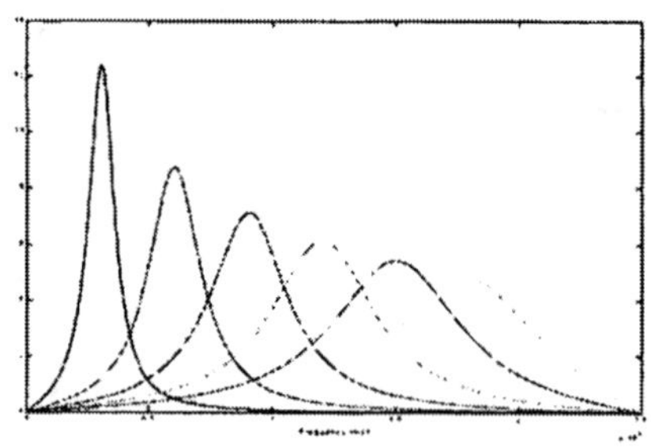

Fig. 6 Frequency responses of the wavelets

As stated above, WT reveals the contribution of different frequency components over the full spectrum from one instant to the next, and the signature of a damaged bearing consists of periodic structural ringing of some resonances. Thus, it is very likely that one or more frequency bands would be influenced by some of these resonances. To detect bearing localized defects, one should observe periodic variations of say, magnitude or phase, in WT of different dilations. If a periodic variation is found at the rate of a certain characteristic defect frequency, it can be concluded that periodic structural ringings due to repetitive impacts, generated upon the passing of each rolling element on the defect, are present — and the bearing is damaged.

There are several ways to determine whether there is a periodicity in the variations of the magnitude of a signal. In this study, magnitude frequency spectrum is used and the values around the characteristic frequencies are examined. A local peak above the noise floor around these frequencies is regarded as evidence that the bearing may be damaged.

Case Study: The experimental setup consists of a spindle driven by a DC variable speed motor and a stand-by bearing test rig (Fig. 7). The test bearings were mounted on a mechanical platform which enables the adjustment of the following: the thrust load, the angular misalignment, and the position of the center of the bearing. Vibration signals were picked up in the radial direction by a

PCB 303A02 accelerometer. The output signals of the sensor were bandpass (5 kHz - 25 kHz) filtered and sent to an A/D converter for data sampling.

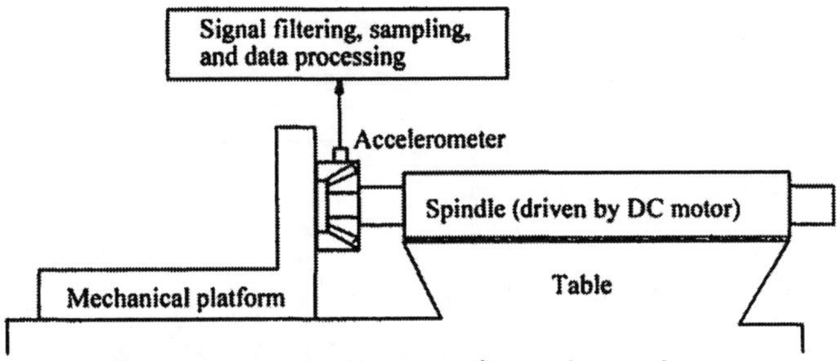

Fig. 7 Schematic diagram of experimental setup

Simulated localized defects were produced on outer-race or roller of bearings by electric-discharge machine to keep their size and depth under precise control. The size of the artificial defects is 60/1000 inch in diameter by 5/1000 inch in depth which is typical. The bearings used in experiment are (i) a bearing with a single outer-race defect, (ii) a bearing with a single roller defect, and (iii) a normal bearing. The characteristic defect frequencies for these bearings can be calculated using the following formulae (Appendix A):

$$f_{or} = 6.854 \cdot f_r \quad (13)$$

$$f_{ro} = 6.24 \cdot f_r \quad (14)$$

where subscripts "*or*" and "*ro*" represent "outer-race" and "roller", respectively, and f_r is the shaft speed of the bearing in Hz.

The zero-phase wavelet transform is first applied to a vibration of the outer-race damaged bearing shown in Fig. 8 along with its envelope spectrum. Each roller-defect impact produces a very remarkable ringing in this vibration. The envelope spectrum obtained using HFRT shows a peak at 195 Hz. Considering that the frequency spacing between neighboring spectra lines is 24.37Hz, it is logical to conclude that this comes from the outer race defect which has a theoretical characteristic defect frequency of 206 Hz. The magnitude WT is shown in Fig. 9 and the ringings can be clearly seen in wavelet transforms #5 and #6. The spectra of the magnitude and phase of the wavelet transforms are shown in Fig. 10. The spectra of magnitude transform of wavelets 4, 5, and 6 all clearly show a peak around 195 Hz while the one from 6 is the largest. The spectra of phase transform show (to a lesser extent) the same periodicity with wavelets 4 and 2. This example basically validates the zero-phase wavelet transform in the sense that its results are consistent with that of the state of the art—envelope spectrum.

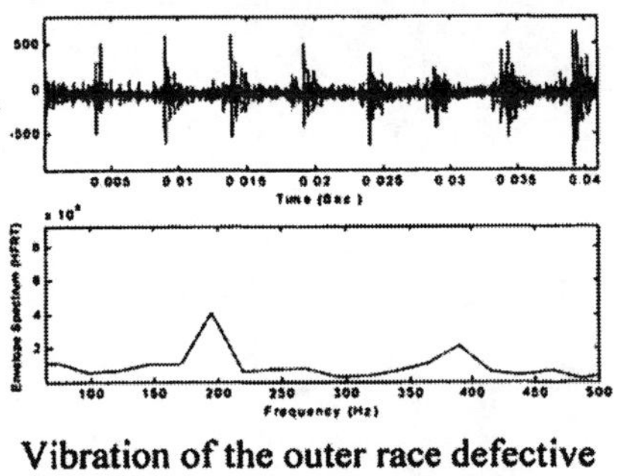

Fig. 8 Vibration of the outer race defective bearing and its envelope spectrum

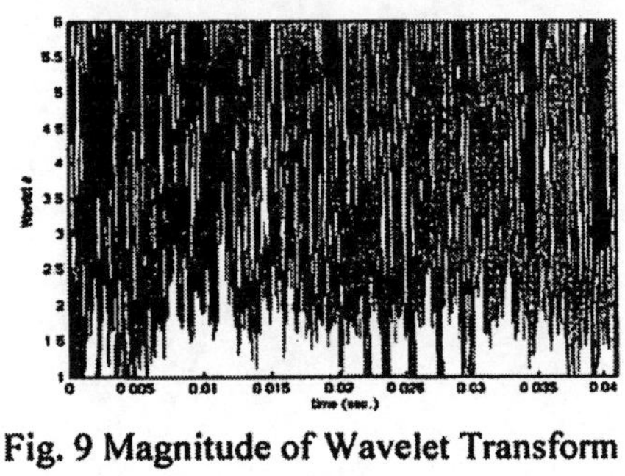

Fig. 9 Magnitude of Wavelet Transform

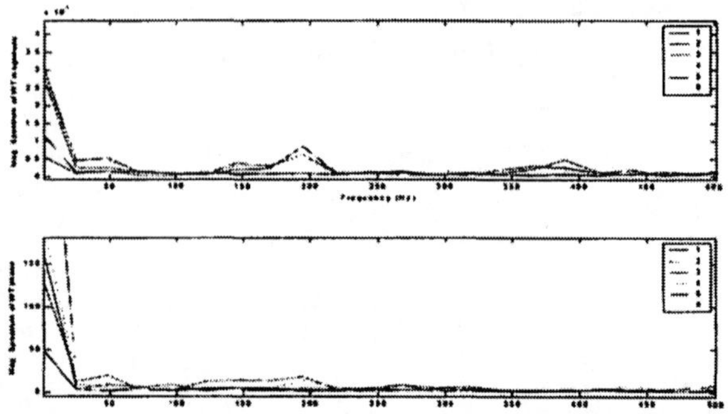

Fig. 10 Spectra of the magnitude and phase parts of the wavelet transform

The zero-phase WT was then applied to a more challenging bearing vibration. Fig. 11 shows a vibration from the roller damaged bearing and its envelope spectrum. In the case of a damaged roller, the transmitting path between the roller and the transducer varies as the damaged roller orbits around. This resulted in a much less remarkable signal. In fact, the envelope spectrum failed to indicate the presence of the fault as the magnitude around the characteristic defect frequency, i.e., 211.34 Hz does not rise well above the noise floor. Fig. 12 shows spectra of the magnitude and phase of the wavelet transform. The magnitude one shows increased magnitude at 219Hz for wavelet 6 and the phase one shows increased magnitude for wavelets 1 and 2.

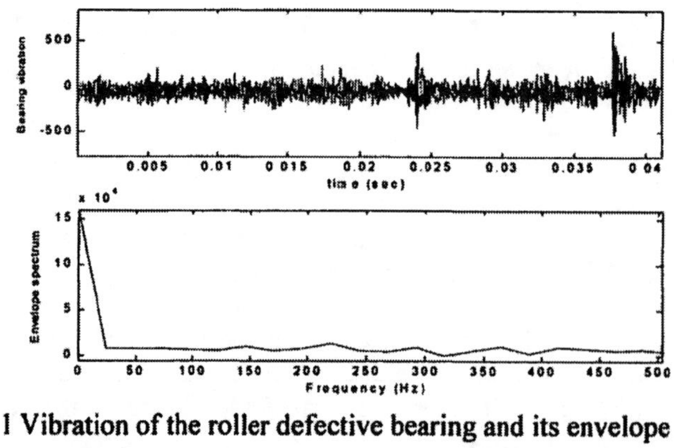

Fig. 11 Vibration of the roller defective bearing and its envelope spectrum

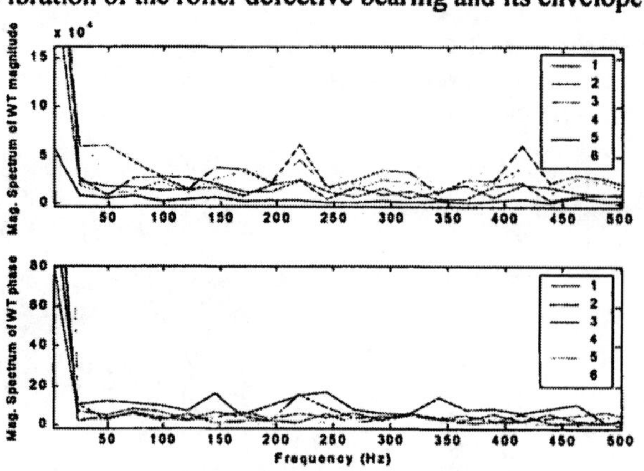

Fig. 12 Spectra of the magnitude and phase parts of the wavelet transform
219 peak/211.34 theoretical

Conclusions: The study designed a zero-phase wavelet transform for bearing localized defect detection and compared its effectiveness with the most popular state of the art—envelope spectrum or HFRT.

1. The zero-phase wavelet transform is as effective as envelope spectrum in detecting outer-race defect which generated a very marked vibration.

2. The zero-phase wavelet transform is more effective in detecting the more challenging roller defect case. Both zero-phase WT magnitude and phase offered useful indications with the magnitude being slightly better.

3. It was noticed that outer race defect and roller defect were revealed by different wavelets. This is because a defect at different bearing parts may excite different resonant modes of the bearing and its nearby structure. Furthermore, it is possible that different modes are excited through the course of defect development. Therefore, it is imperative to have a bearing localized defect detection scheme that requires no band selection such as the proposed one.

4. It was noticed that for the same bearing, WT magnitude and phase revealed the bearing condition with different dilations. This is probably due to the fact that while the magnitude WT yields the best result with a wavelet whose passing band largely overlaps with the ringing, the phase WT yields the best result with a wavelet that offers a passing band that covers both a low frequency constant background noise and bearing ringings.

REFERENCES

[1] Widner, R. L., and Littmann, W. E., "Bearing Damage Analysis", National Bureau of Standard special publication No. 423, April 1976

[2] Downham, E., and Woods, R., "The Rationale of Monitoring Vibration on Rotating Machinery in Continuously Operating Process Plant", ASME paper No. 71-Vibr-96

[3] Weichbrodt, B., and Bowden, J., "Instrument for Predicting Bearing Damage", GE Co. Report, S-70-1021 AD 869633, March 1970

[4] Dyer, D., and Stewart, R. M., "Detection of Rolling Element Bearing Damage by Statistical Vibration Analysis", J. of Mechanical Design, Vol. 100, 1978, pp. 229-235

[5] Gustafsson, O. G., and Tallian, T., "Detection of Damage in Assembled Rolling Bearings", Trans. ASLE, 5, 1962, pp. 197-209

[6] Braun, S., and Datner, B., "Analysis of Roller/ball Bearing Vibrations", J. of Mechanical Design, Vol. 101, No. 1, 1979, pp. 118-125

[7] Ehrich, F. F., "Sum And Difference Frequencies in Vibration of High Speed Rotating Machinery", ASME J. of Engineering for Industry, Vol. 94, No. 1, 1972, pp. 181-184

[8] Eshleman, R. L., "The Use of Sum and Difference Frequencies in Rotating Machinery Analysis", Proc. of Machinery Vibration Seminar IV, Cherry Hill, NJ, Vibration Institute, Nov. 1980, pp. 1-18

[9] Mcfadden, P. D., and Smith, J. D., "Vibration Monitoring of Rolling Element Bearings by the High Frequency Resonance Technique-A Review", Tribology International, Vol. 17, 1984, pp. 3-10

[10] Broderick, J. J., Burchill, R. F., and Clark, H. L., "Design and Fabrication of Prototype System for Early Warning of Impending Bearing Failure", NASA CR-123717, Jan. 1972

[11] Burchill, R. F., "Resonant Structure Techniques for Bearing Fault Analysis", National Bureau of Standards NBSIR 73-252, 1973

[12] Darlow, M. S., Badgley, R. H., and Hogg, G. W., "Applications of High Frequency Resonance Techniques for Bearing Diagnostics in Helicopter Gearboxes", USAAMRDL-TR-74-77, 1974

[13] Li, C. J., and Wu, S. M., In-line Detection of Localized Defects in Bearing by Pattern Recognition Analysis", ASME J. of Engineering for Industry , Vol. 111, 1989, pp. 331-336

[14] C. James Li and J. Ma, "Wavelet Decomposition of Vibrations for Detection of Bearing Localized Defects," NDT&E International, Vol. 30, No. 3, June 1997, p 143-149.

[15] C. James Li and K. McKee, "Diagnostics: Bearing (roller & Journal) vibrations and diagnostics," Encyclopedia of Vibration, Editors: S. G. Braun, D. J. Ewins and S. S. Rao, Academic Press, Aug. 2001.
[16] Grossmann, A., and Morlet, J., "Decomposition of Hardy Functions into Square Integrable Wavelets of Constant Sharp." SIAM J. Math. Anal., Vol. 15, No. 4, 1984, pp. 723-736
[17] Grossmann, A., "Wavelet Transforms and Edge Detection", in Stochastic Processes in Physics and Engineering, S. Albeverio et al., ed., D. Reidel Publishing Co., Dordrecht, Holland, 1988, pp. 149-157
[14] Li, C. J., Ma, J., Hwang, B., and Nickerson, G. W., ?attern Recognition Based Bicoherence Analysis of Vibrations for Bearing Condition Monitoring", Proceedings of Symposium on Sensors, Controls and Quality Issues in Manufacturing, ASME 1991 WAM, Atlanta, Georgia, Dec. 1991, pp. 1-11
[17] Nawab, S., and Quatieri, T., Short-Time Fourier Transform", in Advanced Topics in Signal Processing, J. Lim and A. Oppenheim ed., Prentice Hall, Englewood Cliffs, New Jersey, 1988, pp. 289-337
[18] Rioul, O., and Vetterli, M., Wavelets and Signal Processing", IEEE Signal Processing Magazine, Oct. 1991, Vol. 8, pp. 14-38

APPENDIX A: Bearing characteristic defect frequencies [1]

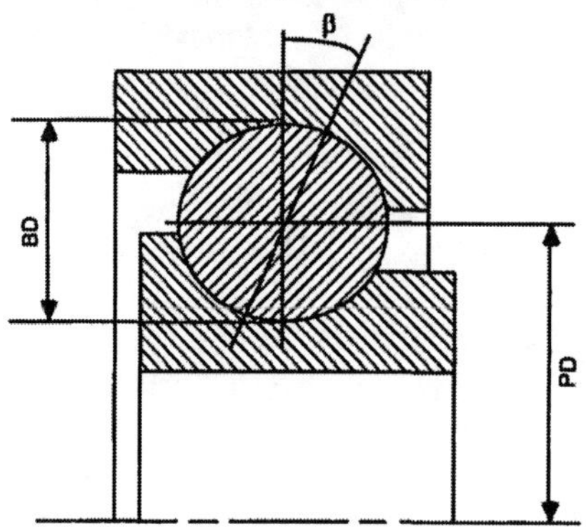

$$f_{or} = \frac{n}{2} f_r \left[1 - \frac{BD}{PD} \cos\beta \right] \tag{A1}$$

$$f_{ir} = \frac{n}{2} f_r \left[1 + \frac{BD}{PD} \cos\beta \right] \tag{A2}$$

$$f_{ro} = \frac{BD}{PD} f_r \left[1 + \left(\frac{PD}{BD} \right)^2 \cos\beta \right] \tag{A3}$$

where
f_{or} = characteristic defect frequency for defect on outer-race
f_{ir} = characteristic defect frequency for defect on inner-race
f_{ro} = characteristic defect frequency for defect on roller
BD = roller diameter
PD = bearing pitch diameter
f_r = bearing rotating speed
β = contact angle between race and ball
n = number of rollers

A KERNEL BASED REPRESENTATION AND DISCRIMINATION FEATURE EXTRACTION FOR GAS TURBINE FAULT DETECTION AND DIAGNOSIS

Venkataramana Kini B., Joydeb Mukherjee, Sunil Menon*, Lalitha E. M.

Honeywell Technology Solutions Lab, Bangalore, India 560076
*Honeywell ES&S Laboratories, 3660 Technology Drive, Minneapolis, MN 55418
venkataramana.bantwal@honeywell.com

Abstract: Accurate gas turbine fault detection and diagnosis (FDD) are essential for improving airline safety, reducing airline costs associated with delays and cancellations, and scheduling predictive maintenance. We present the maximum representation discrimination features (MRDF) algorithm, a kernel-based nonlinear method of FDD that uses features for maximum representation and discrimination. Examples for training from other classes may not be available or are sparse and one must rely on representation features. When adequate examples for all classes are available, the method will use discriminatory as well as representation features for classification. By varying a single parameter one can judiciously decide the extent to which the features will be used. This method improves upon the existing MRDF classifier by introducing a different between-class distance metric. This metric considers the local neighborhood of points rather than all pairs of points between two classes. This method does not require a nonlinear optimization procedure for its solution; only eigen-decomposition of the kernel matrix is required to arrive at the solution. During operation, the KMRDF classifier will classify new sensor data and make decisions about whether the data indicate presence of fault or not. The developed method was tested with snapshot data, both normal and faulty, collected at takeoff from a turbofan propulsion gas turbine and the results are presented.

Key Words: Kernel methods; maximum representation & discrimination features; fault detection; fault diagnosis; fault identification

INTRODUCTION:

Accurate turbine engine fault detection and diagnosis (FDD) is critical for improving aircraft safety and reducing airline operating costs. Incipient fault detection in turbine engines can reduce costs significantly by preventing in-flight shutdowns, delays and cancellations, unscheduled engine removals, or take-off aborts. Accurate incipient fault detection is necessary for a more effective preventive maintenance program.

Turbine engine FDD involves early detection of incipient and sudden faults. When a fault is detected, it will be diagnosed as a sensor, actuator, or system component fault. FDD approaches can be broadly classified into model-based [1] and data-driven approaches. Model-based FDD methods use a physics-based or empirical models of the system to generate residuals (difference between the model outputs and actual system measurements). Residuals are passed through a decision logic system to arrive at a diagnosis. Data-driven methods use historical data from the system to construct quantitative (neural networks, statistical models) or qualitative (trends- or rule-based) methods. In this paper we adopt data-driven approach for FDD.

The primary objectives in data-driven FDD are to reduce variations in the estimated model coefficients and providing compact representations. The compact representation, in terms of lesser number of transformed variables or features, not only saves storage space and processing need, but also skirts a problem with statistical analysis of multidimensional data known as the *Curse of Dimensionality*. Particularly for faulty cases, high dimensionality models using the historically available sparse data points may not yield reliable models.

Principal Components Analysis (PCA) and its nonlinear versions NLPCA (nonlinear PCA [6]), Kernel PCA, Principal Curves and Surfaces [2] have been used to implement *representation-based FDD*. Fisher's linear discriminant analysis (LDA) and its nonlinear counterparts quadratic LDA, kernel-based discriminant functions, etc. [3] have been used to develop *discrimination feature-based FDD*. It is desirable to build a classifier that uses both types of features for the reasons mentioned in the MRDF sub-section. The Maximum Representation and Discrimination Features (MRDF) algorithm exactly addresses this need [4]; however, it has restricted itself to using only linear features. Nonlinear feature extractors are needed to uncover existing nonlinear relationships pertaining to turbo engines present in the data. The algorithm proposed and implemented in this paper attempts to address this shortcoming by introducing a kernel-based nonlinear feature extraction method (Kernel MRDF or KMRDF) and combining both representation and discrimination features judiciously.

The following section briefly discusses linear feature extraction methods PCA, LDA, and MRDF. In the succeeding section we outline successful nonlinear feature extraction methods—locally linear PCA, auto-associative multilayer perceptrons, principal curves, and kernel PCA. The next section goes into core of the paper, explaining the KMRDF algorithm which makes use of the concepts used in the previous sections to extract nonlinear representation and discrimination features. We then compare relative merits of classifiers that use nonlinear representation and discrimination features quantitatively in terms of their turbo engine fault classification performance. Results obtained using KMRDF are compared against PCA, LDA, and MRDF results. A summary and conclusions are provided in the final section.

LINEAR FEATURE EXTRACTION METHODS:

Principal Component Analysis: Among the traditional *representation* feature based algorithms PCA is the most popular. It essentially uses eigenvalue decomposition of the covariance matrix of the data. The problem is formally posed as the maximization of Rayleigh quotient—see Eq. (1):

$$\max\left[v^T C v\right]^{-1}\left[v^T v\right]; \qquad s.t. \quad \|v\|=1 \tag{1}$$

Its solution is obtained with the following eigenvalue decomposition:

$$Cv_i = \lambda_i v_i \qquad for\ \lambda_1 \geq \lambda_2 \geq \ldots \lambda_d \tag{2}$$

where C is the covariance matrix formed from data points $x_1 \ldots x_n$ of dimensionality d. λ_i represent eigenvalues and v_i eigenvectors. Projections along principal subspace, i.e. eigen directions corresponding to the first few eigenvalues, are the principal components. They capture most of the variability in the data. The discarded components are assumed to represent residual subspace corresponding to the noise in the data.

Linear Discriminant Analysis: Components good for representation may not be good for discrimination between different classes. Fisher's linear discriminant analysis strives to find a hyperplane along which, when the data is projected, maximum separability between different classes is achieved. Moving the hyperplane around, we might find an orientation for which the projected samples are well separated. Formally, LDA tries to maximize the ratio of inter-class variance S_B to intra-class variance S_W. This optimization problem is posed as a maximization of the generalized Rayleigh quotient—Eq. (3):

$$\max\left[v^T S_W v\right]^{-1}\left[v^T S_B v\right]; \qquad s.t.\ \|v\| = 1 \tag{3}$$

Once again, its solution is the following generalized eigenvalue decomposition:

$$S_W^{-1} S_B v = \lambda v \tag{4}$$

where, $S_W = \sum_{i=1}^{c}\sum_{j=1}^{n_i}(x_j - \mu_i)(x_j - \mu_i)^T$ and $S_B = \sum_{i=1}^{c}\sum_{j=1}^{c}(\mu_i - \mu_j)(\mu_i - \mu_j)^T$, μ_i is the mean of ith class for c number of classes and n_i is the number of points belonging to class j.

Maximum Representation & Discrimination Features (MRDF): The previously described formalisms work well for their specified purposes. Representation-based fault detection involves extraction of features that ensure larger spread of projections of nominal data. On the other hand, fault identification or classification involves extraction of discriminatory features that ensure separation of samples in different classes while constraining samples in each class to be tightly packed. Generally, for turbo engines the faults rarely occur, so except for one or two faulty classes, different faulty class data will not be available. It is necessary to develop models having simultaneous detection as well as identification capabilities. Hence the need to club representation and discrimination features and make the most of both of them. Maximum representation and discrimination [4] is a framework wherein we can capture both representation and discrimination features.

Measure for best representation: The MRDF representation measure is mainly derived from linear PCA. Given a d dimensional data, linear PCA is used to represent it in fewer dimensions: $p<d$. For each class covariance matrix (C_1, C_2) and its corresponding means

μ_1, μ_2 are found. According to linear PCA, d-dimensional transform, $y = V_p^T x$, on a d-dimensional input x such that the input data is transformed into a lower dimensional (p-dimensional) feature space y that optimally (in the sense of minimal sum of squared reconstruction error) represents the input data. Here $V_p = [v_1, \ \dots \ , v_p]$ is a $d \times p$ matrix containing p basis vectors v_i. Each output feature y_i of the feature vector y is the projection of the input x onto a basis vector v_i. The basis vectors are the solution to the optimization problem in Eq. (5).

$$\max E_R = \max\left[\sum_{i=1}^{p} v_i^T C_2 v_i + \sum_{i=1}^{p} v_i^T C_2 v_i\right]; \qquad s.t. \ \ \|v_i\| = 1 \tag{5}$$

Measure for best discrimination: For discrimination, we want the projections (i.e. transformed values) for each class to be well separated. Talukder [4] suggests using the average squared difference of all projection values for two classes (i.e. all the pair of points between two classes). He argues that it is a better measure of separation, since it does not use the means of the class projections alone. For discrimination, we find basis vectors $V_p = [v_1, \ \dots \ , v_p]$ such that $y_1 = V_p^T x_1$ and $y_2 = V_p^T x_2$ are well separated from each other. Once again, the basis vectors are the solution to the optimization problem in Eq. (6).

$$\max\left[\sum_{i=1}^{p} v_i^T R_{12} v_i\right]; \qquad s.t. \ \ \|v_i\| = 1 \tag{6}$$

where, $R_{12} = (1/n_1 n_2)\sum_{i=1}^{n_1}\sum_{j=1}^{n_2}(x_i - x_j)(x_i - x_j)^T$ and n_i, n_j are the number of points belonging to two classes.

This measure does not separate the means of the two classes; instead it separates all pairs of samples in the two classes. Thus, it can handle classes with multiple clusters. For the discrimination measure in the MRDF algorithm, we also minimize the spread of the transformed samples of each class, i.e. we minimize:

$$E_R = \sum_{i=1}^{p} v_p^T (C_1 + C_2) v_p \tag{7}$$

To find the best set of discrimination measure we use:

$$E_D = \sum_{i=1}^{p}\left\lfloor v_i^T (C_1 + C_2) v_i \right\rfloor^{-1} \left\lfloor v_i^T R_{12} v_i \right\rfloor \tag{8}$$

Linear MRDF Solution: The general solution for maximum representation and discrimination is found by combining representation and discrimination measures. Combining E_R in Eq. (7) and E_D in Eq. (8) we need to solve the optimization problem in Eq. (9).

$$\max E_{RD} = \max \sum_{i=1}^{p} \frac{v_i^T \left[k(C_1 + C_2) + (1-k)R_{12}\right] v_i}{v_i^T \left[kI + (1-k)(C_1 + C_2)\right] v_i}; \quad s.t. \quad \|v_i\| = 1 \tag{9}$$

where, $0 \le k \le 1$ and I represent an identity matrix. When $k = 0$, E_{RD} reduces to E_D, only discriminatory features are obtained and when $k = 1$, E_{RD} reduces to E_R, only representation features are obtained. When an intermediate value of k is used, simultaneous representation and discrimination are obtained. In practice, the solution to Eq. (9), an optimization problem, is arrived at by solving an equivalent eigenvalue problem: $[kI + (1-k)(C_1 + C_2)]^{-1}[k(C_1 + C_2) + (1-k)R_{12}]v_i = \lambda_i v_i$. MRDF gives a good framework for combining representation and discrimination features by choosing the value of k judiciously.

NONLINEAR FEATURE EXTRACTION METHODS:

We briefly describe major approaches in this section instead of giving a full review of this large field and refer readers to Diamantaras and Kung [5] for more details.

Locally Linear PCA: Where a linear PCA fails because the dependences in the data vary nonlinearly with the region in input space, one way to address the problem is to capture nonlinearity using piecewise linear PCA. Claims that using linear PCA locally can be fruitful do not consider global nonlinearity in the data.

Auto-associative Multilayer Perceptrons: Linear PCA can be realized using a linear three layer perceptron with a hidden layer containing a smaller number of nodes than the input. When it is trained to reproduce the input values as outputs (i.e., used in auto-associative mode), the hidden unit activations form a lower-dimensional representation of the data closely related to PCA. To generalize to a nonlinear setting, nonlinear activation functions and additional layers are used [6]. Unlike linear PCA, the resulting network training solves a hard nonlinear optimization problem with the possibility of getting trapped in local minima, and thus the outcome depends on the training's starting point. Moreover, in neural network implementations, there is often a risk of over-fitting.

Principal Curves: This is an approach with a clear geometric interpretation in input space [2] that iteratively estimates a curve (or surface) capturing the structure of the data. The data are mapped to the closest point on a curve, and the algorithm tries to find a curve with the property that each point on the curve is the average of all data points projecting onto it. It can be shown that the only straight lines satisfying the latter are principal components, so principal curves are indeed a generalization of the latter. Again, to compute principal curves, a nonlinear optimization problem has to be solved.

Kernel PCA: Kernel PCA [7] is a nonlinear generalization of PCA as it performs linear PCA in a *feature space* of arbitrarily large (possibly infinite) dimensionality. The input data is mapped to feature space by a function $\phi : x \to \phi(x)$. Compared to the previous approaches, Kernel PCA has the main advantage that its operation involves no nonlinear optimization. It essentially resorts to linear algebra as simple as standard PCA. In comparison with Auto-associative Multilayer Perceptrons-based PCA approaches, Kernel

PCA could be disadvantageous if we need to process a very large number of observations, as this results in a large kernel matrix K (explained below). As for ease of explanation, Kernel PCA is harder to interpret in input space as opposed to Principal Curves; however, at least for polynomial kernels, it has a very clear interpretation in terms of higher-order features.

Although we are mapping $\phi : x \rightarrow \phi(x)$ to arbitrarily higher dimensional space, we need not incur high computational cost. As long as we are able to express our operations on the higher dimensional feature space vector in terms of inner product of feature vectors it amounts to performing operations on the kernels (specifically, inner product kernels) constructed using lower dimensional input data vectors [8]. Because of these advantages the nonlinear correlations are captured using a modified kernel approach. The theory behind Kernel PCA is explained briefly below.

After performing mean-centering appropriately, the covariance matrix in this feature space is given by:

$$\bar{C} = \frac{1}{n} \sum_{i=1}^{n} \phi(x_i)\phi(x_i)^T \tag{10}$$

We must now find eigenvalues $\lambda \geq 0$ and eigenvectors V satisfying Eq. (11) and (12).

$$\lambda V = \bar{C}V \tag{11}$$

$$V = \sum_{i=1}^{n} \alpha_i \phi(x_i) \tag{12}$$

The eigenvectors satisfying the equations lie in the span of $\{\phi(x_j)\}_j^n$. Combining Eq. (10), (11), and (12), after some algebraic manipulation we arrive at Eq. (13):

$$K\alpha = n\lambda\alpha \tag{13}$$

where, K is the kernel matrix (polynomial kernel of degree d) the elements of which are given by Eq. (14):

$$K_{ij} = \left(\phi(x_i) \cdot \phi(x_j)\right) = \left(1 + x_i^T x_j\right)^d \tag{14}$$

KERNEL-BASED MAXIMUM REPRESENTATION AND DISCRIMINATION FEATURE EXTRACTION:

Even though MRDF allows us to simultaneously capture representation and discrimination features, it doesn't capture nonlinear correlations in the data. This work describes a feature extraction procedure referred to as the kernel maximum representation and discrimination feature (KMRDF) extraction method. This method can extract nonlinear features useful for discrimination, representation, or joint representation and discrimination. Nonlinear features for representation [7] try to ensure that the spread of the projections of the samples in one class is large. Features for discrimination [8] try to separate samples in different classes while constraining samples in each class to be tightly packed. Features for representation and those for discrimination are complements of each other. Features for representation typically do not ensure good classification between different classes. The discrimination function R_{12} separates all pairs of samples in the different classes. Taking a cue from support vector machines (SVM) [8], we developed a new formulation for capturing discrimination in terms of boundary points instead of using all pairs of points.

Measure for best nonlinear representation: We apply kernel function ϕ (polynomial or Gaussian) to the input space, where nonlinear correlations will be uncovered [8]. In the feature space the data is represented by using kernel matrices K_1 and K_2.

$K_{1ij} = \phi(x_i) \bullet \phi(x_j) \quad \forall x_i, x_j \in class1$ and $K_{2ij} = \phi(x_i) \bullet \phi(x_j) \quad \forall x_i, x_j \in class2$

The representation measure is given by: $E_R = K_1 + K_2$

Measure for best nonlinear discrimination: For discrimination, we compute the subspace corresponding to the eigenvectors of the average local between-sum of squares matrices in the feature space.

$$\tilde{\phi}_j(x^i) = \bar{\phi}_j(x^i) - \bar{\phi}(x^i) \tag{15}$$

$$E_D = \frac{1}{cn} \sum_{i=1}^{n} \sum_{j=1}^{c} \tilde{\phi}_j(x^i) \tilde{\phi}_j(x^i)^T \tag{16}$$

where, c is the number of classes, n the total number of data points, and $\bar{\phi}_j(x^i)$ is the mean of data points in feature space in the local neighborhood of the point $\phi(x^i)$ with respect to class j, $\bar{\phi}(x^i)$ is the mean of data points in feature space in the local neighborhood of the point $\phi(x^i)$. The kernel MRDF formulation is the optimization problem Eq. (17):

$$\max E_R + E_D; \qquad s.t. \ \|v_i\| = 1 \tag{17}$$

Its solution is found in the eigenvalue decomposition in Eq. (18).

$$\left[kI + (1-k)K \right]^{-1} \left[kK + (1-k)\tilde{K} \right] v_i = \lambda_i v_i \tag{18}$$

where, $\tilde{K} = K_1^T K_1 + K_2^T K_2$ and $K = K_1 + K_2$ assuming K_1 and K_2 are compatible for addition; and k is the weighting factor for discrimination versus representation as in the case of MRDF.

APPLYING KMRDF TO TURBO ENGINE FDD:

Engine Details: The specific turbine engines for which KMRDF-based FDD techniques are developed are two-spool, high-bypass turbofan gas turbine engines. Based on operating pressure, the engine system can be divided into two main components: low pressure and high pressure. The low-pressure section uses a gear reduction unit to drive a single-stage fan off the two-stage, low-pressure turbine. The high-pressure system consists of a double compressor and a gearbox, both driven by the high-pressure turbine.

The available engine measurements (and measurement units) include: total air temperature (deg. F), altitude (feet), Mach number (no units), engine oil temperature (deg. F.), and fuel flow (lbs/hr), low-pressure spool speed (percentage of full speed), high-pressure spool speed (percentage of full speed), and the exhaust gas temperature (deg. C). The engine faults considered in this work are high-pressure power section degradation, combustor liner rupture, bleed band leakage and EGT sensor abrupt failure.

Results: Figure 1 shows the flowchart for turbo engine FDD using KMRDF extraction.

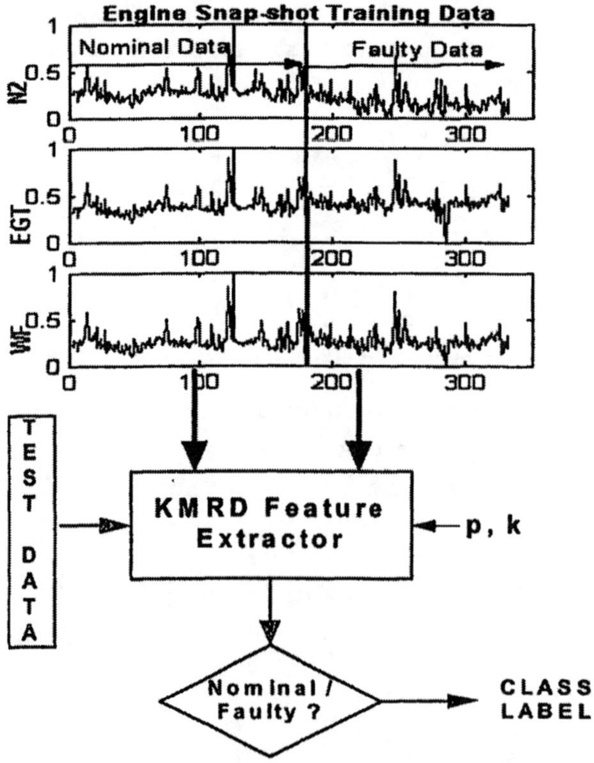

Figure 1: Engine FDD Using KMRDF

Engine data are preprocessed by passing them through a semi-empirical polynomial model of the engine to correct and normalize them for varying operating conditions and to account for engine idiosyncrasies. The outputs from this semi-empirical model are input to the KMRDF. Faulty and nominal regions are identified and segregated from the available historic dataset. These datasets are further partitioned into two parts for training and testing the KMRDF extractor. The training data set is used to develop the KMRDF extractor. The test data is for validating the KMRDF extractor.

To classify the test data, they are projected along KMRDF directions Eq. (18); feature values obtained as projected components determine the location of the data points in the KMRDF space and are then classified according to their *closeness* to the class centers in KMRDF feature space.

DISCUSSION:

Principal component analysis performed on pooled data or a single linear discriminant function (LDA) cannot separate many configurations of data; for example, data from a non-convex set. In such cases, locally linear PCA can be used to form a piecewise linear approximation to the data for representation feature-based classification. Many linear discriminant functions together can be used to form a discrimination feature-based classifier. However, using many piecewise linear PCA or LDA functions increases computation and design costs. Nonlinear transformations lead to more compact representations with fewer components or discriminant functions. Nonlinear transformations formed using iterative NLPCA algorithms suffer from some drawbacks [4] that are circumvented using kernel-based nonlinear transformations. Talukder [4] proposed a nonlinear version of MRDF—nonlinear eigen feature (NLEF)—which operates in higher dimensional feature space by artificially creating quadratic features. Our KMRDF is a more systematic approach using kernel transformation. It can be used to introduce and experiment with. Also, by using the *kernel trick*, one does not have to compute inner products of large dimensional vectors in the expanded higher dimensional feature space. Note that if a linear transformation can capture the spread in the data, the kernel-based method will reduce to PCA or LDA as the case may be. Nonlinear eigen feature-based MRDF is a special case of the more generic KMRDF.

Our experiment shows that higher order correlation information exists in real data and nonlinear KMRDF can efficiently locate such information. PCA and LDA cannot do this as effectively. The results in terms of percentage false positive rates (FPR) and percentage true positive rates (TPR) are shown in the Table I. For comparison, the results are also shown for PCA, LDA, and MRDF. KMRDF outperforms other methods and deals more consistently with the four types of faults we considered. The k values used in the KMRDF to judiciously blend representation and discrimination features for the four cases reported here are 0.5, 0.01, 0.5 and 0.01

Table I: Performance comparison of different algorithms

Fault	PCA		LDA		MRDF		KMRDF	
	%FPR	%TPR	%FPR	%TPR	%FPR	%TPR	%FPR	%TPR
HP Deterioration	6.5	89.0	0.0	85.0	0.0	100.0	0.0	99.0
BB Leakage	5.0	82.0	15.0	65.0	0.0	68.0	3.0	94.0
Combustion Liner Failure	4.0	91.0	12.0	97.0	0.0	100.0	1.0	99.0
EGT Sensor Failure	6.0	79.0	10.0	75.0	15.0	74.0	15.0	100.0

SUMMARY AND CONCLUSIONS:

A new, kernel-based nonlinear maximum representation and discrimination features (KMRDF) algorithm for FDD of turbo aircraft engines was proposed and experimented with using snapshot data from a fleet of propulsion engines. We attempted to detect and classify both progressive and abrupt types of faults such as HP deterioration, combustor liner failure, and abrupt failure in EGT sensors. A test of our technique on real data showed reasonably good results. We compared the results obtained with our KMRDF algorithm with other well-known feature extraction procedures such as PCA, LDA and MRDF.

The KMRDF method improves upon the existing MRDF classifier by introducing a different between-class closeness measure that considers the *local neighborhood* of points rather than all pairs of points between two classes. Our method affords a more systematic way to introduce nonlinear features, as opposed to the nonlinear version of MRDF proposed by Talukder [4].

In this work, a polynomial kernel with a degree of two was used. It remains to be seen what kinds of features (Gaussian, tan hyperbolic or other kernels) emerge and the consequent performance of the classifiers.

REFERENCES

[1] *Fault Diagnosis in Dynamic Systems: Theory and Applications*, R. Patton, P. Frank, and R. Clark, eds., Prentice Hall, Englewood Cliffs, N.J., 1989.

[2] Hastie T. and W. Stuetzle, "Principal curves," *JASA*, vol. 84, pp. 502 - 516, 1989.

[3] Baudat G., Anouar F. "Generalized Discriminant Analysis Using a Kernel Approach" *Neural Computation*12, 2385-2404, 2000.

[4] Talukder A., "Nonlinear Feature Extraction for Pattern Recognition Applications" PhD thesis-CMU, 1999.

[5] *Principal Component Neural Networks: Theory and Applications.* Diamantaras and Kung, John Wiley & Sons, Inc. New York, NY, USA, 1996.

[6] M.A. Kramer. "Nonlinear principal component analysis using autoassociative neural networks". *AIChE Journal*, 37(2):233–243, 1991.

[7] Bernhard Scholkopf, Alexander Smola, Klaus-Robert Muller "Nonlinear component analysis as a kernel eigenvalue problem," *Neural Computation* 10, 1299–1319, 1998.

[8] Bernhard Scholkopf, Alexander J. Smola. *Learning with Kernels: Support Vector Machines, Regularization, Optimization, and Beyond* The MIT Press; 1st Edition, 2001

[9] *Pattern Classification*, R. O. Duda, P.E. Hart and D.G. Stork, John Wiley and Sons, 2nd Edition, 2000.

FEATURE DIMENSIONALITY REDUCTION FOR PARTIAL DISCHARGE DIAGNOSIS OF AIRCRAFT WIRING

Weizhong Yan, Kai F. Goebel

GE Global Research Center
Niskayuna, NY 12309
{yan, goebelk} @ crd.ge.com

Abstract: Partial discharge (PD) analysis has been successfully used as a means to evaluate the integrity of insulation systems, especially high voltage electrical systems such as generators, transformers, and capacitors. In recent years, PD has also gained increasing attention for its application in low voltage insulation systems such as aircraft wiring systems. Besides detecting and acquiring discharge signals/pulses with sufficiently high measuring sensitivities, one of the challenges of using PD for insulation fault diagnosis is to accurately interpret the acquired signals. That is, to associate the PD signals with real physical phenomena and specifically the states of the insulation that is being monitored. Like in any diagnostic/classification system, the key to an accurate and reliable PD diagnosis is a set of high quality features/attributes that represent/capture the characteristics of PD signals. More importantly, these features must posses strong discriminant power so that the classifier designed based on those features gives desired performance. This paper is concerned with the application of linear feature transformation techniques for reducing the number of features that are extracted by collectively using different feature extraction methods. More specifically, this paper investigates the effectiveness of the two well-studied linear feature transformation methods, namely, principal component analysis (PCA) and linear discriminant analysis (LDA), in improving the classification performance of PD diagnostic systems. We apply these techniques to experimental partial discharge data from an ongoing study of aircraft wiring diagnostics. We also compare the results of this study with those from using other feature selection techniques.

Keywords: Dimensionality reduction; Feature selection; Partial discharge; PD diagnosis; Classifier; Principal component analysis; Linear discriminant analysis

1. Introduction: Partial discharges (PD) are a local, partial breakdown event that occurs for example, on the surface or inside insulation of electrical products due to possibly minute defects in insulation structure. These minute defects may be the result of the manufacturing process and/or the result of ageing and mechanical damage of the products. While normal (healthy) condition of insulation gives a baseline level of partial discharge activities, increase of partial discharge activities indicates insulation degradation or faults. Diagnosing insulation defects/faults based on partial discharge activities, generally known as "PD diagnosis" [Gulski (1995)], has played a critical role in condition-based maintenance of insulation systems, especially high voltage insulation systems, such as generators, transformers, and capacitors.

PD diagnosis, a typical classification problem, is to classify measured PD activities into the underlying insulation defects or sources that generate PDs. PD diagnosis is a complex classification problem because 1) PD is inherently a stochastic process and the occurrence of PD very much depends on many factors, such as temperature, pressure, applied voltage, and the test duration [Gulski (1995)], and 2) PD signals contain noise and interference signals. Consequently, like in any diagnostic/classification systems, the key to an accurate and reliable PD diagnostic system is to identify a set of high quality features/attributes. These features should represent/capture the characteristics of PD signals. More importantly, these features must posses strong discriminant power so that the classifier designed based on those features has the desired performance.

Motivated by finding salient features for PD diagnosis, researchers have introduced several different methods for extracting features out of PD pulses, including the widely used statistical analysis of phase-resolved PD patterns and more modern methods, such as, fractal analysis [Gulski (1996)] and textual analysis [Rahman et al (2000)], among others. However, so far no single feature extraction method has been proved to be effective for *all* problems. In fact, effectiveness of features from those individual feature extraction methods on classification is highly problem-dependent. Therefore, in designing PD diagnostic systems, the designer still faces a great challenge in deciding which feature extraction method(s) is most appropriate for the given problem at hand and, ultimately, in how to find a set of features that is optimal for the problem concerned in terms of classification performance. To tackle this challenge, in our previous paper [Yan & Goebel (2005)], we proposed a novel approach that consists of two steps. First collectively utilize a plurality of feature extraction methods, without discerning how good they are for the problem, to generate a feature pool; then apply feature selection to the feature pool to choose the optimal subset of features. The two-step approach not only gives an optimal set of features, but also alleviates the designer from the difficulties of identifying appropriate feature extraction methods. In this paper, we follow the two-step approach; however, we investigate effectiveness of using linear feature transformation in step 2 to find a small number of features that are most discriminative. Linear feature transformation has been widely used in various fields for feature extraction and feature dimensionality reduction. However, linear feature transformation, especially, linear discriminative analysis (LDA), has not been actively studied by the researchers in the field of PD diagnosis. This paper demonstrates through a design example that the linear feature transformation can be effective in improving performance of PD diagnostic systems.

This paper is organized as follows. Section 2 gives an introduction to linear feature transformation methods (principal component analysis and linear discriminant analysis). Section 3 describes the exemplar case that we used for the demonstration and the feature extraction methods used for extracting features from PD signals. Classification results of the example case study using PCA and LDA are presented and compared in Section 4. Section 5 concludes the paper.

2. Feature dimensionality reduction: Feature dimensionality reduction provides several important benefits in classifier design, including: 1) reducing the computational complexity of classifiers; 2) providing a remedy to "curse of dimensionality"; and 3) improving the generalization of classifiers. As a result, feature dimensionality reduction is almost always a necessary step in the design of classification systems. While there are several different methods for feature dimensionality reduction [Fodor (2002)], linear feature transformation is regarded as the most attractive one since it is "simple to compute and analytically tractable" [Duda et al (2000)]. Linear feature transformation reduces feature dimensionality by representing the original features with a smaller number of new features that are the linear combinations of the original ones. Graphically it projects the data from a high dimensional feature space to a low dimensional one.

Mathematically, linear transformation has the general form of $\mathbf{Y} = \mathbf{W}^{\mathrm{T}}\mathbf{X}$, where $\mathbf{X} = [\mathbf{x}_1, \mathbf{x}_2, \ldots \mathbf{x}_n] \in \Re^p$ is a data matrix formed by n p-dimensional vectors, $\mathbf{Y} = [\mathbf{y}_1, \mathbf{y}_2, \ldots \mathbf{y}_n] \in \Re^d$ is the transformed data matrix containing n d-dimensional vectors in the new feature space, and $\mathbf{W} \in \Re^{p \times d}$ is the transformation matrix. For feature dimensionality reduction, $d \leq p$ and preferably $d \ll p$ if the original feature dimension (p) is high. The transformation matrix, $\mathbf{W}$, is obtained by optimizing a specific objective/criterion function. It is the objective/criterion function that distinguishes one transformation method from another. In this paper, we are specifically interested in the two most popular linear transformation methods, namely, principal component analysis (PCA) and linear discriminant analysis (LDA). From the standpoint of the mathematical form, both PCA and LDA are the same. The difference of the two methods comes from the criterion functions used for determining the transformation matrix, $\mathbf{W}$. A brief explanation of the two methods is given as follows.

PCA is a classical statistical technique and has been widely used in a range of applications. For feature dimensionality reduction, PCA attempts to find a smaller set of variables that best describe/represent the given data in terms of representation error. Following the above-defined notations, the criterion function for obtaining the transformation matrix, $\mathbf{W}$, in PCA is $J_{PCA}(\mathbf{W}) = E(\|\mathbf{WY} - \mathbf{X}\|^2)$, where E is the expectation operator. It has proved that the transformation matrix, $\mathbf{W}$, that minimizes $J_{PCA}(\mathbf{W})$ is formed by the eigenvectors corresponding to the d largest eigenvalues of the scatter matrix, $\mathbf{S} = E[(\mathbf{x}_i - \mathbf{m})(\mathbf{x}_i - \mathbf{m})^{\mathrm{T}}]$, where $\mathbf{m} = \frac{1}{n}\sum_{i=1}^{n}\mathbf{x}_i$.

LDA projects the variables to a feature space in such directions that provide optimal separation among classes. Class separability is typically quantified as a function of the between-class scatter, $\mathbf{S}_B$, and the within-class scatter, $\mathbf{S}_w$, as described by Duda et al.

(2000). LDA attempts to determine the transformation matrix, $\mathbf{W}$, by maximizing the Fisher's criterion function defined as $J_{LDA}(\mathbf{W}) = \frac{|\mathbf{W}^T \mathbf{S}_B \mathbf{W}|}{|\mathbf{W}^T \mathbf{S}_W \mathbf{W}|}$ [Duda et al. (2000)].

It has been proved that finding $\mathbf{W}$ that maximizes $J_{LDA}(\mathbf{W})$ is equivalent to solving a generalized eigenvalue problem $\mathbf{S}_B \mathbf{W} = \lambda \mathbf{S}_W \mathbf{W}$, which yields a maximum of $c-1$ (c is the number of classes) eigenvectors if $\mathbf{S}_W$ is nonsingular. Those eigenvectors corresponding to the d largest eigenvalues form the columns of the transformation matrix, $\mathbf{W}$.

From the above-defined objective functions, we can see that LDA finds new features by maximizing discriminating power between classes, while PCA attempts to best represent the entirety of the data without paying any attention to the underlying class structure. As the result, it is intuitive and generally accepted that for classification problems, LDA should perform better that PCA does. Recently, however, some empirical studies, e.g., Martinez and Kak (2001), have shown that when the training data set is small, PCA can outperform LDA. Since LDA requires class information for all data examples, it is a supervised method. On the other hand, PCA is an unsupervised method.

3. A case study: The case studied in this paper is the design of a PD diagnostic system for aircraft wiring fault diagnosis. It is part of a multiple-generation initiative project funded by ONR. Presented here as an example case study is the initial design of an on-ground, de-energized prototype system.

Experiment test for generating PD pulses: PD measurements for aircraft wiring are generated and recorded through laboratory tests. Damage to wiring insulation generally takes two forms: material degradation due to aging or thermal/electrical environment, and chafing that may occur during maintenance and mechanical abrasion during operation-induced vibration. The lab tests conducted in this paper focus on the latter, i.e., wire chafing. For classifier design purpose, two wiring conditions are tested. One is for normal condition wires and another for wires with artificial chafing. Two different ways are considered in producing artificial defects. The first type represents defects occurring in twisted pair wires. For that a small piece of the upper insulation layer was removed from both wires in the twisted pair. This type of defect is the most reproducible one since the thickness of the remaining insulator layer was not changed. Samples for this type were prepared from two aircraft grade wires, types M22759/90-22-95 and M22759/81-22-52. The second type of defect simulated a chafed wire touching the shielding of the cable bundle or a metal part of the aircraft. The wire was chafed in a short length and a piece of tinned copper wire was twisted around the chafed area. The tinned copper wire was also pressed into the chafed section in order to touch the remained insulation layer. Such samples were prepared from three aircraft grade wires, types M22759/90-22-95, M22759/81-22-52 and M81044/6-22-9.

A total of nine wire samples are tested with maximum of 10 repetitions for each sample. After data cleansing (including the removal of noise and incomplete data sets), 596 PD sequences are used for designing the PD diagnostic system. Out of the 596 PD sequences, 225 are for normal wires and 371 for chafed ones.

Feature extraction: We use the following five different methods for extracting features from the PD measurements/signals.

A) Features from statistical analysis of phase-resolved PD patterns. Phase-resolved PD pattern analysis is the most commonly used method for feature extraction in PD diagnosis. Given a sequence of PD pulses and the recorded voltage phase angles at corresponding pulse peaks, a 3D PD pattern is generated, where the number of pulses (pulse count) is plotted as a function of magnitude and phase of the PD pulses. A typical 3D PD pattern is shown in Figure 1. 3D PD patterns are a good representation/summary of all PD pulses recorded within a specified time window and should show different characteristics for different PD activities, thus different PD sources. For the convenience of statistical analysis, the 3D patterns are decomposed into two 2D distributions by projecting it into the two axes - phase and magnitude. Statistical analysis is performed separately for those two distributions. Also, statistical analysis is performed separately for phase angles from 0° to 180° ("positive" PDs), for phase angles from 180° to 360° ("negative" PDs), and on the difference between positive and negative PDs. For each of the distributions, two types of statistics, namely amplitude statistics and shape statistics, are calculated. The statistical descriptors are mean, standard deviation, skewness and kurtosis. In addition, overall maximum magnitudes of positive and negative PDs and correlation between positive and negative PD patterns are also calculated as features.

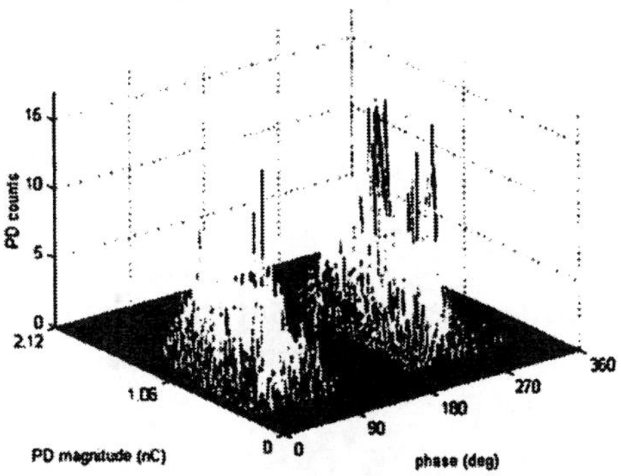

Figure 1: An example of 3D PD patterns

B) Features from PD height distribution analysis. Heights (peaks) of a sequence of PD pulses can be represented in a histogram that shows number of pulses as a function of their magnitude. According to Cacciari et al. (2002), PD pulse height distribution tends to fit well with the two-parameter Weibull function defined as: $F(q)=1-\exp(-(q/\alpha)^{\beta})$, where q is the pulse height, α and β are the scale and shape parameters of the Weibull function. Cacciari et al. determined that the scale and shape (especially shape) parameters differ with different PD sources, thus can be used as features for PD identification or classification.

C) Features from "classification map". PD pulses are different in wave shape depending on the location and nature of the underlying defect that generates PD. One way to capture the different wave shapes is to use so-called "equivalent time-length", T^2, and "equivalent bandwidth", W^2 [Contin et al (2000)].

In the $T^2 - W^2$ plane (also called the "classification map" by [Contin et al (2000)]), each PD pulse is presented as a point and each sequence of PD pulses, which are similar in shape, fall into a well-defined area (cluster) in the $T^2 - W^2$ plane. The location and shape of the clusters in the $T^2 - W^2$ plane differ corresponding to different PD sources [Contin et al (2000 & 2002)]. In this paper, characteristics of the clusters are extracted through statistical analysis and used as features for classification purpose. The features extracted include overall mean, means and standard deviations in both T^2 and W^2 directions, respectively, 1st through 4th orders of moments of distributions in both T^2 and W^2 directions, respectively, direction of the 1st eigenvector of the cluster, and ratio of the first two eigenvalues of the cluster.

D) Features from spectrum analysis. Frequency spectrum of a PD pulse indicates frequency components of the PD pulse. Thus the shape or distribution of frequency spectra should be correlated with different PD sources. In this paper, the first 3 frequencies corresponding to the highest three magnitudes, the three highest magnitudes themselves, the difference between the three frequencies, and the difference between the three magnitudes are used as features.

E) Features from raw PD signals. They are the maximum and minimum peaks of PD pulses, mean and standard deviation of peaks of PD pulses, inception voltage, and PD rate.

The five feature extraction methods yield 86 features in total. Including the temperature and humidity measurements, we have 88 features for each of the 596 data samples, which we can use for feature transformation study.

4. Classification results: PD diagnosis is a classification problem where the extracted features from PD measurements are the inputs and the sources of PD or condition status of the wire monitored are the class targets. There are a great number of classifiers available, ranging from traditional statistical methods to more modern methods (such as neural network classifiers and support vector machines (SVMs)). To study the effectiveness of feature transformation on improving classification performance of PD diagnostic systems, in this paper, we use support vector machines as the classifier for diagnosing aircraft wiring faults. SVMs are a recently developed learning system originated from the statistical learning theory [Vapnik (1995)]. One distinction between SVMs and many other learning systems is that its decision surface is an optimal hyperplane in a high dimensional feature space. The optimal hyperplane is mathematically found by solving a properly formed convex quadratic problem with optimization theory, which is well studied in the field of mathematical programming and can be solved in a relatively straightforward way. For the classification problem concerned in this paper, two of these unique properties, namely, better handling of sparse data and good average performance over a wide spectrum of different classification problems, make SVM classifiers a better choice here.

To evaluate classification performance, the data set is randomly split into two disjoint subsets: one for training and another for evaluation. Also, to improve the robustness of evaluation, the SVM classifier is trained and evaluated 10 times and each time the data is randomly split into two disjoint subsets. The mean, the standard deviation, the minimum, and the maximum of accuracies of the 10 evaluations are reported here. Additionally, 1-sigma random noise is added to the original feature values in order to account for the facts that only limited PD examples are obtained in well-defined lab environment and that PD measurements in real aircraft wires will be much more noisy.

The PCA results: First, we apply PCA to the 88-feature data set described in the previous section. Figure 2 shows the variance explained by each of the principal components (PCs). For illustration convenience, only the first 30 PCs with largest variance explained are shown in the figure. The vertical bars show the percentage of the total variance explained by each individual principal component, while the line with dots shows the cumulative percentage of the total variance explained by all principal components up to the specific one.

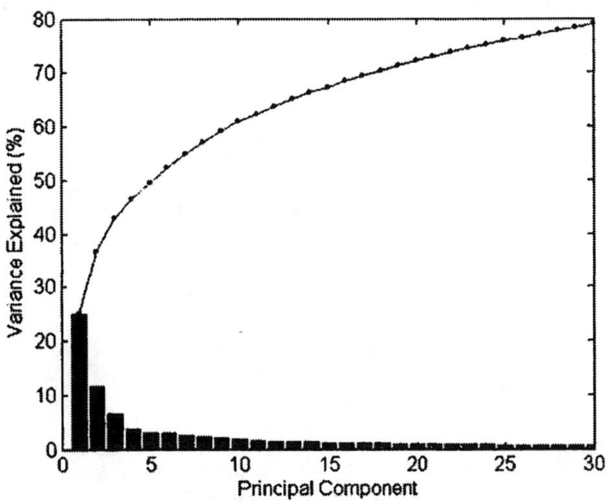

Figure 2: Variance explained by the first 30 principal components

As we can see from the figure, the first 10 PCs explain approximately 60% of the total variance of the data whereas the first 30 PCs explain about 80% of the total variance. Classification accuracies for the designs using the first 3, 10, and 30 PCs as classifier inputs are shown in Table I, respectively. For comparison, also included in Table I are the classification accuracies for the design with all 88 features extracted from the PD signals as its inputs. From Table I, we can see that using only the first 3 PCs as the classifier inputs already provides a reasonable performance – better than that from using all 88 features. Using the first 10 PCs leads to a classification performance improvement. However, further increasing the number of PCs from 10 to 30 does not gain anything in classification performance.

Table I: Classification accuracies summary

Methods		Classification accuracy			
		Mean	Std. Dev.	Min.	Max.
All 88 features		0.9566	0.0206	0.9224	0.9811
PCA	3	0.9644	0.0055	0.9560	0.9706
	10	0.9839	0.0044	0.9748	0.9874
	30	0.9813	0.0036	0.9748	0.9874
LDA		0.9906	0.0048	0.9832	0.9958

The LDA results: The PD diagnostic system concerned in this paper is a 2-class classification problem, with the 2 classes being either normal or chafed condition of aircraft wires. As discussed in Section 2, the maximal number of new features in LDA space is equal to the number of classes minus one. Thus applying LDA to the 88 features extracted in Section 3 results in a single feature in the LDA space. The distributions of the single LDA feature over the 2 wiring conditions (normal and chafed) are show in Figure 3. It clearly shows a good separation between the 2 conditions/classes, which means that the LDA feature carries high discriminant power, thus expects to give a good classification performance.

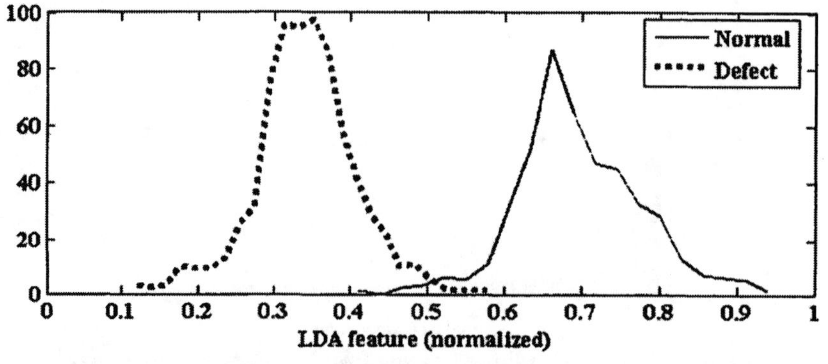

Figure 3: Feature value distribution of the LDA feature

The classification accuracies for the design using the LDA feature are also shown in Table I (last row). Using the single LDA feature achieves the highest classification accuracies.

From Table I, we can see that both PCA and LDA not only dramatically reduce the number of features needed by the classifier, but also improve the classification performance for the PD diagnosis problem. Between the two linear transformation methods (PCA and LDA), LDA seems to be more effective not only in reducing feature dimensionality, but also in improving classification performance. This observation, i.e., LDA is better that PCA in classification, is in accordance with the expectation based on different objective functions used. The observation from Table I that both PCA and LDA use a few features while showing better performance than the design using all 88 features indicates that there exist a great number of redundant features among the original 88 features. This is expected since the 5 different feature extraction methods used do not necessarily produce completely independent features.

5. Conclusions: In designing PD diagnostic systems or any other diagnostic systems, finding an optimal feature set with a small number of features is important. This paper is concerned with using linear feature transformation to reduce feature dimensionality. More specifically, we investigate the effectiveness of linear feature transformation on improving the classification performance of PD diagnostic systems. By using an example of designing a PD diagnostic system for aircraft wiring fault diagnosis, this paper demonstrates that both PCA and LDA can be effective not only in reducing feature dimensionality, but also in improving the classification performance, in the design of PD diagnostic systems for aircraft wiring. Therefore, linear feature transformation (PCA and LDA), especially when working together with our previously proposed 2-step feature extraction approach, can be potentially useful in design of PD diagnostic systems. This paper also demonstrates that LDA outperforms PCA for PD diagnosis because the underlying objective for LDA is specifically to maximize class separability.

Acknowledgment: The authors gratefully acknowledge the guidance and support of Nicole Evers, Janos Sarkozi and Karim Younsi at GE Global Research Center and of Sean Field, Charles Fulbright, and Greg Nepini at NavAir. This work was funded by ONR under contract #: N00014-02-C-0402.

References:

1. Gulski, E. (1995), "Digital analysis of partial discharges", IEEE Transactions on Dielectrics and Electrical Insulation, Vol. 2, No. 5, pp822-37.
2. Gulski, E., Burger, H.P., Zielonka, A., and Brooks, R. (1996), "Classification of defects in HV components by fractal analysis of PD measurements", IEEE Annual Report Conference on Electrical Insulation and Dielectric Phenomena, San Francisco, October 20-23, 1996
3. Abdul Rahman, M.K., Arora, R., and Srivastava, S.C. (2000), "Partial discharge classification using principal component transformation", IEE Proceedings of Scientific Measurement Technologies, Vol. 147, No. 1, pp7-13

4. Yan, W. & Goebel, K. (2005), "Feature selection for partial discharge diagnosis", Submitted to SPIE Smart Structures 2005, March 6 –10, San Diego, CA
5. Fodor, I.K. (2002), "A survey of dimension reduction techniques", Lawrence Livermore National Laboratory (LLNL) technical report, June 2002, UCRL-ID-148494
6. Duda, R. O., Hart, P. E., and Stork, D. G. (2000), *Pattern Classification*, John Wiley & Sons, Inc., New York, NY
7. Martinez, A.M. and Kak, A.C. (2001), "PCA versus LDA", IEEE Transactions on Pattern Analysis and Machine Intelligence, Vol. 23, No. 2, pp228-233.
8. Contin, A., Cavallini, A., Montanari, G.C., Pasini, G., and Puletti, F. (2000), "Artificial intelligence methodology for separation and classification of partial discharge signals", Proceedings of IEEE Conference on Electrical Insulation and Dielectric Phenomena", Victoria, Canada, pp522-26.
9. Contin, A., Cavallini, A., Montanari, G.C., Pasini, G., and Puletti, F. (2002), "Digital detection and fuzzy classification of partial discharge signals", IEEE Transactions on Dielectrics and Electrical Insulation, Vol. 9, No. 3, pp335-48
10. Vapnik, V. (1995), *The nature of statistical learning theory*, Springer-Verlag, New York, 1995

TRACKING BEARING SPALL SEVERITY THROUGH INVERSE MODELING

C. James Li
Department of Mechanical, Aerospace, and Nuclear Engineering
Rensselaer Polytechnic Institute
110 8th St.
Troy, NY 12180
lic3@rpi.edu

Hyeongceol Shin
Department of Mechanical, Aerospace, and Nuclear Engineering
Rensselaer Polytechnic Institute
110 8th St.
Troy, NY 12180
shinhc@rpi.edu

Abstract: The ability to track the severity of a spall in a bearing in operation is desirable for bearing condition monitoring and remaining life estimation. Existing technology estimates spall severity from bearing vibration r.m.s which can be affected by many factors such as operating conditions, and other nearby machine components, and therefore is of little practical use. The proposed method is a model based one and therefore is much more robust. When a spall is present, the impact between rollers and the spall produces a train of impulses which then excite the bearing and its nearby structure. The proposed method identifies a bearing dynamic model from a measured bearing vibration using a system identification method. This bearing dynamic model is then properly inverted to create a stable high fidelity inverse model that recovers the impact impulse behind a measured bearing vibration. Given the vibration of the bearing measured at a different time and therefore a possible different spall size, the inverse model estimates the impulse that excites the bearing vibration. A modeling error compensation scheme then refines the recovered impulse. Subsequently, the spall severity is estimated from the magnitude of impulse by removing the effect of bearing speed. This study also conducted experimental validations to ascertain the effectiveness of the proposed scheme in assessing spall severity.

Key Words: bearings; diagnosis; dynamic model; inverse model; spall;

Introduction: Rolling element bearings are found in most rotating machinery such as aircraft engines, automobiles and electrical motors, and their failures frequently have great consequence. For example, 40% of the failed differential bearings in aircraft engine result in the loss of equipment. Condition monitoring of bearings, therefore, has gained attention in the last two decades as demand for reliability and safety increases. To date, the majority of bearing diagnostic algorithms [1] is designed for the detection of on-set of localized defects such as a spall due to fatigue resulted from repeated contact between rollers and raceways. However, to determine damage severity, predict remaining useful life and spall propagation trajectory, one must be able to track spall severity from measureables such as bearing vibrations.

There is few methods, vibration-based or not, available for tracking the severity of a spall in a rolling element bearing in operation. Reference [2] described a method to identify bearing resonance frequencies excited by roller-defect impacts and track their energy level for severity assessment. However, it did not consider the effect of varying bearing speeds and had limited experimental data. Reference [3] relates bearing vibration r.m.s. to spall size. The method's deficiency lies in that r.m.s. is affected by many factors that have nothing to do with the bearing. Reference [4] reported that bearing stiffness reduces as the size of bearing spall grows and the stiffness reduction decreases rotor-bearing natural frequency and increases vibration amplitude at the frequency. It then finds the relationship between stiffness reduction and remaining life using damage mechanics. It does not estimate the spall severity per se. Additionally, rotor stiffness is a function of many factors such as speed, temperature, load, etc. Therefore, the method will probably not generalize well beyond a bearing testing rig which maintains constant operating condition. Reference [5] describes the utility of a high gain displacement sensor that looks at the outer-race deflection directly. It reported that the spike to deflection amplitude ratio is a good index of spall severity and is insensitive to the operating condition. However, drilling an access hole in the bearing housing to look at the raceway deflection directly is just not a practical solution except for some limited applications.

The objective of this study is to find a vibration-based method that can track the severity of a bearing spall regardless of the operating condition and noise from nearby machine components.

Proposed Method: The proposed method is summarized in Figure 1. When a spall is present, the impact between rollers and the surface defect produces a train of impulses, which then excites the bearing and its nearby structure. Therefore, the bearing signature consists of periodic ringings of resonances as shown in the upper right corner of Fig. 1. An impact model can be used to estimate the impact impulse from spall geometry, and bearing geometry, speed and load. A bearing dynamic model is subsequently identified from the impulse and measured bearing vibration by a system identification method. This bearing dynamic model is then properly inverted to create a stable high fidelity inverse model that estimates impact impulse from a measured bearing vibration. Given the vibration of the bearing measured at a different time and therefore a different spall size, the inverse model estimates the impulse that excites the bearing vibration. The spall severity is then estimated from the impulse magnitude.

The Impact Model: The emphasis of this study is about recovering roller-defect impact impulses from bearing vibration, and deriving a spall damage index from recovered impulses. The issue of selecting an impact model and establishing the utility of it is not part of this study. In the mean time, a simple impact model is examined to understand which factors may affect the impact impulses. Reference [3] describes an impact model suggesting that, for a given bearing, the impulse magnitude is proportional to the bearing rotation speed and the spall size.

$$F\Delta t = \frac{m\pi d_m n(1-\gamma)^2}{60D}x \tag{1}$$

where F is the mean impact force, Δt the impact duration (sec.), D the roller diameter (mm), m mass of the roller (kg), d_m the pitch diameter of the bearing (mm), n shaft rotating speed (rpm), γ the ratio between D and d_m, and x the spall size.

Identification of Bearing Dynamic Model: A rolling element bearing is a complicated system in terms of its dynamic interactions between its parts, i.e., rolling elements, inner race, outer race and cage, in the presence of lubricant. A comprehensive bearing dynamic model is computationally expensive. Such a model is frequently used by bearing manufacturers to understand a bearing's dynamic behavior under various operating conditions. Examples of such a model include BEAST developed by SKF [6], ADORE by Dr. Gupta [7], and BRAIN by NSK [8].

They typically demand much computing power. For example, BEAST runs on a super computer and ADORE runs on workstations. Furthermore, these model's contact force and moment calculations are mainly based on elastohydrodynamnic theory and therefore are not appropriate for describing what happens in the presence of a spall which produces impacts.

An impact between a rolling element and a spall is a wide-band excitation that can excite resonances in the bearing and even its nearby structure. This presents an opportunity to identify the bearing dynamics. Assuming one knows the size of the spall, an impact model can be employed to estimate the impact induced impulse from it, bearing geometry and speed. Having the impulse and bearing vibration excited by the impulse, one can employ a broad array of system identification tools to identify an input-output model of bearing dynamics.

This study, being the first taking this approach, will keep the dynamic model simple to control the computational cost and maintain the model invertability. Specifically, the Auto-Regressive with eXogenous (ARX) model was employed.

An ARX model has the form of:

$$\frac{Y(z)}{U(z)} = \frac{z^{-d}B(z^{-1})}{A(z^{-1})} \tag{2}$$

where Y(z) is the output transform, U(z) input transform, $A(z^{-1}) = 1 + a_1 z^{-1} + ... + a_n z^{-n}$ $B(z^{-1}) = b_0 + b_1 z^{-1} + ... + b_m z^{-m}$, z^{-d} a d-step delay, and z z-transform operator.

MATLAB system identification toolbox was used to select and identify such models. More sophisticated models can be employed in the future as appropriate.

Inverting the Bearing Dynamic Model: Inverting a model like (2) is not as simple as flipping the fraction on the right hand side because the model may contain non-minimum phase zeros which become unstable poles after inversion and therefore an unstable inverse model. Let's rewrite the model as follows where $B^a(z^{-1})$ includes invertible zeros $B^u(z^{-1})$ includes non-invertible zeros.

$$G(z)=\frac{Y(z)}{U(z)}=\frac{z^{-d}B^{a}(z^{-1})B^{u}(z^{-1})}{A(z^{-1})} \tag{3}$$

where, e.g.,

$$B^{u}(z^{-1})=b_0^{u}+b_1^{u}(z^{-1})+\dots \tag{4}$$

An inverse model [9] is

$$\hat{G}^{-1}(z)=\frac{B^{u}(z)}{B^{u}(1)}\frac{z^{d}A(z^{-1})}{B^{a}(z^{-1})B^{u}(1)} \tag{5}$$

where,

$$B^{u}(z)=b_0^{u}+b_1^{u}(z)+\dots \tag{6}$$

Multiplying G and $\hat{G}^{-1}$, one has

$$\frac{B^{u}(z)}{B^{u}(1)}\frac{B^{u}(z^{-1})}{B^{u}(1)} \tag{7}$$

whose frequency response function, which can be obtained by letting $z=e^{jw}$ is real. This means that the phase of the original impulse will be preserved by the proposed scheme.

Error Analysis and Compensation: In Fig. 1, let's call the z-transform of the impact impulse $U_a(z)$. Then the bearing fault signature $C(z)=U_a(z)G(z)$. For the moment, let's assume that the impact impulse estimated by the impact model is virtually identical to the true impact impulse. The system identification yields a bearing model $\hat{G}(z)$ and a residual R(z). Therefore,

$$\hat{G}(z)=\frac{C(z)-R(z)}{U_a(z)} \tag{8}$$

Then, assuming a direct inverse, the inverse model is

$$\hat{G}^{-1}(z)=\frac{U_a(z)}{C(z)-R(z)} \tag{9}$$

When a different bearing vibration C'(z) due to an unknown impact impulse $U_a'(z)$ is applied to the inverse model, the inverse model output is

$$
\begin{aligned}
&C'(z)\hat{G}^{-1}(z) \\
&= (U_a'(z)G(z))\hat{G}^{-1}(z) \\
&= (U_a'(z)G(z))\frac{U_a(z)}{C(z)-R(z)} \\
&= U_a'(z)\frac{U_a(z)G(z)}{C(z)-R(z)} \\
&= U_a'(z)\frac{C(z)}{C(z)-R(z)}
\end{aligned} \tag{10}
$$

which is not the desired $U_a'(z)$. Rather the output is $U_a'(z)$ multiplied by C(z)/[C(z)-R(z)] which came from nonzero modeling error or the residual. To compensate for this, a non-parametric transfer function is constructed to approximate [C(z)-R(z)]/C(z) and then multiplied to the output of the inverse filter. This is possible because both the [C(z)-R(z)] and C(z) are readily available.

Experimental Setup and Experiments: The bearing test rig is shown in Fig. 2. It consists of a spindle driven by a 40 HP variable speed AC motor, and a bearing housing riding on top of an x-y stage. The upper limit of the spindle speed is about 2,000 rpm and can be increased by changing the pulleys. The motor controller can take speed command from a computer through its digital I/O ports. The bearing cone, which includes the inner-race, rollers and cage, is at one end of the spindle while the cup of the bearing is inside the bearing housing. Torquing the handle of the x-y stage produces static radial and axial loads up to 1,000 lbf, respectively. Currently, the setup is instrumented with two accelerometers measuring axial and radial bearing vibrations and a tachometer that generates once-per-revolution pulses. The setup is controlled and monitored by an NT Workstation for unmanned data acquisition.

Bearing vibrations were taken from two bearings with different spall sizes under 5 different speeds because spall size and bearing speed affect the magnitude of impact impulse according to equation 1. The size of the smaller spall is roughly 85% of the larger one widthwise (Due to the irregular spall shape, it is difficult to have a precise number.) The speeds employed were 630, 750, 870, 930, and 990 rpm. Altogether, ten bearing vibrations were collected from 2 bearings at the 5 speeds. Visual inspection of measured bearing vibrations revealed that the ringing pattern is less obvious at the lowest speed due to weaker impact between rollers and the defect. This is expected to have a negative impact on the effectiveness of the proposed scheme.

Results: The impulse is assumed to have a shape of half sine and duration of 4.1e-4 sec (This is the time it takes the roller to travel 1 degree at 1,000rpm). In the absence of a proven impact model, one cannot predict a magnitude i.e., the area under the curve, of the impact impulse from a given spall size, bearing speed and geometry. However, the magnitude is not critical because linear models are used and everything is therefore scalable. One just has to be contented with a relative spall severity assessment instead of an absolute one such as the physical size until a good impact model becomes available. To obtain a bearing dynamics model and then inverse model, an impulse of somewhat arbitrary magnitude (2.5894e-004 Nt-sec.) was used along with the bearing vibration measured at 990rpm from the bearing with the large spall to obtain an ARX model and subsequently the inverse model. Fig. 3 shows the impulse and the bearing vibration along with the impulse recovered by the inverse model. The impulse recovered by the inverse model is very similar to the impulse used to obtain the model and this suggests that the inverse model is of good quality. (Note that the impulse is magnified by 5x to make it more visible.) To test the generality of the inverse model, the ten bearing vibrations were then presented to the inverse model to recover the impulses that excite the bearing vibrations. Fig. 4 shows an example of such a bearing vibration and the impulse recovered by the inverse model.

The 2nd and 4th rows of Table 1 give the impulse magnitudes recovered from all the ten vibrations by the inverse model. These magnitudes are then divided by the bearing speed because Equation 1 suggests that the magnitude of impulse is proportional to speed. Then, they are normalized by the impulse/speed ratio that was used to obtain the ARX model. The normalized values are shown in the 3rd and 5th rows of the table. To be a good spall severity index, the normalized values in the 3rd row and 5th row should equal to 100% and 85%, respectively, regardless of the speeds. It is virtually the case except at the lowest speed because of the aforementioned poor signal to noise ratio. The mean of the 3rd row, i.e., normalized severity index values for the bearing with the large spall, is 101.3% with a standard deviation of 4.9% when the lowest speed data is excluded. The same mean of the 5th row is 88.2% with a standard deviation of 2.2%. Both means agree well with their respective theoretical values, i.e., 100% and 85% and the small standard deviations suggests that the effect of bearing speed on the severity index is not significant.

Conclusions: A model based bearing spall severity tracking method was developed. The method entails a system identification method to identify bearing dynamics between spall induced impact impulse and bearing vibration, and a model inversion method to obtain an inverse model that recovers the impact impulse from a bearing vibration. An error compensation method was also developed to compensate for the modeling error and improve the quality of the recovered impulses. Experimental validation showed that the relative severity index derived from the recovered impulse magnitude offers an accurate and consistent assessment about the spall severity regardless of bearing speeds with the exception of the lowest speed. The inverse model is general enough to be used on a bearing other than the one the model was obtained for (The two bearings used in the experiment have the same model number though.) When an impact model becomes available, the proposed method can provide an estimate of the physical size of the spall.

REFERENCES

[1] C. James Li and K. McKee, "Diagnostics: Bearing (roller & Journal) vibrations and diagnostics," Encyclopedia of Vibration, Editors: S. G. Braun, D. J. Ewins and S. S. Rao, Academic Press, Aug. 2001.
[2] C. James Li and S. M. Wu, "On-line Bearing Damage Severity Assessment via Defect Sensitive Resonance Identification and Matched Filtering," Mechanical Systems and Signal Processing, Vol. 2, No. 3, July 1988, p 291-303.
[3] Zhang, C., Kurfess, T.R., Danyluk, S., Liang, S. Y., "Dynamic Modeling of Vibration Signals for Bearing Condition Monitoring," The 2nd International Workshop on Structural Health Monitoring, Stanford, 1999, p 926-935.

[4] Qiu, Jing, Zhang, Cheng; Seth, Brij B.; Liang, Steven Y. "Source Damage mechanics approach for bearing lifetime prognostics," *Mechanical Systems and Signal Processing*, v 16, n 5, September, 2002, p 817-829
[5] Yu, J.J. (Bently Rotor Dynamics Res. Corp.); Bently, D.E.; Goldman, P.; Dayton, K.P.; Van Slyke, B.G., "Rolling element bearing defect detection and diagnostics using displacement transducers," J. of Engineering for Gas Turbines and Power, v 124, n 3, July, 2002, p 517-527
[6] Lawrence Kren, Bearing Simulation Gets Real: experiments confirm analysis software accurately reflects actual hardware. (Basics of Design Engineering: Bearings). Machine Design, July 11, 2002, p 84-88.
[7] http://www.pradeepkguptainc.com/
[8] Hirotoshi Aramaki, "Rolling Bearing Analysis Program Package "BRAIN", Motion & Control, No. 3, 1977, NSK, p 15-24.
[9] Tomizuka, Masayoshi, "Zero Phase Error Tracking Algorithm for Digital Control," J. of Dynamic Systems Measurement and Control, 1987, V. 105, p 65-68..

Table 1 Impulse magnitudes and severity index values

	Bearing speed	630 rpm	750 rpm	870 rpm	930 rpm	990 rpm
Large Spall	Impulse Magnitude (10^{-4} Nt-sec)	1.2650	1.9577	2.1706	2.6052	2.6636
	Normalized Severity (%)	76.77	99.80	95.39	107.10	102.87
Medium Spall	Impulse Magnitude (10^{-4} Nt-sec)	1.2079	1.7544	2.0372	2.1593	2.1987
	Normalized Severity (%)	73.30	89.43	89.53	88.77	84.91

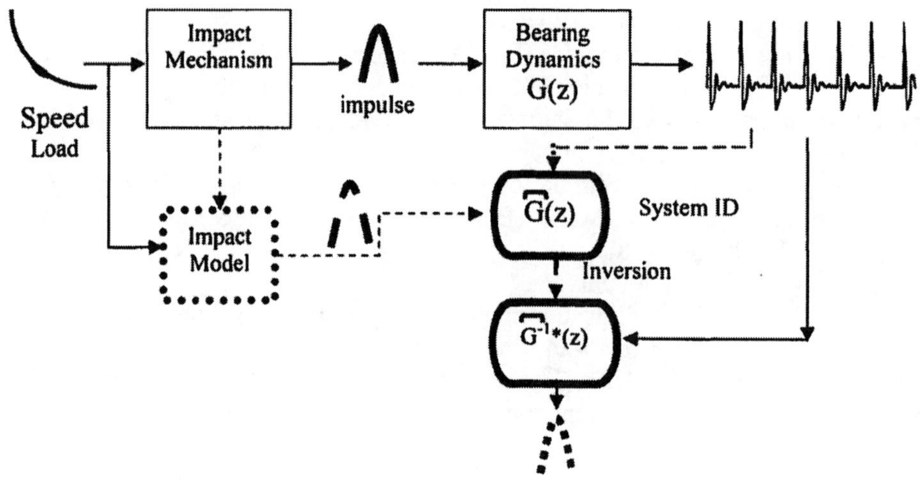

Figure 1. The proposed method

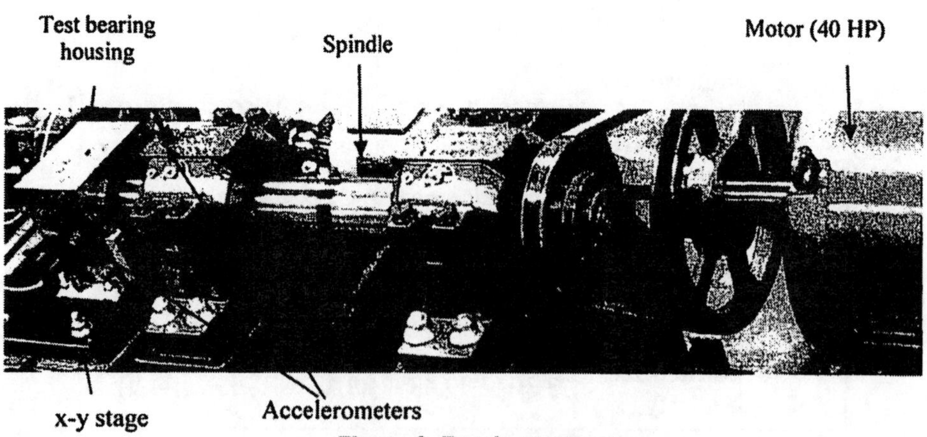

Figure 2. Bearing test setup

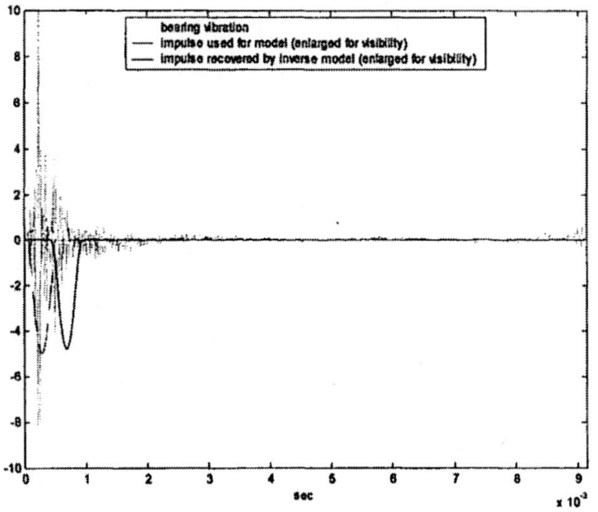

Figure 3. The impulse and bearing vibration used for modeling

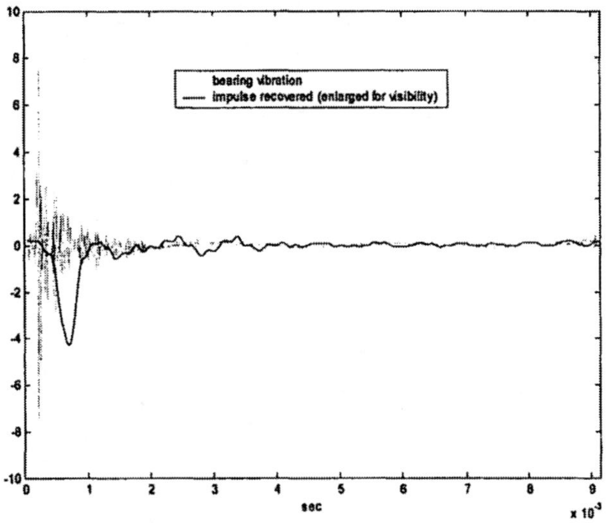

Figure 4. Bearing vibration and impulse recovered from it

Non-Linear Response Effects on Vibration Signatures

Sidney J. Booksh
Machinery Technical Support
200 W. 34th Ave #979
Anchorage, AK 99503
mts.ak@att.net

Abstract- The vibration time-base signal and its associated spectra are functions of the forces in the system and how the structure responds to them. This paper will discuss how that response can make the vibration signals extremely complex through system non-linearity. It discusses two classes of non-linearity- force response and time response. It will demonstrate that beating and modulation are one and the same phenomena and are the result of time response non-linearity. The power series approximation and simple trigonometry do a creditable job of modeling non-linear systems. Computer simulation can help distinguish between the real and the artifact in complex spectra.

Key Words: Beating, comb filter, misalignment, modulation, Non-linearity, power series,

Literature review: Ehrich [1] described the interaction of two signals close together in frequency through a non-linear system. Although he did not use the terms modulation and beating, he appears to have been the first to document and describe the use of sidebands and an oscillating envelope in the diagnosis of a mechanical system. Eshleman[2] expanded on Ehrich and discussed the possibility of misalignment, looseness, rubs, and other phenomena caused by system response non-linearity. He also described the generation of sidebands in the vibration spectra. He raised the subject of using this characteristic as a quantitative diagnostic tool as an area of fruitful future research. Both Ehrich and Eshleman used the Fourier transform to determine the frequency content of truncated signals. Eisenmann [3] and Taylor [5] present explanations of the characteristics of simple modulation and beating. Several case histories are given that add depth to the explanation. Taylor [5] also very briefly touches on the non-linear aspects of misalignment. Other authors, Mitchel [4] and Crawford [6], have discussed the appearance of sidebands in spectra and oscillating waveforms and their use in diagnosing machinery faults.

How mechanical systems respond to forces : The frequency response plot that we are all familiar with, Figure 1, comes with the understood assumption that it only applies to linear systems. Figure 2 is the frequency response map of a system whose response characteristic changes with increasing force.

Here we will look at the effect of the force response characteristic of a system on the vibrations that we observe. We shall assume that at any given force the form of the frequency response curve remains internally proportional, as appears to be the case in the response surface of Figure 2.

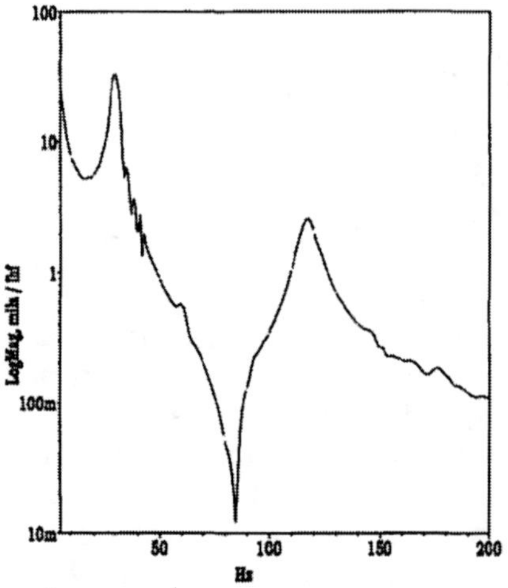

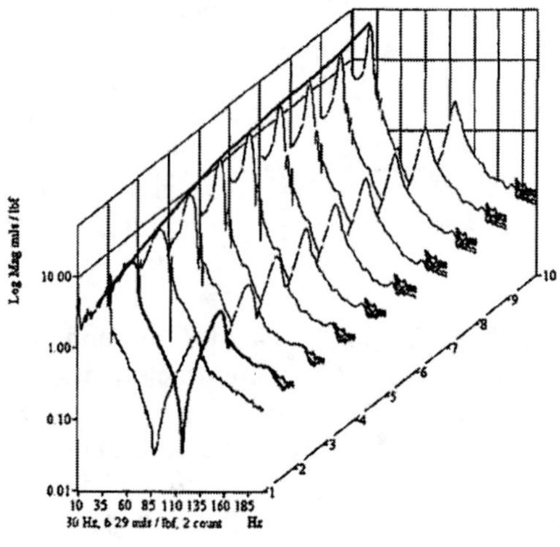

Figure1- Frequency Response **Figure 2 - Response as a function of force**

Figure 3 shows the force response characteristic that might be associated with misalignment, a rub, or looseness. The soft and progressive change in restraint of misalignment is more subtle. A non-linear model for misalignment can do a creditable job of explaining its apparently inconsistent characteristics[7].

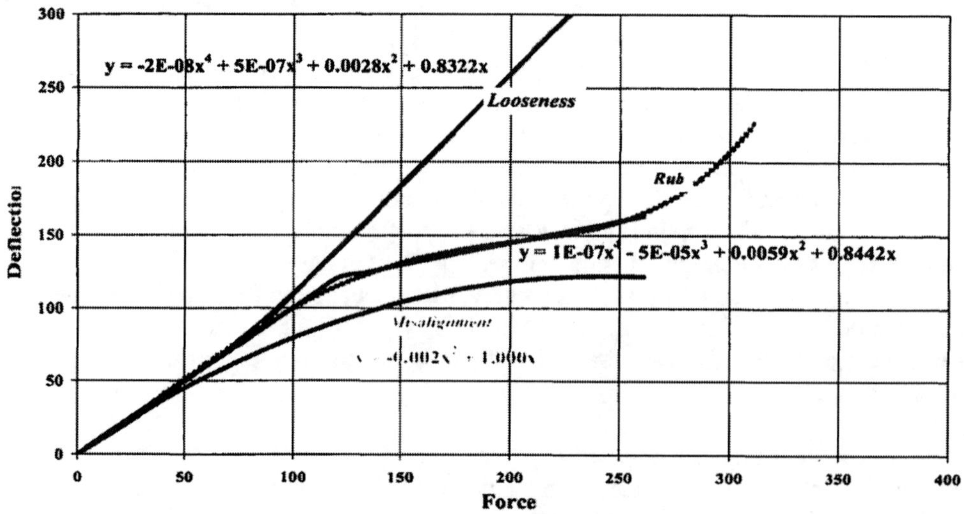

Figure 3- Malfunction deflection response

Modeling a non-linear system : A power series can be used to describe the force response characteristic of the progressively more non-linear systems, as illustrated in Figure 3. It has the form:

$$A + BX + CX^2 + DX^3 + \ldots\ldots MX^N \qquad \text{EQ(1)}$$

Where A,B,C,D,M, and N are arbitrary coefficients; X is the independent variable. In this case, force is the independent variable and may be a single force or the sum of independent forces. Also, the constant term 'A' would imply a static deflection with no force input, so it is assumed to be zero and will be dropped in future equations. The advantage of a power series is that it is has the flexibility to accurately describe some really complex systems and it is easy to handle in a computer. The disadvantage is that the results are only valid within the range of the data used to generate the model. The 'Rub' equation in Figure 3 is extended past its data to illustrate this. There are other equation forms that can be used. It is not uncommon for the form of the equation to drastically affect the insights gained form a model. In Figure 3 the dotted lines show how closely a power series can describe even discontinuous plots. A better fit would be obtained if more terms were used for the looseness and rub curves.

Another form of non-linearity happens when one forcing frequency effects the system's response to other forcing frequencies. The most common forms are the modulation of gear mesh frequency with gear shaft speed and AFB resonances with bearing defect passing frequency.

Single frequencies

Linear Response: Any system can be described by the simple relation

$$\text{Vibration} = \text{Response*Force} \qquad \text{EQ(1)}$$

The forces associated with rotating equipment are traditionally written as Force*Cos ωT. To keep things tractable it will be simplified to Cos F, where F is the forcing frequency. Therefore, the vibration would be B*Cos F. By inspection, the only frequency seen in the time signal (Figure 4, Top) and the spectrum (Figure 5, Top) is the forcing frequency, F. The format of the following plots is linear on top, slightly non-linear in the middle and highly non-linear in the bottom.

Non-Linear Response: When the system gets a little non-linear the response can be described by:

$$\text{Vibration} = \text{B* Force} + \text{C* Force}^2 \qquad \text{EQ(2)}$$

Any progressive restraint might produce such a response. Since the force is described by Cos F, the equation becomes:

$$\text{Vibration} = \text{B*Cos F} + \text{C*Cos}^2\text{ F} \qquad \text{EQ(3)}$$

Since $\text{Cos}^2\text{F} = \frac{1}{2}(1+\text{Cos 2F})$, EQ(3) becomes:

$$\text{Vibration} = \text{B'*Cos F} + \text{C'*Cos(2*F)} \qquad \text{EQ(4)}$$

Henceforward, changing the constant symbols with primes and other devices to indicate a change in value as a result of algebraic manipulation will be dropped. A single frequency force, like unbalance, acting on a slightly nonlinear system produces the fundamental and the second harmonic (Figures 4 and 5, Middle). If the system response is very non-linear, the response might be:

$$\text{Vibration} = B^{*}\ \text{Force} + C^{*}\ \text{Force}^{2} + D^{*}\text{Force}^{3} + E^{*}\text{Force}^{4}\ \text{ad infinitum} \qquad EQ(5)$$

After applying the above process, EQ(5) becomes:

$$\text{Vibration} = B^{*}\ \text{Cos}\ (F) + C^{*}\ \text{Cos}\ (2^{*}F) + D^{*}\ \text{Cos}\ (3^{*}F) + E^{*}\ \text{Cos}\ (4^{*}F) \qquad EQ(6)$$

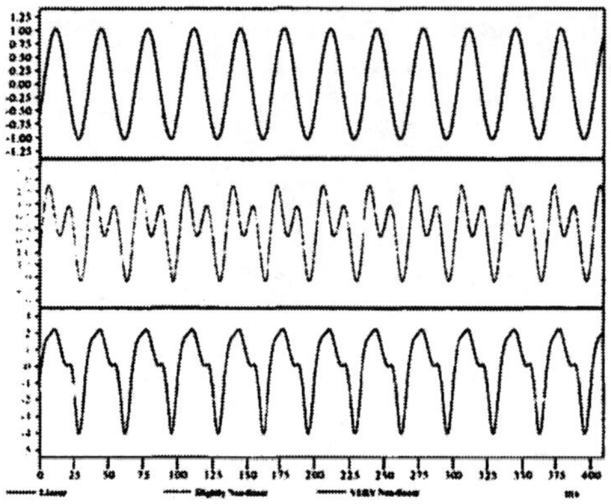

Figure 4- Single Frequency Timebase

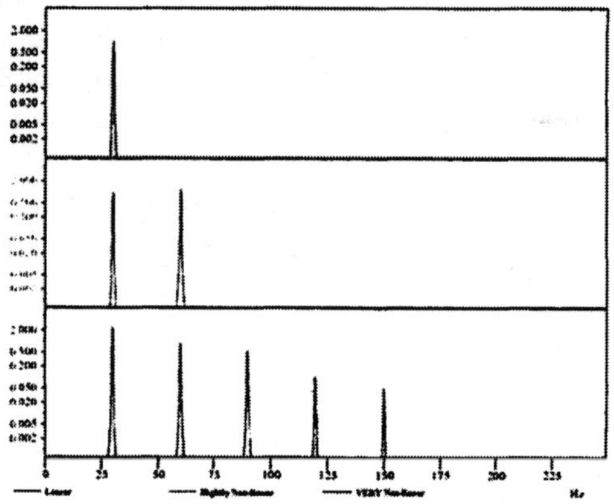

Figure 5- Single Frequency Spectrum

This is demonstrated in the bottom plots of Figures 4 and 5. While the individual frequency components may be discernible in the time-based plot for the highest amplitude harmonics, the waveform is usually too complex to see any others.

Several observations follow from the above progression. First, a second harmonic is neither necessary nor sufficient for a diagnosis of misalignment, conventional wisdom notwithstanding. All that can be said is that the system is rather non-linear. The frequencies seen would be a function of the equation describing the system force response. The assumption of a misaligned coupling behaving as a Hook's joint is not needed to account for the 2X. Second, it would appear that an equation for the system response to a force could be developed from the vibration spectrum. If the trigonometric coefficients of the harmonics are backed out and the frequency response is known or can be reasonably approximated, then one is left with the force response of the system. It would seem that a plot of this function would unambiguously discriminate between looseness and a rub. Perhaps, it would be found that different types of couplings would have unique and consistent force response functions. This may be a way to get a quantitative measure of the amount and direction of misalignment directly from vibration spectra.

Effect of multiple paths: This is one of the things that contribute to the untidiness of real vibration signals. It is especially relevant to gearbox, blade-pass and rolling element bearing signals. If the lengths of the individual paths from the source to the transducer is sufficiently different, then a frequency component can arrive at different times. That is, the signal will be added to itself with a phase shift. If that shift is around 180° then the signal will be lowered; if it is around 360°, it will be elevated. Figure 8 is a plot the effect of several path lengths with frequency. Since the speed of sound in engineering materials is so high, this phase offset would only have a practical effect at the very high frequencies of local structural and component resonances.

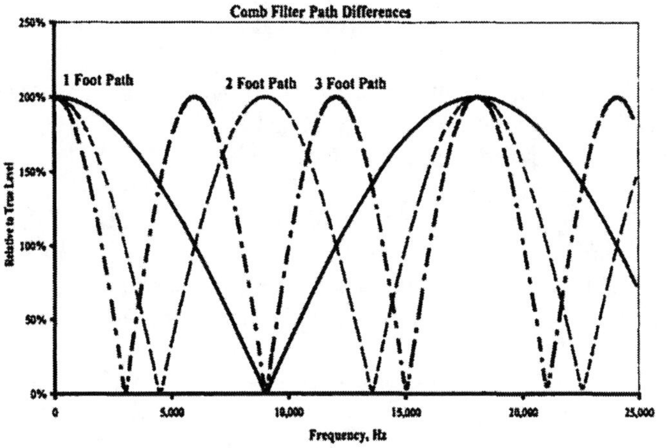

Figure 8- Comb filter

This is called a Comb filter in the acoustics community. Its equation is:

$$\text{Response} = |2 * \text{Cos}(\pi * \text{Frequency, Hz} * \text{Time delay, seconds})| \qquad \text{EQ(7)}$$

It makes designing room acoustics a challenge. This phenomenon has different effects at different frequencies; that is, it filters out and amplifies several widely separated frequency bands. Therefore, it is wise to record the vibrations of complex systems at several locations in order to mitigate this effect. This is one of the reasons why it is difficult to come up with a proper place to monitor gearbox vibrations.

Our outer ear acts as a comb filter and makes it possible for us to determine the direction of a sound source [9]. It may be possible to use the differences in the whole spectrum at several locations to derive the comb filter characteristics then calculate the path length difference and finally, divine the source of a signal. Also, one should also be able to derive the true signal level once the comb filter characteristic is removed.

Sum of two frequencies

Linear Response: Obviously, multiple forces act upon real systems. Given a linear response, B* Force, the sum of two signals, F1 and F2, has a time base like Figure 9 Top and a spectrum like Figure 10 Top. The equation will be :

$$A * \text{Cos } F1 + B * \text{Cos } F2 \qquad \text{EQ(8)}$$

This is generally called a 'Beat'. Neither frequency will be seen in the time base. The plots, below, use 30 and 32 Hz. Measuring from local peak to local peak produces the average of the two or 31 Hz. Measuring from macro-peak to peak or zero to zero results in 2 Hz, which is the difference frequency. Beating can result in sidebands in a non-linear system. Since :

$$\text{Cos } A + \text{Cos } B = 2\ \text{Cos}\tfrac{1}{2}(A+B) * \text{Cos } \tfrac{1}{2}(A-B) \qquad \text{EQ(9)}$$

EQ. 8 becomes

$$2 * \text{Cos}\{\tfrac{1}{2}(F2-F1)\} * \text{Cos}\{\tfrac{1}{2}(F2+F1)\} \qquad \text{EQ(10)}$$

which is generally called 'modulation'. This implies that with the proper choice of frequencies, beating and modulation are indistinguishable. Therefore, the diagnostician needs other independent information to distinguish between them.

The top plot of Figures 9 and 10 is the sum of two signals, 30 Hz and 32 Hz; The second plot is the product of 1 Hz ([32 Hz-30 Hz]/2) and 31 Hz ([32 Hz + 30 Hz]/2. This is called DSSB (Double Sideband Suppressed Carrier) in the communications industry. To better show that the signals are identical, one was subtracted from the other and the result is in the bottom plot of Figure 9.

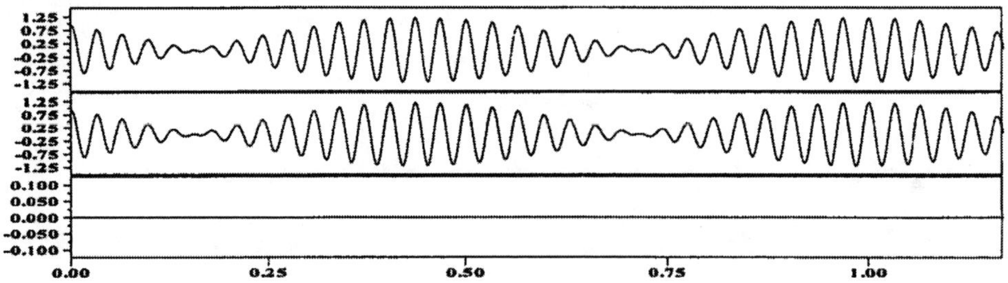

Figure 9– Sum,product, difference timebase

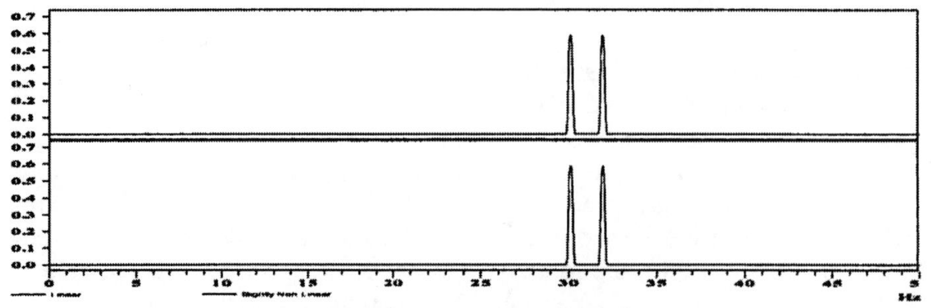

Figure 10- Sum and product spectrum

Non-Linear Response : If we apply EQ(3) to the sum of two frequencies

$$(B * \text{Cos } F1 + C * \text{Cos } F2) + (B * \text{Cos } F1 + C * \text{Cos } F2)^2 \qquad \text{EQ(11)}$$

and apply the trigonometric identities EQ(9) and

$$\text{Cos } A * \text{Cos } B = \tfrac{1}{2}\ \text{Cos}(A+B) + \tfrac{1}{2}\ \text{Cos}(A-B) \qquad \text{EQ(12)}$$

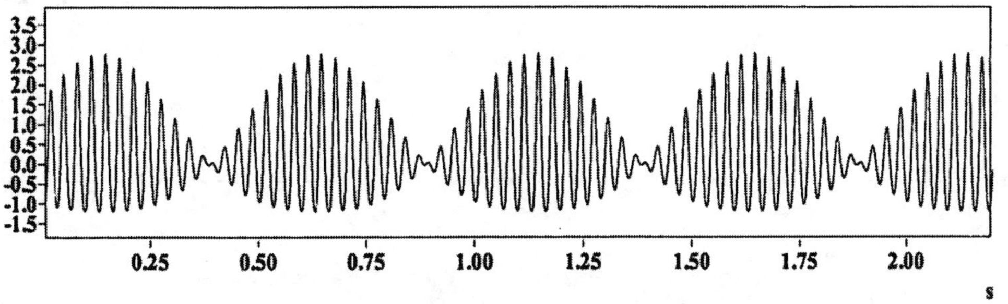

Figure11 - Non-linear force response timebase

we get EQ(13) and plots of Figures 11 and 12.

$$\mathrm{Cos}\,A + \mathrm{Cos}\,B + \mathrm{Cos}\,2A + \mathrm{Cos}\,2B + \mathrm{Cos}\,(A-B) + \mathrm{Cos}(A+B) \qquad \text{EQ(13)}$$

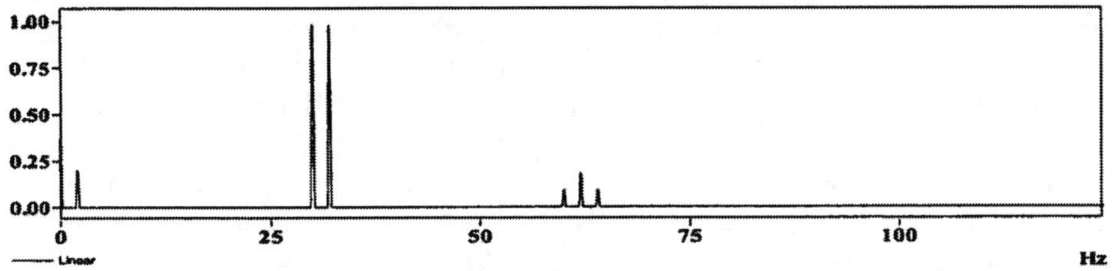

Figure 12- Non-linear force response spectrum

The forcing frequencies for these plots are 30 Hz and 32 Hz.
If, however, the system is non-linear in time, i.e., one of the time varying forces controls the system's response to the other force(s), the result is entirely different as seen in Figures 13 and 14. The input frequencies are also 30 Hz ans 32 Hz, but they appear in neither the timebase nor the spectrum. Figures 13 and 14 show the very familiar looking, but unexpected result. From the trigonometry identity of:

$$\mathrm{Cos}\,A^* \,\mathrm{Cos}\,B = ½\,\mathrm{Cos}\,(A+B) + ½\,\mathrm{Cos}\,(A-B) \qquad \text{EQ(14)}$$

only the sum and difference frequencies are seen in the timebase and the spectrum.

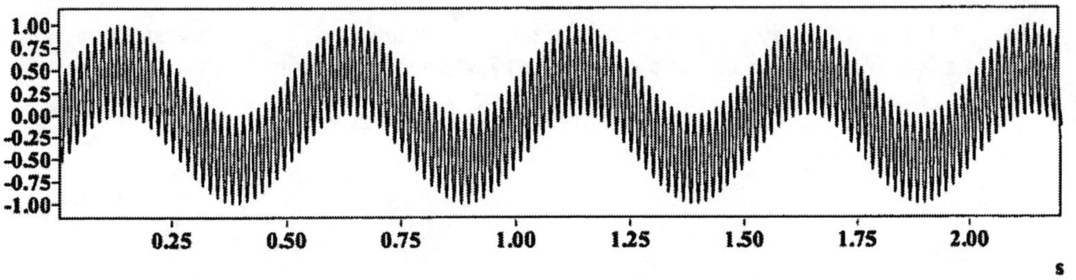

Figure 13 - Non-linear time response timebase

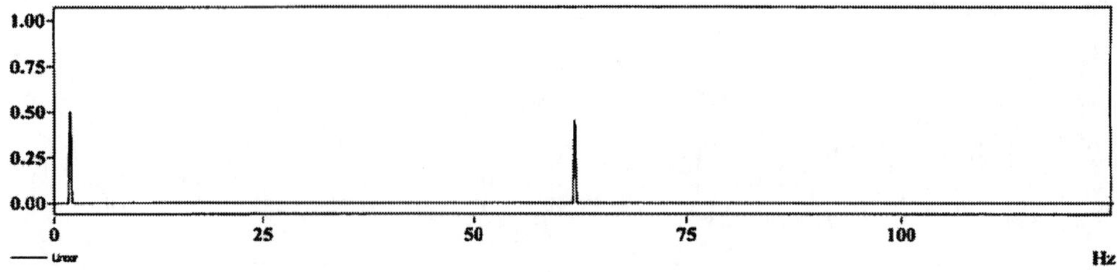

Figure 14- Non-linear time response spectrum

There is not a hint of the true forcing frequencies. The analyst would have to know the characteristics if the system, or be lead on a merry chase of phantoms.

It is often instructive to simulate a system to gain insight into what drives it. This would clearly be the case in the preceding example.

A simulation using a power series to describe a gas turbine's response characteristics made it possible to recognize a low frequency spectral component as spectral (in the paranormal sense) and not real. Figures 15 and 16 are from the GG turbine end of a gas generator. The physical forces in the system are known to be the GG and pump unbalance, 175 Hz and and 106 Hz respectively, and a half-X GG bearing parametric stability problem [8]. The rest are sum and difference frequencies caused by the system's non-linearity. The journal is riding dead center of the bearing so the response is highly non-linear. The vibrations were simulated by using the sum of three frequencies as the variable in a third degree equation, EQ(15). Just a few iterations of the equation coefficients produced reasonably good agreement.

$$\text{Vibrations} = \text{Forces} + 0.1*\text{Forces}^2 + 0.02*\text{Forces}^3 \qquad \text{EQ(15)}$$

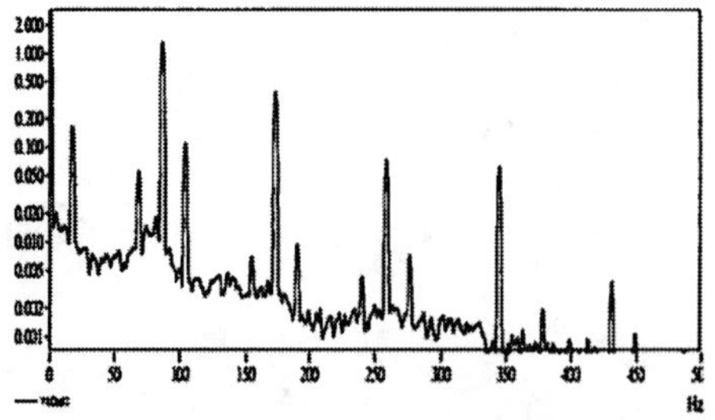

Figure 15 - Gas Turbine Vibrations

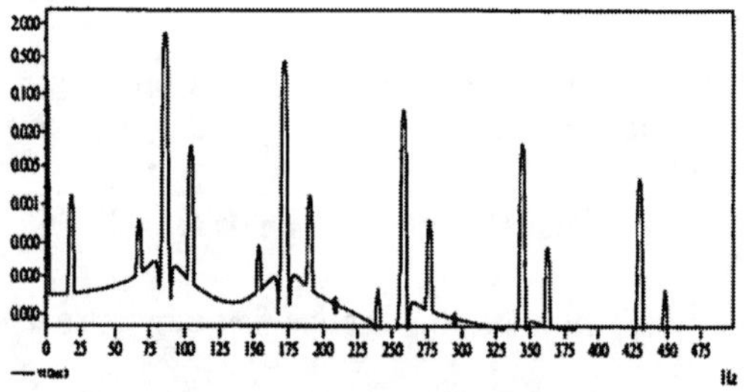

Figure 16- Simulated Vibrations

The 18.5 Hz is not a driver modulating the half-X and harmonics, but the result of system nonlinearity. The force response plot, Figure 17, demonstrates a softening response.

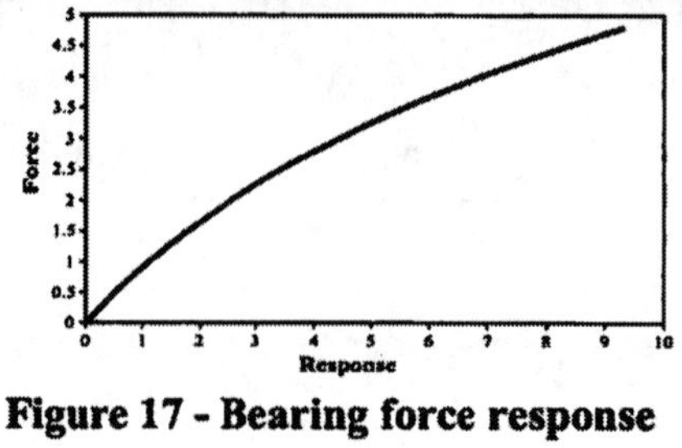

Figure 17 - Bearing force response

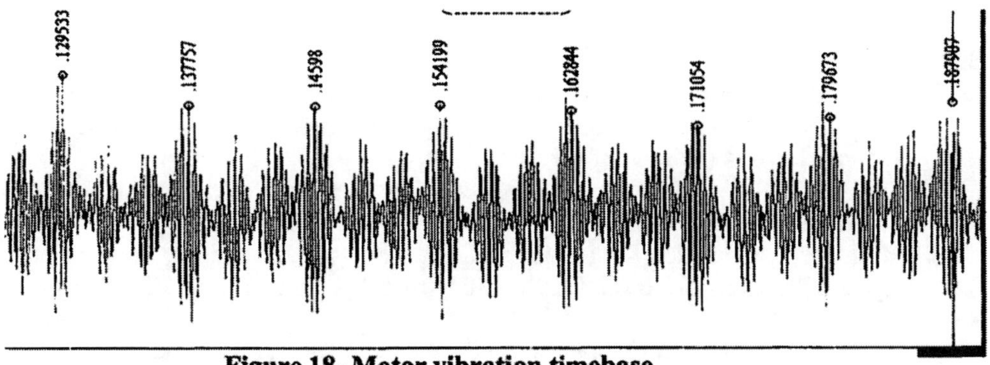

Figure 18- Motor vibration timebase

Another example is the plot of Figure 18 from an electric motor housing and Figure 19, the simulation. In this case, the vibration waveform was the result of the inter-modulation of three drivers – 60 Hz, the line frequency; 20 Hz, the motor's speed; and 800 Hz, the stator slot pass frequency, that is, $F_{60Hz}*F_{20Hz}*F_{800Hz}$.

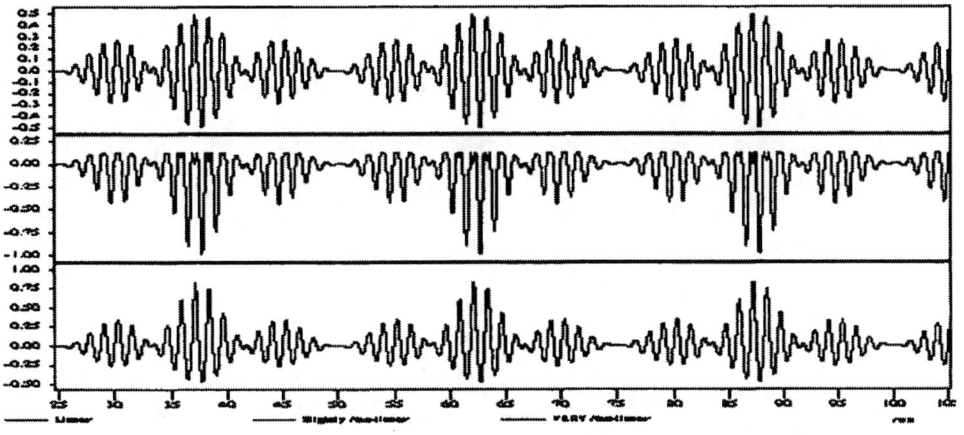

Figure 19- Simulated timebase

Conclusion: The breadth and complexity of the effects of non-linearity are far too great to be able to cover adequately in the space available. Multiple forcing frequencies, multiple signal paths, system frequency response and non-linear force and time response can produce a bewildering and often unexpected array of interactions. Simulating the system responses is one way to separate the phantoms from the real. Power series approximations of force non-linearities with simple trigonometry can make this simulation somewhat easier. It may be a means to quantitatively describe the characteristics of system components like flexible couplings and, therefore, allow quantization of misalignment directly from field vibration data or a trending of changing system physical properties as opposed to vibration levels. Virtual instrument software was used in this paper to simulate the forcing frequencies, non-linear responses, and multiple signal paths. It is hoped that this brief and inadequate summery will encourage others to more fully catalog these interactions.

Bibliography

1. Sum and Difference Frequencies in Vibration of High Speed Rotating Machinery, F.F. Ehrich, Journal of Engineering for Industry, Transactions of the ASME, Vol. 94, pp181 to 184 (Feb 1972)
2. The Role of Sum and Difference Frequencies in Rotating Machinery Fault Diagnosis, R.L. Eshleman
3. Machinery Malfunction Diagnosis and Correction, R.C. Eisenmann, Sr., Hewlett-Packard Professional Books, ISBN0-130240946-1 pages 347 to 364 and 508 to 510.
4. Machinery Analysis and Monitoring, J.S. Mitchell, PennWell Books, ISBN 0-87814-401-3
5. The Vibration Analysis Handbook, James L. Taylor, Vibration Consultants Inc., ISBN 0-9640517-0-2
6. Simplified Handbook of Vibration Analysis, A. Crawford, Computational Systems, ISBN 9994616935
7. Why Shaft Misalignment Continues to Befuddle and Undermine Even The Best CBM and Pro-active Maintenance Programs, John Pitrowski, P/PM Technology, April '97
8. Forced Subrotative Speed Dynamic Action of Rotating Machinery, D.E. Bently, ASME Paper 74-PET-16
9. Handbook for Sound Engineers, Glen M. Ballou, SAMS, ISBN 0-672-22752-5
10. On Period Doubling Bifruncations in Duffing's Oscillator With Negative Linear Stiffness, K.R. Asfar and K.K. Masoud, Journal of Vibration and Acoustics, Vol 114 Pgs 489 to 494, October 1992
11. Structural Modal Multifruncation With Internal Resonance Parts 1 & 2, R.A. Ibrahim, A.A. Afaneh, B.H. Lee, Journal of Vibration and Acoustics, Vol 115 Pgs182 to 201, April 1993
12. Chaotic Vibration of a simple model of the Machine Tool-Cutting Process System, M. Wiercigroch, Journal of Vibration and Acoustics, Vol 119, Pgs 468 to 475, July 1997

PROGNOSTICS III
Expeditionary Fighting Vehicle Program

Chair: Dr. David D. Thurston
General Dynamics Amphibious Systems

IMPLEMENTATION OF A BATTERY PROGNOSTICS SYSTEM FOR THE USMC EXPEDITIONARY FIGHTING VEHICLE

James D. Kozlowski, Chris Rogan
Applied Research Laboratory, The Pennsylvania State University

Abstract: A Battery Prognostics System (BPS) that is capable of accurately assessing the state-of-charge (SOC) and state-of-health (SOH) of lead acid starting-lighting-ignition (SLI) batteries can provide significant operational and logistical benefits when implemented onboard ground combat vehicles. In this paper, the design of a BPS for the USMC Expeditionary Fighting Vehicle (EFV) will be presented. The prognostics hardware utilizes proprietary current sensing circuitry to measure the wide-band impedance of the vehicle batteries in situ. Impedance spectroscopy provides a methodology for assessing the condition of the battery being monitored by comparing the wide-band impedance characteristics to a priori developed electrochemical models. This approach allows for the early identification of potential failure mechanisms and hence determination of the remaining useful life of the battery (i.e., state-of-life), a critical step for prognostics. The focus of this paper will be on the design and implementation of the BPS. Critical design elements will be presented along with performance results from initial system testing, and an analysis of the potential benefits that may be accrued from an operational system.

Key Words: Battery Prognostics; Impedance Spectroscopy; Silent Watch; State-of-Charge; State-of-Health.

1. Introduction: The United States Marine Corps' Expeditionary Fighting Vehicle (EFV) is an advanced amphibious armored vehicle currently under development by defense contractor General Dynamics Land Systems. The battery system onboard the EFV is used to provide the energy necessary to crank the vehicle's engine during startup. When the vehicle is running, the electrical energy required to power onboard electronics is supplied by the vehicle's alternator. It is also often necessary to power onboard electronics while the vehicle is not running. This condition is typically referred to as "silent watch". During silent watch, the vehicle's battery system must provide all the required electrical energy to power the onboard electronics.

Due to the criticality of batteries to EFV mission readiness, it is highly desirable that an advanced battery monitoring system be used. The EFV battery system consists of eight Optima Red Top Lead Acid batteries. Two of the batteries are dedicated to power the onboard computer during startup, and the other six are dedicated to providing the necessary cranking amps required to start the vehicle under a wide range of environmental conditions. In addition, all eight batteries can be used to power onboard electronics during silent watch. A battery monitoring system can potentially offer a number of benefits to a ground combat vehicle. Often times, one or two batteries in a farm may fail prematurely or become weak. If degraded performance is recognized, maintenance personnel will typically not know which of the batteries are weak and will therefore (unnecessarily) replace all eight batteries. If degraded performance is not recognized, the weak batteries will "pull down" the good batteries, with the result being that the entire battery system will fail prematurely. A battery monitoring system would be useful as a troubleshooting aid for vehicle maintenance personnel as it could identify faulty or weak batteries and facilitate removal of those batteries.

An accurate battery monitoring system is also imperative during silent watch. While powering onboard electronics during silent watch, the batteries are being depleted of energy that is also required to start the vehicle at the conclusion of silent watch. The batteries cannot be allowed to drain to the point that sufficient energy is not available to start the vehicle. This can be particularly catastrophic in a combat vehicle where troops may be stranded in harm's way. It is thus imperative that a battery monitoring system be capable of accurately estimating the remaining energy in the vehicle's batteries. In addition to the aforementioned benefits, a battery monitoring system with prognostic capabilities could provide additional benefits. Having the ability to identify the current health of a battery and match that health assessment to the projected use pattern of the vehicle can allow one to make a prediction as to how long the currently installed batteries might last. For example, knowing the health of the batteries and current environmental conditions such as temperature, one could predict how many more times the engine can be cranked using the currently installed batteries.

Penn State's Applied Research Laboratory is under contract to develop the Battery Prognostics System (BPS) for the EFV. The remainder of this paper will discuss the developed system. Section 2 of this paper will provide a short summary of the battery prognostics technology developed at Penn State. Section 3 will give an overview of the BPS designed for the EFV. Section 4 will give a summary of preliminary test results, and conclusions will be stated in Section 5.

2. Battery Prognostics Technology: For overall performance, the capacity and voltage of a cell are the primary specifications required for an application. The capacity is defined as the time integral of current delivered to a specified load before the terminal voltage drops below a predetermined cut-off voltage. The present condition of a battery is described nominally with the state-of-charge (SOC), which is defined as the ratio of the remaining capacity and the initial or rated capacity. Thus, the service history of a cell and its nominal capacity impact the assessment of SOC. Secondary batteries are observed to have a capacity that deteriorates over the service life of the cells. The term state-of-health (SOH) is used to describe the physical condition of the battery, ranging from external behavior such as loss of rated capacity, to internal behavior such as severe corrosion. The remaining life of the battery (i.e. how many cycles remain, usable charge, etc.) is termed the state-of-life (SOL), the prognostic metric [1].

A novel impedance measurement methodology has been developed at Penn State's Applied Research Lab which utilizes electric current sensing technology to measure wideband electrical impedance in electrochemical cells and batteries [2]. In addition, a model-based predictive diagnostics tool was developed which utilizes the aforementioned impedance measurement technology to estimate SOC, SOH, and SOL of primary and secondary batteries ranging from small primary single cells to large banks of rechargeable batteries [3]. This new technology is based upon two important new innovations with regards to battery measurement and diagnostics:

- **Current-Divider Impedance Measurement Methodology**
  The impedance of typical battery cells is quite small, typically just a few milliohms. The standard approach to measuring impedance (driving current through device under test and measuring the voltage drop across it) becomes impractical at such low impedances because prohibitively large currents are required to obtain a large enough voltage drop such that an accurate measurement can be made. Penn State's current-divider circuitry is used to sense the small-signal current present in an equivalent (known) impedance branch, thereby eliminating the need to drive large amounts of current through the battery. Current levels on the order of 20mA are typically all that is required to achieve accurate results.

- **Model-Based Predictive Diagnostics**
  Penn State's Model-Based approach uses the results from the wide band impedance measurement in order to construct an equivalent electrochemical model of the battery. Using the wide band impedance (as opposed to the impedance at one or even a few frequencies) enables the building of a complete model that provides an accurate description of the electrochemical condition of the battery being tested. This model can then be used to extract useful information such as SOC, SOH and SOL.

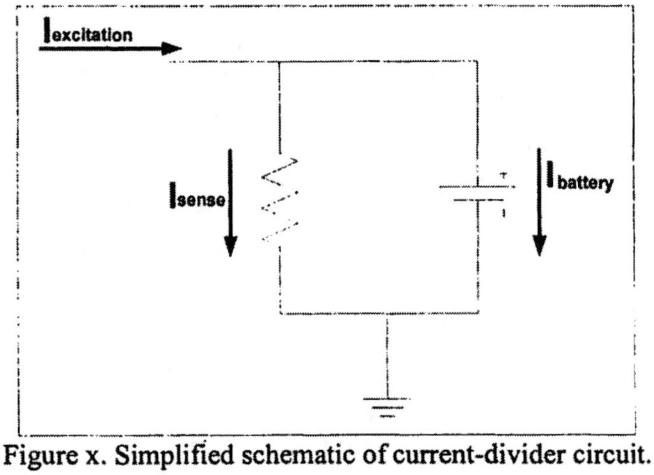

Figure x. Simplified schematic of current-divider circuit.

To obtain the wide band impedance, a series of tones are injected into the battery. The frequencies of interest for lead acid batteries range from approximately 10Hz through 10kHz. This excitation current is divided between the battery under test and a known sense resistor as shown in Figure x. Z_{sense} is known and I_{sense} is measured directly using a novel circuit developed by Penn State ARL (patent submission is progress). $I_{excitation}$ is also measured directly and $I_{battery}$ can be calculated using:

$$Ibattery = Iexcitation - Isense \qquad (1)$$

Once battery current is known, battery impedance can be calculated using:

$$Zbatt = Zsense * \frac{Isense}{Ibattery} \qquad (2)$$

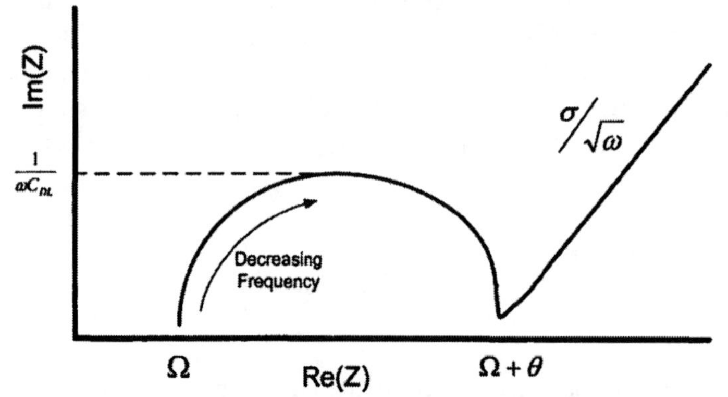

Figure x. Nyquist plot of battery impedance.

If this calculation is repeated over the entire frequency range, a Nyquist plot for the battery can be obtained (see Figure x.). The resulting impedance information relates to the governing equations of the physical electrochemical processes within the battery. To retrieve the electrochemical process information from the impedance data, a Randle's electrical circuit analog is used (see Figure xx). It is from the Randle's circuit model parameters that useful health information can be extracted. For example, lead-acid batteries tend to age from passivation, which has shown to be associated with an increase in charge transfer resistance. Drying out of the electrolyte manifests itself in the Randle's circuit as an increase in the ohmic resistance. Corrosion of the electrode changes the porosity of the electrode and reduces the slope of the linear leg of the plot [1].

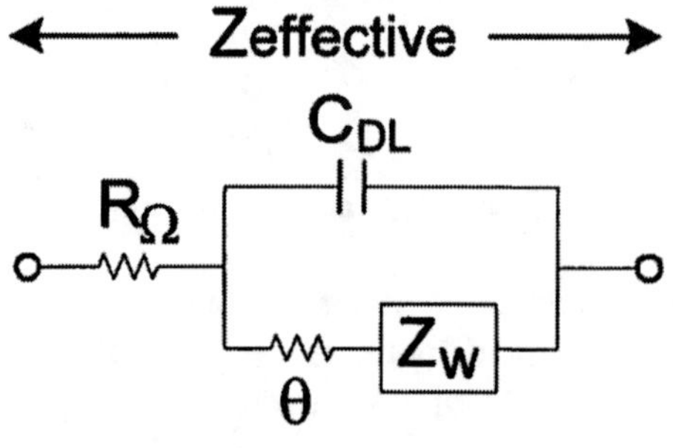

Figure x. Randle's circuit for electrode-electrolyte processes.

Once the model parameters have been determined, classifier algorithms are used to map the parameters to SOC and SOH algorithms that were trained using test data collected during comprehensive laboratory testing. In order to improve robustness, three separate algorithms are run in parallel and results are compared in order to determine the confidence of the final results. SOC estimation is performed using ARMA, fuzzy logic and neural network algorithms, and SOH estimation is performed using fuzzy logic, neural network and linear discriminator algorithms.

The technology has been used successfully in investigations on a wide array of battery chemistries and capacities, both primary and rechargeable, including NiCd, NiMH and Li Ion as well as lead acid batteries. Accuracy of the prognostics has been verified to be < 5% error in all cases in laboratory testing. The testing included wide variations in age, loading and temperature.

3. Design of the EFV Battery Prognostics System: The EFV battery system consists of eight Optima Red Top Lead Acid starting batteries. Therefore, eight measurement circuits or channels are required in order to assess the health of all eight batteries. To accommodate the spatial diversity of the individual batteries within the battery compartment, as well as to maximize system modularity and flexibility, it was decided to implement each measurement channel as its own module that can be affixed to the battery that it will monitor. This approach is also important because the BPS utilizes impedance spectroscopy to assess the health of the batteries. Since the impedance of lead acid starting batteries is quite small (on the order of 10-100mΩs), it is important that the impedance measurement circuitry be physically located close to the battery being monitored since excessively long wire lengths can add significant impedance to the measurement and degrade the result.

For a battery monitoring system to be useful, data collected by all the battery modules must be processed, analyzed and transmitted to the vehicle's onboard computer. This leads to the question of whether to process the measurement data locally at each module, or to do it centrally within the onboard computer. The latter approach requires the modules to transmit large volumes of raw data to the onboard computer, and take up valuable computing resources on the computer. For these reasons, the approach taken was to embed the processing inside each module. The processed results from an individual battery assessment is only a few bytes of data, thus a low bandwidth link can be used to collect system data. For this application, all of the battery modules are linked via a Controller Area Network (CAN) communication bus. One of the eight nodes is integrated to an embedded web server and an Ethernet link is used to relay battery system information to the onboard computer. The system architecture is shown in Figure x.

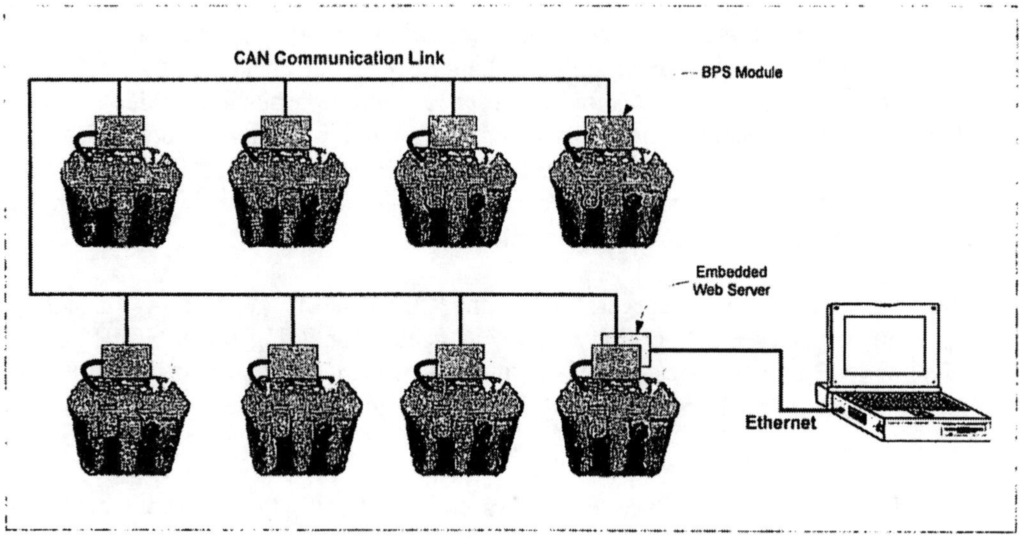

Figure x. BPS System Architecture.

One note regarding the CAN communication link: the vehicle batteries are configured as four parallel strings of two series-connected batteries as shown in Figure x. Since the negative terminals of the two batteries in each series string are 12V apart, and the negative terminal serves as the reference point for the electronics in each module, galvanic isolation is required for the CAN communication link so as to prevent circuit reference points that are 12V apart from being shorted together.

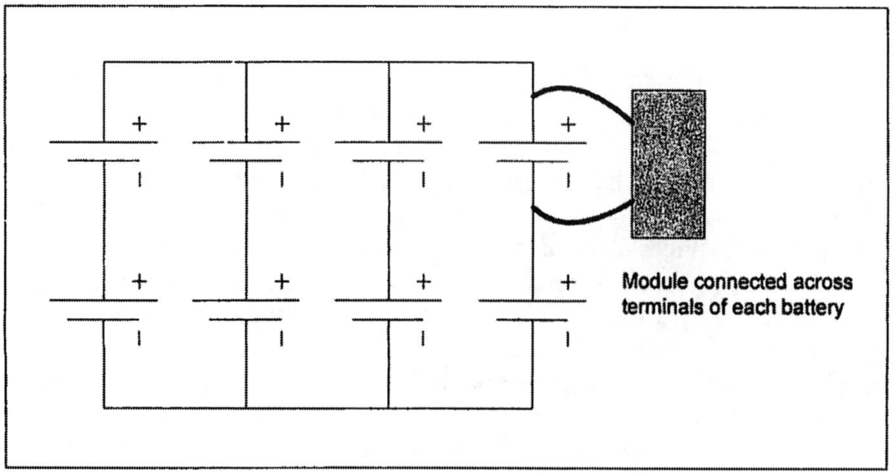

Figure x. Schematic representation of EFV battery system.

The frequency range of interest for lead acid batteries is approximately 10Hz to 10kHz. Therefore, battery module hardware must be capable of generating an excitation signal over that frequency range in order to do an impedance scan of the battery. Additionally, the hardware must be capable of acquiring the sense signal over the same frequency range. The battery modules utilize an 8051-based microcontroller with integral D/A, A/D and CAN communications peripherals. The processor runs at a bus speed of 25MHz and uses 32Kbytes of RAM and 64Kbytes of Flash memory.

Another design issue that was faced was that of providing power for the BPS. The BPS is monitoring the vehicle's battery power system, but the modules themselves require power in order to perform the measurement and to communicate the results. With the current design, each module consumes approximately 1.5W of power during normal operation. There are two basic choices for powering the modules: use a central supply, i.e., the vehicle's main electrical bus, or provide power for each module from the battery it is monitoring. There are advantages and disadvantages to each approach. From a modularity and ease of installation standpoint, it is preferable to have each module powered directly from the battery it is monitoring. The issue with this approach is that if the battery being monitored is degraded to the point that it is not capable of powering the battery module, then that module will essentially be "dead". It will then be left to the user to determine if the problem is the battery or a faulty module. Alternatively, if the central supply is used, additional wiring will be required to be run to each module for power. The advantage of this approach is that as long as one of the four parallel battery strings is good, then all eight battery monitors should be operating. Given the pros and cons of each approach, and the fact that operator and maintainer preference has not yet been determined, the current design allows for either power option.

Another design issue to be considered is when to provide power to the BPS. Obviously, the BPS cannot continue to draw power from the batteries while the vehicle is parked for extended periods, as the BPS would drain the very batteries it is supposed to be monitoring. On the other hand, there are many instances when knowing the current status of the batteries would be advantageous even though the vehicle may not be running (eg. Do I need to replace any of the batteries before embarking on a new mission?). Since current vehicle design only allows for battery system information to be displayed on the vehicle computer, it is reasonable to assume that the "ON" switch for the computer also be used to turn on the BPS. To accomplish this functionality, a semiconductor switch latching circuit was implemented in each module. The input to the latching circuit is the switched power used by the vehicle computer. The purpose of the latching circuit is to turn the BPS module on when voltage is present on the latch signal (i.e., computer "ON" switch is closed).

The current hardware is capable of being installed on the EFV and integrated with the vehicle's electrical and computer systems. Additional work may need to be performed pending results of upcoming onboard vehicle validation testing. In addition, for extended in-the-field testing, ruggedization of the hardware will also be required.

4. Preliminary Testing: Extensive testing of Optima batteries using BPS hardware has been conducted at Penn State laboratory facilities. Test data has been used to train the prognostics algorithms under a wide range of operating and environmental conditions. In addition, validation testing has been conducted at Penn State ARL facilities to verify accurate performance of BPS hardware and software (algorithms). Results of the SOC validation testing performed on Optima batteries is given on Table 1.

Temperature	New	Aged
65°C	4.5%	5.0%
25°C	3.9%	4.8%
-20°C	4.9%	5.4%

Table 1. Error in predicted SOC

Prediction error when estimating SOC is approximately 5% over a wide temperature range, using both new and old batteries. Similar validation testing was performed on faulted batteries. Results are listed in Table 2 for each of the three SOH algorithms. In addition, it should be noted that the accuracy of the SOH prediction algorithm increases as the severity of the fault increases. This is due to the fact that as the severity of the fault increases, the electrochemical indications of the fault become more obvious. A plot of error in SOH predictions vs. fault severity is shown in Figure x. Please note that we characterize 'severity' on a scale from 1 to 10, with 10 being the most severe fault.

Failure Mode	Neural Network	Fuzzy Logic	ARMA
Impending Short	4.1%	6.3%	5.4%
Impending Open	5.5%	8.3%	5.9%
Passivation	4.5%	6.1%	5.3%
Corrosion	5.3%	5.9%	6.0%

Table 2. Error in predicted SOH

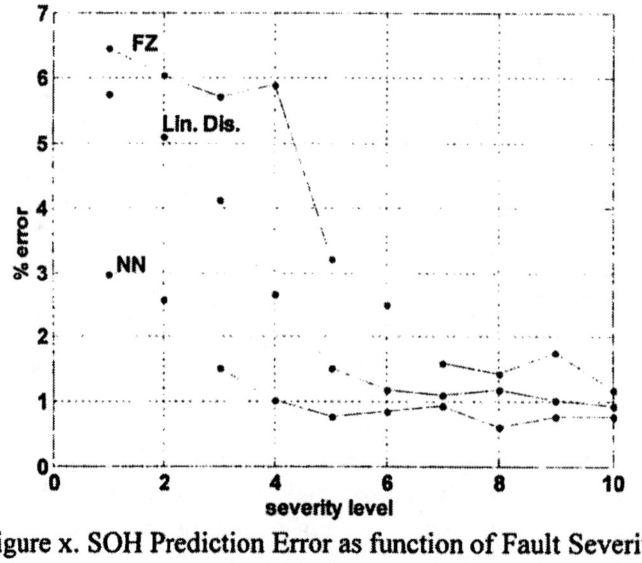

Figure x. SOH Prediction Error as function of Fault Severity

5. Conclusions: A prototype Battery Prognostics System utilizing impedance spectroscopy technology has been developed for the USMC's Expeditionary Fighting Vehicle. The system employs a battery module for each of the vehicle's eight batteries. Each module is equipped with data acquisition hardware, an embedded processor, and a CAN communication link. One of the modules is integrated to an embedded web server to facilitate Ethernet communications with the onboard vehicle computer. The system has been successfully integrated with the EFV's battery system.

The EFV BPS is currently in prototype status. Diagnostic and Prognostic capabilities of the hardware has been verified accurate in the laboratory, and will be installed onboard an EFV in March of 2005 for on-vehicle validation testing. A ruggedized version of the system will be deployed onboard several operational vehicles for extended in-the-field testing in January of 2006.

Acknowledgement: This work was funded by General Dynamics Amphibious Systems under Contract Number EAV002406 for the EFV Prognostics Program managed by Dave Thurston.

References:

[1] J. D. Kozlowski, "Electrochemical Cell Prognostics using Online Impedance Measurements and Model-Based Data Fusion Techniques," *Proceedings of the 2003 IEEE Aerospace Conference*, March, 2003.

[2] J. D. Kozlowski, *U.S. Patent No. 6,307,378 B1*, "Method and Apparatus for Measurement of Electrochemical Cell and Battery Impedances," granted October 2001.

[3] J. D. Kozlowski, C. S. Byington, A. Garga, T. Cawley, M. J. Watson, and T. Hay, *U.S. Provisional Patent No. 60/358,544*, "Model-Based Predictive Diagnostic Tool for Primary and Secondary Batteries," filed February 2002.

PLANETARY GEAR MONITORING ON THE EXPEDITIONARY FIGHTING VEHICLE

Darryl Busch, Philip Zumsteg, George D. Hadden, Ph.D., Sunil Menon, Ph.D.

Honeywell Laboratories, 3660 Technology Drive, Minneapolis, MN 55418
Darryl.Busch@Honeywell.com

Abstract: Honeywell is currently developing a Drive Train Prognostic System (DTPS) for the Expeditionary Fighting Vehicle (EFV). Two mission critical components monitored by the DTPS are the planetary gear sets in the marine drive train that transfer power to the twin water jets. This paper will describe our approach to monitoring the planetary gears using vibration sensors, time synchronous averaging, feature extraction, diagnosis, trending and prognosis. We describe some of the challenges and considerations for planetary gear monitoring on the EFV within the framework of the DTPS architecture.

Key Words: Planetary gear monitoring; prognostics; distributed data acquisition; time synchronous system.

Introduction: The Drive Train Prognostic System (DTPS) is an on-board system that monitors the marine and automotive drive trains on the Expeditionary Fighting Vehicle (EFV). Its role is to detect and provide advanced warning of impending failures several missions in advance, support go/no-go decisions for the crew, and support maintenance decisions. The automotive and marine subsystems monitored by the DTPS are the final drive assemblies, the automotive transmission, and the power transfer module (PTM) that sends power to the transmission in land mode and power to the water jets in water mode. The DTPS effort was awarded to Honeywell by General Dynamics Amphibious Marine Systems with completion scheduled for 2007.

The focus of this paper is our approach for monitoring the two sets of planetary gears in the PTM that drive the two water jets. Our approach uses vibration monitoring of signal averaging synchronized to shaft rotation. Signal averaging is a common technique in gear monitoring and has proved to be very effective. Planetary gears, however, introduce two significant challenges: 1) there are multiple gear contacts meshing at the same frequency, thereby hampering frequency-based separation of the signals from individual gears. 2) the planets move with respect to the sun and ring gears, introducing an amplitude modulation not found in other gear types.

DTPS Overview: The DTPS (Figure 1) consists of a central small form-factor computer called the Subsystem Condition Monitoring Engine (SCME), a suite of Sensors with Embedded Processing (SEPs) and a communications and power distribution network. The SEPs are placed at strategic locations throughout the drive train elements and communicate information to the SCME. SEPs monitor the drive train using sensors and perform the majority of the signal processing in the DTPS, including preprocessing of the acquired data and extraction of salient features. Each SEP consists of one or more sensors, analog circuitry, and a small computer. The current suite of sensors includes accelerometers, pressure sensors, and tachometers.

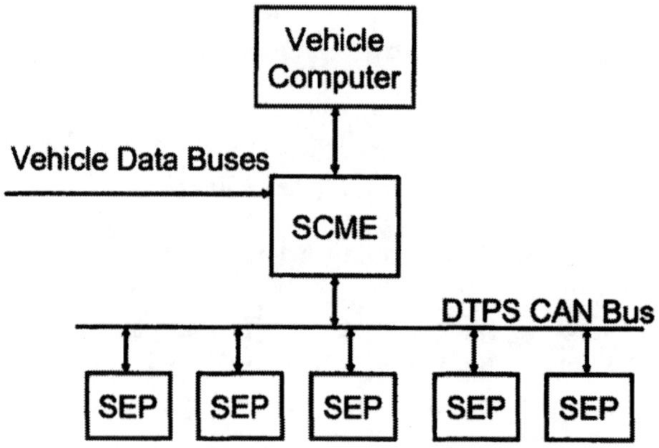

Figure 1DTPS System Architecture

The SCME is the primary prognostic activity center and central brain of the DTPS that produces diagnostic and prognostic results. The SCME coordinates the activities of the SEPs and performs the higher-level processing of information, including; detecting impending faults using the feature evidence provided by the SEPs, isolating the faults, trending, and estimating the remaining operational life of the monitored component. The SCME also gathers and processes information from the vehicle data buses. The SCME provides prognostic conclusions, including current drive train health and remaining life, to the EFV's computer.

Some monitoring tasks require sharing data and results between SEPs. An example is the planetary gear monitoring algorithm described in this paper, which requires that tachometer data acquired and processed by one SEP be transmitted and used by two other SEPs performing vibration monitoring.

Communications and Messaging: Interconnecting the SCME and SEP components is a bi-directional serial communications bus, using the SAE J1939 protocol [1], using the Controller Area Network 2.0B (CAN, 29-bit identifier field) bus physical layer (ISO 11898). While J1939 supports both master-slave messaging for SCME to SEP communications, and peer-peer messaging between SEP components, most DTPS traffic is master-slave, with the SCME acting as bus master and the SEP components as slave devices.

Specified with a maximum signaling rate of 250 Kbits/Second for robust performance, J1939 is best suited for near real-time messages having small protocol data units (PDU). The nominal 8-byte data field (single J1939 packet) may be extended up to 1785 bytes via multiple J1939 PDU packets. Appropriate assignment of J1939 Identifiers to the SCME and SEP components leverages the CAN 2.0B carrier-sense multiple-access, collision-avoidance (CSMA/CA) scheme to ensure the SCME always has highest priority for accessing the J1939 bus.

Within the DTPS, the J1939 Parameter Group Numbers (PGN) are also assigned for the purpose of facilitating seamless communication, while providing the necessary system performance. Thus, short control messages from the SCME to SEP, which may be time-critical (see "Synchronization and Multiple Clock Domains") are assigned PGN's in the sub-100 mSec, response range, while less frequently used data exchange messages are assigned PGN's consistent with less stringent response times.

Synchronization and Multiple Clock Domains: The distributed processing architecture of the DTPS is highly efficient for data flow, an essential element when reducing tens of megabytes of raw data per second into a form suitable for eventual prognostic evaluation. Hence, each SEP is responsible for raw sensor data acquisition from sensors associated with that SEP, as well as feature vector extraction from this data. As a result, the volume of data communicated from a SEP to the SCME may be minimized.

However, the distribution of multiple SEP components and associated sensors across mechanically unified equipment (i.e. the EFV drive train) introduces a challenging problem, the correlation of data collected by more than one SEP. For the SCME to accurately perform prognostic evaluation, the SEP data and results supplied to the SCME must be time-correlated. In addition, some algorithms require the sharing of data and intermediate results between SEPs with accurate alignment in time. Since the SCME, and each SEP, are independent (a necessary characteristic of distributed processing systems), each operates in a unique clock domain.

Hence, a report of the "current clock" time at SEP #1 will differ from a similar report (at exactly the same instant on a global clock) from SEP #2. Without a common time base, the SCME is unable to properly correlate the data and results from SEP #1 and SEP #2. One solution to this problem is to implement a distributed clock-synchronization algorithm, an approach used in some high-criticality systems (i.e. among the redundant components in an aircraft flight control systems). The requisite additional complexity to implement such an approach is impractical in the DTPS, since the single SCME presents a single-point of failure, violating the redundancy requirement for such an approach.

For the DTPS, a three-part solution was developed to provide the necessary synchronization:

1) SCME broadcasts master "time-stamp" message as appropriate
2) SEP marks current data acquisition stream upon receipt of SCME time-stamp message
3) SCME (or SEP as necessary) requests copy of data stream containing time-stamp from another SEP, then aligns the data streams at the requesting device.

J1939 is ideally suited for this approach, as the PGN assignments assure the SCME has immediate access to the J1939 bus, and the broadcast message protocol ensures each SEP receives the message at the same time. Since the interrupt service routine for processing

J1939 messages is identical on each SEP, the resulting time-stamps will be quite accurate.

Implementation Issues: In addition to the need for time-synchronizing multiple data streams at corresponding time-stamps, the DTPS also requires a capability for more absolute marking of features on the raw sensor data stream. Hence, the global event trigger (i.e. time-stamp) message issued by the SCME must be accurately positioned in the data stream by each SEP. At issue is the implementation of this approach on the SCME and SEP.

Three primary factors must be considered:

1) latency between SEP delivery of feature vector(s) to the SCME
2) processing time by SCME to transmit a global event trigger
3) latency at SEP to receive global event trigger and mark data.

Each SEP will require (perhaps varying) time to process collected raw sensor data and process feature vector(s) for transmission to the SCME. The SCME must collect feature vectors from each SEP, also with varying time duration. Finally, each SEP must receive the global event trigger message and insert a time-stamp to mark the data stream. The variability of the first two processes would typically be enough to invalidate the accuracy of such an approach for a single sensor data stream on each SEP. However, the nature of the features obtained by the SEP from the raw data stream is quite periodic, and closely related to the rotational speeds within the drive train. Hence, while the time variability of the first two steps is wide, the third step regains the necessary accuracy, since it is applied to a subsequent data stream.

However, the distribution of multiple SEP components and associated sensors across mechanically unified equipment (i.e. the EFV drive train) introduces a challenging problem, the correlation of data collected by more than one SEP. For the SCME to accurately perform prognostic evaluation, the SEP data and results supplied to the SCME must be time-correlated. Since the SCME, and each SEP, are independent (a necessary characteristic of distributed processing systems), each operates in a unique clock domain.

Thus, the third step in the process requires the highest accuracy. On the SEP, the complexity of processing indicates a need for operating system (O.S.) support to provide memory allocation functions, device drivers, interrupt service routines and process scheduling. The high computational requirements of the SEP favor a processing platform capable of supporting various embedded O.S. solutions, including eCos and Linux. Based on the need for very predictable interrupt service response times, eCos has been selected as the SEP O.S.

DTPS Algorithm Architecture: The DTPS algorithm architecture is a distributed, open, extensible software architecture that can host any number and any combination of diagnostic and prognostic algorithms running simultaneously. The DTPS architecture is based on a system previously developed by Honeywell called MPROS [2]. MPROS (for Machinery Prognostics and Diagnostics System) is a distributed shipboard system that performs diagnostics and prognostics on mechanical equipment (e.g. engines, generators, and chilled water systems). The MPROS prototype system incorporated four main algorithm types that addressed hundreds of failure modes in total.

Figure 2 is a diagram of the algorithm architecture, which consists of a set of a suite of OSA-CBM compliant algorithm modules. The outputs of the architecture are diagnostic and prognostic conclusions, which are passed on to the vehicle computer and used by soldiers and maintainers to make decisions regarding the readiness and health of the vehicle. Raw data and intermediate results are also available.

Note that a given algorithm lies partly on the SEP and partly on the SCME. Generally, the portion of the algorithm that lies on the SEP is that which provides sufficient data reduction for efficient communication over the bus – typically up to and including feature extraction. The subsequent stages of the algorithm are generally incorporated on the SCME.

The architecture supports a set of simultaneously active algorithms that can have overlapping areas of expertise. The advantage of this design is that more algorithms generally increase the chance of reaching a correct conclusion. The disadvantage of this design, however, is that sometimes there is disagreement among the conclusions. To handle situations of agreement and disagreement between overlapping algorithms, the system incorporates a Knowledge Fusion (KF) software module that uses a technique called the *Dempster-Shafer Rules of Evidence* to interpret the various conclusions. This technique can combine "beliefs" expressed by any of the algorithms in their conclusions to reach a combined belief. The output of this subsystem is a set of conclusions (e.g. identity of the root failure mode) and associated probabilities

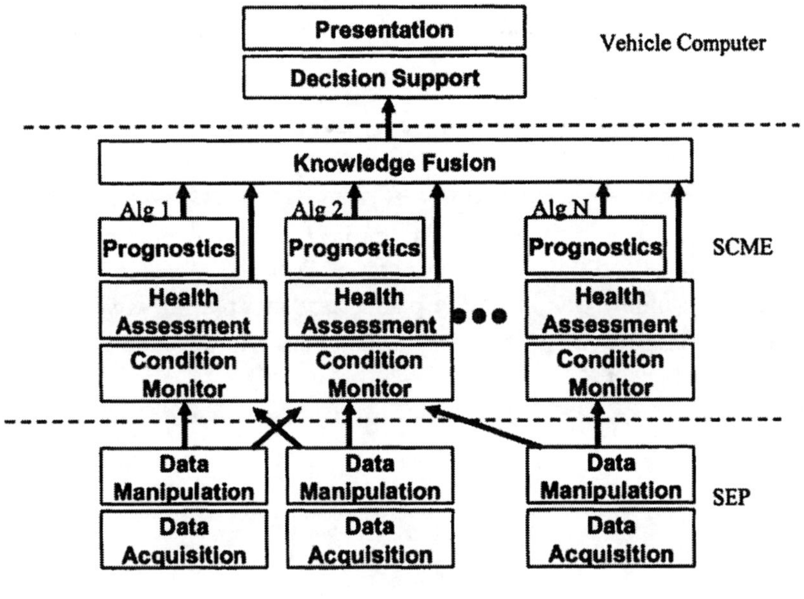

Figure 2 Algorithm Architecture

In addition, the KF subsystem performs a further merging based on remaining life estimates. If two conclusions from separate algorithms are providing the same prognostic root cause for the existing symptoms, KF combines their remaining life estimates. In order to do this, the algorithms are allowed to deliver "prognostic vectors", a data structure which can express complex remaining life estimates (e.g. the probability of failure is 50% in six months AND 80% in twelve months).

Planetary Gear Monitoring Algorithm: Our algorithm for vibration monitoring of planetary gears sets is based on the synchronous averaging and separation technique developed by Forester [5] and covered under Australian patent number 40638/95. The method provides a separate synchronously-averaged vibration signal for each of the individual planets based on an estimate of each planets instantaneous contribution to the total vibration signal.

The essential requirement for the separation technique is to place the vibration sensor so that the contribution from each planet is strongly amplitude modulated at the carrier frequency. We refer to this as "planet pass modulation". Ideally we place the sensor on the ring gear such that the vibration is strongest as the planet passes near the sensor and weakest when the planet is on the far side of the ring. The modulation envelopes for the planets peak at different points in time as the planets pass by the sensor in succession and this provides a means for separation. Other locations tend to reduce the degree of modulation and thus separability.

The algorithm uses windowing to selectively divide the vibration signal between the planets during synchronous averaging; more signal is allocated to the average corresponding to the planet closest to the sensor, and less to planets far from the sensor. Various windows are described by other researchers in the literature. The key innovation in Forester's technique is to use a window train of the form

$$\beta_p(t) = \left(1 + \cos\left(2\pi f_c t - p2\pi / P - \phi\right)\right)^n \qquad (1)$$

for each planet, p, where P is the number of planets, f_c is the rotational speed of the planet carrier, t is time, and n is the window order that allows one to adjust the duty cycle of the window train and thus the discrimination between planets. The offset angle ϕ is set to align the window train with the planet pass modulation envelope. Figure 3 illustrates a vibration signal with planet-pass modulation, and how the window function isolates a portion of the signal.

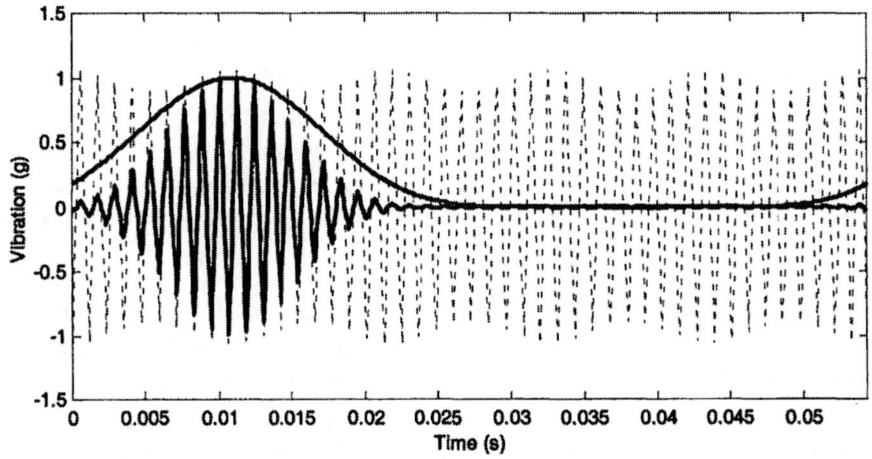

Figure 3 Planet-pass modulation and algorithm window

The benefits of this window function are:

a) It has no discontinuities, which can introduce noise in the averaged signal.

b) It takes no longer to perform than conventional averaging.

c) At all times 100% of the vibration signal is divided among the planetary gears. That is, $\sum_{p=0}^{P-1} \beta_p(t) = \text{constant}$ provided that $1 \le n \le P-1$. The window order is typically set to $n = P-1$ to obtain the maximum separation.

d) The averaged signal for each planet will have the planet pass modulation and window train modulation averaged out, provided two conditions are met. The first condition is that the average is made over N ensembles where NN_p / N_r is an integer (each ensemble is one relative rotation of the planet gear). Here N_p and N_r are the number of teeth on the planet and ring gear, respectively. The second condition is that kN_p / N_r is not an integer for any $1 \le k \le n$.. If it is, the value of n must be reduced until this condition is met.

There are two techniques for setting the window offset angle ϕ [3]. The more direct method is to monitor the carrier with a positional reference (e.g. a shaft encoder) and then relate the reference signal to the planet positions. One sets ϕ so that the window function peaks when one of the planets is closest to the vibration sensor. The other method, which we use, is to estimate the planet carrier position through software analysis of the planet pass modulation. One sets ϕ so that the window function peaks at the same time as the planet-pass modulation envelope. We perform the analysis to find the planet-pass modulation using the synchronously averaged signal of the ring gear.

With this planet separation technique, we are able to apply other commonly used gear prognostic approaches. We perform the signal averaging in the angular domain [4] rather than the time domain signal in order to accommodate modest variations in speed. We chose to only monitor the planetary gears under certain vehicle conditions in order to improve measurement-to-measurement consistency and obtain better trends. Namely we monitor at higher speeds when the EFV is up on plane and traveling at a relatively steady speed and the gears are under high load. The monitoring bandwidth is nominally 4 to 6 times the gear mesh frequency. From the averaged signal, we extract several commonly used statistical features and trend their values over time.

We perform the average over as many as $N = 10 \times N_r$ ensembles (14 seconds) to obtain a good signal-to-noise ratio (SNR), but we are exploring using half that. The number of ensembles is a trade-off between SNR and data acquisition time, and thus opportunities for acquisition during the desired steady-state, on-plane conditions. Since 14 seconds of steady state may be a rare occurrence in the field, working with many smaller ensembles could ultimately produce better trends, depending on how infrequently the 14 seconds of steady state occur.

Several unique aspects of our monitoring technique arise due to particulars of the DTPS system architecture and/or the EFV. Firstly, the complete planetary gear monitoring algorithm is distributed across the SCME and several SEPs. The SCME monitors the built-in sensors via the vehicle data buses (Figure 1), and then determines if the conditions are appropriate for one of the algorithms to run, such as the planetary gear algorithms. When the SCME triggers the planetary gear algorithms, one SEP acquires and processes the tachometer signal, and then transmits the results to the two other SEPs that are performing vibration monitoring of the port and starboard planetary gears. These two SEPs perform synchronous averaging on their respective vibration signals and extract features. The SEPs transmit their features to the SCME which uses the features to detect, isolate and prognose faults.

Another unique aspect is that the algorithm essentially runs on buffered data from the immediate past. At any given time, the SCME is analyzing the vehicle conditions over the last 20 seconds (nominally) to determine if the right conditions were present for sufficient time to run any of the DTPS algorithms. At the same time, the SEPs maintain a rolling buffer of the last 20 seconds of sensor data in anticipation of a trigger from the SCME to run the algorithms on this data. The rationale driving this approach is it allows the DTPS to make best use of the available opportunities. The SCME uses sophisticated triggering logic to ensure that each algorithm gets its share of run time, and that any rare opportunities to run a less used algorithm will be utilized.

Summary: In this paper we described the Drive Train Prognostic System Honeywell is developing for the Expeditionary Fighting Vehicle, and our approach to monitoring the planetary gear systems. We described some of the many unique solutions we are employing, including a distributed network of smart sensors, a proven algorithm architecture, and a method for performing synchronous averaging on planetary gears. The work is ongoing, and much of the development is still ahead of us. The first installation of the DTPS on EFVs is planned for Q3 2005, and the planetary gear monitoring software is planned for Q3 2006.

References:

[1] SAE J1939 protocol "Recommended Practice for a Serial Control & Communications Network – Surface Vehicles".

[2] George D. Hadden, "Shipboard Machinery Diagnostics and Prognostics/Condition Based Maintenance: A Progress Report", with Bonnie Holte Bennett, Peter Bergstrom, George Vachtsevanos, and Joe Van Dyke, *Proceedings of the 2000 IEEE Aerospace Conference* (Aero2000), 2000.

[3] McFadden, P. D., "Examination of a technique for the early detection of failure in gears by signal processing of the time domain average of the meshing vibration", *Mechanical Systems and Signal Processing*, Vol. 1(2), pp. 173-183, 1987.

[4] McFadden, P. D. "Interpolation techniques for the time domain averaging of vibration data with application to helicopter gearbox monitoring", *Aero Propulsion Technical Memorandum 437*, Department of Defense, Aeronautical Research Laboratory, September 1986.

[5] Forrester, B. D., "Advanced Vibration Analysis Techniques for Fault Detection and Diagnosis in Geared Transmission Systems", PhD thesis, Swinburne University of Technology, February 1996.

BEARING INCIPIENT FAULT DETECTION: TECHNICAL APPROACH, EXPERIENCE, AND ISSUES

Patrick Kalgren, Carl S. Byington P.E., Sanket Amin, Kallappa Pattada

Impact Technologies, LLC
2571 Park Center Blvd
State College, PA 16801
Patrick.Kalgren@impact-tek.com

Abstract: This paper describes a technical approach for detection and prediction of incipient bearing faults in complex machinery applications. The authors have developed and adapted the ImpactEnergy™ suite of vibration signal processing algorithms that are sensitive to bearing fatigue and wear events but minimize sensitivity to machinery noise generated by gearing. The ImpactEnergy™ algorithm suite develops an array of time and frequency domain features generated through an adaptive enveloping and demodulation scheme to provide quality diagnostic outputs for advanced machinery health monitoring. The technique reviewed here provides early detection and severity assessment of bearing faults by utilizing multiple regions of the vibro-acoustic energy spectrum. The multiple bandwidth strategies and feature extraction algorithms begin with a foundational understanding of mechanical fault propagation in drive train bearings. Experience with these algorithms and correlation of the indicators within seeded and transitional tests is offered in this paper. The authors discuss specific issues of observability and application of these techniques to the Expeditionary Fighting Vehicle.

Key Words: Bearing Diagnostics; Demodulation; DTPS; Statistical Analysis; Vibration Analysis

Introduction: Awareness of current and projected asset health, through machinery condition monitoring, enhances military operations and planning. A unique Drive-Train Prognostic system (DTPS) in development for the Marines Expeditionary Fighting Vehicle (EFV) will utilize a distributed, high-fidelity, processing architecture to provide state of the art machinery monitoring and prognostic capability in high criticality areas of the multi-mode drive-train. Analysis of machinery vibration provides data useful for condition assessment and fault diagnosis.

Mechanical systems are fitted with acceleration transducers strategically located to record vibrations excited by various mechanical components and sub-systems. This recorded information commonly passes through signal conditioning, pre-processing, and feature extraction stages to analysis algorithms, which compare the conditioned signal with baseline test data and evaluate differences that may indicate problems in the mechanical system. However, traditional bearing analysis techniques fall short of providing the early fault detection capability essential for prognostics on complex machinery systems. The authors have researched and developed theImpactEnergy™ feature extraction technique, based on traditional signal processing and vibration feature extraction principles, that generates time and frequency domain features through an advanced enveloping and demodulation scheme. ImpactEnergy™ increases the bearing fault detection horizon when compared with traditional techniques and provides severity assessment of bearing faults by utilizing multiple regions of the vibro-acoustic energy spectrum. This technique provides the detection, diagnostic, and prognostic foundation for bearings monitored by the EFV DTPS.

High Frequency Resonance Principle: Vibro-acoustic data continues to provide some of the most reliable quantitative indicators of bearing, gear, and rotating member fatigue and wear. The indicators typically spread throughout the vibro-acoustic regime. Monitoring the high frequency range of the spectrum for increasing levels of vibrations is an effective method of identifying and tracking bearing and rolling element incipient fault condition (1, 2, and 3). Early material distress and incipient faults, usually masked by machinery noise at low frequencies, are most detectable at higher frequencies. Indication of developing failure at this point provides the greatest detection horizon. Figure 1 shows the conceptual basis of early fault detection involving bearing components.

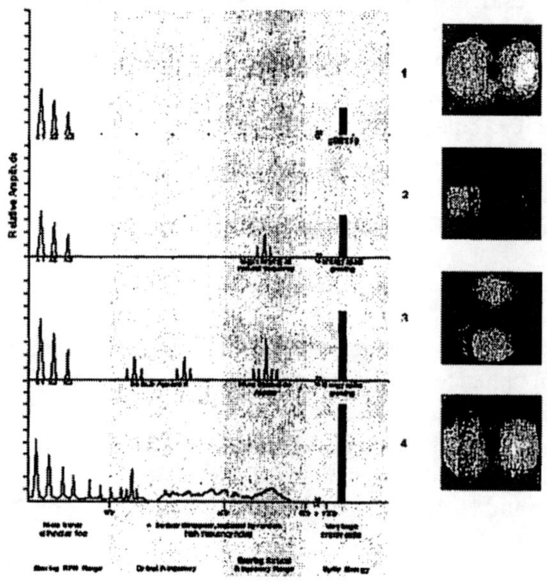

Figure 1 Incipient (early) failure symptoms are most apparent at the highest frequencies. [1]

[1] Adopted from *Illustrated Vibration diagnostics Chart* by Technical Associates of Charlotte Inc. 1994

In failure stage 1, problems appear in the ultrasonic frequency range correspondingly with an increase in spike energy. During failure stage 2, slight defects begin to ring bearing and structural natural frequencies (fn) with the appearance of sidebands around fn. A developed fault in a mechanical component such as a rolling element will create an impulse every time the defect rolls over the race causing the natural frequencies of the bearing and the housing structure to be excited. At failure Stage 3, bearing defect component frequencies and harmonics appear predominantly if the overall machinery noise is not too high. The result will be an increase in the vibration energy at these high frequencies, usually greater than 10 kHz. As wear progresses more harmonics appear with stronger sidebands around defect frequencies and fn. Wear becomes visible to the unaided eye. At the very end of life, depicted as Stage 4, introduction of additional harmonics and high frequency random noise disguises defect frequencies hindering their visibility in the spectrum.

The impulse produced from a defect has an extremely short duration compared to the interval between the pulses. Furthermore, the energy of the pulse distributes itself at very low-levels and over a broad spectrum. This wide distribution of energy makes bearing defect detection difficult through conventional spectrum analysis and more profoundly so when in the presence of higher energy vibrations from gears and other machine components. Fortunately, the defective impact excites a resonance in the system at much higher frequencies than the vibration generated by other components. This structural energy is usually concentrated into a narrow band that is easier to detect than the widely distributed energy of the bearing defect frequencies. Demodulating this narrowband frequency region reveals modulating components with far less ambiguity, leading to an extended fault detection horizon and enhanced detection confidence.

Technique: ImpactEnergy™ is a signal processing and feature extraction technique, compatible with commercial, off the shelf (COTS) sensors, that exploits the resonance principle discussed in the previous section. Incorporating a unique demodulation process, this algorithm detects mechanical fault frequencies that serve as early warning indicators of emerging failure. This is typically in the range of 10-70 kHz in most applications, although in some applications it may range as high as 100 kHz. An enveloping and demodulation process, depicted in Figure 2, takes advantage of the information available in the resonance excitations. The process begins with use of analytically pre-determined filter bands that permit analysis of multiple signal paths. The pass bandwidth of these filters permits several of the relevant sidebands on the carrier signal to pass through. Next, the band-passed filtered signal is full wave rectified to extract the envelope. This rectified signal then passes through a low pass filter to remove the high frequency carrier signal and final processing removes DC content. Overall, the process produces several vibration features for use in diagnostics and prognostics. Primary figures of merit include the peak amplitude of the component bearing fault frequencies in the power spectral density of the enveloped data. These features are described next.

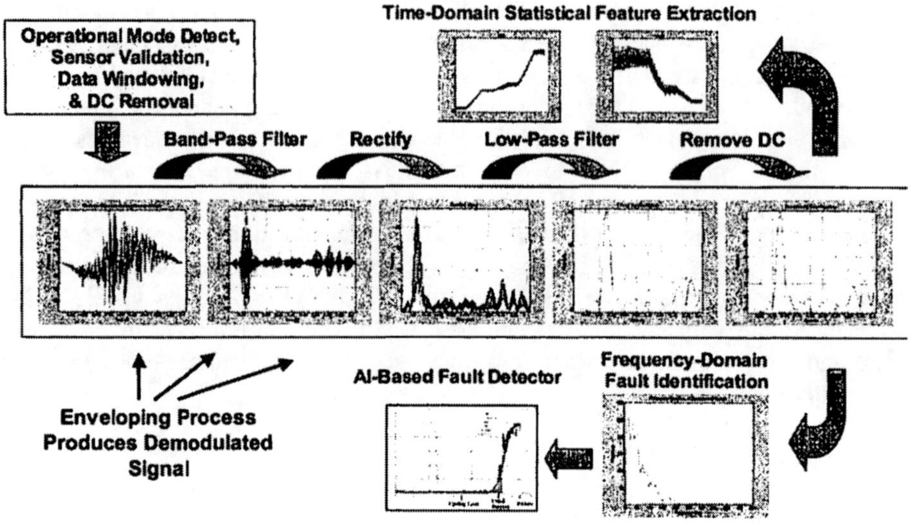

Figure 2 – Flow Chart of the Demodulation (Enveloping Process)

Frequency Domain Features: The fundamental fault frequencies of bearing components qualify as effective indicators of health for the respective component. Specifically their FFT extracted frequency magnitudes are in direct relation to the level of damage. A large magnitude in the frequency spectrum at any of component frequencies would indicate a potential fault occurring in that component of the bearing. The equations for calculating the bearing component frequencies are well known and simple to calculate and provided below for review. These equations assume no skidding or sliding of the elements along with a stationary outer raceway and rotating inner raceway.

Ball Spin Frequency (BSF):

$$f_{BSF} = \frac{P_d}{2B_d} fn\left(1 - \left(\frac{B_d}{P_d}\right)^2 \cos^2\beta\right) \text{ (Hz)} \quad (1)$$

Ball Pass Frequency Inner Race (BPFI):

$$f_{BPFI} = \frac{z}{2} fn\left(1 + \frac{B_d}{P_d}\cos\beta\right) \text{ (Hz)} \quad (2)$$

Ball Pass Frequency Outer Race (BPFO):

$$f_{BPFO} = \frac{z}{2} fn\left(1 - \frac{B_d}{P_d}\cos\beta\right) \text{ (Hz)} \quad (3)$$

Cage Frequency (CF):

$$f_{CF} = \frac{fn}{2}\left(1 - \frac{B_d}{P_d}\cos\beta\right) \; (Hz) \quad (4)$$

Where: fn = the rotational speed in units of Hz

z = number of rolling elements

B_d = individual ball diameter

P_d = pitch diameter

β = contact angle

Figure 3 and compare two frequency domain plots from an example test set of vibration data generated near the end of a test, only minutes before failure and test shutdown. Figure 3 shows bearing fault frequencies extracted through conventional FFT analysis. Indicated are the four fault frequencies, as well as the first four harmonics of the BPFI. However, the ImpactEnergyTM extracted FFT (Figure 4) improves the visibility, and subsequently the ability for detection, of the fault frequencies by providing a much lower and cleaner noise floor. At the test conclusion, a visual inspection indicated a failure of the inner raceway, which correlates to the higher magnitude BPFI frequencies compared to the shaft and other bearing frequencies. In a complex machinery environment, such as the EFV, the machinery noise floor can easily mask component frequencies limiting the ability for damage detection only prior to critical failure.

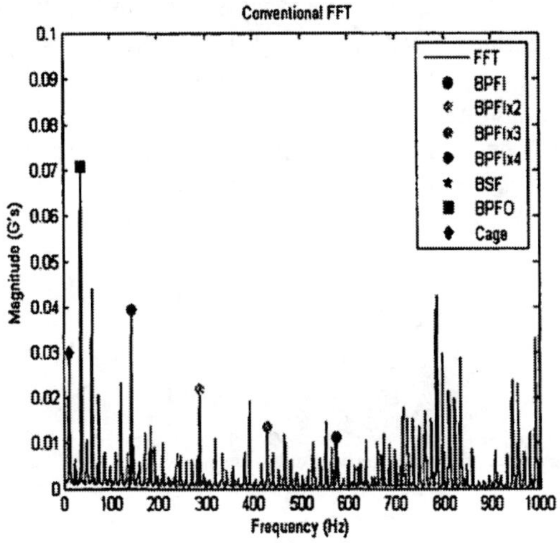

Figure 3 – Conventional FFT

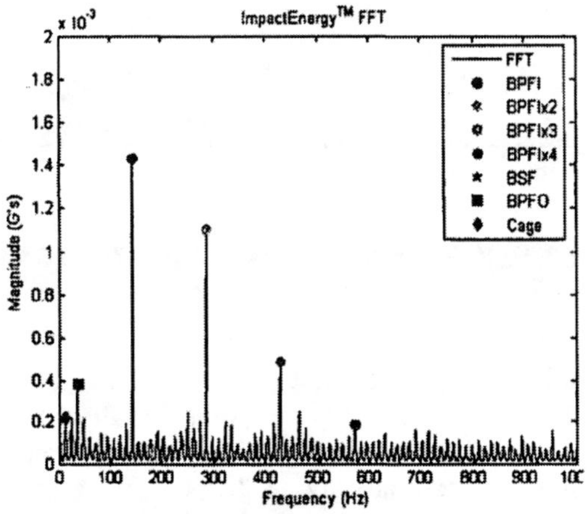

Figure 4 ImpactEnergy™ FFT

Statistical Detection Comparison: As exhibited above, the ImpactEnergy™ technique enhances the ability to detect and distinguish defect frequencies from surrounding machinery noise. This not only increases the detection horizon for incipient faults but also produces diagnostic features that increase confidence in final prognostic estimates, which reduces the probability of false alarms and nearly eliminates opportunity for missed detection. Statistical detection analysis, using probability distribution plots as the one shown in Figure 6 can be used for selecting and implementing optimal thresholds for "calling out" specific fault conditions based upon separability of features in differing fault regimes. The Probability of False Alarm P(FA) and the Probability of Detection P(D) are related because both are measured with respect to a particular threshold applied to the reduced/fused features. Raising the threshold to decrease P(FA) results in a consequential decrease in P(D) and an increase in the probability of missed detection P(MD) [4]. For effective diagnostics, balancing the P(FA) and P(D) using available data sets, simulations, and models as appropriate, is an effective method for setting thresholds. Maximizing feature separability and reducing the variance as much as possible is key to achieving good performance, which can be accomplished through proper mode detection, enhanced signal processing, and fusion algorithms.

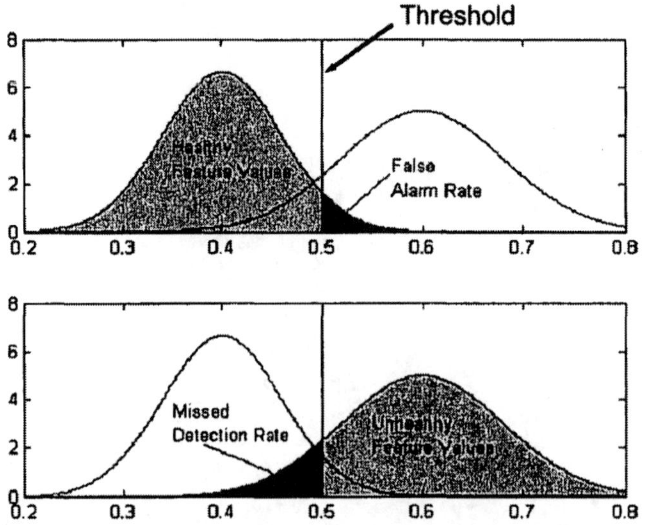

Figure 6 – Relationship of Statistical Detection Metrics

A statistical study conducted by the authors incorporates three different vibration processing methods to compare their ability to distinguish three separate bearing health regimes. The same dataset used for all three techniques, is a compilation characterizing the three stages of bearing life from healthy, to incipient fault, and finally failure spall. The developing fault occurs on the outer raceway of the bearing so the BPFO frequency feature serves as good comparison metric among the three methods. The illustrations below each exhibit three bearing conditions: healthy bearing seeded fault, and progressing spall. Figure 7 shows separability results for the BPFO feature utilizing a legacy bearing fault detection method. This low-fidelity approach, widely implemented in commercial systems, samples accelerometer data with an effective bandwidth of 0-1kHz. A damage alarm threshold chosen at the intersection of the healthy feature PDF right edge and the damage PDF left edge represents a common choice that minimizes false alarms while maximizing detection. The low-fidelity technique provides low separability of healthy and seeded fault data and can result is a high false alarm rate.

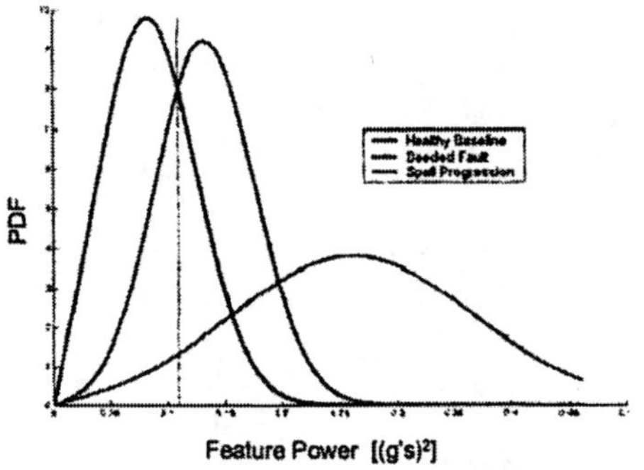

Figure 7 – Legacy System Bandwidth (0-1 kHz)

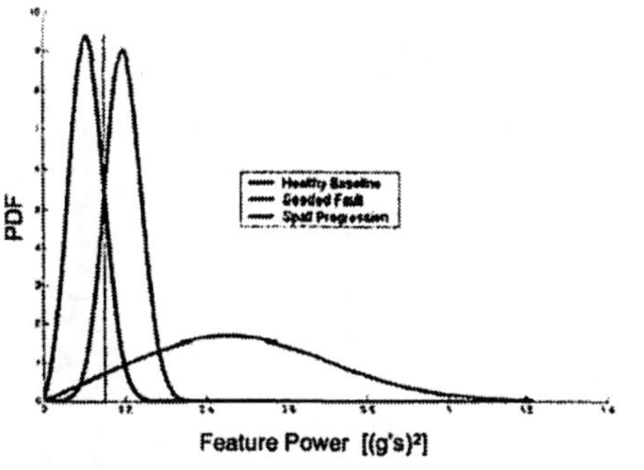

Figure 8 – Traditional Processing Higher Bandwidth (0-25kHz)

The next technique compared (Figure 8) is similar to the previous technique with the exception of an increased effective bandwidth (0-25kHz). While the technique reduces the variance for two of the three fault regimes, further minimizing the P(FA) and P(MD), there still exists a significant overlap due to high variance of the third region, spall progression, which adds a complicating factor in computation of final health estimates. The final plot illustrates results obtained using ImpactEnergy™. The results in Figure 9 depict a notable decrease in variance for the healthy and seeded fault regimes while the distance between means of the three fault distributions has increased substantially reducing P(FA) and P(MD) along with significant reduction in overall of the spall progression regime.

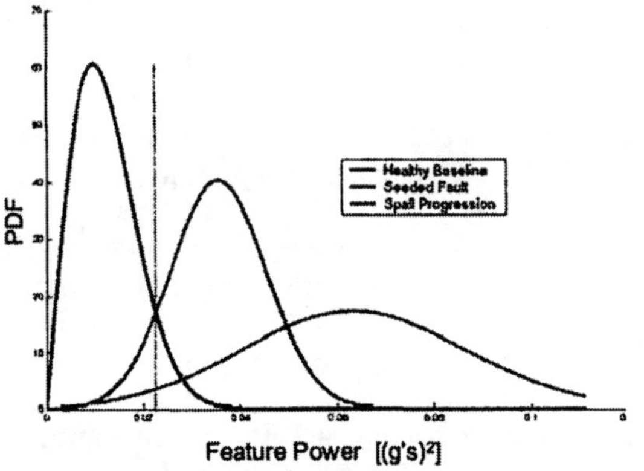

Figure 9 – ImpactEnergy™ Separability Results

Table I presents a summary of the statistical detection comparison of the three techniques. The table shows P (FA) and P (MD) for corresponding thresholds. Values are provided for separation between both healthy and seeded fault and healthy and spall progression regimes. For this study, a comparison is made of a single feature, however, in critical applications; fusion of multiple features on multiple levels would provide the highest separation and detection ability.

Table I - Quantitative Statistical Detection Comparison Results

Technique		Seeded Fault	Spall
ImpactEnergy™	Threshold	0.0227	0.0282
	P(FA)	7.62%	1.90%
	P(MD)	9.23%	5.52%
Freq. Domain (25kHz)	Threshold	0.1524	0.2003
	P(FA)	15.40%	1.72%
	P(MD)	5.50%	9.60%
1 kHz Engine Order	Threshold	0.1101	0.1564
	P(FA)	28.18%	4.83%
	P(MD)	29.33%	14.45%

Time Domain Features: Time domain features developed during the process play a dual role in the overall bearing PHM approach. Algorithm configuration and mechanical system customization utilizes time domain features in the initial identification of significant resonant bands. Time domain features also play a significant role in an overall approach to condition monitoring within a data-driven, multi-level feature fusion process. Some figures of merit from multiple broad and narrow filtered bands include features such as maximum amplitude, crest factor, root-mean-square (RMS), and kurtosis. The maximum absolute vibration amplitude measured during a specified time interval can indicate the onset of bearing damage. The RMS value of a vibration signal is a time analysis feature signifying the measure of power content in the vibration signature. The feature escalates significant when bearing damage is present but also is good for tracking the overall machinery noise level. The simplest approach to measuring defects in the time domain is using the RMS merit. However, the RMS level may not show appreciable changes in the early stages of bearing damage. An additional measure is the "crest factor," which is defined as the ratio of peak level of the input signal to RMS level. Therefore, peaks in the time series signal will result in an increase in the crest factor value. This feature is useful in detecting changes in the signal pattern due to impulsive vibration sources such as tooth breakage on a gear or a defect on the outer race of a bearing. An additional extracted feature is kurtosis, the fourth statistical moment of broadband and specific narrowband vibration signals. A changing kurtosis value potentially signifies the formation of a spall or other incipient fault conditions.

Customization for EFV: Adaptation to different mechanical systems results in a processing technique dedicated and optimized for the intended use. A highly dynamic vehicle such as the EFV requires a sophisticated health monitoring system. The DTPS software utilizes techniques well understood and established. A high degree of customization for the specific application ensures maximal effectiveness. Using data obtained during seeded fault and operational characterization tests, the DTPS team has begun the process of honing the technologies. The ImpactEnergyTM customization process begins with analysis of data obtained from the tests described above. Time and frequency domain features provide insight to isolate effective target bands and develop initial threshold settings. Selection of primary and secondary sensors and development of higher order data fusion specific to the EFV drive train requires substantial characterization data. Real time operational data provides the final linkage required to ensure system reliability. During the verification and validation stage, operational data will provide ongoing feedback, permitting analysis of realistic data and refinement of the deployed algorithms.

Summary: This paper provided a brief description and results from a specialized bearing processing technique. The authors have demonstrated the portability of the algorithm and discussed adaptation for a health monitoring system specific to the Expeditionary Fighting Vehicle. Common vibration analysis techniques pose risk in application for complex mechanical systems like EFV because machinery noise typically overwhelms bearing defect indictors. ImpactEnergy™ exploits structural resonance frequencies that carry incipient defect information from bearing components. Much like a stethoscope may isolate and focus on a local signal, the algorithm described here focuses on the monitored bearing. Demodulating the information rich resonance region provides a much cleaner broadband energy spectrum for detection analysis when compared to traditional techniques. This simplifies the extraction of diagnostic features and provides a higher confidence prognostic. Ongoing related developments provide possible future enhancements to ImpactEnergy™ including automated signal validation and automated adaptive band selection. The incorporation of these additional algorithms and techniques into a prognostic health monitoring system can increase accuracy of the health estimate. The First Check™ algorithm examines raw vibration data to diagnose the validity of the information provided by sensing transducers. Operation of this algorithm in a pre-processing stage ensures sensor signal. Automated Band Selection (ABS) is another technique under development for pre-processing to ImpactEnergy™. System resonance identification is critical in determining the effective bandpass filter ranges for producing accurate and effective demodulation results. Preliminary tests of the ABS process show benefits in decreased variance results, permitting greater fault detection with fewer false alarms. An effective implementation of ImpactEnergy™ will produce an array of diagnostic features upon which classifiers and inference engines may build by incorporation of additional physics of failure models and usage-based methods to deliver a robust overall health prognosis. More significantly, the technique described herein provides a much earlier failure detection capability compared to conventional vibration analysis techniques. In an application such as the EFV, this is critical for maximal vehicle utilization, mission completion, and crew safety.

References:

1. Bagnoli, S. Capitani, R., Citti, P., "*Comparison of Accelerometer and Acoustic Emission Signals as Diagnostic Tools in Assessing Bearing Damage,*" Proc. 2nd Intl. Conf. Condition Monitoring, London, pp 117-125, May 1988.
2. Braun, S., Datner, B., "*Analysis of Roller/Ball Bearing Vibrations,*" Transactions of the ASME, Vol. 101, pp 118-125, Jan 1979.
3. D'Amato, E. and Rissone, P., "*Using the Envelope Method to Monitor Roller Bearings,*" Proc. 1st Intl. Mach. Monitoring Diagnostic Conf., NV, pp 560-566, Sept 11-14, 1998.
4. Dowling, Martin "*HUMS METRICS: Monitoring the Monitor,*" Rotorcraft Industry Technology Association, Inc. Final Report., Jan 2000.
5. Kalgren, Patrick W., Byington, Carl S., Pattada, Kallappa., "*An Intelligent Ultra High Frequency Vibration Monitoring System for Turbomachinery Bearings,*" Proc. 2004 ASME/STLE International Joint Tribology Conference, CA, Track 8, TRIB2004-64317, October 24-27, 2004.
6. Byington, C., Watson, M.., Kalgren, P., "*Metrics Evaluation and Tool Development for Health and Usage Monitoring System Technology,*" Proc. 59th AHS Forum
7. Harris, Tedric A., "*Rolling Bearing Analysis,*" John Wiley & Sons, Inc., New York, pp 993-1000.

STRUCTURAL HEALTH MANAGEMENT

Chair: Dr. Robert W. Bolton
Texas A&M University

A Time and Frequency Domain On-line Crack Identification of Stiffened Plate Panels

Agung Budipriyanto **A.S.J Swamidas** **M.R. Haddara**
Graduate Student **Professor of Civil Engg.** **Professor of ONAE**
e-mail : agungbp@engr.mun.ca

Faculty of Engineering and Applied Science
Memorial University, St. John's, NL, Canada

Abstract:

This paper describes a methodology for on-line crack identification of a stiffened plate structure. The methodology is based on the measured/computed root mean square (rms) amplitude response obtained from random vibration testing and analysis. The random excitation force simulated the excitation force exerted on the structure due to the modeled wave spectrum (with a dominant frequency of 2 Hz). The fundamental frequency of the structure was 584.3 Hz (flexural-torsional frequency of the main stiffener), whereas the first bending frequency of the stiffened plate system was 1136.7 Hz. The basic idea is that when a crack is present in a structure, the computed/measured rms amplitude of response would change; and by monitoring the rms responses at several locations for several crack cases, the crack location and extent could be identified. A one-twentieth scale model, of a stiffened side shell panel of a tanker ship structure, was tested in the experimental study; a specimen was tested for different crack development scenarios. Since the methodology does not require the measurement of the excitation forces, the measured structural responses were normalized by the computed rms response amplitude [computed for the excitation frequency range (0 to 5 Hz)]. Results of testing carried out on the model and analysis of test results for predicting various damage cases are reported.

Key Words: Crack identification; Experimental testing and analysis; The root mean square of response; Stiffened plate panel; Vibration based inspection in time domain

Introduction:

Identifying crack in large structures such as ships or bridges is a challenging task. Earlier studies have reported that fatigue cracks in side shell of ship structures mostly occurred at the intersection of longitudinals with web frames or bulkhead (Sucharki [1]). The crack starts at a region of high stress concentration in the longitudinal connection to bulkhead or web frame; it then propagates across the flange and the web of the longitudinal. Even though under normal operating conditions the cracks rarely lead to a catastrophic failure of the ship structure, if the crack reached the side shell plate it might lead to leakage of tanker ship contents. The spillage would prove to be harmful to the environment.

Investigations on uni-directionally stiffened plate models representing the side shell of tanker ship structures were conducted by Zubaydi et al. [2]. Experimental and numerical studies were carried out. The contributions from non-structural modes generated from the test bed were investigated and they found the contributions to be negligible. A crack detection methodology using a function having damping force and non-linear restoring force terms as damage indicator was proposed. Back propagation neural network was employed to compute the damage indicator. They stated that the results were encouraging. Budipriyanto et al. [3] reported on the development and fabrication of a stiffened plate model of a tanker ship side shell structure for crack identification. The model was made of aluminum; it consisted of a plate with two web frames, which stiffened the plate vertically and four horizontals, which stiffened the plate horizontally. The best sensor locations were determined from numerical simulation using finite element analysis. Based on the above study, six accelerometers and six strain gages were mounted on the model to monitor the response changes due to cracks. The natural frequencies obtained from experimental tests and numerical investigations using finite element were in very good agreement.

Identification of cracking in time domain has drawn the attention of a number of researchers [4-7]. Lee et al. [4] proposed a scheme for detection of faults in heating, ventilating and air-conditioning (HVAC) systems using residuals and changes in model parameters as damage indicators. The residual values were computed by subtracting measured values under fault conditions to their values under normal conditions. The parameters examined in the study were supply air temperature, cooling coil control signals, supply of duct static pressure and flow difference between fans. To investigate the change in model parameters the systems were modeled using ARMAX (Autoregressive moving average with external input) and ARX (Autoregressive with external input) models; then the model parameters were calculated using a Kalman filter. It was assumed that physical changes in the systems would result in changes of model parameters. The systems were considered to be damaged when the damage indicators were exceeded by three standard deviations. They reported that the scheme worked successfully when it was applied to laboratory experimental data obtained from systems under damaged and healthy conditions.

Worden et al. [5] carried out numerical simulation and experimental investigations on damaged and undamaged structures. A statistical analysis called outlier analysis was proposed for damage detection using Mahalanobis squared distance as a damage indicator. The results of the investigations showed that the proposed damage indicator was able to discriminate between damaged and undamaged structures. Andrade et al. [6] proposed a crack identification technique using the Kolmogorov-Smirnov test; the technique was based on the assumption that time signatures of a structure under cracked and intact conditions had different probability distribution functions. An experimental investigation on gears under cracked and intact conditions was performed to investigate the performance of the proposed technique. The gear under investigation was operated at a rotational speed of 5 Hz. An accelerometer was utilized to monitor the responses which were recorded at 5.12 kHz frequency sampling rate. Choi and Stubbs [7] developed a damage detection methodology using the mean strain energy computed from displacement response; a damage indicator was defined as the ratio of the element stiffness of intact and cracked elements. The methodology was verified by utilizing numerical data computed from a continuous I beam subjected to impact excitation. They showed that the proposed methodology was able to detect the damage location and quantify its crack.

In the present paper, the results of numerical and experimental investigations of a stiffened plate model of a tanker ship's side shell are reported. The root mean square responses under intact and four different crack cases were determined using computed and measured responses. Acceleration and strain responses of the model due to cracks occurring at one and two locations of the intersection between a horizontal and bulkheads were examined. A damage identification scheme is presented.

Numerical and Experimental Investigation:

To investigate the root mean square response of a 1/20 model of a tanker ship's side shell structure numerical and experimental investigations were carried out. The model was developed and fabricated for crack identification investigation using its vibratory response. It was made of aluminum; it had two web frames and four horizontals (Budipiyanto, et al. [3]). The 1/20 model was tested under intact and various damage conditions. The damage cases considered in this investigation are presented in Table 1. In the numerical investigation the root mean square (rms) responses were computed using a finite element analysis computer program [8]. The numerical investigation results were compared with the experimental results.

Table I: Damage cases

Case No	Crack Locations	
	Location 1	Location 2
1	10 mm (0.394")	none
2	20 mm (0.787")	10 mm (0.394")
3	20 mm (0.787")	20 mm (0.787")
4	25.4 mm (1.000")	30.8 mm (1.212")

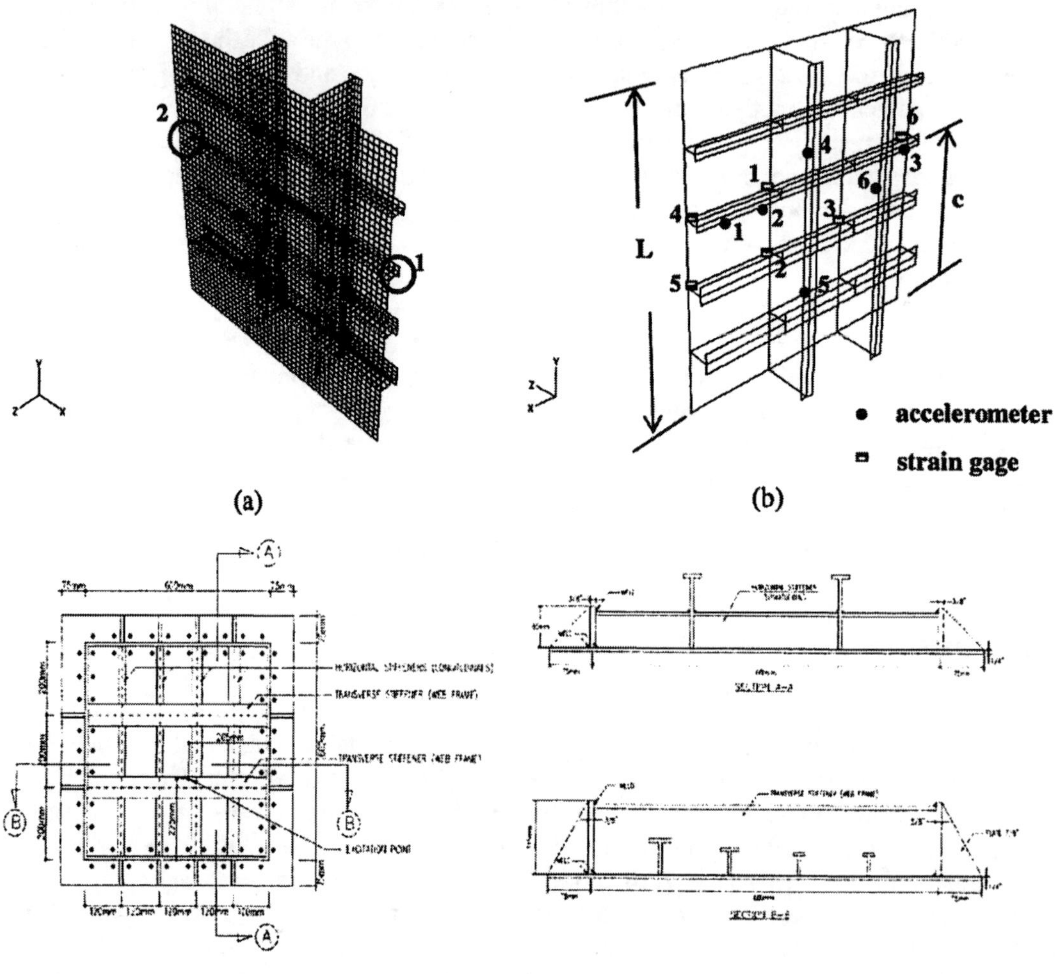

Figure1. (a) Finite element model and crack locations; (b) Accelerometer and strain gage locations and number and notations used in Eq. 2; (c) The physical model of a 1/20 side shell structure

Numerical Investigations:

A finite element compute package program ABAQUS [8] was used to compute the root mean square responses of the 1/20 model of the side shell structure under intact and damage cases investigated in this study. The finite element mesh of model is shown in Fig. 1(a). The model was tested using an excitation force which simulated wave forces following Pierson-Moskowitz (P.M) wave spectrum. To obtain the wave force spectrum, random wave height time history of the spectrum was computed and the wave forces exerted on the prototype of a side shell structure were calculated using the method presented in Haddara [9]. It was found that the dominant frequencies of the force were around 0.1 Hz. Based on these results, the 1/20 model was excited by random excitations whose dominant frequency was 2 Hz.

The rms at points, where accelerometers and strain gages were located [see Fig.1(b) for these sensors' locations], were calculated in the frequency range of 200 Hz to 600 Hz and then normalized by the largest value. The rms values showed significant changes around the frequency of 600 Hz (for detailed results see Budipriyanto et al. [10]). The package program assumed that the system response has a zero mean. Therefore the mean of responses under investigated damage cases could not be obtained.

Experimental set up and experimental investigations:

The 1/20 stiffened aluminum plate model was fabricated and tested to investigate the change of the mean and the root mean square response of the model under intact and damaged conditions. Figure 1(c) shows the fabricated model. The experimental set up is shown in Fig. 2. The model was excited by random loads generated from a Wavetek signal generator; the input signal was filtered by a low pass Krohn-Hite filter with a cut off frequency of 2 Hz to simulate wave force dominant frequency regime. The signal was amplified and then employed to drive a B&K model 4812 vibration exciter which was connected to the model thorough a stinger. To monitor the excitation force a Kistler model 912 load cell was attached to the stinger and it was connected to the plate of the tested model. The dynamic responses of the model were measured using MM EA-06-062AP-120 strain gages and accelerometers. The transducer signals (load cell, strain gages and accelerometers) were passed through amplifiers prior to acquiring data using LabView [11] data acquisition system which recorded data at a frequency sampling of 12 kHz for about 8 to 10 seconds for one measurement. Ten measurements were carried out for each case. The recorded data were stored in a personal computer for further analysis.

The response data were calibrated. To obtain the root mean square (rms) of the model's response under intact and damaged conditions the response was converted into frequency domain for obtaining the response autospectrum and then the rms was calculated in frequency range of 570 Hz to 600 Hz, by integrating the auto spectrum of response. The frequency range was selected since in that range the analytical rms values showed significant changes; it would be due to the fact that there were two dominant frequencies in that frequency range. In addition it was carried out to minimize the effect of noise, which would be present in the measured data. Although the frequency range selected was different from the range used for the numerical investigations, it would not matter since only relative values of the rms were used (those rms values were normalized by the largest value). It should be noted that the response data used for computing the rms were different from the data used earlier for obtaining the amplitudes presented in Budipriyanto et al. [3]. Moreover the response data were not filtered for obtaining the rms.

The mean of the response and the rms of excitation forces from different measurements under the same conditions as well as under different damaged cases were also examined. It was observed that the mean and the rms of the excitation forces obtained from ten measurements for each case changed from one measurement to the other for the same conditions. Therefore it was decided that mean of the response would be zeroed and the measurement response was normalized by selected rms amplitude of response in the range of excitation frequency (0 to 5 Hz). The detailed results could not be presented in this paper due to limitation of space.

Numerical and Experimental Results:

The normalized rms obtained from numerical and experimental investigations are shown in Fig. 3 (a), (b), (c) and (d). These values were the average values of ten measurements. The experimental acceleration rms values show the same trend as the numerical values. The rms values obtained from accelerometer #1, #2 and #3 changed consistently due to the increase of crack lengths (see Fig. 1(b) for sensor numbering); in addition the accelerometers #4, #5 and #6 (located on two vertical web frames) decreased consistently. The strain rms values measured from strain gages #1, #4 and #6 show the same trend as the numerical values while results obtain from strain gages #2, #3 and #5 do not. This might be due to the very small strain changes measured from these gages and the residual errors from the sensing instrument.

Crack identification scheme:

The rms values obtained from the numerical and experimental investigations suggested the use of a damage indicator, e_j that can be expressed as

$$e_j = e_{j-1} + \left| rms_j - rms_{j-1} \right| \qquad j = 1,2,3,4 \tag{1}$$

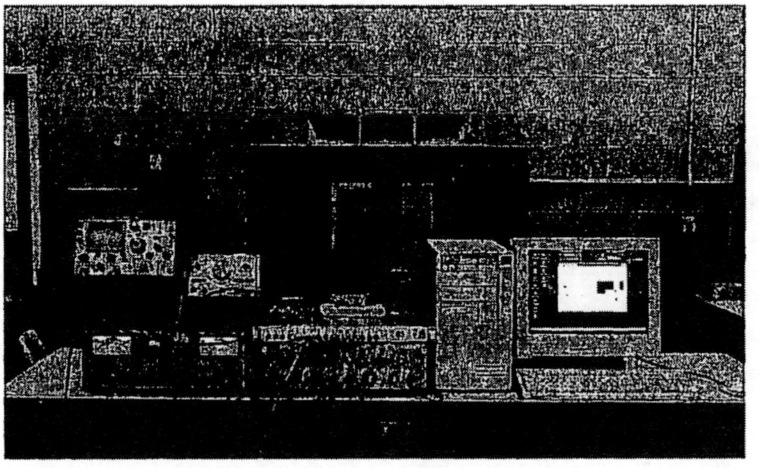

Figure 2. Experimental set up

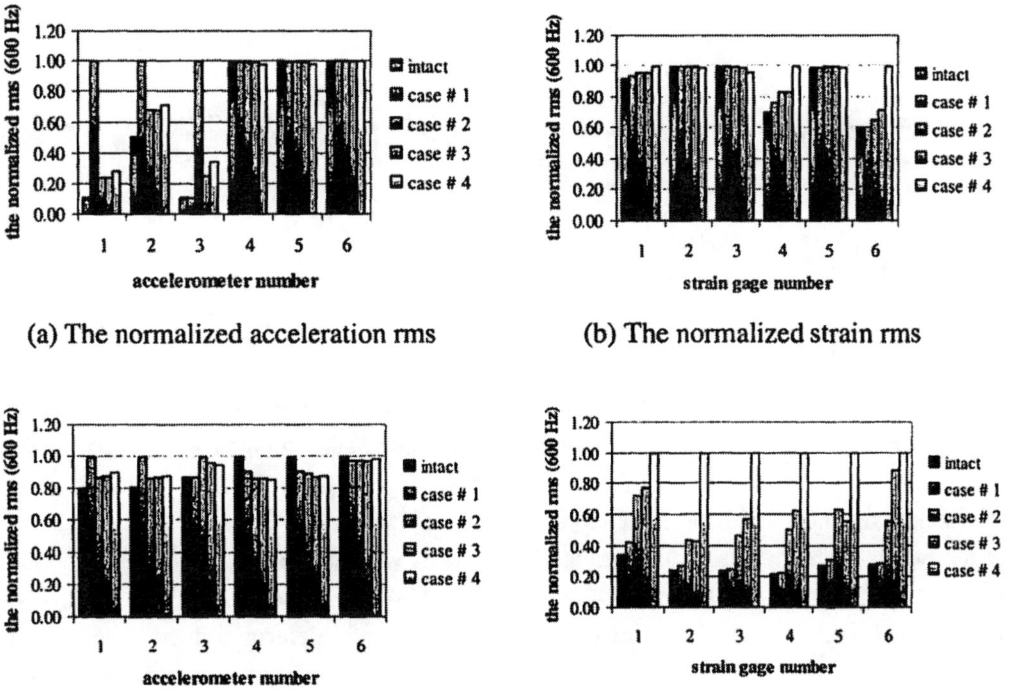

(a) The normalized acceleration rms

(b) The normalized strain rms

(c) The normalized acceleration rms

(d) The normalized strain rms

Figure 3. The normalized rms of acceleration and strain responses under intact and damage conditions: (a) and (b) numerical results (c) and (d) experimental results

Table II The e_j values calculated from numerical and experimental acceleration responses

Damage case	e_j values computed from responses of accelerometer number					
	4		5		6	
	numerical	experimental	numerical	experimental	numerical	experimental
intact	1.000000	1.000000	1.000000	1.000000	0.999149	1.000000
1	1.005380	1.094260	1.005297	1.092061	0.999505	1.024879
2	1.010845	1.136756	1.010715	1.109315	1.000000	1.025507
3	1.011046	1.140343	1.010911	1.133327	1.000712	1.031954
4	1.024998	1.145407	1.024775	1.142847	1.001302	1.041472

where j is the damage case number, $|\,|$ denotes the absolute value and e_0 is taken equal to rms_0 which is the rms values for intact condition. Table II presents the e_j values computed from responses of accelerometers # 4, #5 and #6 under intact and damaged cases. The e_j under intact, damage cases #1 and #2 were then fitted by following expression

$$e_j = \beta_0 + \beta_1\left(\frac{c_1}{L}\right)\left(\frac{a_{j1}}{d_1}\right)^2 + \beta_2\left(\frac{c_2}{L}\right)\left(\frac{a_{j2}}{d_2}\right)^2 + \beta_3\left\{\left(\frac{c_1}{L}\right)\left(\frac{a_{j1}}{d_1}\right)^2 + \left(\frac{c_2}{L}\right)\left(\frac{a_{j2}}{d_2}\right)^2\right\}^{1/2} \quad (2)$$

where $j = 0,1,2$ is the damage case number ($j = 0$ is the intact case), c_1 and c_2 are distances measured from the bottom side of the side shell to the horizontal where cracks were occurring [in the present case $c_1 = c_2 = c$, see Fig. 1(b)], L is the side shell plate length, a_{j1} and a_{j2} are crack lengths at crack location number 1 and 2 for jth damage case and d_1 and d_2 are the maximum possible crack lengths at locations 1 and 2 repectively; the values of d_1 and d_2 were used for normalizing crack lengths so that Eq. (2) becomes dimensionless. Total possible crack length in this paper was taken as the horizontal's flange width plus the horizontal's web depth for the values of d_1 and d_2 repectively. Since the cracks investigated were located in the same horizontal, the value of d_1 was equal to d_2; it was 49.4 mm. For known values of L, c_1, c_2, d_1, d_2, a_{i1}, a_{i2} and e_j ($j = 0,1,2$) the constants β's in Eq. (2) were determined in least square sense.Then Eq. (2) was utilized to estimate the rms amplitude for damage cases #3 and #4. The estimated damage indicator amplitudes are presented in Table III. The error between the actual and estimated amplitudes is also shown in the same table. It can be seen from the table that the numerical e_j amplitudes were estimated accurately using the extrapolation function. The error was found to be less than 0.5% for damage cases 3 and 4 for accelererometer #4. The e_j amplitudes for accelerometer #4 calculated from

Table III Actual and estimated values of the e_j amplitudes and the error in computation

e_j values	Case No	Values computed from accelerometer number					
		4		5		6	
		numerical	experimental	numerical	experimental	numerical	experimental
actual	3	1.0110	1.1403	1.0109	1.1333	1.0007	1.0320
	4	1.0250	1.1454	1.0248	1.1428	1.0013	1.0415
estimated	3	1.0144	1.2174	1.0142	1.1963	1.0002	1.0504
	4	1.0198	1.2731	1.0196	1.2329	1.0006	1.0573
error (%)	3	0.3363	6.7614	0.3264	5.5590	-0.0500	1.7829
	4	-0.5073	11.1489	-0.5074	7.8841	-0.0699	1.5170

measured responses and Eq. (2) showed a much larger error of 11.15% maximum. The error obtained for accelerometers #5 and #6 were much less. Hence it seems to be possible to utilize Eq. (2) for estimating the subsequent damage indicator amplitudes for different crack lengths. The effort is now being made to estimate the amplitudes computed using reponses of accelerometers #1, #2, #3 and strain gages and to calculate the e_j amplitude using Eq. (2) with modified β's . In addition estimating the damage indicator amplitudes for cracks occurring at two different horizontals is also being carried out. Efforts are also being made to predict the crack sizes from measured reponse.

Conclusions:

Results of numerical and experimental investigations carried out on the root mean square response of a stiffened plate model of side shell of tanker ship structure are reported. The panel consisted of a plate which was stiffened by four horizontals in the horizontal direction and two web frames in the vertical direction. The model was excited using random forces having a dominant frequency of 2 Hz. The rms of the response obtained from numerical and experimental investigation were compared and used to compute a damage indicator. Using the first two damage cases, a simple equation was proposed to estimate the damage indicator values for the subsequent damage cases. The difference between actual and estimated damage indicator values obtained from numerical investigations were less than 0.5 % and the difference was less 11.15% for the values obtained from experimental measurements. More studies are being carried out to find a more accurate equation for estimating the response of accelerometers located in the horizontals and the responses of strain gages as well as those of different damage cases.

Acknowledgements:

This study was financially supported by National Sciences and Engineering Council of Canada through Discovery Grant program awarded to Drs. A.S.J. Swamidas and M.R. Haddara.

References:

[1] Sucharski, D., 1995. "Crude oil tanker hull structure fracturing: An operator perspective," Proceedings of a Conference on Prevention of Fracture in Ship Structure, Washington D.C., pp. 87-124

[2] Zubaydi, A., Haddara M.R. and Swamidas, A.S.J., 2002 "Damage Identification in a Ship's Structure Using Neural Networks", Ocean Engineering, 29(10), pp. 1187 – 1200.

[3] Budipriyanto, A., Haddara, M.R. and Swamidas, A.S.J., 2004. "Analysis and Testing of Orthogonally Stiffened Plate Panels for Crack Identification," Proceedings of The 58th Meeting of the Society for Machinery Failure Prevention Technology, pp. 223-232.

[4] Lee, W.Y., Park C. and Kelly, G.E.,1996. "Fault Detection in Air-Handling Unit Using Residual and Recursive Parameter Identification Methods," ASHRE Transactions, 102(1), pp. 1 –12.

[5] Worden, W., Manson, G., Fieller, N.J.,2000. "Damage Detection Using Outlier Analysis," Journal of Sound and Vibration, 229(3), 647-667

[6] Andrade, F.A., Esat, I. and Badi, M.N., 2001."A New Approach to Time-Domain Vibration Condition Monitoring: Gear Tooth Fatigue Crack Detection and Identification by the Kolmogorov-Smirnov Test," Journal of Sound and Vibration, 240(5), pp. 909-919

[7] Choi, S. and Stubbs, N., 2004. "Damage Identification in Structures Using the Time-domain Response," Journal of Sound Vibration, 275(3-5), pp. 577-590

[8] ABAQUS Theory and User's Manual ver. 6.4, 2002, Hibbitt, Karlsson and Sorensen, Inc.

[9] Haddara, M.R., 1971, "On the Stability of Ship Motion in Regular Oblique Waves," International Ship Building Progress, 18(207), pp. 416-434.

[10] Budipriyanto, A., Haddara, M.R. and Swamidas, A.S.J., 2004. "Crack Identification in Cross Stiffened Plate System Using the Root Mean Square of Time Domain Responses," submitted for publication to Canadian Journal of Civil Engineering.

[11] LabVIEW Reference Manual, 1998, National Instruments.

THE EFFECT OF SOLUTION TREATMENTS, COOLING RATES, AND NATURAL AGING TIMES ON THE MECHANICAL PROPERTIES OF 6061-T4 ALUMINUM

P. Cullip-Brennan, A. Ordonez-Chu, E. Villalobos
E. Lee*, J. McLennan**
R. Clark, J. Ogren, and O.S. Es-Said

National Science Foundation
Research Experience for Undergraduate Program
Mechanical Engineering Department
Loyola Marymount University
Los Angeles, CA 90045

*NAVAIR
Code 4342, MS Building 2188
Patuxent River, MD 20670-1908

**Boeing Satellite Systems Inc.
2260 E. Imperial Hwy
El Segundo, CA 90245

Abstract:

The influence of varying the thermal processing parameters on the physical and mechanical properties of 6061-T4 aluminum alloy was studied. The variables altered were solution treatment temperatures, quenching media and natural aging durations. The influence of varying these parameters on the tensile strength, electrical resistivity and hardness of the alloy is discussed and correlated.

Keywords:

Al 6061-T4; Correlation of strength and hardness; Faulty heat treatments

Introduction:

AL 6061 alloy is used in automotive and aerospace applications in both the T4 (solution treated and naturally aged) and the T6 (peak aged) tempers. The composition of the alloy is given in Table I, [1].

Table I: Chemical Composition of Al 6061, [1]

Alloying element	Cu	Si	Fe	Mn	Mg	Zn	Cr	Ti
Weight Percent	0.15-0.4	0.4- 0.8	0.7	0.15	0.8-1.2	0.25	0.35-0.4	0.15

In the T4 temper, the alloy is more corrosion resistant although it has less strength as compared to the T6 temper.

It has been discovered recently that some aluminum parts, including Al 6061-T4 alloy, may have experienced faulty heat treatments prior to installation to aircraft. The objective of this study is first to map out the effect of possible pitfalls in the solution treatment temperatures, quenching media and natural aging durations on the mechanical tensile properties. Accordingly, a wide range of solution treatment temperatures were chosen from 470°C (low) to 570°C (high). Quenching is chosen to represent a range of cooling rates from high (water quenching) to medium (air quenching) to low (mixture of furnace cooling and air quenching or delayed air quench) cooling rates.

A second objective is to quantitatively correlate the hardness measurements and electrical resistivity measurements to the tensile behavior. Electrical resistivity measurements and Rockwell E hardness measurements are conducted on a routine basis by aircraft technicians. High resistivity measurements indicate a poor corrosion resistance and a high strength, (2-4) and high hardness numbers indicate a high strength, (5). This would provide a predictive tool for aircraft maintenance crews to evaluate aluminum components in aircraft structures.

Experimental Procedure:

A 22 ft long extruded aluminum 6061-T6 panel was cut into 21" length plates. The thickness of the panel was 0.12". These plates were milled into 1" strips. The strips were then either cut into 3" strips that were used for hardness and electrical conductivity testing, or they were cut into 10" strips that were later milled into tensile bars. Forty-five hardness/electrical conductivity bars and 135 tensile bars were machined. Three tensile specimens and one hardness/electrical conductivity bar were grouped and averaged to form each condition out of the forty-five total conditions. The forty-five conditions consisted of heat treating and natural aging the samples according to Table II. The thermal processing included five solution treatment temperatures three quenching media and three natural aging durations, (5 x 3 x 3 = 45).

Table II: Thermal processing of the 6061 aluminum alloy

Solution Heat Treatment (30 min)	Quenching Media	Natural Aging
470°C	Air	1 day
500°C		5 days
530°C		
550°C	Water	34 days
570°C	Delayed Air	

The "Delayed Air" quench describes a condition where the furnace is turned off, furnace door is opened and samples are kept inside the furnace for two minutes before being air-cooled. Accordingly the highest cooling rate was the water quenching followed by air cooling and then by delayed air cooling.

Tensile testing was performed according to ASTM E8 standards and superficial 15T and Rockwell E hardness scales were used in hardness testing. Measurements of electrical resistivity were conducted by using an Eddy Current apparatus.

Results and Discussion:

Figs. 1 – 3 represent the stress-strain characteristics (yield strength, ultimate strength and percent elongation) of samples solution treated at different temperatures and naturally aged at 1 day (Fig. 1), 5 days (Fig. 2) and 34 days (Fig. 3) respectively.

For water quenched samples irrelevant of natural aging at one day, 5 days or 34 days, the strength values generally increase with the increase in the solution treatment temperature.

For air quenched and delayed air quench samples the strength values increase with the increase in solution treatment temperature after 1 day natural aging and until 530°C with 5 days natural aging. The strength values are not affected by the solution treatment temperature after 34 days of natural aging.

In Fig. 4, the ultimate tensile strength is plotted vs. solution treatment temperature for different quenching media. In the water quenched samples, the strength values are higher after 5 days, followed by 34 days at low solution treatment temperatures, 470° - 570°C. However, these values are similar at higher temperatures, 530 - 570°C. Nevertheless, these values are always higher than the values of the one day naturally aged samples. In the air quenched condition, the 5 days naturally aged samples have the highest strength values and in the delayed air quench samples, due to the loss of super-saturation, the

Figure 1 – Strength Values vs. Solution Treatment Temperatures for One Day Natural Aging

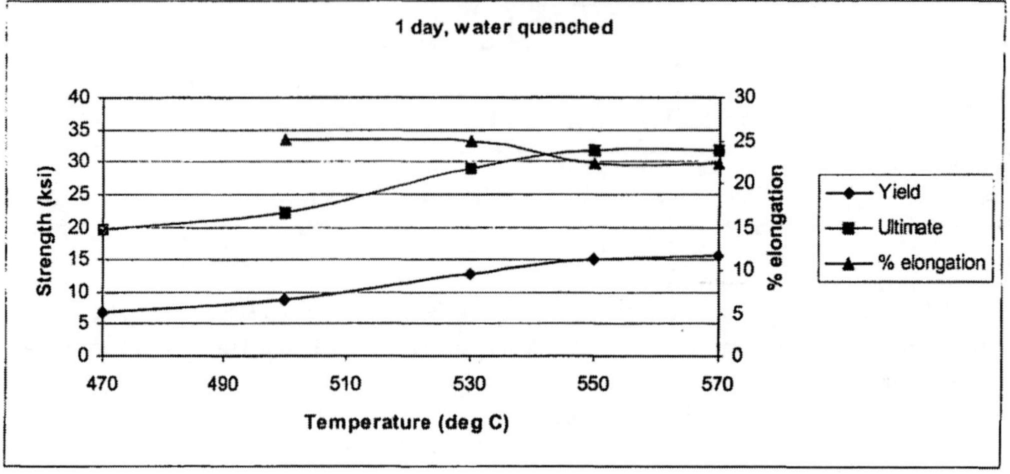

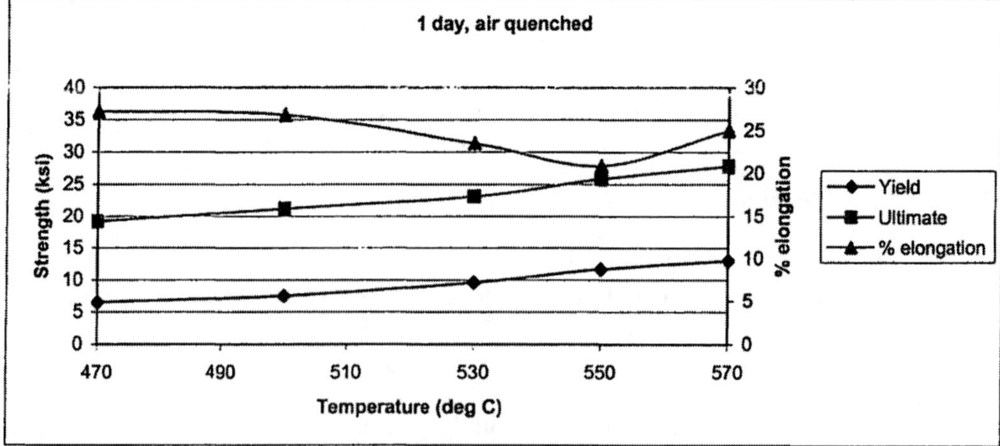

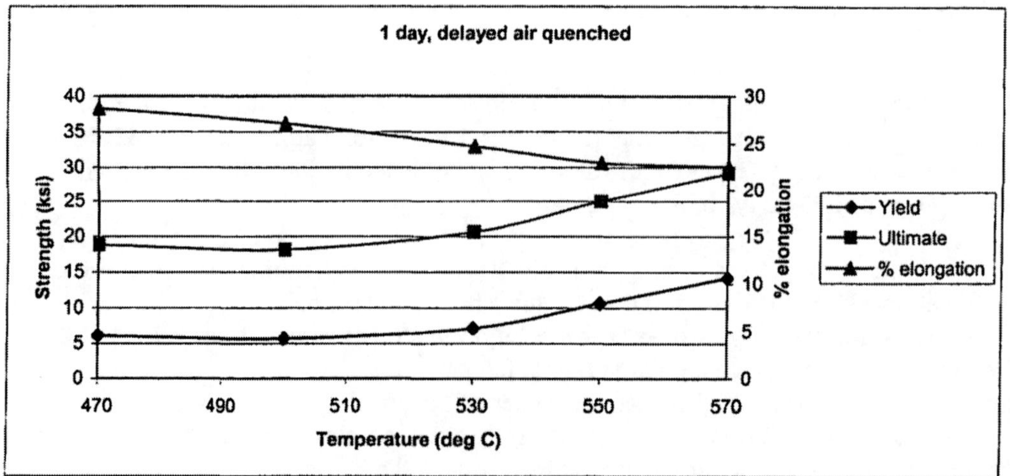

Figure 2 – Strength Values vs. Solution Treatment Temperatures for 5 Days Natural Aging

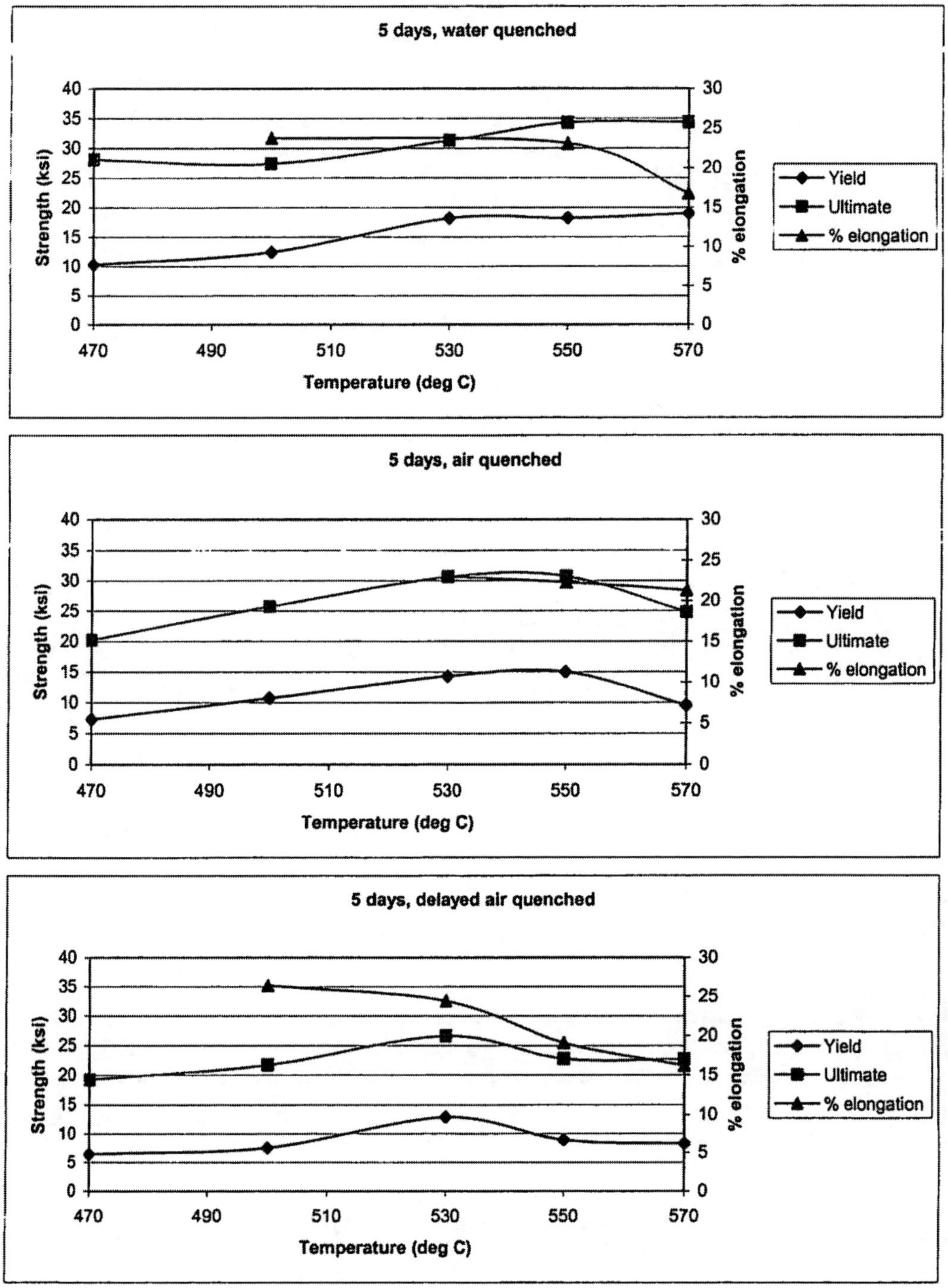

Figure 3 – Strength Values vs. Solution Treatment Temperatures for 34 Days Natural Aging

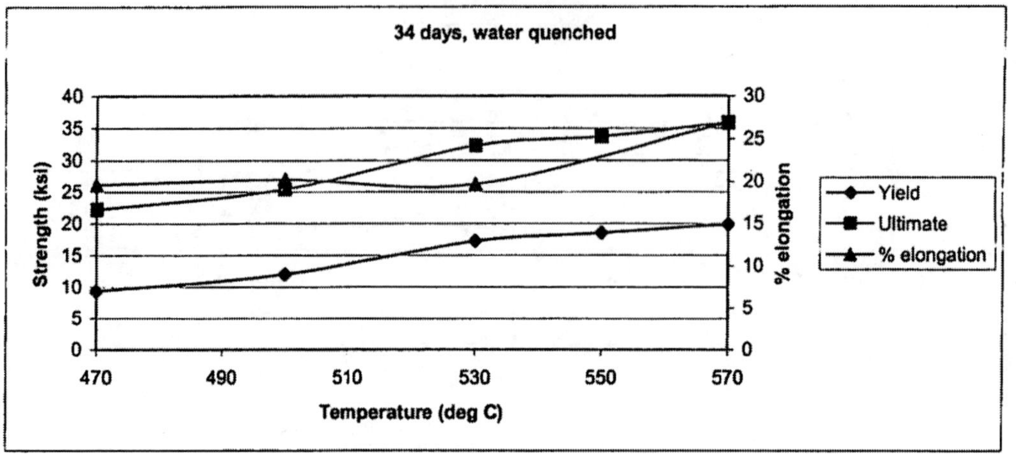

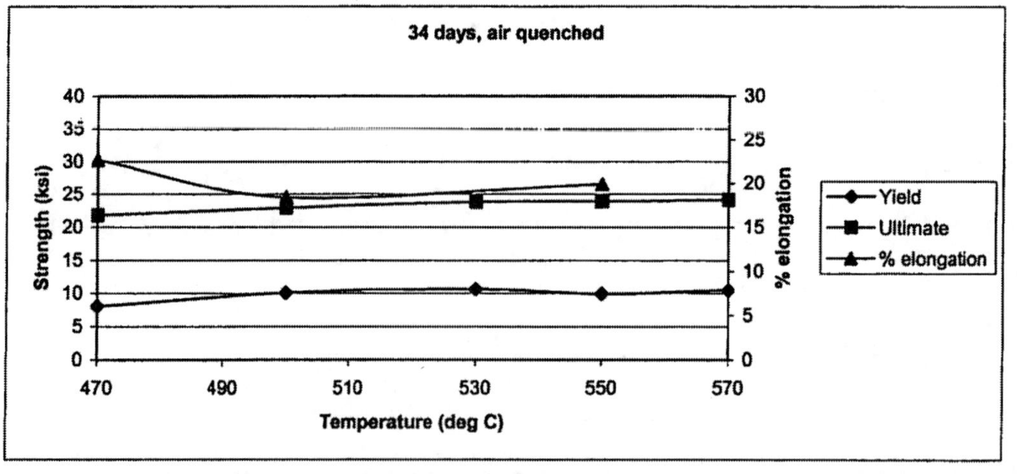

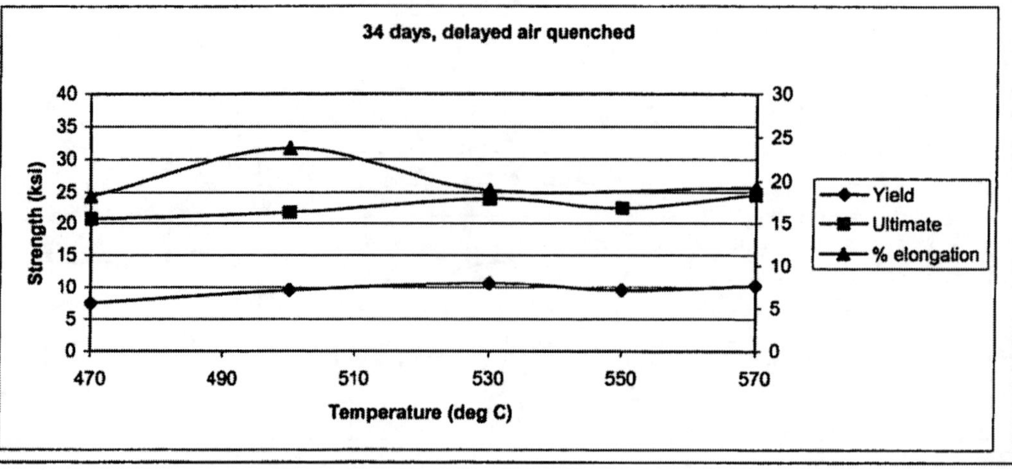

Figure 4 – Ultimate Strength vs. Solution Treatment Temperature for Different Quenching Media

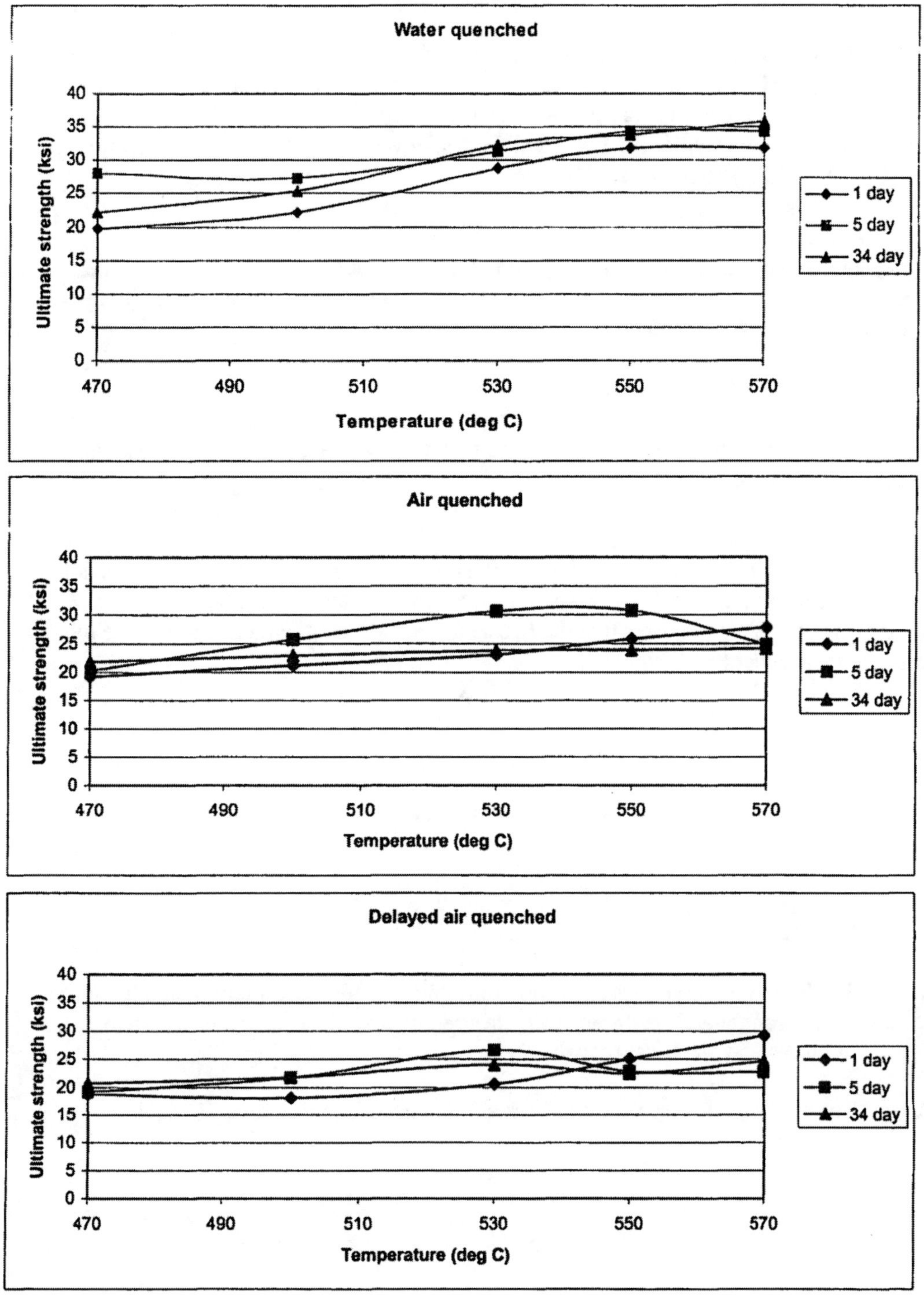

duration of natural aging time has minimal effects on the strength values, i.e., they are similar.

In Fig. 5, the ultimate tensile strength is plotted vs. solution treatment temperature for different aging times. For all natural aging times, water quenched samples have the highest strength values. The air cooled samples has higher strength values than the delayed air quench samples after one and 5 days, however these values become similar and low (20-25 K_{si}) after 34 days.

Correlations between ultimate strength and yield strength values with (15T) hardness measurements are shown in Figs 6 and 7, a relatively good direct proportionality relation was found with $R^2 = 0.6783$ and 0.7221 respectively.

Correlations were also found between ultimate strength and yield strength values with electrical resistivity measurements. A good inverse proportionality relation was found with $R^2 = 0.7681$ and 0.7943 respectively, Figs. 8 and 9.

Finally, an excellent correlation ($R^2 = 0.9998$) was found between the (15T) superficial hardness scale and the Rockwell E scale which 6061-T4 aluminum alloys are customarily measured with. This indicates that the two scales are almost identical and the use of either scale is acceptable, Fig. 10.

Conclusions:

1. Water quenched samples produced the highest strength values for all solution treatment temperatures and natural aging times.

2. Overall, the higher the solution heat treatment temperature, the stronger the alloy becomes.

3. There were relatively good correlations between ultimate strength and hardness, yield strength and hardness, ultimate strength and electrical resistivity and hardness and electrical resistivity.

4. The 15T hardness scale could be used in place of the Rockwell E scale.

Figure 5 – Ultimate Strength Values vs. Solution Treatment Temperatures for Different Natural Aging Times

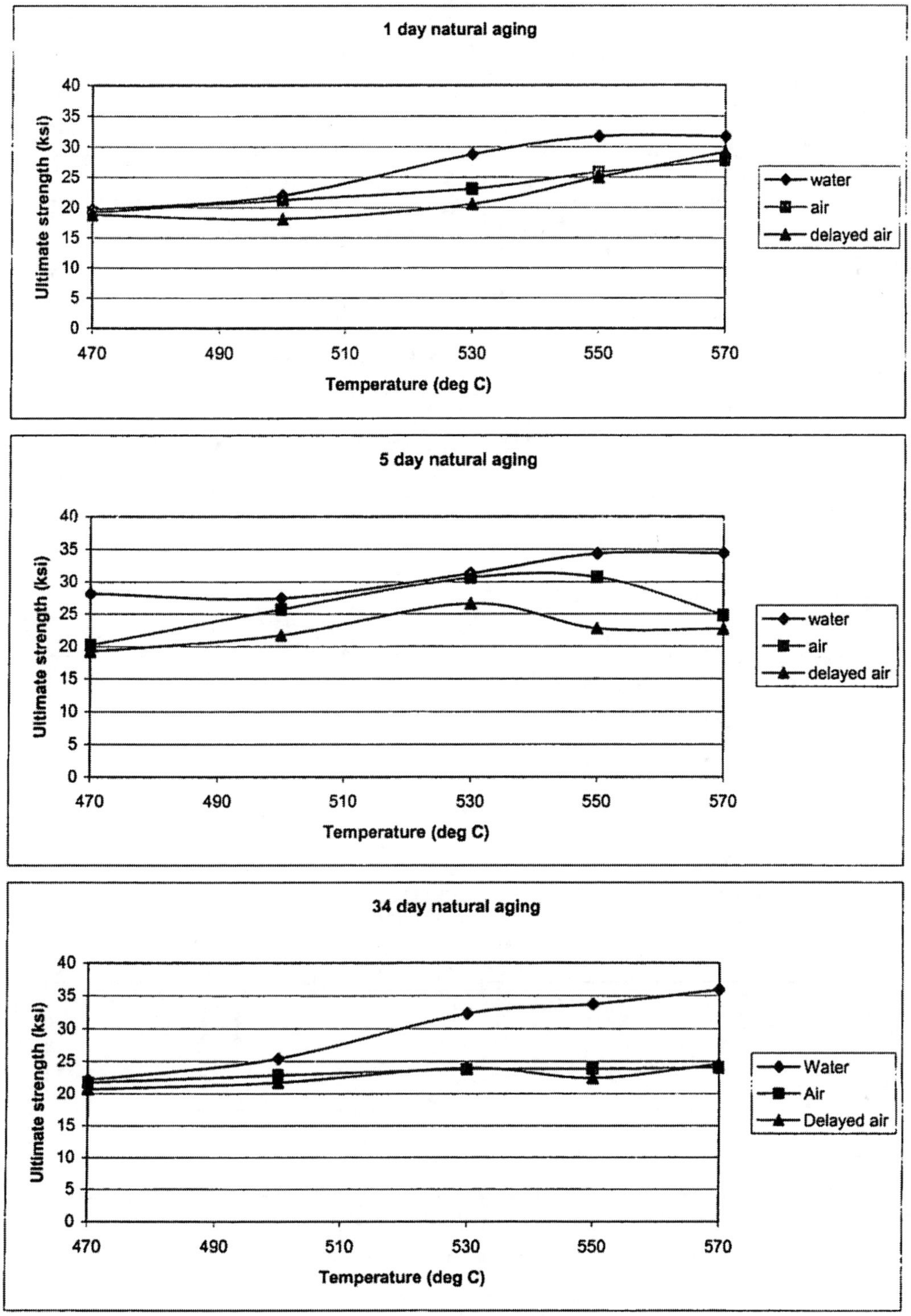

Figure 6 - Correlation between Ultimate Strength and Hardness Values

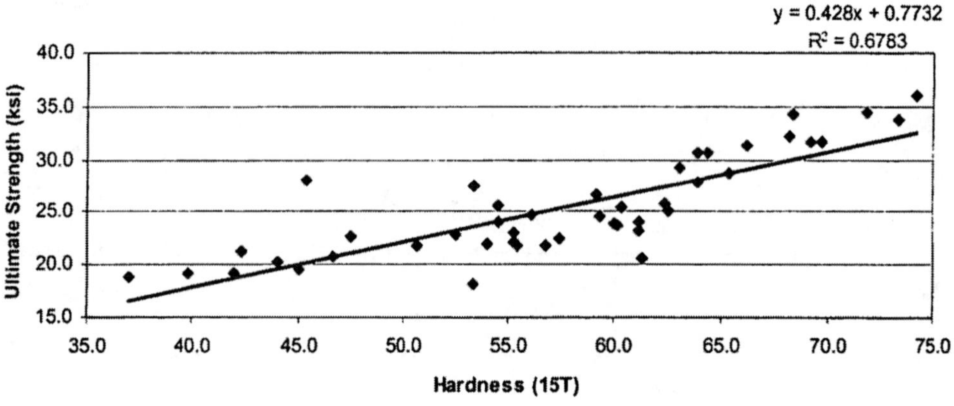

Figure 7 - Correlation between Yield Strength and Hardness Values

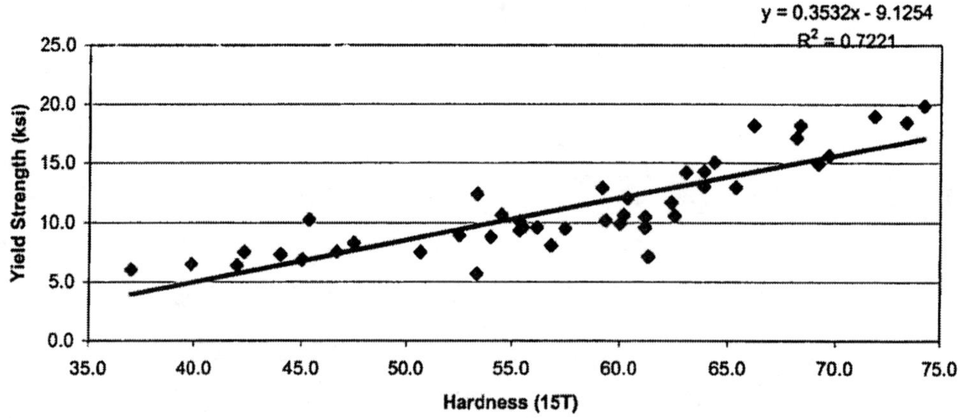

Figure 8 - Correlation between Ultimate Strength and Electrical Resistivity Values

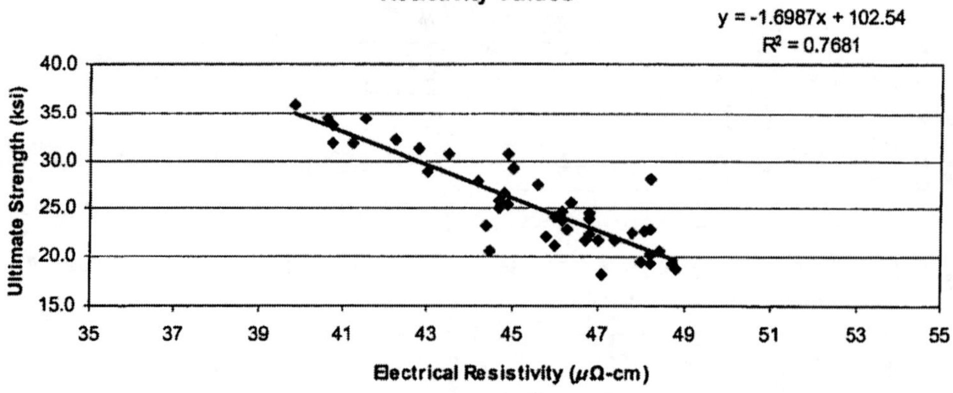

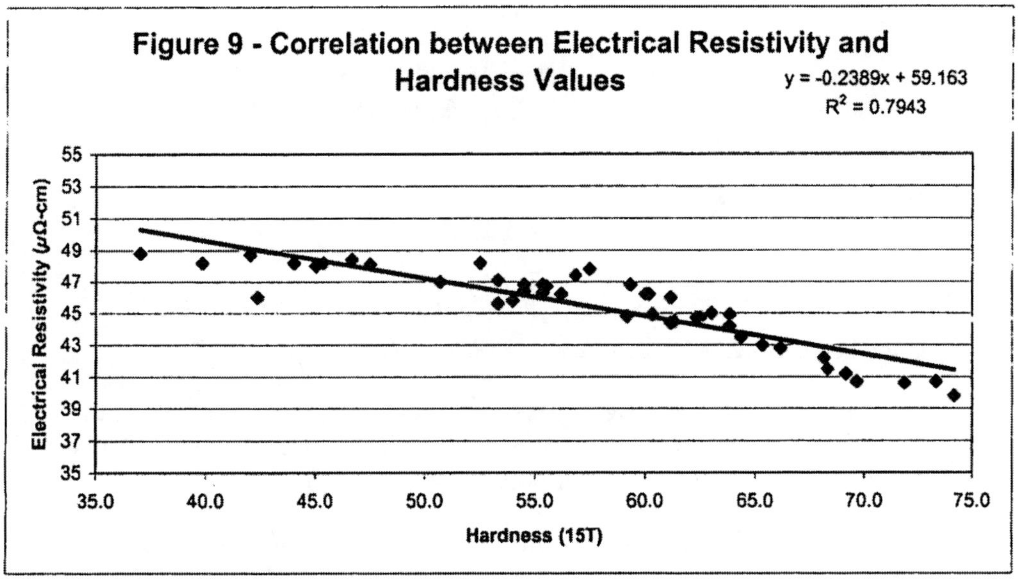
Figure 9 - Correlation between Electrical Resistivity and Hardness Values
y = -0.2389x + 59.163
R² = 0.7943
Electrical Resistivity (μΩ-cm)
55
53
51
49
47
45
43
41
39
37
35
35.0
40.0
45.0
50.0
55.0
60.0
65.0
70.0
75.0
Hardness (15T)

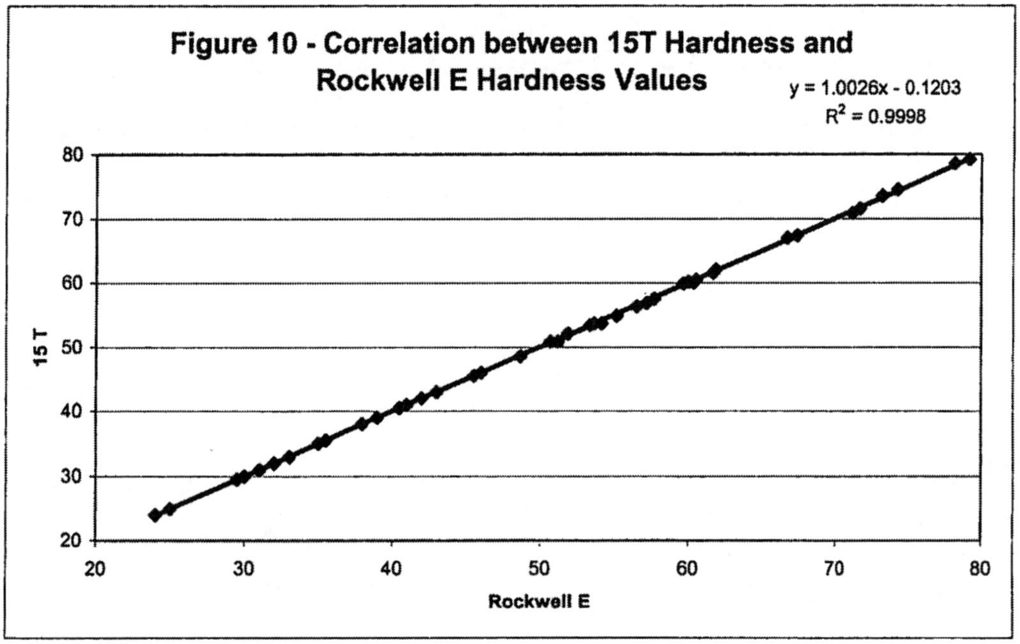
Figure 10 - Correlation between 15T Hardness and Rockwell E Hardness Values
y = 1.0026x - 0.1203
R² = 0.9998
15 T
80
70
60
50
40
30
20
20
30
40
50
60
70
80
Rockwell E

Acknowledgments:

This work was funded by the National Science Foundation (Grant # EEC-0353668), Research Experience for Undergraduates. The authors wish to thank Ms. Maria Guerrero for the preparation of the manuscript.

Reference:

[1] Mat Web: Material Property Data, "Aluminum 6061-T6; 6061-T651," http://matweb.com/search/specificMaterial.asp?bassnum=MA6061

[2] M.U. Islam and W. Wallace, "Stress Corrosion Crack Growth Behavior of 7475-T6 Retrogressed and Reaged Aluminum Alloy." Metals Technology, 11 (1984), pp.320-322.

[3] M. Talianker, B. Cina. "Retrogression and Reaging and the Role of Dislocations in the Stress Corrosion of 7000-Type Aluminum Alloys." Metall. Trans. A, 20A (1989), pp.2087-2092

[4] O.S. Es-Said, W.E. Frazier and E.W. Lee, "The Effect of Retrogression and Reagaing on the Properties of the 7249 Aluminum Alloy." Journal of Materials (JOM), 55(1), January 2003, pp.45-48.

[5] M. Iskandar, D. Reyes, Y. Gaxiola, E. Fudge, J. Foyos, E.W. Lee, P. Kalu, H. Garmestani, J. Ogren, O.S. Es-Said, "On Identifying the Most Critical Step in the Sequence of Heat Operations in 7249 Aluminum Alloy." Engineering Failure Analysis Journal (EFA), 10 (2003), pp.199-207.

MECHANICAL LOADS IDENTIFICATION AND DIAGNOSTICS IN LAND VEHICLE SUSPENSIONS THROUGH INTEGRATED SENSING AND FORCE STATE MAPPING

Muhammad Haroon, Timothy J. Johnson and Douglas E. Adams

Purdue University
School of Mechanical Engineering
Ray W. Herrick Laboratories
140 S. Intramural Drive West Lafayette, IN 47907-2031
Ph: (765) 496-6033 Fax: (765) 494-0787
deadams@purdue.edu
http://widget.ecn.purdue.edu/~FEND2

Abstract: Commercial and defense vehicle systems operate in highly variable loading environments. Variability in static and dynamic loading can cause damage to components and lead to excessive warranty costs and overly aggressive maintenance schedules. Manufacturers are utilizing data interrogation methodologies for identifying loads and damage in suspensions, tires and other vehicle components to reduce the total cost of vehicle ownership. The application of loads and damage identification techniques for mechanical vehicle systems is discussed here. Operating, passive response data from suspension systems in passenger cars are examined. Mechanical damage in these components, i.e., degradation to shock, is identified experimentally using force state maps and transmissibility functions, which are generated using acceleration data in conjunction with fundamental mechanics models. The force state maps also provide the suspension restoring force estimates, which must be used to prognosticate remaining life and predict trends in the damage indices. Diagnostic challenges associated with nonlinear operational characteristics of these components and their sensitivity to loading levels and spectra are also discussed.

Key Words: Loads identification; force state maps; vehicle diagnostics

Introduction: There is an increasing demand in the consumer, commercial and defense sectors for integrated suspension systems. Suspension components and subsystems have historically been supplied by different manufacturers. This component-wise approach leads to components that are over designed to compensate for possible deficiencies in companion components, causing a significant increase in the weight of the suspension system. Similarly, life of some components may be significantly reduced if unanticipated interactions accelerate degradation. For example, unexpected interactions between struts and torsion bars can considerably reduce bushing life. When integrated system modules are manufactured by one company with design authority over all interconnected components, it is possible to design higher performing, longer lasting systems. Integrated suspension modules incorporate isolation, steering (knuckles and dampers) and other subsystems including shock absorbers, struts, torsion bars, mounts, etc., leading to complex loading and degradation mechanisms in bushings, linkages, struts etc. Industry is producing integrated suspensions but better understanding is still required of: 1) complex loading due to interactions between components; and 2) degradation mechanisms and rates at which degradation escalates in each component. To achieve this understanding, loads and damage identification algorithms are required for quantifying loads and their directionality in experimental data acquired from suspension components (e.g., lateral, thrust, torsional etc.), and detect, locate and quantify degradation mechanisms, i.e., faults in suspensions (e.g., bushing rupture, joint preload, leaky shock, warped torsion bar). This better understanding of loading and degradation mechanisms will also lead to reduced maintenance and warranty costs.

In this paper, two techniques are presented that use response acceleration data to identify the loads and degradation in a system. Restoring force plots characterize and identify the internal loads that are set up in the suspension system under different amplitude and frequency inputs at the tire patch of the vehicle. These plots also capture the changes caused in the internal forces by the degradation of the system components. Transmissibility functions isolate the dynamics between two response points and are thus sensitive to the local changes caused by component degradation. Restoring forces work better for inputs with narrow bandwidths (e.g., sinusoidal) and transmissibility functions are more suited to broad band inputs (e.g., random). This trade-off motivates the selection of an algorithm in an event-driven manner based on the bandwidth of the excitation. The utility of both techniques is demonstrated by applying them to experimental vehicle data.

Restoring Force: The restoring force is an internal force that opposes the motion of an inertial element within a system. It can be derived using Newton's Second Law applied to a body with constant mass, *m*, and acceleration vector, *a*: $\Sigma \mathbf{F}=m\mathbf{a}$. The stiffness and damping in a system resist the motion of a given inertia; consequently, the forces in the stiffness and damping elements are referred to as components of the restoring forces. Therefore, restoring forces can be used to identify the internal forces setup in a system under different loading. In addition, individual nonlinearities have particular restoring forces; therefore, the nonlinear nature of the internal forces can be characterized by the restoring forces within a system.

A major advantage of the restoring force technique is that it only requires that the output accelerations of a system be measured. Consider the two degree-of-freedom quarter car model shown in Fig. 1. The equations of motion for this system are given by,

$$M_1\ddot{x}_1+(C_1+C_2)\dot{x}_1-C_2\dot{x}_2+(K_1+K_2)x_1-K_2x_2+N_1[x_1(t),x_2(t),\dot{x}_1(t)\dot{x}_2(t)] \quad (1)$$
$$+N_2[x_1(t),x_b(t),\dot{x}_1(t)\dot{x}_b(t)]=C_1\dot{x}_b+K_1x_b$$

$$M_2\ddot{x}_2-C_2\dot{x}_1+C_2\dot{x}_2-K_2x_1+(K_2+K_3)x_2=N_1[x_1(t),x_2(t),\dot{x}_1(t)\dot{x}_2(t)] \quad (2)$$

where $x_k(t)$ are the displacements of the unsprung and sprung masses, M_k, $x_b(t)$ is the displacement of the tire patch, C_k and K_k are the damping and stiffness in the tire and suspension, $N_1[x_1(t),x_2(t),\dot{x}_1(t)\dot{x}_2(t)]$ denotes the nonlinear forces in the suspension and $N_2[x_1(t),x_b(t),\dot{x}_1(t)\dot{x}_b(t)]$ denotes the nonlinear forces in the tire.

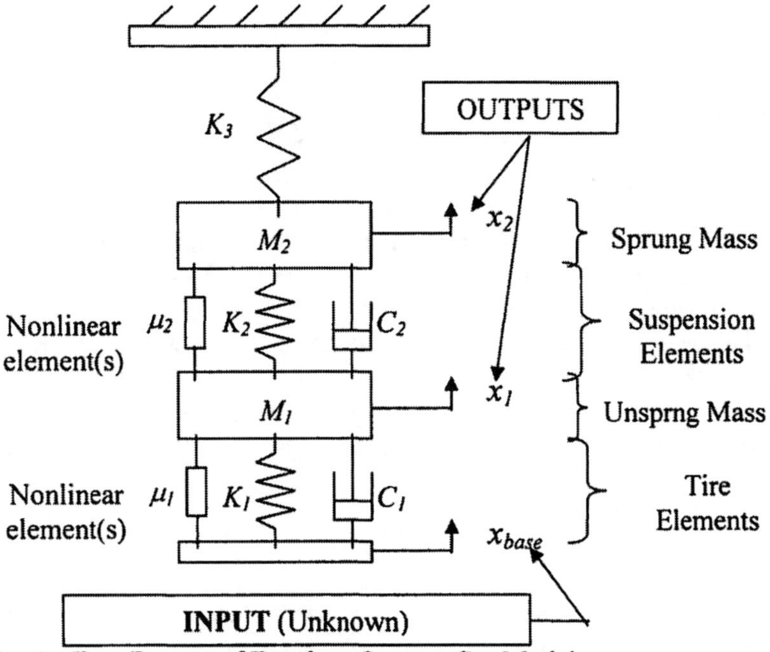

Fig. 1: Two Degree of Freedom Quarter Car Model.

The equation for the sprung mass, M_2, can be rearranged to give the restoring forces as,

$$M_2\ddot{x}_2 = -C_2(\dot{x}_2 - \dot{x}_1) - K_2(x_2 - x_1) - K_3 x_2 + N_1[x_1(t), x_2(t), \dot{x}_1(t), \dot{x}_2(t)] \qquad (3)$$

The plot between the acceleration of the sprung mass and the relative velocity between the sprung mass and the unsprung mass allows the damping restoring force in the suspension to be characterized. Similarly, the plot between the acceleration of the sprung mass and the relative displacement between the sprung mass and the unsprung mass allows the stiffness restoring force in the suspension to be characterized. Haroon and Adams [1] showed that restoring force plots can be used to characterize the nonlinear forces in vehicle suspension systems, using only response measurements.

Transmissibility: Johnson and Adams [2] showed that transmissibility could be used to effectively detect and locate structural changes. Transmissibility functions only contain the zeros, and not the poles, of discrete frequency response functions (FRF), and therefore, contain information about more localized regions of structures. For example, a two degree-of-freedom system has an FRF matrix

$$\mathbf{H}(\omega) = \frac{1}{\Delta(\omega)} \begin{bmatrix} A_{11}(\omega) & A_{12}(\omega) \\ A_{21}(\omega) & A_{22}(\omega) \end{bmatrix} \qquad (4)$$

where $\Delta(\omega)$ is the characteristic polynomial whose roots are the system poles. The entries $A_{ij}(\omega)$ contain the system zeros, which are a function of a few of the total degrees of freedom in the system. Transmissibility is defined as the ratio of two response degrees of freedom. The following equation shows that the characteristic polynomial is cancelled in the transmissibility calculation, and all that remains is the ratio of system zeros.

$$T_{ij(k)}(\omega)=\frac{X_{ik}(\omega)}{X_{jk}(\omega)}=\frac{\dfrac{X_i(\omega)}{F_k(\omega)}}{\dfrac{X_j(\omega)}{F_k(\omega)}}=\frac{H_{ik}(\omega)}{H_{jk}(\omega)}=\frac{\dfrac{A_{ik}(\omega)}{\Delta(\omega)}}{\dfrac{A_{jk}(\omega)}{\Delta(\omega)}}=\frac{A_{ik}(\omega)}{A_{jk}(\omega)} \tag{5}$$

Thus acceleration response measurements can be used to directly access information about the zeros of the system FRF. This property means that a transmissibility measurement across two discrete structural degrees of freedom will primarily contain information about the structural path between those two degrees of freedom. Damage and other structural changes can hence be detected and located within a sensor array.

No measurement of the input is required for computing the transmissibility, which can therefore be applied for vehicle diagnostics because data can be taken during vehicle operation and no information about the road surface or vehicle speed is necessary to make diagnostic decisions.

Experimental Setup: Response data was collected from the left-front McPherson Strut suspension system of an Isuzu Impulse using a two-post hydraulic shaker apparatus. A picture of the experimental setup is shown in Fig. 2. The MTS® hydraulic shakers with a maximum dynamic pressure of 3000 psi, an input frequency range of 0-100 Hz and a maximum stroke of approximately 8 inches, were used to excite the front tire patches of the car with different types of inputs in the vertical direction. Table I lists the details of the inputs used with corresponding signal processing parameters. Tri-axial accelerometers (PCB® T356B18), with nominal sensitivity 1 V/g, were used to measure vertical, lateral and longitudinal accelerations at four different locations on the suspension system; a) Steering knuckle-strut connection (unsprung mass, $\ddot{x}_1$), b) Body-strut connection (sprung mass, $\ddot{x}_2$), c) steering knuckle-control arm connection (above mount, $\ddot{x}_3$) and d) steering knuckle-control arm connection (below mount, $\ddot{x}_4$).

Fig 2: Shaker Test Setup.

The signals were recorded with an IOTECH® Portable Data Acquisition System and converted into '.mat' files for further analysis in MATLAB®. The IOTECH® system allowed a wide range of sampling frequencies and the application of high and/or low pass filters to remove noise and aliasing from higher frequency components.

Table 1: Experimental Inputs and Corresponding Signal Processing Parameters

Input	Frequency Range (Hz)	Sampling Frequency (Hz)	Low Pass Filter Cut-off (Hz)	High Pass Filter Cut-off (Hz)	Number of Time Pts., N_t
Random	0-35	500	100	0.1	150,000
Sine Sweep	0-15 @ 0.025 Hz/s	500	100	0.1	300,000

Loads Identification: Restoring force curves were used to identify the loads between response locations. A very slow chirp input to avoid abnormal transitions through nonlinear resonances (i.e., lingering bifurcations or other effects) from 0-15 Hz, at a rate of 0.025 Hz/s, was used as the base excitation to generate the restoring force. The acceleration response measurements were taken and then integrated offline to estimate the velocity and displacement responses. A 100 Hz bandwidth low pass filter was used to reduce aliasing back into the frequency range of interest from approximately 3 to 35 Hz.

Velocity (damping) and displacement (stiffness) restoring force curves were generated in the vertical, lateral and longitudinal directions for different input amplitudes and frequencies between four different response locations: a) x_1 and x_2, b) x_1 and x_3. c) x_3 and x_4, d) x_2 and x_3. As can be seen from the two degree-of-freedom model in Fig. 1, the restoring force curves between x_1 (unsprung mass) and x_2 (sprung mass) will give the forces in the strut. Figures 3 and 4 show damping and velocity restoring forces, respectively, between the four response locations in the vertical direction for an input amplitude of 1mm. Figure 3(a) shows that the damping restoring force in the strut has a piece-wise linear characteristic at 3.7 Hz. The damping forces between the other three response locations have a mainly linear characteristic with hysterisis at the respective input frequencies. The displacement curve in Fig. 4(a) shows a nonlinear force with primarily hysterisis (i.e., backlash) in the strut. The stiffness force across the steering knuckle-control arm mount (Fig. 4(c)) shows an odd-powered polynomial characteristic.

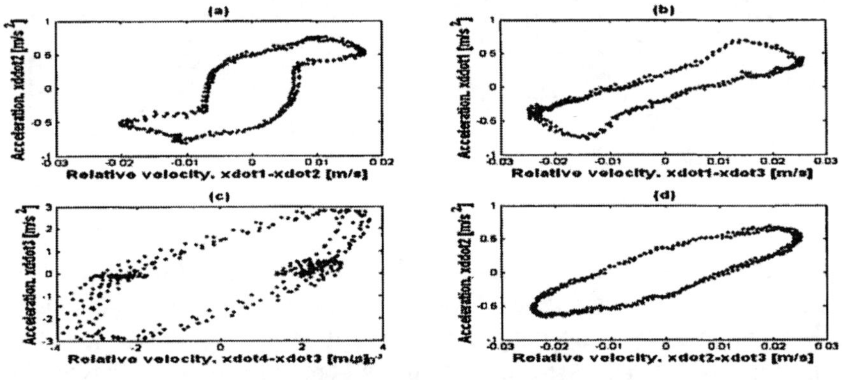

Fig. 3: Velocity (damping) Restoring Forces between Four Response Location Pairs for a 1 mm Input; (a) x_1 and x_2 at 3.7 Hz, (b) x_1 and x_3 and 3.45 Hz, (c) x_3 and x_4 at 13 Hz, (d) x_2 and x_3 at 3.5 Hz.

Similar restoring force curves were generated in the longitudinal and lateral directions along with the forces for the cross-coupling between the three directions. Figure 5 shows four such restoring force plots. In the same manner, restoring force plots can be generated in multiple directions and for multiple response pairs to identify the loading mechanisms in different components of the vehicle suspension.

The amplitude and frequency characteristics of the nonlinear forces were also characterized using restoring force plots. The vertical damping restoring force in the strut was generated for multiple input amplitudes and frequencies. Figure 6 shows the change

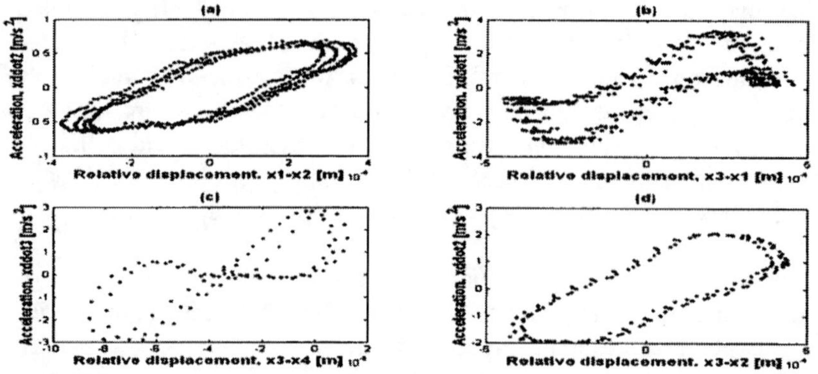

Fig. 4: Displacement (Stiffness) Restoring Forces between Four Response Location Pairs for a 1 mm Input; (a) x_1 and x_2 at 3.5 Hz, (b) x_1 and x_3 and 13.5 Hz, (c) x_3 and x_4 at 13 Hz, (d) x_2 and x_3 at 14 Hz.

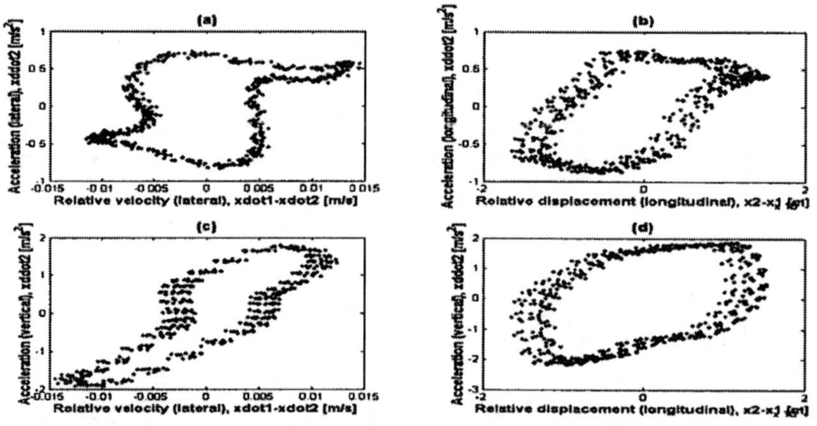

Fig. 5: Restoring Forces in the Strut : (a) Velocity Restoring Force in the Lateral Direction (13.9 Hz), (b) Displacement Restoring Force in the Longitudinal Direction (14.95 Hz), (c) Velocity Restoring Force for Coupling between Vertical and Lateral Directions (13.5 Hz), (d) Displacement Restoring Force for Coupling between Vertical and Longitudinal Directions (14.5 Hz) .

in the nonlinear damping force in the strut with amplitude at a constant frequency of 3.8 Hz. The damping characteristic changes from a saturation type in Fig. 6(a) to a piecewise-linear form in Fig. 6(d), in which positive relative velocities suffer more saturation than negative velocities, as anticipated in a typical shock absorber.

Figure 7 shows the change in nonlinear strut damping force with frequency at a constant 1 mm input amplitude. The trends in curves Fig. 7(a)-(d) indicate that some of the nonlinear effects decrease as the frequency increases. This trend occurs because an increase in frequency results in an increase in stroke velocity and, hence, a gradual movement of the system away from the central hysteresis loop in Fig. 6(a) to the outer relatively linear portion of the restoring force curve in Fig, 7(d). The multiple loops in the restoring force curves, as shown in Fig. 7(d), are due to the presence of harmonics of the fundamental frequency in the data, which recall was taken for a slowly varying chirp displacement input at the base of the tire.

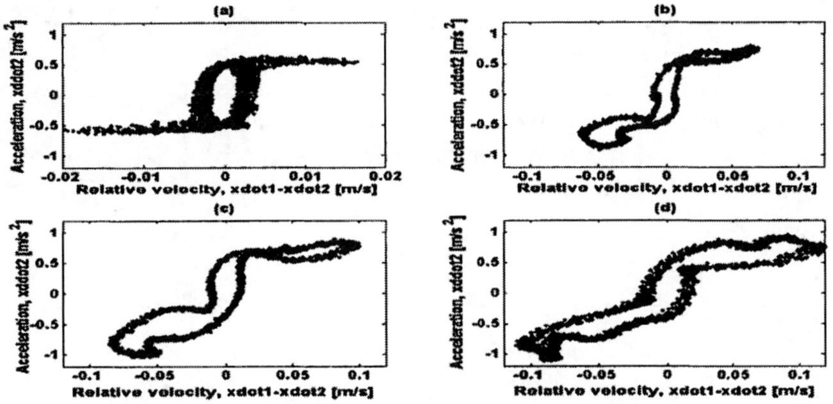

Figure 6 – Velocity Restoring Force in Strut Showing Change in Nonlinear Damping Force with Amplitude for Constant Frequency Input at 3.8 Hz at the Tire Patch: (a) 0.5 mm Base Excitation; (b) 2.0 mm; (c) 3.0 mm; and (d) 4.0 mm.

Thus, the amplitude and frequency characteristics of the vertical damping forces in the strut have been characterized. A similar characterization can be done for other components and in all three directions. In this manner, the loads in the vehicle suspension system components can be identified.

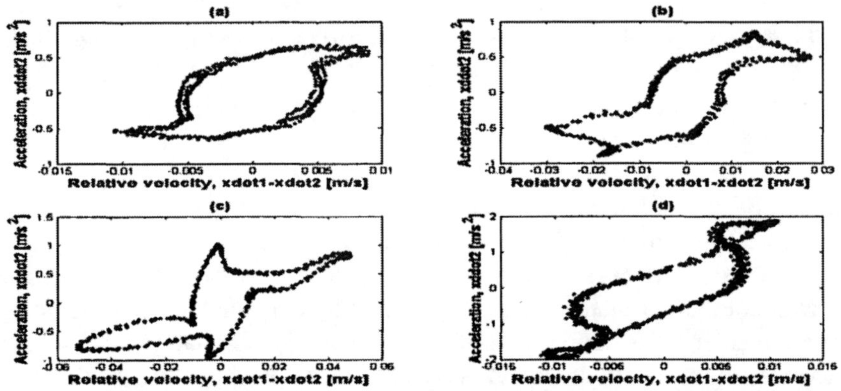

Figure 7 – Velocity Restoring Force in Strut Showing Change in Nonlinear Damping Force with Frequency for Constant Amplitude Input of 1 mm at the Tire Patch: (a) 3.5 Hz Base Excitation; (b) 4.0 Hz; (c) 6.8 Hz; and (d) 14.5 mm.

Damage Identification: Damage was simulated in the system by means of a wooden piece wedged between two coils of the left-front suspension spring (Fig. 8). The expected effect of this damage is an increase in the suspension stiffness. Restoring force curves were generated for the damaged system as in Fig. 3 and Fig. 4. Figure 9 shows velocity restoring force curves between the four response location pairs before and after damage.

The obvious changes in the restoring forces indicate that there is damage in the system, but it cannot be said with any certainty that the damage is in the strut. The restoring force in the strut (Fig. 9 (a)) does not show a significant change while that between x_1 and x_3 (Fig. 9(b)) shows a significant change. Hence, although the presence of damage can be detected using restoring forces, some other information is needed to properly locate and identify damage. Transmissibility functions can serve this purpose.

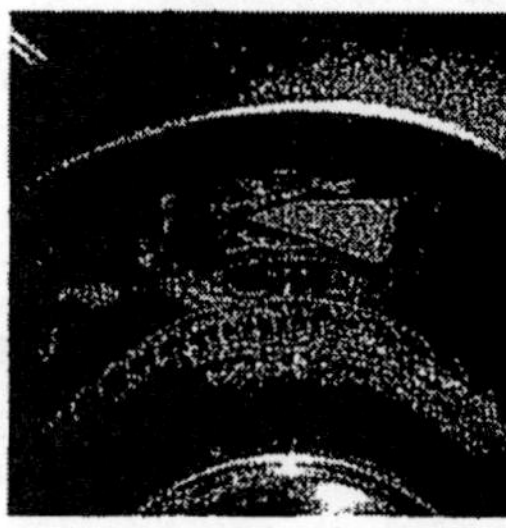

Fig. 8: Damage.

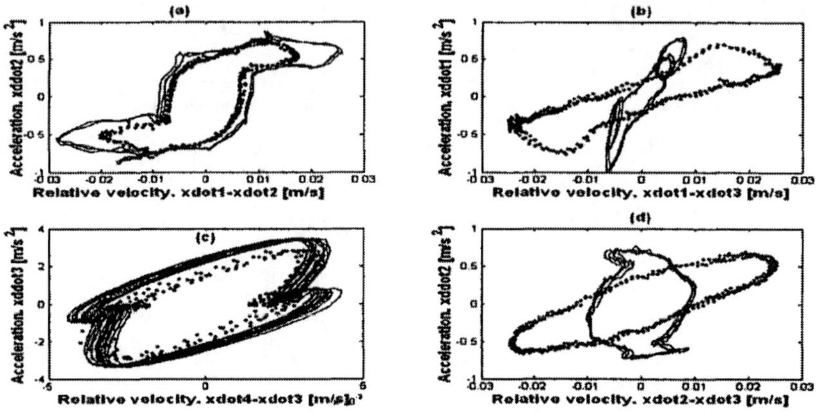

Fig. 9: Damping Restoring Forces, Damaged (——) and Undamaged (····).

Transmissibility functions between the previously mentioned response locations were generated using random inputs with a Gaussian distribution for multiple amplitudes, before and after damage. The vertical transmissibility functions between the response locations that have the damage location in the path between (Fig. 10((a) and (d)) them show a significant change, while the other functions show no significant change. This result is to be expected because transmissibility functions capture the local dynamics between two points. As stated earlier, the induced damage is expected to cause am increase in the stiffness of the suspension. Consequently, the natural frequencies (peaks in the transmissibility functions) are expected to move up. This movement is observed in the first two peaks (near the sprung mass resonance) of the transmissibility functions in Fig. 10 ((a) and (b)). But the peak around 15 Hz (near unsprung mass resonance) actually shows a decrease in frequency and its level goes down. This result is unexpected, as of now. In order to better understand what dynamic changes result from the damage, the two degree-of-freedom model in Fig. 1 was created in Matlab® Simulink® with linear and nonlinear elements. The vertical transmissibility functions were generated for three different cases: (a) nonlinear cubic stiffness and coulomb friction damping in suspension, (b) increase in linear stiffness and friction; and (c) increase in stiffness and decrease in friction. The results are shown in Fig. 11. Note that the simulated transmissibility function has just one peak while the measured function has multiple peaks. The reason for this result is that the actual system has more than two degrees of freedom. For the transmissibility X_2/X_b, an increase in the linear stiffness plus an increase in the coulomb friction damping causes the first peak (sprung mass resonance) to shift forward and down, while the second peak (unsprung mass resonance) moves backwards and down (matching the experimental results). On the other hand, if the linear stiffness increases and the coulomb friction damping decreases, the first peak moves forward and upwards (as seen in the experimental results) and the second peak also shifts forwards and upwards.

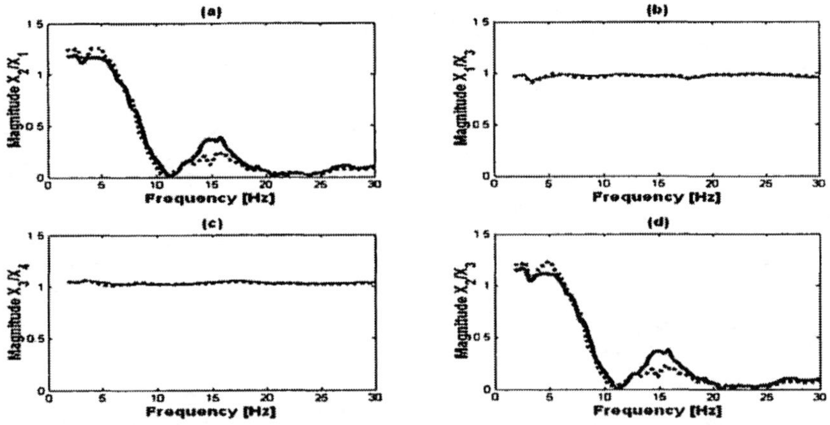

Fig. 10: Transmissibility Functions between Response Location Pairs for Unamaged (——) and Damaged (····) Cases: (a) X_2/X_1, (b) X_1/X_3, (c) X_3/X_4, (d) X_2/X_3.

This shift leads to the following explanation for the trends observed in the experimentally estimated transmissibility functions. Damping causes the response peaks to flatten and also governs the damped natural frequency according to the relation $\omega_d = \sqrt{1-\zeta^2}\omega_n$, where ω_n is the undamped natural frequency and ζ is the damping coefficient. Coulomb friction is more severe for low amplitudes and frequencies. At the sprung mass resonance (around 4 Hz), as the relative motion is almost nonexistent (further reduced by increase in stiffness), Coulomb friction damping effects decrease and hence the peaks increase in amplitude. On the other hand, at the unsprung mass resonance (around 14 Hz) the relative motion is very high (coulomb friction has little effect) and is reduced by the increase in stiffness. This decrease in the relative amplitude across the strut causes the coulomb friction effects to increase and results in a corresponding decrease in the peak amplitude and frequency. Thus, damage in the system can be located and identified by using transmissibility functions.

The technique to be used for diagnosis depends on the bandwidth of the input to the system. Restoring force plots are easier to generate for narrow band inputs (e.g., sinusoidal) and transmissibility functions require broad band inputs (e.g., random). Hence, depending on the bandwidth of the input to the system, one of the two techniques presented here can be used for damage diagnosis.

Conclusions: Restoring force curves and transmissibility functions provide a means of measuring structural changes between two sensors using passive vibration measurements.

Restoring force curves contain information about the internal forces setup in a system between two response points and transmissibility functions isolate the dynamics between two response locations. Hence, these two techniques are well suited to identifying loading and degradation mechanisms in systems. Either technique can be used in an event driven manner depending on the bandwidth of the input to the system. Restoring forces can be used for narrow band inputs and transmissibility functions for broad band inputs.

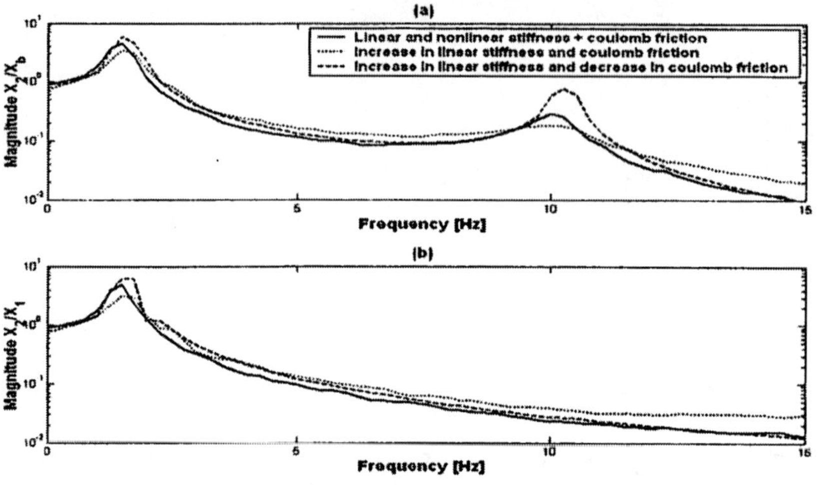

Fig. 11: Simulated Transmissibility Functions between Sprung and Unsprung Mass for Changes in Stiffness and Coulomb Friction Damping: (a) X_2/X_b, (b) X_2/X_1.

References

[1] Haroon, M., Adams, D.E., Luk, Y.W. and Ferri, A.A., "A Time and Frequency Domain Approach for Identifying Non-Linear Mechanical System Models in the Absence of an Input Measurement," 2004, Proceedings of IMECE, Anaheim, CA.

[2] Johnson, T. J. and Adams, D. E., "Transmissibility as a Differential Indicator of Structural Damage", 2002, American Society of Mechanical Engineering Journal of Vibration and Acoustics, Vol. 124, No. 4, pp. 634-641.

Bibliography

- Afolabi, D., "An Anti-Resonance Technique for Detecting Structural Damage," Apr. 1987, Proceedings of the 5th International Modal Analysis Conference, pp. 491-495.
- Masri, S.F., Caughey, T.K., Miller, R.K., and Saud, A.F., "Identification of Nonlinear Vibrating Structures: Part I – Formulation," 1987, Journal of Applied Mechanics, 54, pp. 918-922.
- Masri, S.F., Caughey, T.K., Miller, R.K., and Saud, A.F., "Identification of Nonlinear Vibrating Structures: Part II – Applications," 1987, Journal of Applied Mechanics, 54, pp. 923-929.
- Zhang, H., Schulz, M. J., Naser, A., Ferguson, F., and Pai, P.F., "Structural Health Monitoring Using Transmittance Functions," 1999, Mechanical Systems and Signal Processing 13 (5), pp. 765-787.

IMPLEMENTATION OF AMBIENT RESPONSE MONITORING INSTRUMENTATION FOR A BRIDGE ASSESSMENT TEST BED

Robert Bolton*, Norris Stubbs*, and Charles Sikorsky**

*Texas A&M University
Departments of Engineering Technology and Civil Engineering,
College Station, TX. 77808
bolton@tamu.edu and n-stubbs@tamu.edu

California Department of Transportation
Office of Earthquake Engineering
Sacramento, CA 942874
charles_sikorsky@dot.ca.gov

Abstract: The initial hardware and instrumentation implementation phase of a permanent ambient response (output-only) bridge monitoring system is describe in this paper. The system was installed on a short two-span monolithic slab reinforced concrete bridge and consisted of a distributed accelerometer network connected in a star configuration to a local field acquisition system. Valid high speed point-in-time response data was automatically recorded three times per day and stored on-site. A remote off-site personal computer was used to automatically collect data for further processing and storage from the field instrumentation via an analog modem connection. The point-in-time datasets consisted of multiple near synchronous (burst mode) acceleration time traces from twenty-eight single and bi-axial response locations on the underside of the bridge deck. Experimental and numerically simulated dynamic characteristics of the bridge are also presented along with selected response characteristics collected over a three month startup period.

Key Words: bridge monitoring; dynamic response; structural assessment; remote sensing

Introduction: Study of the changes in dynamic structural properties that may indicate trends in the condition of structural systems has been an active research area over the last twenty years [1-10]. Several modal parameter extraction methods [12-21] in the literature rely on output-only data derived from unmeasured structural excitation by surrounding environmental sources such as passing traffic, wind, and seismic activity. Utilizing output only methods to determine the structural properties of civil systems can significantly decrease the difficulty and cost of collecting point-in-time data from remote structures by eliminating the need for active measured excitation equipment. A Level IV Non-destructive Damage Evaluation (NDE) method [11] (a method which locates and sizes damage, and evaluates the effect of the damage on the structural capacity of the damaged system) based on the Damage Index Method (DI) [6-8,10] is being used in an ongoing research effort to monitor changes in the structural properties of two new highway bridges. SIMO modal tests were conducted on these structures prior to opening of the roadway to traffic to establish baseline numerical models and then the subsequent periodic point-in-time modal data collection process was automated to allow trending of dynamic responses and structural condition.

This paper describes this instrumentation and monitoring system and presents selected results that include a comparison of the extracted modal properties of a two-span reinforced-concrete bridge derived from the SIMO test and the automated output-only method. The remainder of this paper is divided into eight major sections. The first, presents details of the bridge geometry and permanent instrumentation layout. The next section presents results of predicted dynamic properties from a numerical finite element model. Initial SIMO results follow this section. In the fourth section, extracted modal properties from the automated output-only method are presented and in the fifth section point-in-time results are compared. The last two sections are devoted to a summary and conclusion sections.

Structure Description: The reinforced concrete bridge described is a standard Caltrans reinforced concrete monolithic cast-in-place slab deck design. The structure is located in the Southern California high desert where temperature extremes are 120 °F in the summer and near freezing in the winter months. Daily temperature excursions are typically 20 °F to 30 °F. The deck cross-section is 0.43 m (1.41 ft) thick by 13.1 m (42.5 ft) wide with symmetric 10.0 m (33.0 ft) long north and south spans. The deck is monolithically constructed and rigidly connected to 0.77 m (2.5 ft) thick by 2.0 m (6.5 ft) deep abutment walls which rest on five 380 mm (15 inch) round prestressed concrete piles. The slab is supported at mid-span by five uniformly spaced 380 mm (15 inch) round prestressed concrete columns. Figure 1 depicts a profile view of the completed structure. Caltrans Type 25 reinforced concrete barrier rails were cast-in-place along the longitudinal edges of the structure. A down stream view of one rail on a sister structure

Figure 1. Profile view of completed two-span reinforced concrete highway bridge.

Figure 2. View of instrumentation conduits and concrete barrier rail attached to deck.

and two instrumentation conduit boxes are depicted in Figure 2.

Figure 3. View of instrumentation conduits attached to underside of deck slab.

Figure 4. View of instrumentation closet containing conditioning equipment and logger.

Permanent Instrumentation: The permanent vibration monitoring instrumentation and cabling was installed in protected NEMA 4 metal conduit and junction boxes on the under-side of the concrete deck slab. A view of this installation is shown in Figure 3. Deck slab instrumentation consisted of forty-two accelerometer channels divided into twenty-eight vertical acceleration responses and fourteen transverse acceleration responses. Single axis ICSensor 3140 +/-2g MEMs accelerometers mounted on aluminum orientation blocks were installed at each response location. These units have a frequency range of DC-200 Hz and sensitivity of 1000 mV/g. Accelerometer response points were distributed in a grid pattern consisting of four longitudinal sensor lines parallel to the direction of travel along the structure. Each sensor row contained seven accelerometer response points, four of which were at third points of each span and the remaining three located near the abutments and mid-span columns. Outside edge sensor lines contained two accelerometers mounted in a biaxial orientation. The two-bay instrumentation closet installed on one abutment sidewall is shown in Figure 4. This closet housed all instrumentation including filtering and offset adjustment conditioning circuits for each accelerometer channel, a Campbell Scientific CR9000 multiplexed field data logger for collection of response data, DC power supplies with battery backup, and an analog modem for periodic transmittal of response data. Data is collected several time per day and transmitted via a remote modem link late at night. Collection times and characteristics are programmable via the remote modem link. Response data is collected in burst-mode at a 200 Hz scan rate triggered by a mid-span accelerometer.

Numerical Model of Structure: A numerical solids-based dynamic finite element model of the structure was created in Ideas, Version 8m2. These data were used to evaluate the modal properties extracted from subsequent experimental results. The analytic model consisted of a monolithic deck rigidly connected to 0.61 m (2.5 ft) thick by 2.0 m (6.5 ft) high abutment walls and the five central columns. Rigid boundary conditions were specified at the base of the columns and abutment supports. Mass of stiffness properties matched those for the concrete specified for structure (f'c of 3250 psi and nominal weight density of 150 pcf). The first six lower modes identified in the finite element model are depicted in Figure 5. The corresponding modal frequencies are listed in Table I along with results of the SIMO and ambient testing discussed in subsequent sections. Figures 5(a) through Figures 5(e) depict standard beam type global and torsional modes. Figure 5(e) and Figure 5(f) are the first plate modes of the structure. The finite element model of the structure indicates that the plate modes may be very closely spaced.

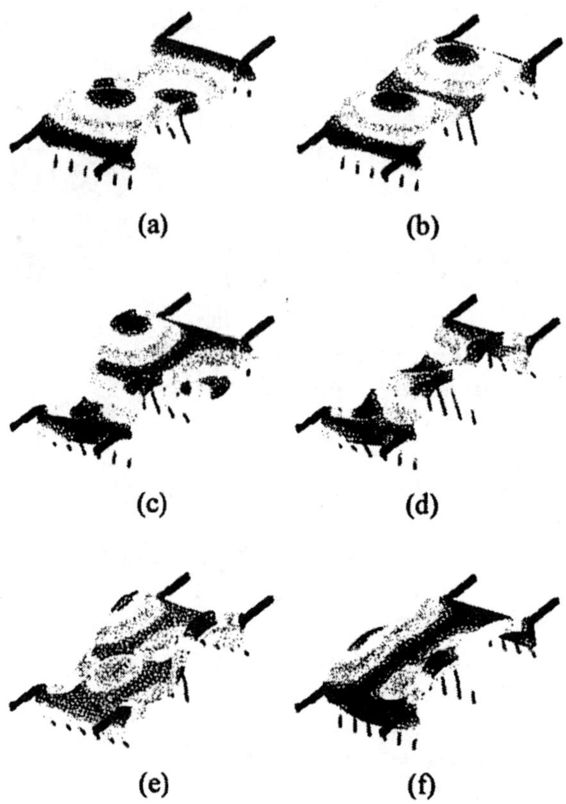

Figure 5 First six finite element bending and torsional modes; (a) 11.2 Hz, (b) 14.1 Hz, (c) 16.7 Hz, (d) 18.2 Hz, (e) 22.4 Hz, and (f) 22.4 Hz.

Table I Modal frequencies from numerical model and experimental results.

Mode Type	Mode Frequency					
	1 (Hz)	2 (Hz)	3 (Hz)	4 (Hz)	5 (Hz)	6 (Hz)
FEM	11.2	14.1	16.7	18.2	22.4	22.4
SIMO	11.2	13.6	16.4	19.5	23.3	21.7
FDD	10.4	12.7	15.6	17.0	19.8	20.0

Experimental SIMO Results: The initial SIMO impact test on the newly constructed reinforced-concrete two span bridge was performed prior to opening of the bridge to traffic. Therefore, the structure was assumed to be in a near pristine undamaged condition when tested. Figure 6 depicts the layout of the measured experimental modal response locations. These points corresponded to the permanent instrumentation layout but were located on the top surface of the deck. An incremental SIMO test method was used to collect the response data. Incremental testing implies that a single impact point was used for the whole structure and response data were collected incrementally in six test setups using repeated impacts of the common input point. The global structural data set included the response of forty-two structural DOFs monitored by the permanent instrumentation system plus additional unmonitored transverse and longitudinal DOFs. Five replications of each structural response were collected. Bolton, et al. [22] presents a more complete description of the SIMO test instrumentation. Average Frequency Response Functions

(FRFs) and coherence functions for each response were computed from the time data using MEScope VES Version 2.0. Figure 7 depicts an overlay of the averaged FRFs for the vertical response of each deck node. Coherence was above 0.95 for all vertical response degrees of freedom which indicated a good signal to noise ratio in the data. The FRF functions presented in Figure 7 display significant interactions of the middle three structural modes. Only the 1st bending, 1st torsional, and 1st plate modes are distinct and well separated from other modes. The extracted natural frequencies from the SIMO test are listed in Table I. The corresponding structural mode shapes for these frequencies are shown in Figure 8. These modes shapes are very similar to the finite element modes shapes and follow a similar modal order.

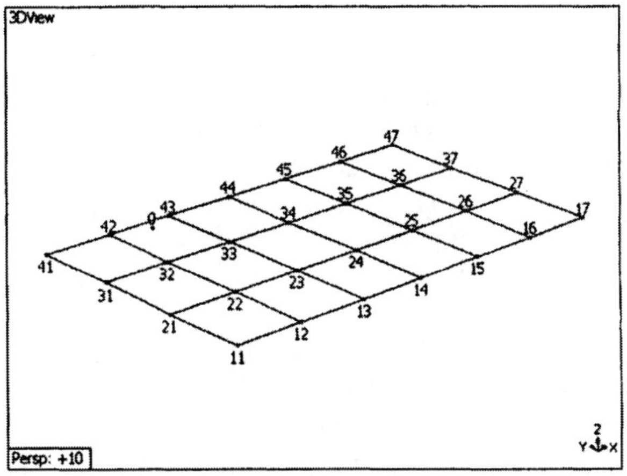

Figure 6 Layout of response locations for the structural system.

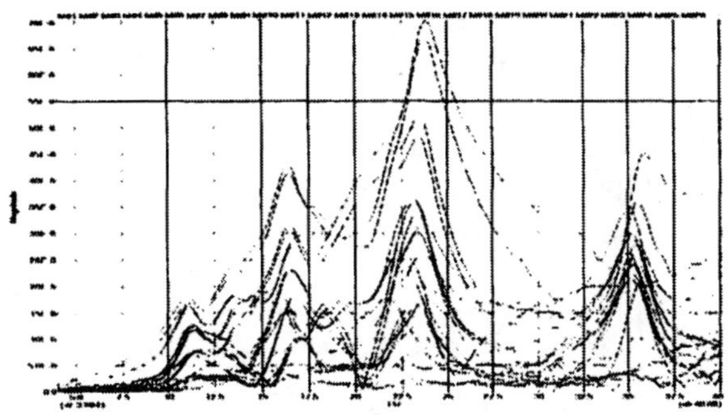

Figure 7 Overlay of SIMO vertical frequency response functions for all deck locations.

Output-only Modal Properties: An automated version of the Frequency Domain Decomposition (FDD) [21] method was implemented to extract natural frequencies and mode shapes from the point-in-time response measurements. A unique feature of permanent instrumentation setup was the collection of near-simultaneous (Used burst mode collection of complete channel list with 200 Hz scan rate.) with response data from all monitored DOFs. This allowed formation of complete system spectral density matrices. A single routine was developed to preprocess the time data, generate the spectral density matrix, and perform Singular Valued Decompositions (SVD) of the normalized spectral matrix at each spectral line in the range of valid structural responses. Local maxima in the SVDs were then located and the frequencies and corresponding mode shapes extracted from the decomposed SVD matrices. Figure 9 presents an overlay of several mid-span vertical DOF diagonal elements of the spectral density matrix from a typical point-in-time data set. The original time data set contained approximately twenty vehicular impact events. The range of significant excitation levels in Figure 9 spans the 10 Hz to 20 Hz range containing the first six vertical bending, torsional, and plate modes of the structure. The specialization of the time response processing and FDD decomposition routine to this structure results in a straightforward extraction of modal properties for any combination of point-in-time response data sets. Average modal frequencies for the six lower modes extracted from the point-in-time response data using the FFD method are presented in Table I. The lower natural frequencies extracted using the FDD method are discussed later in the conclusions of the paper. The mode shapes associated with each extracted FDD mode are depicted in Figure 10.

Point-in-Time Data Analysis: The large quantity of collected point-in-time data allowed for a more detailed investigation of the daily fluctuations of modal properties extracted for various combinations of the data sets. A preliminary comparison of results for a five-day response dataset follows. This dataset consisted of 20 vehicular impact subsets recorded at eight-hour intervals starting at midnight local time each day.

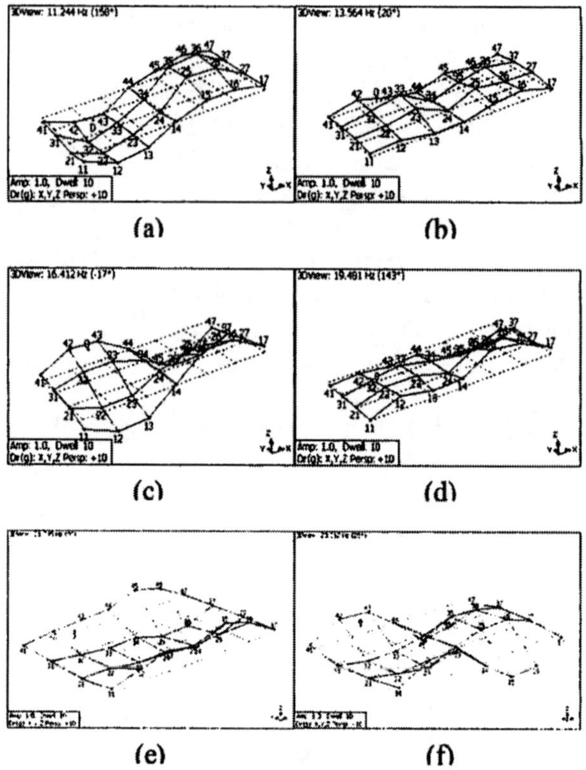

Figure 8 First six SIMO experimental global modes; (a) 11.2 Hz, (b) 13.6 Hz, (c) 16.4 Hz, (d) 19.5 Hz, (e) 21.7 Hz, and (f) 23.3 Hz.

Repeatability of response data: A comparison of five mean daily frequencies and corresponding Modal Assurance Criteria (MAC) for the six lower models identified in the finite element and SIMO analyses is presented here. Mean daily frequencies for these modes are tabulated in Table II through Table VII. MAC calculations are also included in these tables to establish the similarity of extracted modes shapes. Each tabulated frequency and associated mode shape was derived from sixty daily impact events. The tabulated MACs in Table II through Table VII indicated that the 1st and 2nd bending modes and the 1st plate mode show a high degree of repeatability with most MAC ratios well above 0.90. Other modes generally showed good repeatability but not to the degree of the modes noted above. The 1st torsional mode and the 2nd plate mode were

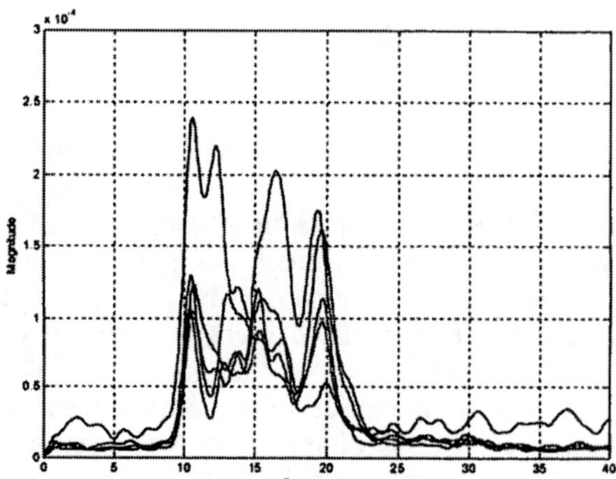

Figure 9 Overlay of autospectra of mid-span vertical acceleration response DOFs.

particularly difficult to extract as indicated by the lack of data for several days in Table IV and Table VII. The average variability in mean daily frequencies was 1.9% and is summarized in Table VIII for each mode.

Daily variation in frequencies: The same five-day data set was used to evaluate the daily variation in extracted frequencies and mode shapes. This data is presented in Table IX through Table XIV for the six lower modes identified in earlier analytic results. The variations in modal frequencies in these results do not indicate a consistent pattern in daily fluctuations across six identified modes. However, the cross-MACs for four of the six modes are very high indicating little daily variation in mode shape characteristics.

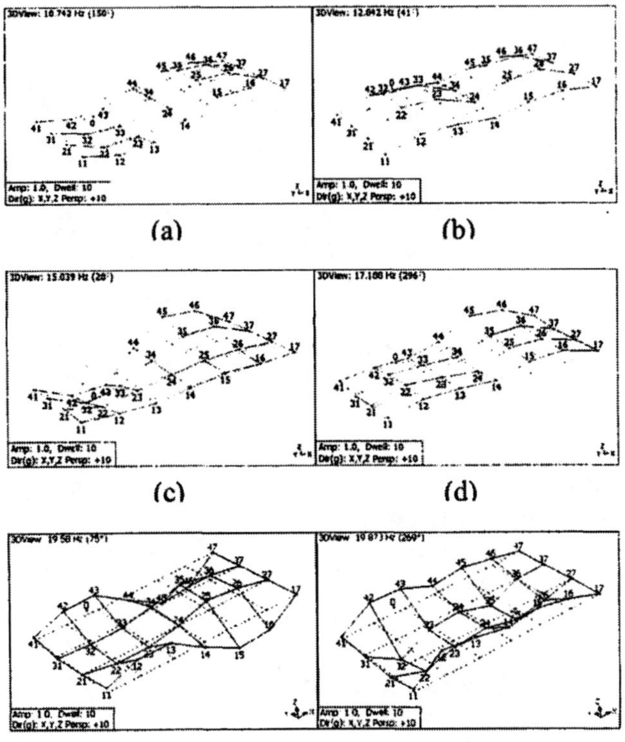

Figure 10 Six FDD experimental global modes; (a) 10.4 Hz, (b) 12.7 Hz, (c) 15.6 Hz, (d) 17.0 Hz, (e) 19.8 Hz, and (f) 20.0 Hz.

Table II Comparison of mean daily frequencies and mode shapes for 1st bending mode.

Mean Freq		Modal Assurance Criteria				
Day	Freq (Hz)	1 10.30	2 10.50	3 10.45	4 10.35	5 10.25
1	10.30	1.00	0.98	0.99	0.95	0.98
2	10.50	0.98	1.00	1.00	0.94	0.99
3	10.45	0.99	1.00	1.00	0.94	0.99
4	10.35	0.95	0.94	0.94	1.00	0.94
5	10.25	0.98	0.99	0.99	0.94	1.00

SUMMARY: This paper presented preliminary results of a research effort focused on the implementation of an automated health monitoring system for a permanently instrumented reinforced concrete highway bridge. The monitoring methodology utilizes extracted point-in-time modal properties of the structure derived from an output-only modal extraction method. A detailed description of the structure and instrumentation was presented and results of an analytical dynamic analysis and initial SIMO test were presented. This was followed by a more extensive presentation of output-only modal results, which included a discussion of the daily variability of modal property characteristics.

CONCLUSIONS: Direct comparisons between SIMO and out-put only results were not presented in this paper because of a significant time lapse between the SIMO testing and collection of the presented output-only data. However, a review of frequency data in Table I indicates a significant difference in extracted frequencies between the SIMO results and the output-only results. This difference is attributed to two possible causes. The first is a break-in of the structural system resulting from several months of traffic on the structure before collection of the output-only data presented here. The second possible cause is the mass effect of traffic on the structure during collection of the output-only data. Although not presented here MACs between SIMO and output-only mode shapes were lower than those between similar output-only modes indicating some difference in extracted mode shapes. Several conclusions can be drawn from the results presented in this paper.

1. The analytic, SIMO, and output-only methods returned similar modal properties for six lower modes indicating that good correlation should be expected between detailed analytic models and experimentally derived properties of the real structure.
2. The SIMO and implemented output-only method had

Table III Comparison of mean daily frequencies and mode shapes for 2nd bending mode.

Mean Freq		Modal Assurance Criteria				
Day	Freq (Hz)	1 12.65	2 12.74	3 12.55	4 12.55	5 12.94
1	12.65	1.00	0.92	0.97	0.91	0.94
2	12.74	0.92	1.00	0.92	0.94	0.97
3	12.55	0.97	0.92	1.00	0.89	0.95
4	12.55	0.91	0.94	0.89	1.00	0.93
5	12.94	0.94	0.97	0.95	0.93	1.00

Table IV Comparison of mean daily frequencies and mode shapes for 1st torsion mode.

Mean Freq		Modal Assurance Criteria			
Day	Freq (Hz)	1 15.48	2 15.67	3 15.04	5 16.02
1	15.48	1.00	0.81	0.70	0.24
2	15.67	0.81	1.00	0.76	0.09
3	15.04	0.70	0.76	1.00	0.05
5	16.02	0.24	0.09	0.05	1.00

Table V Comparison of mean daily frequencies and mode shapes for 2nd torsion mode.

Mean Freq		Modal Assurance Criteria				
Day	Freq (Hz)	1 16.75	2 16.94	3 17.09	4 17.53	5 16.70
1	16.75	1.00	0.89	0.77	0.86	0.87
2	16.94	0.89	1.00	0.84	0.89	0.91
3	17.09	0.77	0.84	1.00	0.92	0.89
4	17.53	0.86	0.89	0.92	1.00	0.94
5	16.70	0.87	0.91	0.89	0.94	1.00

Table VI Comparison of mean daily frequencies and mode shapes for 1st plate mode.

Mean Freq		Modal Assurance Criteria				
Day	Freq (Hz)	1 19.58	2 18.85	3 19.48	4 20.17	5 20.95
1	19.58	1.00	0.91	0.92	0.93	0.87
2	18.85	0.91	1.00	0.89	0.87	0.76
3	19.48	0.92	0.89	1.00	0.92	0.86
4	20.17	0.93	0.87	0.92	1.00	0.94
5	20.95	0.87	0.76	0.86	0.94	1.00

similar difficulty in accurately and consistently identifying several closely spaced modes. More data may improve the results of both methods.

3. Output-only results exhibited lower extracted natural frequencies than SIMO testing.
4. For the structure under test, fluctuation of extracted modal frequencies of approximately 1.9% should be expected. The mode shapes for well-defined frequencies showed very good repeatability.
5. Variations in mode shapes appear to be smaller than the corresponding changes in modal frequency.
6. The datasets consisting of 60 to 100 vehicular impact events used in the presented output-only calculations provided good repeatability of extracted properties.

Aknowledgements: The authors thank the California Department of Transportation, the Texas Engineering Experiment Station, and the Dwight Look College of Engineering at Texas A&M. University for the funding and support of this investigation.

References

1. Aktan, A.E., Lee, K.L., Chuntavan, C., and Aksel, T. "Modal Testing for Structural Identification and Condition Assessment of Constructed Facilities." Proc. of the 12th Int. Modal Analysis Conf, Honolulu, Hawaii, Vol. 1, 462-468, 1994.
2. Biswas, M., Pandey, A.K., Samman, M.M. (1990). "Diagnostic Experimental Spectra/Modal Analysis of a Highway Bridge." Int. J Analytical and Experimental Modal Analysis, 5(l),33-42.
3. Cawley, P., and Adams, R.D. (1979). "The Locations of Defects in Structures from Measurements of Natural Frequencies." J Strain Analysis, 14(2), 49-5 7.
4. Doebling, S., Farra, C., Prime, M., and Shevitz, (1996), "Damage Identification and Health Monitoring of Structural and Mechanical Systems from Changes in Their Vibration Characteristics: A Literature Review," Technical Report LA-13070-MS, Los

Table VII Comparison of mean daily frequencies and mode shapes for 2nd plate mode.

Mean Freq		Modal Assurance Criteria		
Day	Freq (Hz)	1 20.41	2 20.17	3 19.34
1	20.41	1.00	0.66	0.62
2	20.17	0.66	1.00	0.53
3	19.34	0.62	0.53	1.00

Table VIII Variability of mean daily frequencies for six lower modes.

Mode	Sample Size	Mean Frequency (Hz)	STD (Hz)	COV (%)
1	5	10.37	0.09	0.9
2	5	12.69	0.15	1.1
3	4	15.55	0.35	2.3
4	5	17.00	0.30	1.8
5	5	19.81	0.71	3.6
6	3	19.97	0.46	2.2

Table IX Daily variation of 1st bending mode.

Frequency		Modal Assurance Criteria		
Time of Day	Freq (Hz)	12 AM 10.35	8 AM 10.55	4 PM 10.45
12 AM	10.35	1.00	1.00	0.98
8 AM	10.55	1.00	1.00	0.98
4 PM	10.45	0.98	0.98	1.00

Table X Daily variation of 2nd bending mode.

Frequency		Modal Assurance Criteria		
Time of Day	Freq (Hz)	12AM 13.04	8 AM 12.94	4 PM 12.60
12 AM	13.04	1.00	0.98	0.94
8 AM	12.94	0.98	1.00	0.98
4 PM	12.60	0.94	0.98	1.00

Alamos National Laboratory, Los Alamos, NM.
5. Farrar, C., and Jauregui, D. (1996). "Damage Detection Algorithms Applied to Experimental and Numerical Modal Data from the I-40 Bridge. " Report No. LA13074-MS, Los Alamos National Laboratory, Los Alamos, New Mexico.
6. Stubbs, N., and Osegueda, R. (1990a). "Global Non-Destructive Damage Evaluation in Solids." Int. J Analytical and Experimental Modal Analysis, 5(2), 67-79.
7. Stubbs, N., Kim, J.T., and Farrar, C.R. (1995). "Field Verification of a Nondestructive Damage Localization and Severity Estimation Algorithm." Proc. of the]3th Int. Modal Analysis Conf, Nashville, Tennessee, Vol. 2, 210-218.
8. Stubbs, N., Richardson, M., and Sikorsky, C. (1992). "Nondestructive Damage Assessment of a Bridge Using Modal Testing and Structural Reliability." Proc. 3rd Workshop on Bridge Engineering Research in Progress, La Jolla, California, 63-66.
9. Toksoy, T., and Aktan, A.E. (1994). "Bridge-condition Assessment by Modal Flexibility." Experimental Mechanics, 34(3), 271-278.
10. Stubbs, N., Sikorsky, C., Park, S., Choi, S., and Bolton, R., (1999) "A Methodology to Nondestructively Evaluate the Structural Properties of Bridges, Proc. of the 17th Int. Modal Analysis Conf, Kissimmee, Fl., 1260-1268.
11. Rytter, A., (1993). Vibrational Based Inspection of Civil Engineering Structures, PhD. Dissertation, University of Aalborg, Aaloborg, Denmark.
12. Ibrahim, S. and Mikulcik, E., (1977). "A Method for the Direct Identification of Vibration Parameters from the Free Response," The Shock and Vibration Bulletin, 47(4), 183-198.
13. Ventura, C., (1999). "Ambient Vibration Measurements of a Logging Bridge in British Columbia, Canada, Proc. of the 17th Int. Modal Analysis Conf, Kissimmee, Fl.
14. Cooper, J. and Desforges, M., (1997). "Modal Parameter Identification from Ambient Response Data using Advanced Identification Methods," Proc. of DAMAS 1997, Sheffield Academic Press, Sheffield, UK, 409-418.
15. De Stefano, A., Sabia, D., and Sabia, L., (1997). "Structural Identification using Armav Models from Noisy Dynamic Response under Unknown Random Excitation," Proc. of DAMAS 1997, Sheffield Academic Press, Sheffield, UK, 419-428.

Table XI Daily variation of 1st torsion mode.

Frequency		Modal Assurance Criteria		
Time of Day	Freq (Hz)	12 AM 15.52	8 AM NA	4 PM 15.14
12 AM	15.52	1.00	NA	0.86
8 AM	NA	NA	NA	NA
4 PM	15.14	0.86	NA	1.00

Table XII Daily variation of 2nd torsion mode.

Frequency		Modal Assurance Criteria		
Time of Day	Freq (Hz)	12 AM 17.19	8 AM 17.19	4 PM 16.50
12 AM	17.19	1.00	0.93	0.94
8 AM	17.19	0.93	1.00	0.90
4 PM	16.50	0.94	0.90	1.00

Table XIII Daily variation of 1st plate mode.

Frequency		Modal Assurance Criteria		
Time of Day	Freq (Hz)	12 AM 19.68	8 AM 21.05	4 PM 19.53
12 AM	19.68	1.00	0.87	0.92
8 AM	21.05	0.87	1.00	0.92
4 PM	19.53	0.92	0.92	1.00

Table XIV Daily variation of 2nd plate mode.

Frequency		Modal Assurance Criteria		
Time of Day	Freq (Hz)	12 AM 19.14	8 AM 19.39	4 PM 19.92
12 AM	19.14	1.00	0.88	0.73
8 AM	19.39	0.88	1.00	0.71
4 PM	19.92	0.73	0.71	1.00

16. Paultre, P., Proulx, J., and Talbot, M., (1995). "Dynamic Testing Procedures for Highway Bridges using Traffic Loads," Journal of Structural Engineering, 121(2),ASCE, New York, NY, 362-376.
17. Asmussen, J., Ibrahim, S., and Brincker, R., (1998). "Random Decrement: Identification of Structures Subjected to Amibient Excitation," Proc. of the 16th Int. Modal Analysis Conf., Santa Barbara, CA, 914-921.
18. Bonato, R., Ceravolo, R., and De Stefano, A., (1997). "Time-Frequency and Ambiguity Function Approaches in Structural Identification," Journal of Engineering Mechanics, 123(12), ASCE, New York, NY, 1260-1267.
19. Fasana, A., Garibaldi, L., Piomibo, B., and Ruzzene, M., (1997). "Application of the Wavelet Estimation Technique for the Analysis of a Motorway Bridge," Proc. of DAMAS 1997, Sheffield Academic Press, Sheffield, UK, 419-428.
20. Ventura, C., Brinker, R., Kirkegaard, P., and Ibrahim, S., (1998). Response Analysis of Large Structures Course, 16th Int. Modal Analysis Conf, Society for Experimental Mechanics, Bethel, CT.
21. Brincker, R., L. Zhang and P. Andersen: 'Modal Identification from Ambient Responses using Frequency domain Decomposition', Proc. of the 18th International Modal Analysis conference (IMAC), San Antonio, Texas, February 7-10, 2000.
22. Bolton, R., Stubbs, N., Park, S., Choi, S., Sikorsky, C., "Measuring Bridge Modal Parameters for Use in Damage Detection and Performance Algorithms," Proceedings 17th International Modal Analysis Conference & Exposition, Kissimmee, FL., 1269-1275, Feb. 1999.

FAILURE ANALYSIS METHODOLOGY

Chair: Ms. Nancey J. Maegerlein
Naval Surface Warfare Center/Crane

ENSURING QUALITY WEAPONS FOR THE NAVAL WARFIGHTER – A MATERIALS ENGINEER'S PERSPECTIVE

Marc Pepi
Naval Air Warfare Center – Materials Division
Code 497000D
China Lake, CA 93555
marc.pepi@navy.mil

Abstract: Defective ordnance not only negates the mission for which the ordnance was intended, but can be a strain on defense budgets, as increased funding is required to assess the nature of the defect and pinpoint how widespread the problem may be. If the weapon defect is severe enough, it may be deemed unworthy for flight, resulting in downtime, lost missions, increased inspection costs, etc. This paper will describe the procurement life cycle utilized for Naval ordnance, which has transformed over the years from "lowest bidder" to "best value". Materials engineering involvement throughout every stage of the procurement process (planning, source selection plan and evaluation, pre-award meetings, post-award meetings, first article testing and quality systems reviews) is crucial in providing the warfighter with the most reliable and highest quality weapons.

Key Words: Best Value, Defense Acquisition, Naval Warfighter

Introduction: The federal government has recently introduced a new way to purchase items such as supplies, equipment, ordnance, etc. The former process was basically referred to as "lowest bidder" procurement, which had numerous problems associated with it. Although there is a chance the lowest bidder may fulfill all contractual needs, very often, choice of the lowest bidder has the potential to lead to program delays, budget overruns, change orders and poor quality. This paper will discuss the advantages of the "best value" acquisition process for precision strike weapons, and the role of materials engineering support throughout the life cycle in reducing the chances that defective weapons and subcomponents make their way into service.

The "Best Value" Process: It should be noted that the author is not an expert in defense acquisition, and has essentially been associated with the Best Value procurement process through a technical support role. That said, the following is presented to lay the groundwork for the importance of the role of the materials engineer throughout the acquisition process. As mentioned, the "lowest bidder" procurement process has the potential for many problems. Although some creative alternatives to "lowest bidder" acquisition have been employed recently (most notably, the Air Force held an online "reverse auction" to find the lowest bidder for weapons systems acquisitions [ ]), the result is still a "lowest bidder". The Navy actually led the way in the Best Value acquisition process in 1989, with the innovation of employing a "greatest value source selection of firm-fixed-price supplies in which cost and past performance were the only award factors [ ]. Two regulations were passed in the 1990's that gave birth to Best Value: the Federal Acquisition Streamlining Act of 1994, and the Clinger-Cohen Act of 1996. These regulations gave the government greater flexibility in the acquisition process, in that Best Value procurements use past performance, technical capabilities, qualifications of key contractor personnel and other quality indicators as source selection factors [ ]. However, the statutes and regulations provide only a procedure to use Best Value procurement; they do not require an agency to use Best Value procurement [ ]. Best Value is intended to provide the government the "best value" for its money. It is now the preferred source selection method. As discussed in reference [ ], contracting officers should not use Best Value when any of the following criteria are applicable:

- There is adequate time available before the commodity is needed
- Price will be the only factor considered for award of the contract
- No bidder discussions are necessary

Best Value has given rise to increased full and open competition during acquisition, and has allowed the government to award contracts to bidders that may have a pricier proposal, yet offer a higher-quality product. The Best Value Handbook [ ] lists the Guiding Principles for Best Value Source Selection, which includes the following steps:

- Acquisition planning
- Soliciting proposals
- The evaluation process
- Evaluation reports
- Competitive range
- Written and oral discussions
- Best and final offers
- Selecting the source
- Award and debriefing

Problems with "Best Value": Some believe that the Federal Acquisition Regulation (FAR) definition is subjective and permits the government agency to exercise broad discretion [6]. There may also be some subjectivity when evaluating the offeror proposals; for instance, two evaluators rating the same proposal may get different results. It was Rep. Cardiss Collins' (D-IL) feeling during congressional debates on efficient competition that "emphasizing subjectivity in awarding contracts creates a breeding ground for prejudice because [contracting officers], given the choice, will usually go with a well-established large firm instead of a small business offering a lower price" [6]. She went on to state that this type of procurement may exclude smaller companies that can do the job and provide a lower price, but do not have the reputation of a large business. Subjectivity might also be possible in the areas of "quality", "value" and "past performance". In addition, this system may punish a new contractor that may not have much of a past performance history, even if, technically, they can meet all the contract requirements.

The Role of Materials Engineering Support: Although the contracting officer, in conjunction with the weapon program manager, is responsible for the Best Value process listed above, support from materials engineering is beneficial in acquisition planning and source selection. Further, it may be necessary for materials support at pre-award meetings, if two or more offerors are very close in Best Value scoring after the evaluation process. Once an offeror has been selected, the government usually requests a post-award meeting, if an offeror is new, or the offeror is to manufacture an item that has not been produced in a long time. Materials engineering support is requested at this meeting, as well as first article testing. Finally, support may be requested at a quality system review, while the manufacturer is in full production.

Acquisition Planning: This phase of the Best Value process is where the source selection plan (SSP) is created. This plan not only outlines how proposals will be solicited and how negotiations will be conducted, but also lists the evaluation factors to be used for rating each of the offerors. In general, these factors include past performance, technical ability, price/cost and small business utilization. Technical ability, depending on the weapon, can consist of critical skills, processes and procedures, quality system and tests and inspections. Materials engineering input is provided in the areas of critical skills, processes and procedures, as well as tests and inspections. Critical skills, processes and procedures outlines the technical areas that will be evaluated on each offeror's proposal. Depending on the weapon, these could include areas such as material, forging, casting, heat treatment, machining, and coatings. The tests and inspections section usually outlines first article testing and lot acceptance testing (FAT and LAT).

Source Selection: The rating of best and final offers (BAFOs) is accomplished through a source selection and evaluation team, consisting of the weapon program manager and as many or few technical support evaluators he or she feels is necessary in making a fair and unbiased decision. For precision strike weapons, teams usually consist of the program manager, quality engineer, the metrologist and the materials engineer. The Material engineer rates each offeror with respect to the required critical skills, processes and procedures, as well as tests and inspections. Narrative statements are utilized to list the strengths and weaknesses of the offeror with respect to the evaluation factors. It is important to note that at this stage, offerors are not rated against each other, but against the evaluation factors set forth in the SSP. Ratings are forwarded to the weapon program manager who, in turn, sends the information to the contracting officer.

Pre-Award Meetings: Pre-award meetings are sometimes necessary during the Best Value process if two or more offerors have equivalent source selection evaluation scores. A meeting can also be held if only one offeror has responded to the request for proposal (RFP). These meetings serve a couple of purposes: to confirm or refute the original evaluation given to the contractor, and to attain a "comfort level" with the prospective contractor's abilities to make the product. Discussions should focus only on those factors within the SSP. It is at these meetings that government personnel can convene with the contractor's upper management and technical personnel, view the manufacturing equipment and production line and make an assessment as to whether the contractor will be able to manufacture the intended weapon. The materials engineer sends recommendations in the form of reports to both the weapon program manager and the contracting officer, listing the advantages and disadvantages of each facility.

Post-Award Meetings: Once the offeror is awarded the contract, a post-award meeting is held at the discretion of the government. At this meeting, the government ensures the contractor understands the technical aspects of the contract. It is also a forum for the contractor to discuss issues it may have with the engineering drawing package, automated data list (ADL), etc. From the outset, a relationship is forged and the government and contractor work as a team to ensure a quality product will be produced.

First Article Testing: After the post-award meeting, the contractor is granted a certain timeframe to produce finished products and subcomponents as part of the FAT. The time frame granted depends on the complexity of the weapon or component. The FAT ensures the contractor has the ability and capability to furnish a product that conforms to all contract requirements for acceptance. First article testing may be appropriate when [7]:

- The contractor has not previously furnished the product to the government
- The contractor previously furnished the product to the government, but:
  - there have been subsequent changes in processes or specifications
  - production has been discontinued for an extended period of time
  - the product acquired under a previous contract developed a problem during its life
- The product is described by a performance specification
- It is essential to have an approved first article to serve as a manufacturing standard

The materials engineer accompanies the program manager, the quality team and the contracting officer on this visit, examining material certifications, process certifications, in-house procedures, equipment calibration, test results, actual testing and on-going production of related commodities, among other tasks. Once it is deemed that the contractor has passed FAT, production of the weapon may begin.

Quality System Review: Sometimes the need may arise to perform a quality system review (QSR) on a commodity that has shown a decline in overall quality. These reviews may come about as a result of noting inadequacies during a FAT for a different product. The materials engineer may have to support the quality engineer on a QSR since, often, the drop in quality may be due to a manufacturing process. The QSR is designed to provide the manufacturer insight on ways to improve the process and increase the quality of the weapon. It is also the materials engineer who may examine a weapon or weapon component that may have failed while in service or inventory.

Minimizing/Eliminating Defects: As discussed, the materials engineer provides input during the life cycle of the weapon. The following represent a number of case histories, typifying the findings noted during FAT, preproduction and full production.

Conical Bomb Fins: Conical bomb fins are attached to general purpose bombs for a more controlled trajectory. Failure of the fin in flight has the potential of sending the weapon away from its intended target. During first article testing, it was noted that the seam welds were failing shear testing (see Figure 1), and a production run of a similar item exhibited flash rusting upon exiting the zinc phosphate booths and a less thick than nominal powder coating. Recommendations were made in each of these areas to ensure both the current commodity and the commodity in production met the governing requirements.

Suspension Lugs: Suspension lugs are utilized to attach general purpose bombs to the underside of aircraft. There are two lugs per bomb, and failure of one of these lugs in flight would most assuredly lead to the failure of the other lug, freeing the bomb from the aircraft at an unplanned location. Failure was noted in one of these lugs as a result of proof testing of bomb suspension lugs. The lugs were required to show no cracks as a result of this testing: however, the part failed through the bail. A metallurgical examination revealed that the failure originated at a forging lap (Figure 2 shows a similar lap). Recommendations were offered to minimize the probability of this occurrence in the future. This forging lap should have been detected through magnetic particle inspection (MPI), so it was necessary to review the MPI equipment and process. It was determined that the inspection technique was inadequate, and hundreds of thousands of these lugs in inventory were re-inspected.

Laser-Guided Bomb Fins: Similar to the conical bomb fins, these fins are utilized to keep the bomb falling along a planned trajectory. The primary contractor requested the use of powder coating over the thin aluminum fins. The thin aluminum sheet was non-heat treatable grade. Subsequent tensile testing showed a dramatic reduction of mechanical properties as a result of the bake process. The powder coating curing temperature was deemed excessive enough to affect the properties of this alloy. Recommendations were made to use infrared curing instead of conventional oven curing.

Figure 1 Macrograph showing a conical fin seam weld specimen that failed shear testing (top), and a specimen that passed testing. Failure is required to occur within the parent material, not the weld.

Figure 2 A forging lap found in the bail of a bomb suspension lug through magnetic particle inspection.

Fuze Cable Assemblies: These components are the means by which an electrical signal from the aircraft is transmitted to the bomb fuze. Failure of this component could lead to a "dud" and a lost mission. Adhesion problems within the fuze plug between polyurethane and electroless nickel-plated brass contacts were noted by the primary contractor (Figure 3). The polyurethane was a new design chosen to enhance adhesion to the coil cable attached to the fuze plugs. After environmental testing, this design failed electrical testing. Recommendations were made to utilize an adhesive epoxy on the metal contacts to improve adhesion.

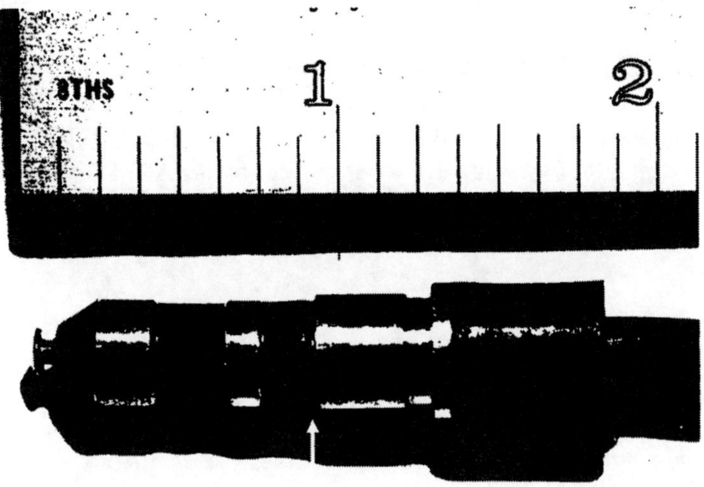

Figure 3 Macrograph of fuze plug from a fuze cable assembly showing metal to polyurethane regions of poor adhesion (arrow shows one interface).

Conclusions:

- Best Value procurement allows the government greater flexibility in the acquisition process, since past performance, technical capabilities, qualifications of key contractor personnel and other quality indicators are used as source selection factors.
- Materials engineering involvement throughout every stage of the acquisition process (planning, source selection plan and evaluation, pre-award meetings, post-award meetings, first article testing and quality systems reviews) is crucial in providing the warfighter with the most reliable and highest quality weapons.

References

[1] Galbraith, M., "Internet Contract Auction Saves Money", Hilltop Times, 25 January 2000, http://www.hilltoptimes.com/archive/20010125/Mainstory.html.
[2] Best Value Overview, http://development.dynsys.com/abm_backup/bpbest.cfm.
[3] Lieberman, R.D and O'Brien, K.R., "Negotiated Procurement", www.mshpc.com/article5.htm.
[4] 48 CFR § 36.3; 10 USC § 2305a (b).
[5] "The Best Value Handbook", Guiding Principles for Best Value Source Selection, June 1996.
[6] Heisse, J.R., "'Best Value' Procurement: How Federal and State Governments are Changing the Bidding Process", Thelen Reid & Priest LLP, 29 April 2002,
[7] Federal Acquisition Regulation System, Subpart 9.3 – First Article Testing and Approval, http://www.arnet.gov/far/current/html/FARMTOC.html

FAILURE PREVENTION IN COMPUTER SYSTEMS

Aleksey Urmanov and Kenny Gross
Sun Microsystems, Inc. San Diego, CA 92121

Abstract: Computer systems are required to be ever more reliable and, at the same time, less complex to maintain. Improved reliability, availability and failure avoidance are becoming major differentiators in the arena of high-performance computer systems. High reliability of computer systems is achieved not only by hardware redundancy but also by adding software intelligence that models the behavior of the system and detects any abnormal deviations from known or expected and assists a system administrator in the decision making process during critical situations. In this paper we present a software intelligence framework for supporting decisions made by a system administrator for prevention of failures in complex computer systems executing business or scientific applications.

Key words. Failure prevention; decision support systems; knowledge representation and manipulation.

Introduction: Complex machine-human systems such as cars, commercial airplanes, nuclear power or chemical plants, computer networks tend to comprise of hundreds and thousands of smaller components interconnected in a way the system could provide its services to its users. The mean time between failure of individual components often cannot be made desirable long earthier due to cost limitation or due to fundamental limitation imposed by technology or by laws of nature. A fault tolerant design is used to make systems tolerate failures of individual components and work around failures [3]. At the system level, failure prevention concerns with sustaining the overall system in a condition in which it continues to provide its services to the users at a prescribed level of availability. The system may have failed components and still has no significant degradation effects. For example, a high end server with hundreds of CPUs can offline some faulty CPU's for which the error rates exceed some specified thresholds and continue to support multiple users and run multiple applications.

Managing data centers and compute ranches with heterogeneous computer systems creates a problem of making decisions quickly in critical situations. Critical situations are caused by component failures, user and administrator mistakes, software bugs and software aging. When a critical situation occurs, a human operator may not be able to make correct decisions quickly enough to avoid possible downtime because of the inability to comprehend or physically obtain necessary information about the environment, user activities, errors, hardware/software states and other necessary information in a short period of time. The solution to this problem can be in the development of intelligent systems supporting the decision-making process at a specified level of availability.

In designing and exploiting a decision support system (DSS) it is important that experts and users of the DSS can easily understand the chosen knowledge representation. Otherwise, the accumulation and evaluation of knowledge becomes impractical. Typical knowledge representation models are logical model [7], product model [2], frame model [5], semantic network model [8]. The logical model is computationally complex, and inferences often lead to wide probabilistic intervals. The product model has very low efficiency of manipulating of knowledge because of long verification of applicability of rules. There is a great interest in representing imprecise and incomplete knowledge and many publications devoted to this topic [1, 6, 10-14]. However, until recently there was no efficient scheme that would allow designing computationally light and fundamentally sound decision support systems.

In the paper we consider a practical problem of managing a datacenter at a prescribed level of availability and demonstrate one of the ways of attaining and sustaining ever increasing levels of availability by developing an intelligent decision support system using the imprecise and incomplete knowledge representation and manipulation framework developed by Sirodzha [9].

The imprecise quanta of knowledge framework: Sirodzha [9] developed a theory for representing and manipulating quanta (portions) of knowledge with applications to designing intelligent decision support systems (DSS) for the human operator assistance.

Representing knowledge: One of the first problems that needs to be solved to be able to design an intelligent DSS is the problem of formal representation of knowledge. Let's assume that knowledge can be treated as data structured in a special manner. The problem of formal knowledge representation then is the representation of knowledge by some data structures and the algorithmic construction of such structures. The knowledge representation and manipulation framework of Sirodzha [9] is build on the notion of imprecise knowledge quanta, which are special structures of data.

The following three definitions specify *terminal* imprecise quanta of knowledge used for the algorithmic construction of various level imprecise quanta of knowledge.

Definition 1: A level-zero *vector* imprecise quantum of knowledge, denoted as $\pi k_0 y$, is represented as

$$\pi k_0 y = [\, a_1^{(1)} \,|\, d_x^{(1)}(a_1^{(1)}), \ldots, a_{p1}^{(1)} \,|\, d_x^{(1)}(a_{p1}^{(1)}) \;:\; \ldots \;:\; a_1^{(n)} \,|\, d_x^{(n)}(a_1^{(n)}), \ldots, a_{pn}^{(n)} \,|\, d_x^{(n)}(a_{pn}^{(n)}) \,] .$$

where $\pi k_0 \in Q_\pi$ with the following semantics: the observed subject of the decision (SD) $\omega \in \Omega$ is characterized by n parameters (or inputs) x_j, $(j=1,\ldots,n)$ of different types. The parameters take values $a_{kj}^{(j)} \in {}^{\pi}X^{(j)}$, $(k_j=1,\ldots,p_j)$ and have associated degrees of correctness (DC) $d_X^{(j)}(a_{kj}^{(j)}) \in D$.

Q_π is the set of symbols coding the semantics of imprecise knowledge quantum structures. Ω is the set of subjects of the decision. D denotes the range of the correctness function. The inputs of imprecise knowledge quanta are separated by a semicolon, and the components within each input are separated by commas.

The degree of correctness is the correctness function $d_X^{(j)}: {}^{\pi}X^{(j)} \rightarrow [-1,+1]$ of the imprecise sets

$$^{\pi}X^{(j)} = \{ \ a_1^{(j)} | d_X^{(j)}(a_1^{(j)}), \ldots, a_{pj}^{(j)} | d_X^{(j)}(a_{pj}^{(j)}) \ \}, \quad (j=1,\ldots,n)$$

A DC of -1 means that the knowledge is absolutely incorrect. A DC of 0 means complete uncertainty. A DC of +1 means that the knowledge is correct. All intermediate values of the DC reflect the degree of trust or distrust to the corresponding knowledge.

Definition 2: A level-zero *selector* imprecise quantum of knowledge, denoted as $\pi k_0 a$, is represented as

$$\pi k_0 a = V_k^{(p)} [\ a_1 | d_A(a_1), \ldots, \ a_p | d_A(a_p) \ ] = a_k | d_A(a_k),$$

where $\pi k_0 \in Q_\pi$ with the following semantics: the value a_k with the DC of $d_A(a_k)$ has been selected from the set ${}^{\pi}A$ of imprecise values of the observed property.

Definition 3: A level-one *indicator* imprecise quantum of knowledge, denoted as $\pi k_1 b$, is represented as

$$\pi k_1 b = \chi_Y^{(j)}(\ a_k^{(j)} | d_Y^{(j)}(a_k^{(j)}) \ ) = \{ \ 1, \text{ if } (a_k^{(j)} | d_Y^{(j)}(a_k^{(j)})) \in {}^{\pi}Y^{(j)} \quad 0, \text{ otherwise} \},$$

where $\pi k_1 \in Q_\pi$ with the following semantics: if $\chi_Y^{(j)} = 1$ then the value $a_k^{(j)}$ of the parameter x_j is observed with the DC of $d_Y^{(j)}(a_k^{(j)}) \in D$ and is not observed if $\chi_Y^{(j)} = 0$. ${}^{\pi}Y^{(j)}$ is the imprecise set of measured values of the property x_j of the SD. D is the range of the correctness function. $b \in S$, the symbolic language of the imprecise quanta of knowledge, consisting of the set of letters, digits and operation symbols of the theory of algorithms.

These three types of quanta of knowledge specify the set of the *terminal* π-quanta of knowledge

$$K_\pi = \{ \ \pi k_0 y, \ \pi k_0 a, \ \pi k_1 b \ \}.$$

Definition 4: Algorithmic constructions generated from the components of K_π by a finite number of application of the superposition operator Π, the string concatenation operator SCON(·), and the matrix concatenation operator MCON(·) are called *various level πk-knowledge* or *approximate* quanta of knowledge.

The superposition operator Π is the procedure of finding function (or algorithm) $h(x_1,\ldots,x_n)$ given functions (or algorithms) $f(y_1,\ldots,y_m)$, $g_1(x_1,\ldots,x_n),\ldots, g_m(x_1,\ldots,x_n)$, according to the formula $h(x_1,\ldots,x_n) = f[g_1(x_1,\ldots,x_n),\ldots,g_m(x_1,\ldots,x_n)]$. The string concatenation operator SCON(·) is the procedure of consecutive concatenation of numeric and symbolic expressions as elements e_i of sequences $\{e_i\}_{i=1}^m$ into a string: $SCON(e_{i=1}^m) = [e_1,\ldots,e_m]$. The matrix concatenation operator MCON(·) is the procedure of consecutive concatenation of numeric and symbolic expressions as the elements of sequences $\{e_i\}_{i=1}^{m}$ into a matrix.

The resulting family of the πk-knowledge is called the *standard family of imprecise models of knowledge representation* for designing intelligent DSS.

Definition 5: *Level-zero imprecise quanta of knowledge* are algorithmic constructions from the terminal π-quanta $\pi k_0 y$ and $\pi k_1 a$, generated by a finite number of application of the superposition operator $\sqcap$. Using this definition various level-zero π-quanta of knowledge can be constructed.

Definition 6: *Level-one imprecise quanta of knowledge* are algorithmic constructions from the terminal π-quanta $\pi k_0 y$, $\pi k_1 a$, and $\pi k_1 b$ generated by a finite number of application of the superposition operator $\sqcap$ and the string concatenation operator SCON(·).

Definition 7: *Level-two imprecise quanta of knowledge* are algorithmic constructions from the terminal π-quanta $\pi k_0 y$, $\pi k_1 a$, and $\pi k_1 b$ generated by a finite number of application of the superposition operator $\sqcap$ and the matrix concatenation operator MCON(·).

The formal definition of πk-knowledge solves the problem of knowledge representation for designing intelligent DSS.

Manipulating knowledge: All approximate statements, for which the degree of correctness (dc) can be assessed, consist of series of steps needed to form a conclusion and to compute the degree of correctness (dc) of the conclusion in a π-quantum of knowledge based on the information in its inputs. All nonzero components in each input domain are related via the logical operation OR and different inputs are combined using the logical operation AND. One of the domains is designated as the output and interpreted as a conclusion followed from the logical combination of the inputs. An approximate statement in a π-quantum of knowledge is formed according to the following implication rule:

$$\cap_{i=1\ldots n}(\cup_{j=1\ldots ni}\ a_j^i) \rightarrow C$$

The following assumptions [9] are made:

A1) The degree of correctness is a number from the interval [-1,+1]. The dc of +1 means that the corresponding event or conclusion is absolutely correct. The dc of -1 means that the corresponding event or conclusion is absolutely not correct. The dc of 0 means that the π-quantum has no knowledge. All other values reflect the degree of correctness of event or conclusion.

A2) All approximate statements are simple implication rules in which individual components and inputs are combined using the logical operations AND, OR, and NOT.

A3) The degree of correctness of a π-quantum of knowledge Y is determined according to the following rule:

```
dc[Y]≡dc[Conclusion]=dc[Combined Input]dc[Implication]
```

A4) Each π-quantum of knowledge is a dynamical part of πk-knowledge. A π-quantum is an active state when all its inputs are specifies along with their degrees of correctness and the output contains the calculated value of the dc of the conclusion.

A simple π-quantum of knowledge contains two inputs and one output. The approximate statement represented by this π-quantum is expressed as

"if e_1 AND e_2 then C"

The dc of the input is taken to be the minimum of the dc of the individual inputs:

$$dc[e_1 \text{ AND } e_2] = \min(dc[e_1], dc[e_2])$$

meaning that the input information is as good as its least reliable piece.

For a simple π-quantum of knowledge with 2 events e_1 and e_2 in an input the dc of the input is defined as the most correct piece of information, i.e.

$$dc[e_1 \text{ OR } e_2] = \max(dc[e_1], dc[e_2])$$

For 3 independent events supporting the conclusion of a π-quantum of knowledge the dc is defined as

$$dc[e_1 \text{ OR } e_2 \text{ OR } e_3] = dc[e_1]+dc[e_2]+dc[e_3]-dc[e_1]dc[e_2]-dc[e_1]dc[e_3]-dc[e_2]dc[e_3]+dc[e_1]dc[e_2]dc[e_3]$$

The reasoning behind this rule is that the dc of the combined input is bigger than the dc of each individual input and at the same time does not exceed 1 to satisfy assumption A1.

For negated events the following rule applies:

$$dc[\text{NOT } e] = dc[e]$$

The computation of the dc of π-quantum Z whose conclusion is supported by two independent π-quanta Y_1 and Y_2 is performed using the following rule:

$$dc[Z]=dc[Y_1]+dc[Y_2]-dc[Y_1]dc[Y_2], \text{ if } dc[Y_1]>=0 \text{ and } dc[Y_2]>=0$$
$$dc[Z]=dc[Y_1]+dc[Y_2]+dc[Y_1]dc[Y_2], \text{ if } dc[Y_1]<0 \text{ and } dc[Y_2]<0$$
$$dc[Z]=(dc[Y_1]+dc[Y_2])/(1-\min(abs(dc[Y_1],dc[Y_2]))), \text{ if } dc[Y_1]<0 \text{ or } dc[Y_2]<0$$

The reasoning behind this rule is that if the conclusion of the π-quantum Z is supported by both π-quanta then the dc of π-quantum Z becomes greater than the dc of each individual supporting π-quanta. If the signs of the supporting π-quanta are different then the result is determined by the biggest in the absolute value dc but its degree of correctness is reduced.

An algorithm was developed in [9] to compute the dc of an arbitrary π-quantum of knowledge, πkY. The following Matlab-like pseudo-code of the algorithm takes as the input a structure representing the π-quantum with all its inputs and the dc of the implication specified, computes the dc of the conclusion in the π-quantum and stores it in Y.Output.dc.

```
if Y.Activate
   % Compute the dc of the combined input
   Y.Output.dc=+1;
   for i=1:Y.NumberOfInputs
      % Compute the dc of each input
      Y.Input[i].dc=-1;
      for j=1:Y.Input[i].NumberOfComponents
         % Compute the dc of input's i components 1 through j
         if Y.Input[i].Component[j].NOT and Y.Input[i].Component[j].dc<0
            % If component's value is preceded by NOT and
            % its degree of correctness is negative
            Y.Input[i].dc=max(Y.Input[i].dc,-Y.Input[i].Component[j].dc)
         else if
            Y.Input[i].dc=max(Y.Input[i].dc,Y.Input[i].Component[j].dc)
         end if
      end for
      % Compute the dc of combined inputs 1 through i
      Y.Output.dc=min(Y.Output.dc,Y.Input[i].dc)
   end for
   % Compute the dc of the conclusion
   Y.Output.dc = Y.Output.dc * Y.Implication.dc
end if
```

Consider an application of the framework to building a decision support system to help a system administrator manage a computer system.

Applications: An experienced computer system administrator can resolve many critical situations in a matter of seconds. However, in his/her absence, a substituting person, who is not familiar with all the details of the local system, can easily take wrong actions and complicate a critical situation further. Even an experienced system administrator can occasionally make mistakes because of incorrect interpretation of some information or lack of proper attention. In order to assist a system administrator to make correct decisions in critical situations when a large amount of information needs to be analyzed in a short period of time and a great variety of possible critical situations exist, a prototype intelligent decision support system based on the formalism presented in the preceding sections was designed.

Consider one of the possible critical situations which is characterized by various hardware failures that do not cause immediate problems but require making a maintenance decision. In an event of loss of power redundancy on a component that executes users jobs, a system administrator must make a quick decision. A loss of one of the two power converters working in a current-sharing mode does not cause immediate problems but increases the risk of having a power loss event. A power loss event causes a domain crash and a possible loss of users data and results. A decision should be made whether to deactivate the affected component immediately after a loss of redundancy or to wait until a planned system reboot or maintenance event. In order to make a correct decision in this situations several factors should be considered. One of them is whether or not the dynamical reconfiguration capability is enabled and what is the current risk associated with executing dynamic reconfiguration commands. Also, the decision maker should evaluate whether the system will be able to handle the current or possibly increased user load if the affected component is deactivated.

The following zero-level imprecise knowledge quantum represents the parameters of the SD (the component with redundant dc-dc converters) and their values

$$
\begin{aligned}
\pi k_0 y = [\, & a_1^{(1)} | d_x^{(1)}(\text{Increasing}), a_2^{(1)} | d_x^{(1)}(\text{Decreasing}), a_3^{(1)} | d_x^{(1)}(\text{Stable}) : \\
& a_1^{(2)} | d_x^{(2)}(00), a_2^{(2)} | d_x^{(2)}(01), a_3^{(2)} | d_x^{(2)}(10), a_4^{(2)} | d_x^{(2)}(11) : \\
& a_1^{(3)} | d_x^{(3)}(\text{PowerON}), a_2^{(3)} | d_x^{(3)}(\text{PowerOFF}) : \\
& a_1^{(4)} | d_x^{(4)}(\text{Down}), a_2^{(4)} | d_x^{(4)}(\text{Up}), a_3^{(4)} | d_x^{(4)}(\text{Idle}), a_4^{(4)} | d_x^{(4)}(\text{Busy}) : \\
& a_1^{(5)} | d_x^{(5)}(\text{Enabled}), a_2^{(5)} | d_x^{(5)}(\text{Disabled}) : \\
& a_1^{(6)} | d_x^{(6)}(\text{None}), a_2^{(6)} | d_x^{(6)}(\text{Low}), a_3^{(6)} | d_x^{(6)}(\text{Moderate}), a_4^{(6)} | d_x^{(6)}(\text{High}) : \\
& a_1^{(7)} | d_x^{(7)}(\text{Deactivate}), a_2^{(7)} | d_x^{(7)}(\text{Wait}) \,] .
\end{aligned}
$$

where the parameters denote respectively:

x_1 = Voltage
x_2 = PowerGood
x_3 = ComponentStatus
x_4 = DomainStatus
x_5 = DynamicReconfiguration
x_6 = DRRisk
x_7 = Action

The following level-one imprecise knowledge quantum is constructed by using the terminal quanta, according to the definition of level-one imprecise quanta of knowledge.

$$
\begin{aligned}
\pi k_1 Y = [\, & (\text{Voltage}=1|d(1),2|d(2)) \cdot (\text{PowerGood}=2|d(2),3|d(3)) \cdot \\
& (\text{Status}=1|d(1)) \cdot (\text{DomainStatus}=3|d(3),4|d(4)) \cdot \\
& (\text{DynamicReconfiguration}=1|d(1)) \cdot (\text{DRRisk}=1|d(1),2|d(2)) \\
& \rightarrow (\text{Action}=1|1)=1|d(Y) \,] .
\end{aligned}
$$

A graph-model of this imprecise quantum of knowledge is shown in Fig. 1. This imprecise knowledge quantum describes the following situation. If a voltage-change detector indicates an increase or decrease in voltage and the power good register indicates that one of the redundant dc/dc converters cut off its output and the converters are powered on and the domain is idle or busy and the dynamic reconfiguration capability is enabled and there's no or low risk of executing a dynamic reconfiguration command then the corrective action is determined exactly by its first value of deactivating the component.

Some of the sources of uncertainty reflected in the degrees of correctness of different inputs are the uncertainty associated with change detection procedures used to track voltage signals, the uncertainty in assessing a current risk of executing dynamical reconfiguration commands, the uncertainty due to possible incorrect operation of the processes reporting registers values such as the power good registers values.

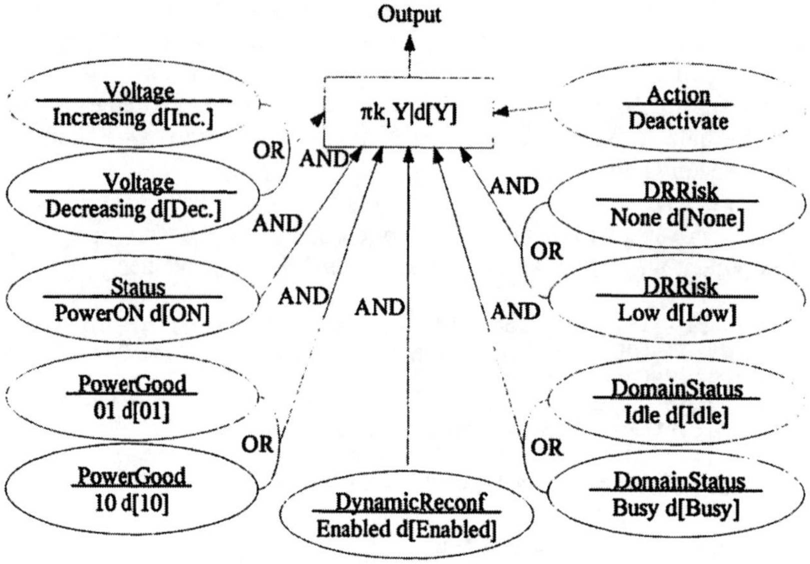

Fig. 1. A graph-model of the imprecise quantum of knowledge.

In summary, when a critical situation, labeled "loss of power redundancy", occurs, the decision support system suggests noncontradictory actions with corresponding degrees of correctness. The system administrator can execute the action with the highest degree of correctness to speed up decision making and avoid mistakes. If needed, the top action recommended by the decision support system can be executed autonomously.

Conclusions: The usage of intelligent decision support system is beneficial as it allows to avoid mistakes, increase availability and help a system administrator have the expertise of most skilled system administrators. Avoidance of mistakes and quick decision making in critical situations helps prevent and avoid system failures. The described framework for representing and manipulating imprecise and incomplete knowledge for decision making in critical situations proved to offer powerful tools for building computationally light decision making aids that have great application potential in complex machine-human and autonomous systems.

References:

[1] Athanassopoulos, A.D. & Podinovski, V.V. (1997). Dominance and potential optimality in MCDA with imprecise information, *Journal of the Operational Research Society*, 48, 142-150.

[2] Björk, B.C. (1989). Basic Structure of a Proposed Building Product Model. *Computer Aided Design*, 21 (2), pp. 71- 78.

[3] Cristian, F. (1991) Understanding fault-tolerant distributed systems. *Communications of the ACM*, Vol. 34, No. 2, pp. 57-78.

[4] Miettinen, K. & Salminen, P. (1999), Decision-aid for discrete multiple criteria decision making problems with imprecise data, *European Journal of Operational Research* 119, 50-60.

[5] Minsky, M. A. (1981). A Framework for Representing Knowledge. In J. Haugeland (Ed.), *Mind Design: Philosophy, Psychology, Artificial Intelligence* (pp. 95-128). Cambridge, MA: MIT Press.

[6] Morrissey, J. M. (1990). Imprecise information and uncertainty in information systems, *ACM Transactions on Information Systems*, Vol. 8, Issue 2, pp.: 159-180.

[7] Nilsson, Martin (1983). A Logical Model of Knowledge, *Proceedings of IJCAI*, pp. 374-376.

[8] Quillian, M.R. (1968). Semantic Memory, In M. Minsky, (Ed), Semantic Information Processing, MIT Press, Cambridge, MA, pp. 216-260

[9] Sirodzha, I. B. (2002) Algorithms for decision-making support in special situations of flight control. *Radio Electronics. Computer Science. Control*, Zaporozhye University, Ukraine, No. 2, pp. 112-117. (In Russian)

[10] Walley, P. (1991). Statistical Reasoning with Imprecise Probabilities, Chapman and Hall, London,.

[11] Weber, M. (1987), Decision making with incomplete information, *European Journal of Operational Research* 28, 44-57.

[12] Yager, R.R. (2002). Veristic Variables, *IEEE Transactions on Sytems, Man, and Cybernetics – Part B: Cybernetics,* Vol 30, No. 1, February, pp. 71-84.

[13] Zadeh, L.A. (1997). Towards a Theory of Fuzzy Information Granulation and Its Centrality in Human Reasoning and Fuzzy Logic, *Fuzzy Sets Syst.*, Vol. 90, pp. 111-127.

[14] Zadeh, L.A. and Kacprzyk, J. (1999). Computing With Words in Information/Intelligent Systems 1 and 2, Springer-Verlag, Telos.

EXTRACTING INFORMATION FROM CONVENTIONAL AE FEATURES FOR ONSET DAMAGE DETECTION IN CARBON FIBER COMPOSITES

Runar Unnthorsson[a] and Niels Henrik Pontoppidan[b] and Magnus Thor Jonsson[a]

[a] Department of Mechanical Engineering, University of Iceland
Hjarðarhaga 2-6, IS-107 Reykjavik, Iceland
[b] Informatics and Mathematical Modeling, Technical University of Denmark
Richard Petersens Plads 321, 2800 Lyngby, Denmark
Corresponding author Runar Unnthorsson, runson@hi.is

Abstract: We have analyzed simple data fusion and preprocessing methods on Acoustic Emission measurements of prosthetic feets made of carbon fiber reinforced composites. This paper presents the initial research steps; aiming at reducing the time spent on the fatigue test. With a simple single feature probabilistic scheme we have showed that these methods can lead to increased classification performance. We conclude that: the derived features of the TTL count leads to increased classification under supervised conditions. The probabilistic classification scheme was founded on the histogram, however different approaches can readily be investigated using the improved features, possibly improving the performance using multiple feature classifiers, e.g., Voting systems; Support Vector Machines and Gaussian Mixtures.

Key Words: Acoustic Emission; Carbon fibres; Data fusion; Fatigue testing; Probabilistic classification; Supervised learning

Introduction: During the design phase of a prosthetic carbon fiber foot at Össur hf., prototype models are built, evaluated, and improved to meet the design criteria. For evaluation several testing methods are used; visual tests, stiffness measurements and fatigue testing. Feet are mainly subjected to dynamic loading, which sometimes requires low loads to initiate and propagate faults [16]. Because of this fatigue testing is extremely important part of the design process. Fatigue tests of composites can take very long time, up to several weeks. It is therefore valuable for an engineer, that wants to test a prototype, to be able to see if it will fail early during the test. Considerable time can be saved by this. By shortening the test time, feet can be; designed and developed in less time and at a lower cost than previously possible. During the fatigue testing of an prosthetic feet, the stiffness is monitored and visual tests are also performed regularly. However, visual tests can only detect faults that extend to the surface of a foot or affect it in some way. In order to be able to predict the fatigue strength of a foot undergoing fatigue testing, other NDT methods can be used for obtaining data.

When microstructural changes occur in composites, energy is released and transient stress waves are generated. These stress waves are called called Acoustic Emissions (AE) [16]. The stress waves travel through the composite and when they reach the surface it will vibrate. The small surface displacements generated by the vibration of the composite can be detected by using appropriate sensors. According to Duesling [3] piezoelectric sensors are most popular. AE can be generated by several types of damage that occurs in the material, e.g. fibre breakage, matrix cracking, delamination etc. Because microstructural damage generates AE the method has the potential of detecting damage in its early stages, much earlier than possible by monitoring deflection or performing visual inspection. Therefore a microstructural damage that starts to formate in the last cycles of a fatigue test can be detected, a damage that would pass undetected by the other two previously mentioned methods.

AE signals are not only generated by damage, they can also come from other sources, such as; the testing machine [16], electrical disturbances [16, 5], friction due to rubbing of parts [12] and the friction generated by opening and closing of matrix cracks [13]. According to Hamstad [7] AE is also generated under loading because of the different material properties of the fibers and the matrix.

Friction generated AE can provide important information about the condition of an composite [1]. Some researchers have attempted to filter the AE signal generated by friction from the total signal in order to better detect AE from damage, but this can be difficult [15]. The fact that a damage only emits AE once but friction many times suggests that approaches based on friction will be more robust.

An AE parameter analysis is the conventional way of performing an AE monitoring. The results from using standard AE parameters, especially amplitude, are contradicting in the literature [5, 2, 8, 3]. Amplitude suffers from attenuation. The main reasons for the attenuation are; geometric spreading, dispersion, internal friction and scattering [11]. Also, Prosser et al. [10] reported that the same type of damage doesn't always produce AE with the same amplitude. This indicates that amplitude is not a very good feature. Godin et al. [6] claimed that conventional AE analysis cannot distinguish between different AE sources and suggested that more advanced methods, such as multivariate analysis and classifiers, should be used. Similar comments were made by Tsamtsakis et al. [15]. They suggested that new parameters, like force or displacement, should be added.

This paper reports the results of using several methods to extract information from conventional AE features. The features were obtained using a commercial AE acquisition system (AESmart 2000 from DECI Inc) and consist of a fixed set of standard AE features. The emphasis was put on generating features which could be used for fast and reliable detection of damage onset. It was also considered important for the detection method to be robust. Robustness was believed to be obtained by basing the features on friction based AE.

The outline of the remaining of the paper is as follows: First the setup of the acquisition system is explained then the methodology is explained. It starts with explaining how new features are derived from the available feature set by both applying sensor fusion and using moving window second order moment. The section ends with a description a simple classification system based on the assumption that faulty specimens have inhibit greater AE activity than normal ones, thus threshold on the features can be used for classification. We compare classification properties of the different features by using using receiver-operator characteristics (ROC) curves. The features are compared and discussed – identifying the useful features. Finally the concluding remarks are made and future work is outlined.

System Setup: The fatigue test specimens were two types of prosthetic feet differing slightly in stiffness and size. Foot no. 1 is stiffer and smaller than foot no 2. The construction of both feet is the same, they are made from unidirectional carbon fiber reinforced epoxy and woven mats are used for the top and bottom layers for nicer look. The test was performed according to ISO 10328 specifications, with a "Foot/Limb" test system at Össur's testing facilities. During the fatigue test, two actuators were used to flex the foot for 2×10^6 cycles at 1.5 Hz. The maximum load for each actuator was kept constant and the deflection was monitored. No change in maximum deflection was measured, which means that no stiffness reduction was observed during the tests. For data collection SE9125-MI data transducers and the AESmart 2000 AE acquisition system, from from DECI Inc., was used.

According to Dunegan [4] the system splits the AE signal into low frequency (LF) and high frequency (HF) signals. The LF signal contains frequencies between 20 to 60 kHz and the HF signal contains frequencies between 100 to 500 kHz. Data is only collected from one sensor at a time, each time the system switches between the sensors it stores the data in an Excel file. The data that is stored is not the actual AE signal but instead the following features that the system extracts from this signal; time of the recorded event, HF counts above threshold (TTL counts), HF peak amplitude, LF peak amplitude, ratio of the HF/LF peak amplitudes, event count, time difference between HF and LF signals. Other features are also stored by the system but, they are irrelevant to the test.

The system was set up to switch between the two sensors, i.e. one on each foot, using 6 second dwell time. The monitoring was performed twice a day for 4 to 6 hours each time. The gain for both the LF and HF gain was set to 60 dB. The threshold setting for the LH and HF was set to 200 mV. The time interval for counting the TTL counts was set to 1 ms.

Data preprocessing: During data acquisition each foot was monitored for periods of six seconds, before the system switched started monitoring the other foot. Each entry in the data file is the result of the following events

1. AE signal crosses the trigger threshold
2. Acquisition board computes the set of features from a 1 ms analysis window
3. Acquisition board goes back to detecting threshold crossing in AE signal. Further the system might also change foot if the 6 s period has ended

<u>Moving window second order moment</u>: We compute the second order moment over a fixed number of consecutive data entries, move the window and repeat. This gives the moving window second order moment that peaks when the local mean value change, and also when

the local variance increase. So this is just a simple "amplification" of changes in a time series. Besides from giving a different measure (variance instead of values) this method also performs filtering since a number of consecutive samples are used for each value.

$$\check{\mu}(n) = \frac{1}{N} \sum_{n'=0}^{N-1} x(n-n') \tag{1}$$

$$\check{\sigma}^2(n) = \frac{1}{N-1} \sum_{n'=0}^{N-1} (x(n-n') - \hat{\mu}(n))^2 \tag{2}$$

Time normalization: Each data entry comes from a 1 ms analysis window triggered by a threshold crossing. So we can imagine several regimes (explaining the data entries in the 6 s periods where the system monitors each foot).

1. few data entries with a few high readings – occasional low activity
2. many data entries with a few high readings – steady low activity
3. few data entries with many high readings – occasional high activity
4. many data entries with many high readings – steady high activity

This means that if we omit the time information and look at the feature values, we cannot differ between the occasional and steady activity. Instead we fuse the features in each 6 s period. We deliberately use the term fuse, even though most features are just summed up during the period. But for the time entry we take the earliest value, for the event count we take the maximum, and for the ratio we recalculate; since the original ratio computation was constrained. Another benefit of time normalization is reduction of data size as several events are combined into one.

We can also apply the moving window second order moment to the time normalized data in order to further enhance changes.

Energy normalization: In the LF/HF peak amplitude we have the peak amplitudes, we already have the number of TTL counts in a 6 s period. Multiplying those two quantities and sum over the events in the period gives an upper bound estimate of the "energy".

$$E = \sum_i TTL_i P_i \tag{3}$$

Filtering: One way of decreasing feature variance is low-pass filtering, essentially this forces the classification system to evaluate the feature value for consecutive examples. This smears out occasional high values, but also the sudden steps in the feature, thus this delays the significant changes. Further notice that this is related to the post classification processing described last in the following section.

Feature processing: We adopt a very simple setup for supervised single feature classification of the specimen condition. Assuming that a damaged specimen generate a "loud" AE signal, it will also generate many TTL counts with high peak amplitudes and many event counts. Thus as a rule of thumb: The feature value in a faulty specimen should be larger than the normal. In order to compute the feature thresholds, we compute the histograms of each feature on two sets of data, a normal and a faulty. Histograms are not directly well suited for classification tasks, so instead we use the normalized cumulated sum of the histogram, corresponding to a sample of the true cumulated density function (CDF). The CDF gives the probability that the feature is less or equal to τ.

$$P(x \leq \tau) = \int_{x=-\infty}^{\tau} p(x) \tag{4}$$

A good feature for classification will behave such that $P_{\text{normal}}(x \leq \tau) \gg P_{\text{faulty}}(x \leq \tau)$ is fulfilled, not everywhere but in an interval. Actually $P_{\text{normal}}(x \leq \tau) = X, X \in [0; 1]$ says that X of the normal examples have a feature value x less or equal to τ. Obviously this means that thresholding with this value of τ corresponds to $1 - X$ false alarms, and that $1 - P_{\text{faulty}}(x \leq \tau)$ of the faulty specimens is correctly detected (see further [14]). Varying τ from the smallest to the largest observed value whilst tabulating $1 - P_{\text{normal}}(x \leq \tau)$ and $1 - P_{\text{faulty}}(x \leq \tau)$ gives the Reciever Operator Characteristics (ROC) shown in Figure 1.

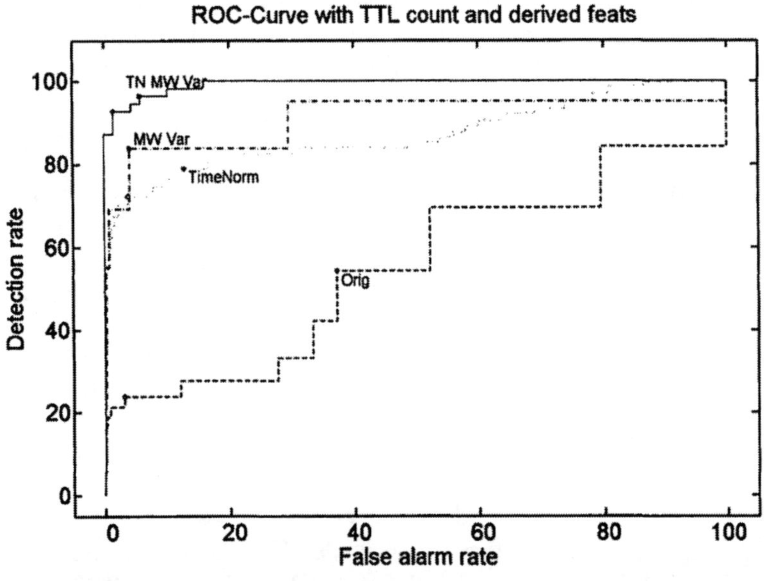

Figure 1: ROC-curves for the TTL count and the three derived features. The o's and stars on the curves indicate the performance of the two optimal thresholds. Notice the consensus between the different optimality criterions wrt. ranking of the features – also the Neyman-Pearson criterion at 5% agrees. The figure also show our conclusion, that the time normalization (data fusion) and moving window variance (processing) increase the performance of the system.

$$\text{Detection rate } D(\tau) = 1 - P_{\text{faulty}}(x \leq \tau)s \tag{5}$$

$$\text{False alarm rate } F(\tau) = 1 - P_{\text{normal}}(x \leq \tau) \tag{6}$$

This approach with a single dimension classifier is very simple, with multiple dimensions, like using several of the obtained features, either pattern recognition or voting systems should be explored.

We adopt two rules for obtaining the optimal threshold:

1. Maximize the difference between the detection rate and false alarm rate
2. Minimize the absolute distance to the optimal point (100% detection, 0% false alarm)

Also applicable is the Neyman-Pearson Criterion[14], where the best classifier/threshold is the one with the highest detection rate given a set false alarm rate, say 5%.

Having decided on a threshold we also have knowledge of the false alarm rate of the classifier under normal conditions. In many cases occasional false alarms will occur, an we could filter out with a moving average low pass filter. The length of the filter and the significant number of alarms for the filtered signal can be obtained through the Binomial hypothesis testing (this is further described for a unsupervised setting in [9]).

In short we can vary the threshold in order to balance the rate of false alarms versus detection, however it is a trade-off. Direct improvement is only available through a better classifier. Another available trade-off is between classification accuracy and delay. If we know that the false alarm rate is 5 out of 100, we could wait until 12 out of 100 was classified as faulty; however this will introduce a delay since we need to gather enough fault-classifications.

Results and Discussion: Table 1 shows the statistical parameters for the ROC curves corresponding to the original features, i.e. those provided by the acquisition system.

Feature	false 1	detect 1	false 2	detect 2	Area
TTL count	3	24	37	54	0.55
HF amp	3	10	18	15	0.15
LF amp	0	2	25	8	0.07

Table 1: Original data

None of these features are able to discriminate well between normal and faulty signals under the assumption that faulty signals are louder. The LF amp with the assumption that normal signals are loud works quite well (area under ROC-curve 0.93), this could fit the "steady low amplitude" regime descried under *time normalization*.

Applying a *moving window variance* on the original features the classification performances is considerably improved. Now take a look at the features generated by taking a moving window variance of the original data, shown in Table 2, then we have features with considerably better detection performance.

This increase in performance is in some cases followed by increased false detection rate. However, these features are generally better suited for classifying as indicated by the much larger area under the ROC curves. The performance of the time normalized data is shown in Table 3.

Feature	false 1	detect 1	false 2	detect 2	Area
TTL count	4	84	4	84	0.91
HF amp	16	87	16	87	0.84
LF amp	36	79	36	79	0.67

Table 2: Moving Window Variance of Original data

Feature	false 1	detect 1	false 2	detect 2	Area
TTL count	4	72	13	79	0.87
HF amp	11	49	22	56	0.60
LF amp	3	8	48	45	0.31
Event count	43	91	26	73	0.78

Table 3: Time Normalized data

These features result in better classification than with the original features (Table 1), however not as good as with the moving window variance. However the combination, by applying moving window variance on those features we hoped for similar improvement as was observed for the original data. Table 4 lists the ROC performance of the new set of features.

Feature	false 1	detect 1	false 2	detect 2	Area
TTL count	1	93	6	96	0.99
HF amp	13	93	13	93	0.92
LF amp	23	84	23	84	0.81
Event count	83	100	54	55	0.52

Table 4: Moving Window Variance of Time Normalized data

According to the Moving window variance of the normalized TTL counts is definitely a good feature.

Feature	false 1	detect 1	false 2	detect 2	Area
TTLxHF	13	56	13	56	0.54
TTLxLF	5	19	40	35	0.29
TTLx(HF+LF)	5	29	42	51	0.43

Table 5: Energy Normalized data

In order to see if the classification performance of the feature created by applying moving window variance on the normalized TTL counts could be improved further, a filtering was applied. As Table 6 shows that the filtering improved the false alarm rate compared to Table 4, but the area below the curve is slightly decreased as the detection rate is also decreased. Figure 2 shows the time series of both features derived from the TTL count as well as their respective classification. The various preprocessing steps reduce the local variance of the classification.

Feature	false 1	detect 1	false 2	detect 2	Area
NTTL variance	0	95	0	95	0.98

Table 6: Filtered moving window variance of the normalized TTL counts

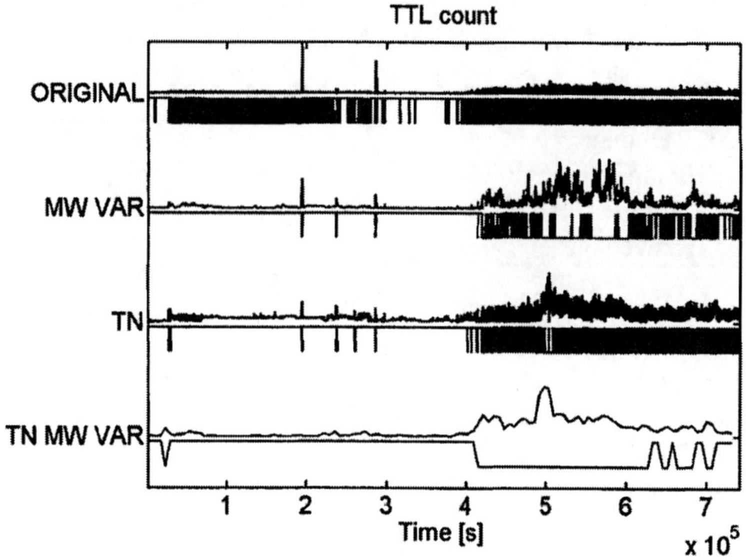

Figure 2: The time series of the TTL count and three derived features. Each entry in the features has been classified; this binary output is plotted just below the feature (the lower value is faulty, upper is normal). It is believed that the specimen turns faulty after $4 \cdot 10^5 s$. The uncertainty using the original TTL count is clearly visible during under both normal and faulty conditions, as the classification switches all the time.

Conclusion: We have analyzed simple data fusion and preprocessing methods on Acoustic Emission measurements of prosthetic feets made of carbon fiber reinforced composites. This paper presents the initial research steps; aiming at reducing the time spent on the fatigue test. With a simple single feature probabilistic scheme we have showed that these methods can lead to increased classification performance. We conclude that: the derived features of the TTL count leads to improved classification and damage detection in a supervised setup.

The probabilistic classification scheme was founded on the histogram, however different approaches can readily be investigated using the improved features, possibly improving the performance using multiple features classifiers, e.g., Support Vector Machines and Gaussian Mixtures.

Acknowledgements: The authors gratefully acknowledge Össur hf. (`www.ossur.com`) for providing samples and access to their testing facility. The research of Runar Unnthorson was supported by a research grant from RANNIS, the Icelandic Research Council (`www.rannis.is`). His visit at DTU was supported by grants from the Helga Kristjánsdóttir and Sigurliði Kristjánsson Memorial Fund and Landsvirkjun's (`www.lv.is`). Niels H. Pontoppidan was supported by the European Commission through Growth Project AE-WATT (`mole.imm.dtu.dk/aewatt/`).

References:

[1] J. Awerbuch and S. Ghaffari. Monitoring progression of matrix splitting during fatigue loading through acoustic emission in notched unidirectional graphite/epoxy composite. *Journal of Reinforced Plastics and Composites*, 7(3):245–64, 1988.

[2] S. Barre and M.L. Benzeggagh. On the use of acoustic emission to investigate damage mechanism in glass-fibre-reinforced polypropylene. *Composites Science and Technology*, 52(3):369–376, 1994.

[3] L.A. Duesing. Acoustic emission testing of composite materials. *Reliability and Maintainability Symposium, 1989. Proceedings., Annual*, pages 128–134, 1989.

[4] H.L. Dunegan. An alternative to pencil lead breaks for simulation of acoustic emission signal sources, August 2000.

[5] M. Giordano, A. Calabro, C. Esposito, A. D'Amore, and L. Nicolais. An acoustic-emission characterization of the failure modes in polymer-composite materials. *Composites Science and Technology*, 58(12):1923–1928, 1998.

[6] N. Godin, S. Huguet, R. Gaertner, and L. Salmon. Clustering of acoustic emission signals collected during tensile tests on unidirectional glass/polyester composite using supervised and unsupervised classifiers. *NDT and E International*, 37(4):253–264, 2004.

[7] Marvin A. Hamstad. Acoustic Emission Primer, Lecture delivered by Dr. Hamstad at the acoustic emission working group meeting on august 4, 2003 (aewg-46). internet, 2004. http://www.engr.du.edu/profile/frame_bio.html.

[8] K. Ono. Acoustic emission behavior of flawed unidirectional carbon fiber-epoxy composites. *Journal of Reinforced Plastics and Composites*, 7(1):90–105, 1988.

[9] Niels Henrik Pontoppidan, Jan Larsen, and Sigurdur Sigurdsson. Non-statrionary Condition Monitoring with the AEWATT Toolbox. Submitted to MFPT 59, 2005.

[10] W.H. Prosser, K.E. Jackson, S. Kellas, B.T. Smith, J. McKeon, and A. Friedman. Advanced waveform-based acoustic emission detection of matrix cracking in composites. *NDT and E International*, 30(2):108, 1997.

[11] William H. Prosser. Applications of advanced waveform-based ae techniques for testing composite materials. *Proceedings of the SPIE conference on Nondestructive Evaluation Techniques for Aging Infrastructure and Manufacturing: Materials and Composites*, pages 146–153, 1996.

[12] S.R. Ravishankar and C.R.L. Murthy. Characteristics of ae signals obtained during drilling composite laminates. *NDT and E International*, 33(5):341–348, 2000.

[13] N. Sato and T. Kurauchi. Interpretation of acoustic emission signal from composite materials and its application to design of automotive composite components. *Research in Nondestructive Evaluation*, 9(3):119–136, 1997.

[14] Clayton Scott and Jeffrey Silverman. The Newman-Pearson Criterion. internet, 2004. `http://cnx.rice.edu/content/m11548/latest`.

[15] D. Tsamtsakis, M. Wevers, and P. De Meester. Acoustic emission from cfrp laminates during fatigue loading. *Journal of Reinforced Plastics and Composites*, 17(13):1185–1201, 1998.

[16] M. Wevers. Listening to the sound of materials: acoustic emission for the analysis of material behaviour. *NDT and E International*, 30(2):99–106, 1997.

EFFECT OF DAMPING ON COAL MILL VIBRATION

Rudolph J. Scavuzzo
Professor Emeritus, University of Akron
Akron, Ohio 44325-3903
Consultant to ECG, RScavuzzo@uakron.edu

Michael T Santucci
Engineering Consultants Group, Inc.
PO Box 1862, Akron, Ohio 44309-1862
mike@ecg-inc.com

Abstract: A coal mill was observed to go into resonance at various coal feeder speeds. As a result of the resonant condition, shaft failures occurred. The dynamic displacements of the three grinding rollers were measured for subsequent analysis using instrumentation techniques developed by Engineering Consultants Group Inc. (ECG).

A finite element model of the mill and grinding rollers (3 each) was developed. These major components of the FEA model were tied together through the coal bed stiffness. Basic measurements results and a modal analysis of the mill were reported at the 2004 meeting of MFPT. In the study, transient forces, which were based on measurement, were applied to the three grinding rollers to determine the dynamic response. The addition of dampers to the roller compression springs was evaluated for the effect on the shaft stresses

Key Words: Machinery Vibration, Coal Mill, Pulverizer, Resonance, Shaft Failure, Dynamic Response, Dampers

Introduction: Coal mills grind pea-sized coal into a powder that is blown into the furnace of a power plant. Normally, there are several mills that supply the ground coal to the furnace. The amount of coal going into the furnace is controlled by the coal feed rate to each mill and by the number of mills on line. In the mill considered, the bowl of the mill turned at a measured speed of 58.5 rpm. Three grinding rollers were forced against the coal bed by compression springs. There are stops on each roller to prevent loading on a bowl with no coal. Once the coal is fed into an operating mill the rollers lift off the stops. For 100% coal feeder loading, the coal bed on the bowl has a thickness of about ¾ inch. The design considered is shown in Figure 1.

In the particular design that was studied, the operation of the mill was normal until the coal feeder reached about 65% of full loading. Then the three grinding rollers began to vibrate with amplitudes that reached an order of magnitude larger at 75% feeder loading than with the coal feeder loading below the vibration threshold. This vibration led to premature mill shaft failures.

This mill grinding roller vibration was measured and analyzed and the entire mill modeled with finite elements. The objective was to determine the cause of the vibration and to seek a means of eliminating the vibration and resulting shaft failures. Based on these measurements, forces acting through the compression springs could be calculated. These forces were applied to the end of the rollers so that the measured displacements amplitudes were obtained. Viscous dampers were added to the model to evaluate the possible reduction in shaft stresses and the resultant increase in fatigue life.

The following is a short review of the experimental work [1] and basic finite element modeling. Then the dynamic loading, the dynamic response and the effects of applying damping are discussed.

Figure 1 Overview of Coal Mill

Experimental Investigation: Probes were mounted so that the displacements of the compression spring shafts shown on Figure 2 could be measured. These transducers recorded the vertical dynamic motions of the three grinding rollers. From these measurements, the depth of the coal bed could be determined. These motions could be correlated with the rotation of the bowl. The coal bed depth increased with the coal feeder loading of the mill with a maximum depth of about ¾ inch. As a result, the stiffness of the coal bed decreases as the loading to the mill increased. As described in Ref [1], this change in stiffness changes the frequencies of the first two modes of vibration and brings the system into resonance with the exciting forces. Measured coal depth with feeder rate for Roll 1 is plotted on Figure 3. Coal bed depths as a function of feeder speed of the other two rollers are similar.

Measured vibration amplitude of Roll 1 is plotted on Figure 5 for feeder speeds of 50%, 70% and 90%. The variation of the vibration magnitude with feeder speed can easily be observed. This amplitude is plotted on Figure 4. Resonance begins when the feeder loading exceeds 65% of capacity and ends at approximately 85% with a dominant frequency of 15.8 Hz. The resonant frequency was determined with a FFT of the measured motion as shown on Figure 5. This same resonant frequency was found in the dynamic measurements of all three rolls. However, this resonant value is not seen in the FFT of the roller motions below 60% feeder loading or above feeder loading of 90%. Two mode shapes of the three grinding rollers were measured at resonance. At times two rollers vibrated out-of-phase with the third roller almost motionless. In the second case, two grinding rollers move in-phase with the third roller out-of-phase. Moreover, the third vibrating roller had higher amplitudes then the two in-phase rollers. Excitation of the system was first though to be caused by bearings [2]. However based on a personal communication with R Eshleman, slip-stick vibration must also be examined.

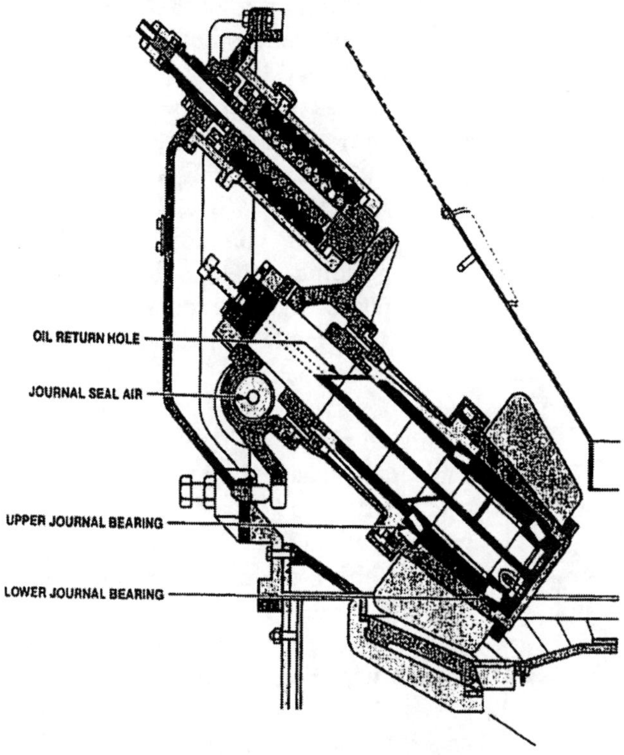

Figure 2 Details of the Grinding Rollers

Finite Element Model: The finite model of the coal mill was developed with the Noran Engineering, Inc version of NASTRAN and FEMAP. The actual modeling was developed with FEMAP, exported to NASTRAN where calculations were run and the results were imported back to FEMAP to view and interpret the results. The mill was modeled in two parts: a FEA model of the three grinding rollers in space was developed and a model of the bowl, shaft and gear. The two parts were merged to develop the full model. The rollers were located in space such that the rollers would fit properly in the bowl. These rollers were basically rigid and pivoted about constraints that had to be located in the local coordinate systems of the rollers. Nodal constraints were applied as follows: 1 Torsional motion of the gear was constrained from rotating at the worm gear, 2 Vertical motion was constrained at the thrust bearing, 3 There are two lateral constraints; one at the lower radial bearing and one at the upper radial bearing. The stiffness of the upper radial bearing was included in the model. Local coordinate systems were used to apply constraints to the grinding rollers. Initially, a coal bed stiffness of 100,000 lb/in was specified. The

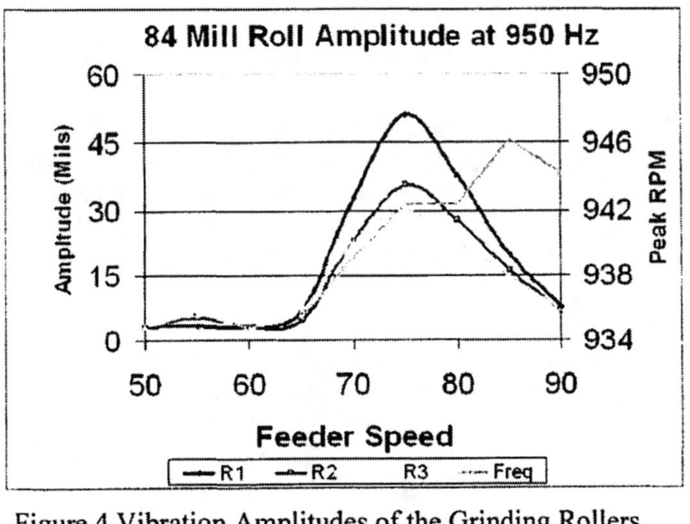

Figure 4 Vibration Amplitudes of the Grinding Rollers

Figure 3 Measured Mill

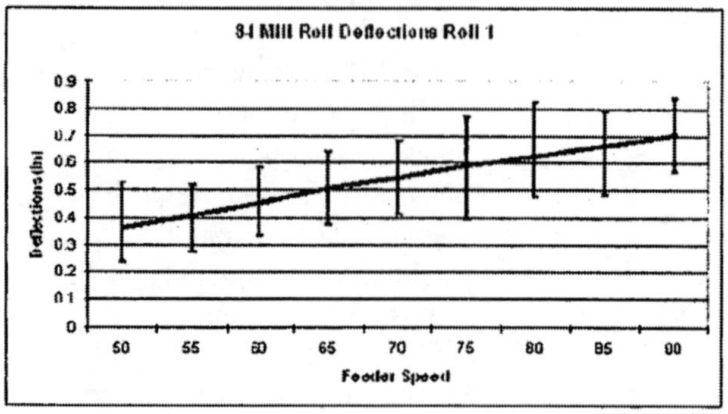

Figure 3 Deflection of Grinding Roller 1 with

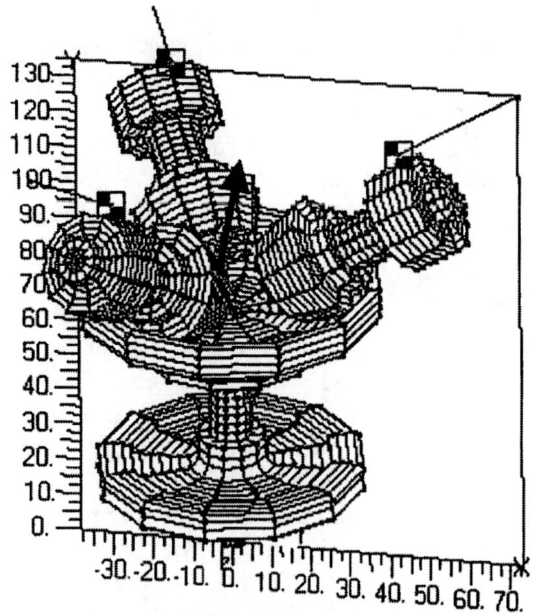

Figure 6 FEA Model: Grinding Rollers, Bowl, Shaft and Gear.

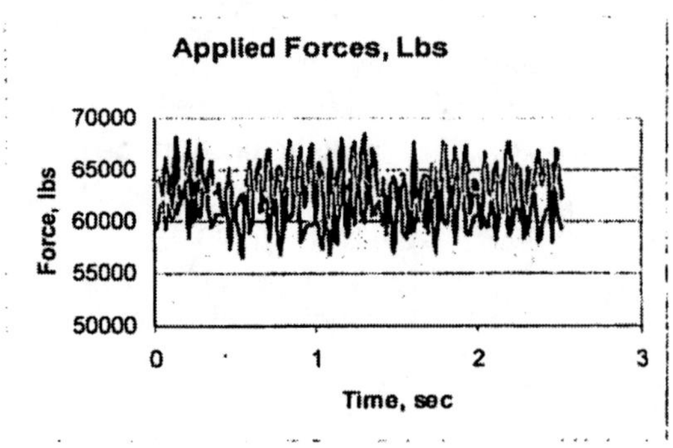

Figure 7 Applied Loads

fundamental modes of vibration were about 14 Hz. This value was increased so that the first and second modes had frequencies of 14.8 and 15.0 Hz., respectively. Transient response calculations were based on this geometry.

In the mode shape of the first mode of vibration, two of the grinding rollers vibrate out-of-phase and one does not move. This mode shape was observed experimentally. However, the most observed mode shape has two rollers in-phase and one out-of-phase. This mode shape corresponded to the calculated mode shape of the second mode of vibration. It should be noted that the third mode of vibration was a torsional mode of the bowl with respect to the fixed gear. This motion did not change with coal stiffness and was not observed experimentally. In this analysis viscous dampers were added in parallel to the compression springs for the 15 Hz model.

Input Forces: Input forces were based on measured compression spring displacements multiplied by the stiffness times a factor. These three input forces (Figure 7) were applied to the center of each of the three rollers. One force is shown on Figure 6. The factor was scaled so the measured displacement range of the springs was approximately obtained (Figure 8). Calculated ng axial stresses in the mill shaft on presented on Figure 9. These stresses are high enough to eventually fatigue the shaft. Then viscous dampers were applied modeled which were parallel to the springs and sized so that the rollers have about 20% damping based on a roller weight of 10,000lbs and a frequency of 15 Hz. Thus, C_c, the critical damping constant is as follows:

$$C_c = 2m\omega = (2)(2)(\pi)(15)(10000)/386 = 4886 lb * \sec/in$$

Thus, the effective viscous damping constant for each of the three dampers is 976 lb-sec/in. for 20% damping. Deflections are shown in Figure 10. By comparing the range deflections, peak to peak, of the damped to undamped calculations, the range values are reduced to 50% of the undamped range deflections. As a result, alternating stresses are reduced by the same percentage and shaft stresses will no longer fatigue the mill shaft.

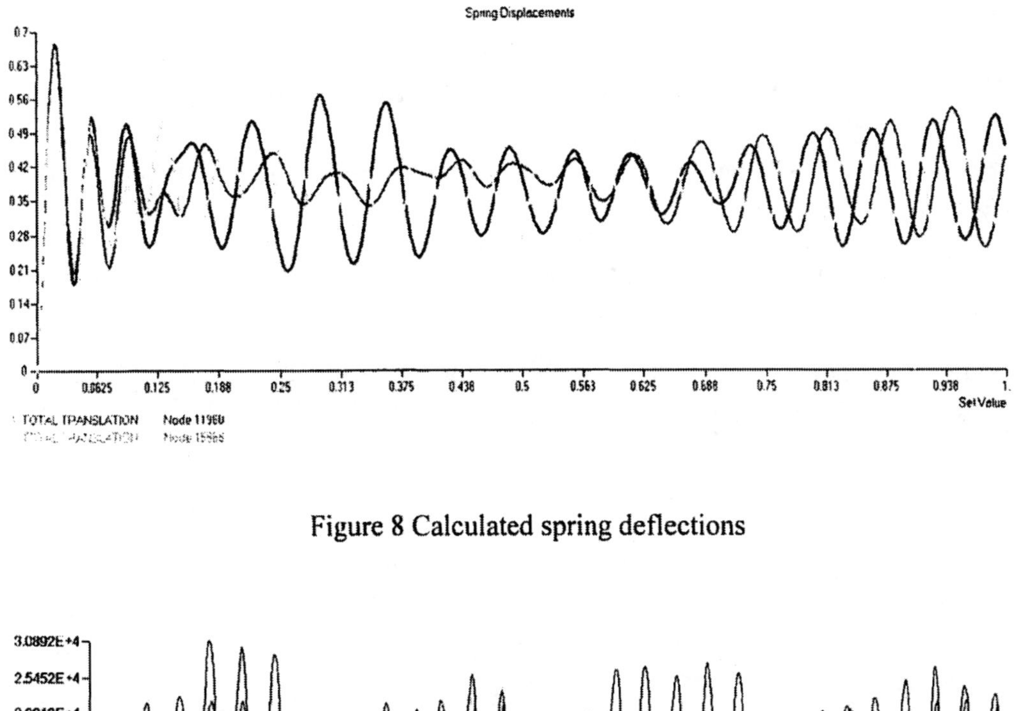

Figure 8 Calculated spring deflections

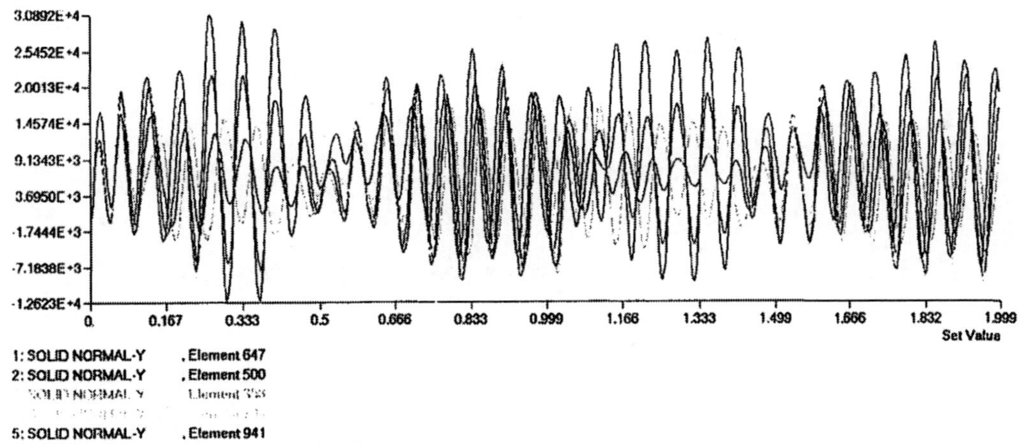

Figure 9 Calculated axial shaft stresses.

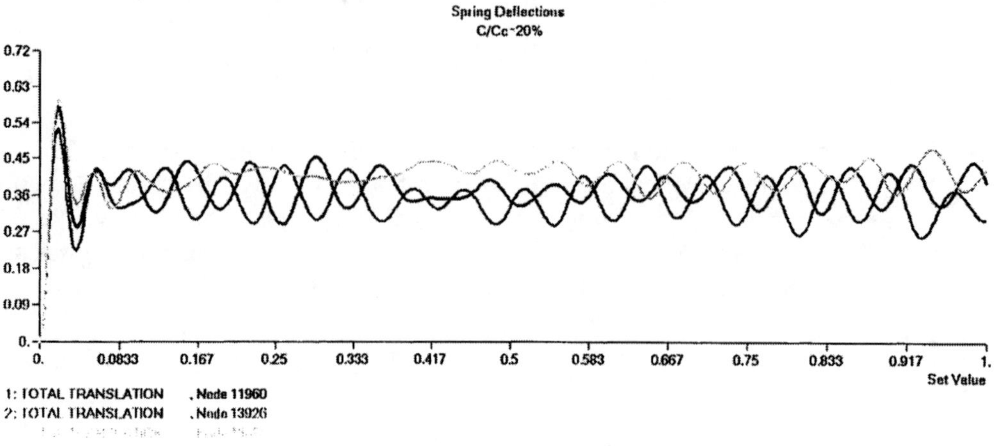

Figure 8 Calculated compression spring dynamic deflections.

Conclusions: By applying a dynamic force to the center of the grinding rollers, which is based on measured spring deflections, the dynamic stresses of the mill shaft were estimated. These applied forces were scaled so that the range deflections were approximately correct. By adding 20% damping to the grinding rollers, stresses are reduced by a factor of two and are no longer predicted to fatigue the shaft.

Acknowledgement: The authors would like to acknowledge the assistance of the technical staff at Noran Engineering, Inc., Los Alamitos, CA during the development of the FEA model and applied dynamic loads.

References:

[1] M.T. Santucci and R. J Scavuzzo, "Experimental And Analytical Comparison Of Machinery Vibration," Proceedings of MFPT, April 2004

[2] Ronald L. Eshleman, Basic Machinery Vibration, VIPress, Inc., Clarendon Hills, IL, 60514, Chapter 4.

REMOTE NDE TECHNOLOGY FOR INACCESSIBLE SHIPBOARD PIPING INSPECTION

John M. Liu
Welding & NDE Branch
Naval Surface Warfare Center Carderock Div.
W. Bethesda, MD 20817-5700
Liujm@nswccd.navy.mil

Christopher P. Nemarich
Engineering and Manufacturing Division
GEO-CENTERS, Inc.
1201 M St S.E. Suite 050
Washington DC 20003
cnemarich@geo-centers.com

Abstract: Piping inside fuel and ballast tanks on US Navy ships requires periodic inspection and maintenance. However, because the cost to open and enter tanks for inspection is considerable the Navy has extended scheduled in-tank inspections by up to 12 years. This has reduced maintenance costs but has increased the risk that piping corrosion goes undetected. This has led to some unexpected piping failures and in a few cases has resulted in oil spills. To prevent these pipe failures the US Navy is evaluating technologies that can enable rapid and inexpensive periodic condition assessment of piping that is inside of fuel and ballast tanks. Pipe failures were investigated to determine the exact nature of the failures and the ship systems most affected. Then a detailed technical performance requirements list was developed on which to base a review of remote inspection equipment capability. Commercially available technologies were identified and candidates selected for laboratory evaluation. The technologies presently undergoing evaluation include those that involve the insertion of sensor(s) into the pipe from outside of the tank as well as non-invasive techniques that use externally placed sensors from outside of the tank. This paper provides an overview of the problem definition, the performance specifications and a description of some potential technology solutions. The capabilities and limitations of each of the technologies under investigation are presented along with some preliminary performance testing results.

Keywords: Condition Assessment; Corrosion and Defect Detection; NDE Requirements; Optical Pipe Inspection; Pipe Inspection; Remote NDE Techniques; Guided Wave Technology; Ultrasonic Inspection

Introduction: Piping inside fuel and ballast tanks on US Navy ships present a maintenance dilemma. This piping, subjected to a corrosive saltwater environment, requires periodic inspection to insure proper coating and structural integrity. However, the cost to empty, clean and gas-free a typical bank of 4 tanks for safe personnel entry can be up to $100K. Personnel must then perform a visual inspection of piping that often runs along the overhead inside of these tanks. This is a time consuming task and involves risk of injury to personnel. With a fleet of over 200 surface ships the collective cost to the Navy to open and inspect the piping inside of tanks is enormous. To reduce maintenance costs, the Navy has extended scheduled tank openings and inspections by up to 12 years. This has increased the risk of piping corrosion going undetected, especially in older ships. Unexpected piping system failures have occurred and have in some cases resulted in the discharge of oil into the water. When an unexpected thru-tank pipe failure occurs, personnel must enter the tanks on an emergency basis to inspect and repair the failed pipe runs. If an oil spill occurs as a result of the pipe failure then there is an additional cost to clean up the spill and to dispose of fuel-contaminated water. If a sewage line inside of a fuel tank ruptures, the fuel-contaminated sewage and waste-contaminated fuel both must be treated as hazardous waste and the associated disposal costs are very high. Preventing oil spills is of high importance to the US Navy so the Naval Sea Systems Command (NAVSEA), with funding provided for under the Oil Spill Prevention Initiatives Program, directed the Naval Surface Warfare Center Carderock Division (NSWCCD) to evaluate and recommend remote NDE inspection technologies to eliminate the need for personnel to enter the tanks to perform pipe inspections. The goal of this effort is to provide the Navy with a low cost means to quickly and safely perform periodic quantitative inspections of thru-tank piping, thus reducing or eliminating the risk of unexpected pipe failures causing fuel oil spills.

First, a thorough investigation of known thru-tank piping failures was performed. Failure types, system service (fuel, water, etc.), pipe size and material and piping configurations were cataloged. Detailed performance requirements were then developed in order to establish measurement and cost objectives. Applicability of existing NDE inspection standards were examined based on observations and measurements made on actual failed pipe sections. Commercial-off-the-shelf (COTS) and developmental NDE technologies were then reviewed and candidates were identified and selected for laboratory evaluation based on their expected ability to detect the predominant types of pipe corrosion that occurs in thru-tank piping systems most vulnerable to fuel oil spills. NSWC is currently evaluating several candidate technologies as presented in this paper. Upon completion of the laboratory evaluation, successful candidate NDE technologies will then be used for shipboard evaluations.

Piping System Configurations and Failure Mechanisms: A review of thru-tank piping failures was conducted. Fluid piping systems were ranked by the number and severity of failures. Ship drawings of these systems were reviewed for piping arrangements and layout, material and size. Pipe service was noted as to whether the pipe carried fuel or seawater. The maximum (110 feet) and typical length of thru-tank pipe runs, typical internal tank arrangements (elbows, bends, hangers) and locations of entry and exit points from the tanks were noted for these systems. The specific failure mechanisms, if known, were noted. Failure locations were also noted if known. Examples of failed pipe sections from two different Navy ships are shown in Figure 1.

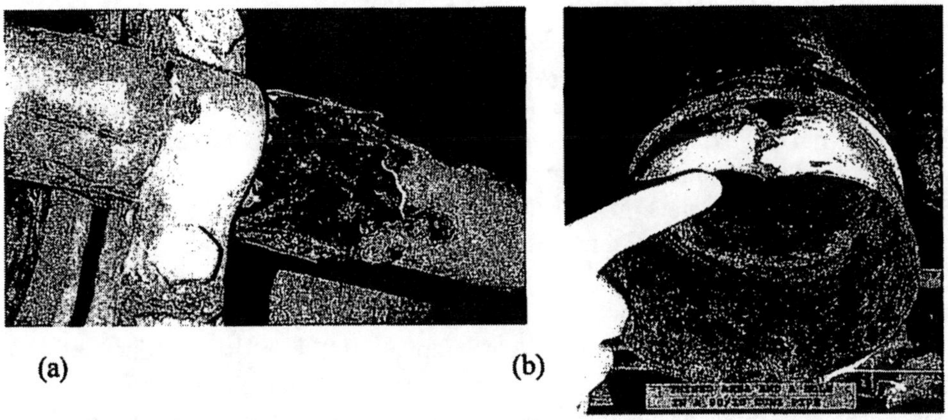

(a) (b)

Figure 1 (a) Corroded steel JP-5 pipe inside a ballast tank (b) Corroded and failed copper nickel (CUNI) sewage pipe inside a fuel tank.

The study revealed that Copper Nickel (CUNI) and Carbon Steel pipe were the two predominant pipe materials used. CUNI piping is sometimes supported by steel hangers that tend to corrode badly. From the study it is concluded that general corrosion has been the main cause of piping failures leading to oil spills. The most common piping system failures are in clean water overboards passing through fuel tanks and fuel lines inside of seawater ballast tanks. The most common pipe sizes ranged form 2 to 12 inches in diameter.

Performance Goals: From the data analysis, a set of performance goals was developed in order to help select the NDE technologies most likely to successfully detect the expected failure types (e.g. wall thinning) and with applicability to the both CUNI and carbon steel pipe. Performance goals were developed to include both physical measurement (e.g. ability to resolve cracks of a minimum size, wall thinning etc.) as well as operational goals. The ability of the equipment to assess the condition of shipboard piping without requiring either entry to tanks or the breaking of seals to the interior of pipes was rated very important since breaking the seals to access the interior of fuel and sewage pipes poses hazards to human operators. The performance goals of equipment was consistent with operational and maintenance standards as specified in NSTM505 [1]. The stated cross-sectional thickness loss, remaining thickness, and defect areas were set to fulfill the minimum thickness requirements for steel and 90/10 CUNI piping in NSTM505. Operational requirements included constraints on the physical size and weight of the type of NDE equipment that can be easily transported from one location to another. The attributes, capabilities, and cost goals developed were:

- Assess the conditions of piping without need to open the seal of the pipe or require human entry into tanks.
- Weight of system less than 50 lbs.
- Dimensions of major equipment to be less than 2 ft x 2 ft x 2ft.
- Space for mounting sensors less than 1 inch surrounding the pipe and 6 inches along the pipe surface.

- Ability to inspect pipe diameters from 2 to 12 inches.
- Ability to inspect both carbon steel and CUNI pipe.
- Inspect straight pipe length of 30 ft.
- Inspect beyond an elbow with radius larger than 5 times the pipe diameter.
- Detect internal and external wall thinning due to corrosion to better than a 10% wall cross sectional loss from up to 30' away.
- From up to 30' away detect internal and external wall loss in an area larger than 8 inch x 8 inch at a loss of wall thickness not exceeding 25%.
- Detect fuel / oil leakage in piping without applying pressure inside the pipe from up to 30' away.
- Accuracy of spatially locating defects to < 1 foot at 30 ft from the transducer.
- Inspect for internal and external defects for wall thickness up to 0.5 inch.
- Operation independent of whether pipe is submerged in fluid inside the tank and whether the pipe is filled with fluid internally.
- Complete the assessment of a 30 ft length of pipe in less than 1 hour
- Detect the presence of bulkheads, pipe supports, hangers, and welds.
- Provide consistent measurement repeatability regardless of equipment operator.
- Provide capability for data / signal recording, archiving, retrieval for condition based maintenance purposes.
- Operate on battery power alone (a plus but not a requirement).
- Acquisition cost of equipment set less than $50K.

NDE Technology Review and Evaluation: Most NDE techniques require point-to-point access, either internally or externally to the pipe. However, several techniques are possible for remotely assessing the condition of thru-tank piping. These so-called "long-range" techniques, allow the conditions of the pipe over long distance to be assessed when a sensor is placed at one location only. Among the later are all remote viewing optical techniques as well as ultrasonic guided wave techniques, both of which will be described in some detail later. Since entry to the tanks is restricted, techniques requiring point-to-point access to the outside surface of the pipe are only described briefly. Techniques generally known as "pigging a pipe" can be applied to a closed pipe if it is empty, can be isolated, and the seal broken to allow insertion of equipment for internal point-to-point evaluations. These techniques were deemed "Intrusive Techniques". The alternatives to these are called "Long-range, Non-intrusive Techniques".

Intrusive Techniques: The inspection of empty piping by accessing from the inside is widely used in the petrochemical and gas pipeline industries, and is known as "pigging' of a pipe. Frequently, sensors are scanned over large distances for point-to-point mapping of the pipe conditions. Widely used are optical imaging techniques and those that are based on the principles of eddy current, magnetic flux leakage, and ultrasound that do not require rigid physical coupling between the sensor and the pipe walls. Internal and external corrosion-induced defects in steam generation tubing and in above and below ground pipelines are successfully inspected by these techniques.

Optical imaging by inserting a miniaturized camera to the pipe interior is widely used in various industries for close examination of the walls of tubes and pipes. The transport of the sensors is achieved by manual push-pull or mechanized crawler systems, accompanied by analog or digital position encoding and image recording capabilities. One NDE system that performs well inside water filled pipe is based on an ultrasound technique known as IRIS [2]. In this technique, waves from a single transmit/receive transducer interrogates the pipe wall via a rotating mirror in water filled pipe, producing typical A, B, C scans for defect characterization.

<u>Long-Range, Non-Intrusive Techniques:</u> There are three viable techniques, which complement one another, to achieve the goals of thru-tank piping inspection without human entry to the tank or breaking the seal to a closed pipe. One is based on propagation of "long-range" ultrasonic waves. Another is based on inserting a specially equipped camera through a manhole to record images of the external surfaces of piping inside the tank. The third is based on the use of a vapor detector delivered near the pipe surface inside a ballast tank by a ROV delivered through a manhole.

Ultrasonic Guided Wave Technology for Piping: Ultrasonic technology is usually applied by launching and receiving pulses of ultrasound through the thickness of a material. This usually requires point-to-point measurement over the region where defects are suspected to exist. In the case of defect detection in piping inside a tank, this mode of operation not only is time consuming but also requires a human operator to have access to the inside a tank. A more viable approach is to take advantage of the fact that the tubular structure of a pipe allows ultrasonic waves to propagate along the axis of a pipe. Such guided waves can travel long distance with little attenuation and generate recognizable signals from corrosion induced wall thinning internal or external at the pipe wall. This technology is currently available commercially, and has thus far been applied mostly in the petrochemical industry. The outstanding features here is that guided waves can be launched in a section of the pipe outside of a tank, and as they propagate along a pipe inside the tank, defects along the pipe can be detected and characterized. In addition, transducers once mounted can be left in place for additional pipe condition monitoring over a long period of time. For a thin-walled pipe (thickness / diameter < 10), ultrasonic wave in the frequency range of tens to hundreds of kilohertz can propagate for long distance along the pipe [3]. Guided wave propagation for NDE is depicted in a sketch in Figure 2a. Many guided wave modes can be excited, each with unique dynamic mechanical displacements distributed through the wall thickness and along the pipe axis. A sketch of mode patterns is shown in Figure 2b. From the standpoint of this project, the most significant difference between the torsional modes and the other modes is that the dynamic displacements associated with torsional modes are of the "in-plane shear" type, and as such are not coupled to any fluid present inside or outside the pipe.

The easiest way to generate a guided wave mode is to place transducers around the circumference of a pipe and excite them simultaneously consistent with the displacements associated with that mode. Currently, piezoelectric, clamp-on, transducer arrays [4] and magnetostrictive ribbon transducers [5] are in use commercially. The later have a very small footprint, and can be easily mounted with double sided adhesive tape or with epoxy and left in-place for periodic pipe condition monitoring over a long period of time. This is shown in Figure 3.

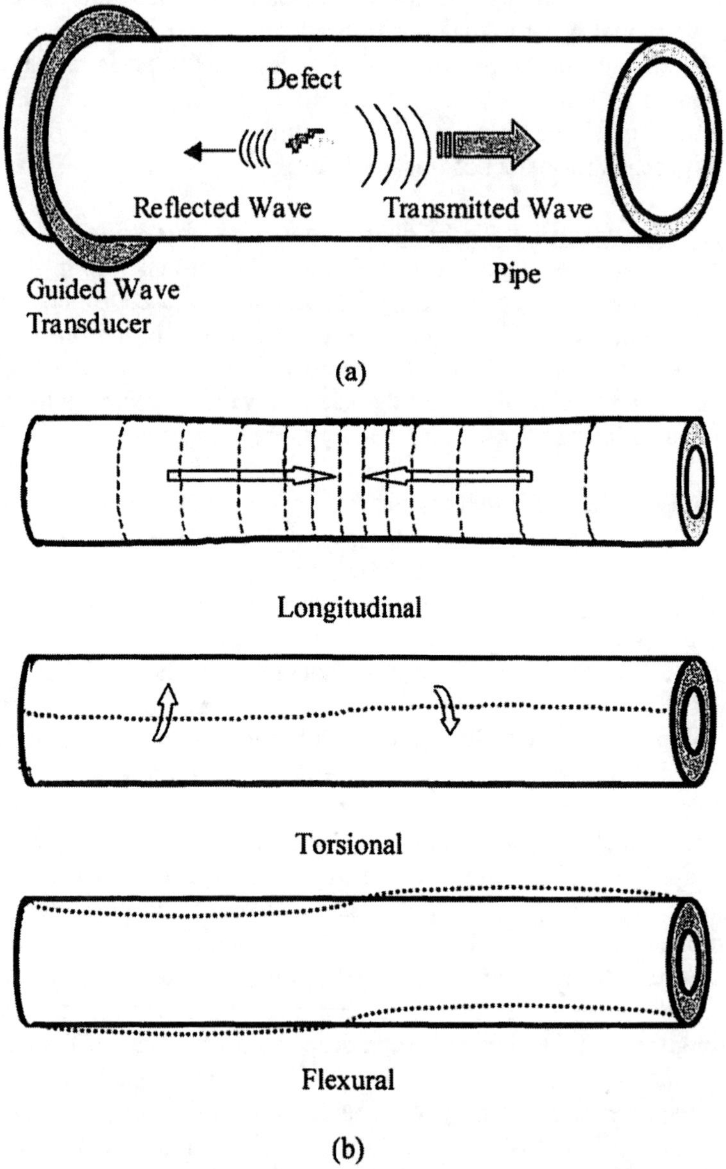

Figure 2 (a) Guided wave propagation (b) Displacements associated with ultrasonic guided waves modes.

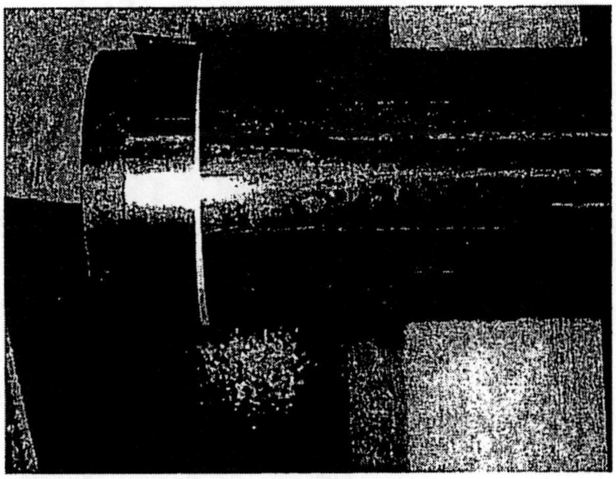

Figure 3. A magnetostrictive ribbon ultrasonic guided wave transducer attached to a 4" steel pipe.

All interfaces existing between the pipe walls and an external member will reflect the guided waves. Thus, a weld between the pipe and a bulkhead, external attachments, hangers, will generate reflections, which can serve as markers in the pulse-echo pattern. The physical location of defects is obtained by their relative position with respect to these markers. Severe distortion of these marker signals at a later time could indicate material deterioration at these welded locations.

To evaluate ultrasonic guided wave systems, an NPS 4" SCH 40 L-shape pipe consisting of two pipes 5 feet long welded together via a 90 degrees elbow was prepared. Three feet from the end of one leg, there was a 1" diameter through hole, followed 1 foot away by an area around the circumference having its thickness reduced to half of the wall thickness to simulate corrosion. Beyond the elbow, there was another annular area simulating corrosion, followed by a 0.5" dia. through hole three feet from the end of the second leg. A sketch of the test pipe is shown in Figure 4a. Piezoelectric, clamped on transducers were used. The pulse-echo pattern is shown in Figure 4b. We note that in addition to most of the features and defects in the pipe, some noise signals referred as 'reverberations" are also seen. These are mostly originated from multiple reflections from pipe ends and the transducer itself. Guided wave ultrasound signals require some explanation. This is related to distinguishing reflections from clamps, weld, and other fixtures from those associated with localized corrosion induced defects. For fixtures that are attached uniformly around the pipe wall, the signals return to different elements in the transducer array mostly in phase and equal in amplitude. Localized defects like corrosion, a hole, or a crack, will have returns that are out of phase or unequal in amplitude at differential locations around the circumferentially mounted transducer array, as a result of mode conversion at these defects. Thus, the measurement system displays two traces, one for the total amplitude without consideration of these phase differences, the other for the mode-converted signals. Note that these mode-converted signals are small at obstructions that are connected to the pipe over the entire circumference.

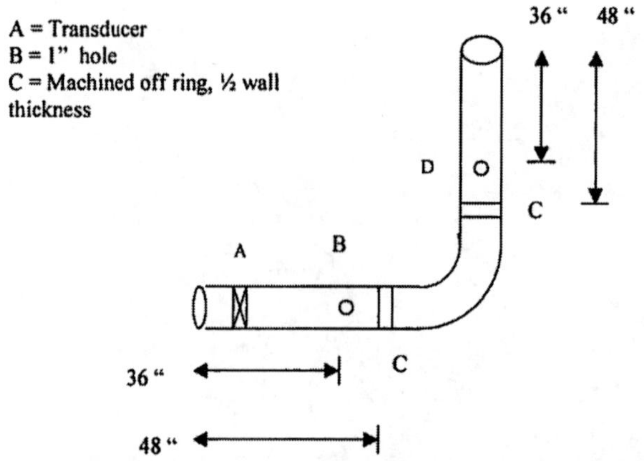

Figure 4 (a) Sketch of L shape pipe specimen with artificial defects

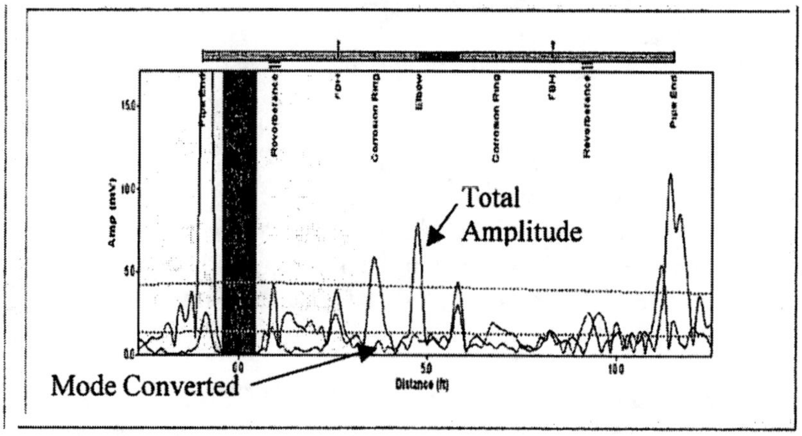

Figure 4 (b) Ultrasonic guided wave pulse-echo data showing reflections from various features and defects.

It has to be emphasized that currently the technology of ultrasonic guided waves is not advanced enough to detect with confidence small defects like pinholes and small surface cracks having an effective wall cross sectional loss of less than about 10 percent. Unless the formation of theses type of defects is preceded by corrosion induced thinning, techniques other than guided wave ultrasound must also be used to mitigate such shortcomings and supplement the results of guided wave ultrasound.

Remotely Controlled Optical Inspection for Oil Residuals: One such technique is based on remotely observing (via a video camera) the fuel or oil residuals from a leaking pipe or on the water surface inside a ballast tank. There are two aspects of this technology. The first one is the ability to mechanically control a video camera through a manhole to acquire images anywhere inside the tank. The Naval Research Laboratory (NRL) has developed this part of the technology for their ISIS tank wall corrosion evaluation system [6]. We are adapting this system for pipe leak detection.

The second aspect is the use of special light sources that would make fuel or oil residuals more visible than water and other background materials inside the tank. Fuel and oil emit light strongly under ultraviolet illumination while water does not emit visible light under ultraviolet illumination. These fluorescence characteristics allow a conventional video camera to remotely detect fuel or oil residuals on the surface of piping or on the top surface of water. To illustrate the detectability of fuel or oil using this concept, we show in Figure 5 the appearance of 5 liquids when illuminated by ultraviolet light. The specimen is a rusted steel plate with 5 flat-bottom-holes of 0.020 inch in depth. The water remains dark, whereas the fuel and oil show their characteristic visible spectra, clearly distinguishable from the steel background. This approach using a variety of light sources and video cameras is being evaluated.

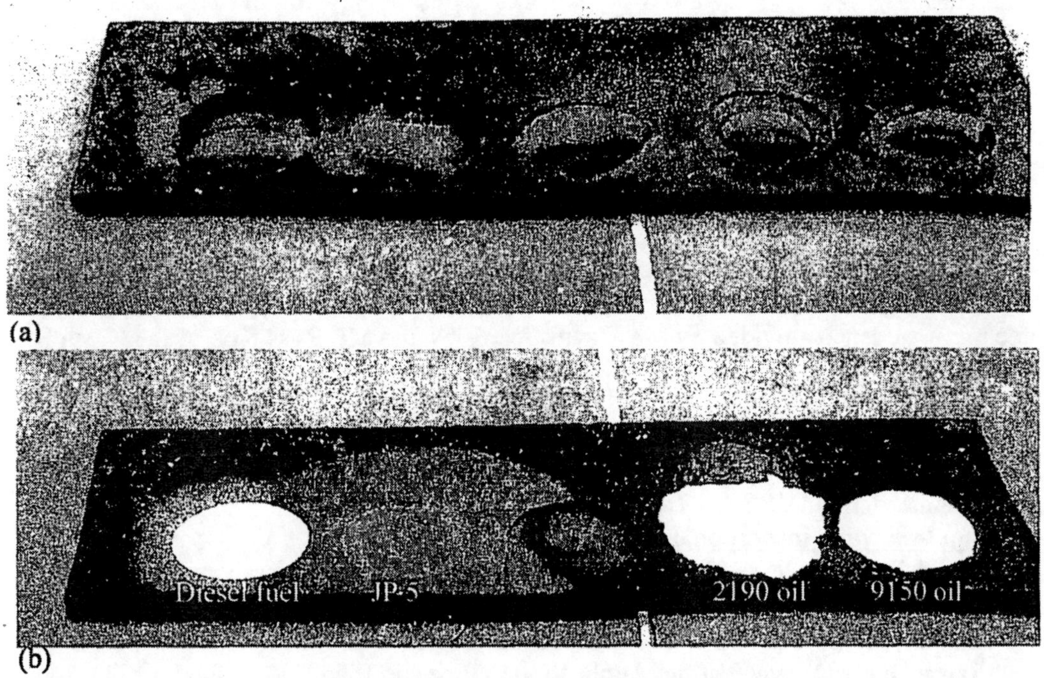

Figure 5. (a) Appearance of water under ultraviolet illumination.
(b) Appearance of fuel and oil under ultraviolet illumination

Remotely Launched Fuel Vapor Detector Inside Ballast Tanks: The third approach that should complement the other two in achieving the goals of NDE for Oil Spill Prevention is based on the detection of fuel vapor using a remotely launched sensor. We envision a chemical sensor will be placed in a ROV, which is to be launched through the manhole to a partially filled tank. The launching, retrieving, and data acquisition technology using a ROV for shipboard application has already been developed by NRL for their ROPI system [7]. This was originally designed for up close inspection of the conditions of paint on tank walls. The transport of a chemical sensor via the ROV to the vicinity of a pipe would require some addition to the electromechanical components and data acquisition systems which should be quite straight forward. The performance of some of these chemical sensors in the high humidity environment in a ballast tank and their ability to distinguish the small amount of residual fuel vapor potentially left inside the tank from true fuel leaks from a pipe is being evaluated. It is envisioned that a truly leaking pipe will produce an increase in response from a sensor as it approaches the source of the leak. In the case of piping existing near the top of a tank wall, robotic, magnetic crawlers can be deployed from the ROV to allow the sensor to be transported to those locations higher up. We are planning the testing of these concepts in collaboration with the NRL team. If successful, this can complement the performance of fluorescence imaging and ultrasound techniques.

Conclusions: We have reviewed current COTS technology for the assessment of the conditions of thru-tank piping without human entry to the tanks. For the purpose of achieving more efficient maintenance to prevent piping failures leading to oil spills, we have identified three techniques that complement one another. These are guided wave ultrasound, remote optical imaging using the fluorescence characteristics of fuel or oil residuals, and fuel or oil vapor detection inside a tank by sensors launched via an ROV. Preliminary data indicate the merits of these techniques. Further evaluation is continuing for potential Fleet applications.

Acknowledgements: The authors gratefully acknowledge the program support and technical guidance provided by Mr. David Beardsley and Mr. Russ Kok of the Naval Sea Systems Command.

References:

1. Naval Ships' Technical Manual Chapter 505 Piping Systems, published by Commander, NAVSEA, 1 Dec. 2001.
2. http://www.iris-inspection.com
3. Joseph L. Rose, Ultrasonic Waves in Solid Media, Cambridge University Press (2004).
4. Cawley, P., M. J. S. Lowe, D. N. Alleyne, B. Pavlakovic, P. Wilcox, "Practical Long Range Guided Wave Testing: Application to Pipe and Rails," Mat. Eval. 61 (1), 2003, pp. 66-74.
5. Light, G. M. and H. Kwan, S. Y. Kim, and R. L Spinks, "Health Monitoring for Aircraft Structures," Mat. Eval. 61 (7), 2003, pp.844-847.
6. Kaznoff, A. I., Brinckerhoff, B., Spadafora, S. J., and Perez, A., "Success Stories: Navy, Naval Assets Present Tough But Achievable Challenges", AMPTIAC Quartery, 7 (4), 2003, pp.66-73.
7. Bill Groeninger and Ed Lemieux, Naval Research Laboratory, private communications 2004.

Materials Selection for Spur Gears by Using CMS Software

R. Chantarat and O.S. Es-Said

Mechanical Engineering Department, Loyola Marymount University, Los Angeles, CA 90045, USA

Abstract:
Materials selection for light, strong, durable spur gears is developed for preliminary spur-gear designs by using Cambridge Materials Selector (CMS) Version 2.01. Spur-gear models are developed from bending fatigue stresses and surface fatigue stresses based on Lewis and Hertz works. A coupling equation of bending fatigue stresses and surface fatigue stresses is determined for optimum materials selection. CMS generates the graphical material database to select the best materials based on a coupling equation.

Keywords:

Cambridge Materials Selector; Materials selection; Materials indices; Spur gear; Optimization

Introduction:

Mating gear teeth acting against each other should have conjugate action. Conjugate action is to have constant angular velocity ratio of gear pairs during meshing. Conjugate action occurs when mating gears have involute profile [1], which relates to spur gear geometries. A change in gear geometry can affect stresses in the gear.

In order to select gear materials, determining failure stresses is as important as determining involute properties. Trade-off between involute properties and failure stresses allow for optimum designs, and maximize gear performance. For example, minimizing gear mass and cost can provide the highest gear performance. However, constraints on involute profiles and design stresses limit gear options.

The performance of mechanical elements such as spur gears can be described by the equation, $p = f_1(F) f_2(G) f_3(M)$, where $f_1(F)$ is a functional requirement (force on spur gear) and $f_2(G)$ is a geometric parameter (involute geometry for spur gears) and $f_3(M)$ is a material property or material index, $\boldsymbol{p}$ describes the performance of mechanical elements. For example, cost, mass or volume is performance of mechanical elements. Optimum design involves the selection of the materials and geometries which maximize or minimize $\boldsymbol{p}$ [2].

In order to select materials with multiple constraints (combinations of performance equations), coupling equations are useful to determine optimum materials. Coupling equations are defined as $M_2 = (C)M_1$. **M₁** and **M₂** are material indices and **(C)** is a constant fixed by the design specification. A material index is a combination of material properties which characterizes the performance of a material in a given application [2].

A search into various literatures and databases produced no reported method for selecting optimum spur-gear materials. In this paper, methods for selecting optimum spur-gear materials are developed by the use of material indices and coupling equations. Methods for developing material indices and coupling equations were presented by professor Ashby in reference [2]. Material indices and coupling equations are applied by constructing a slope in CMS software in order to select optimum gear materials.

Theoretical Discussion:

Identifying design requirements for spur gear designs is important for preliminary design because design requirements are the main gear selection [2]. Table I displays design requirements to select spur-gear materials.

Table I: Design requirements to select spur-gear materials

Function	Spur gear (transmitting constant angular velocity and specific power)
Objective	Minimize mass
Constraints	(a) Fatigue bending stresses are specified
	(b) Surface fatigue stresses are specified
	(c) Pinion and gear shall be made from the same materials

Two important factors contributing to the failure of gear teeth are bending stresses and pitting of gear teeth. A formula for computing the bending stress in gear teeth was introduced in 1892 by Wilfred Lewis, and it still remains the basis for gear design today [1]. Surface fatigue failure of spur gears causes pitting due to many repetitions of high contact stresses. The maximum surface contact stress can be computed from the Hertz stress equation [1, 3].

A spur gear is modeled in Fig. 1 to formulate mathematical equations. In this design, face width (F) is a free variable because it does not affect involute properties. If the involute geometries of the spur gears are specified, only the face width is a free variable because a spur gear is planar mechanism [4].

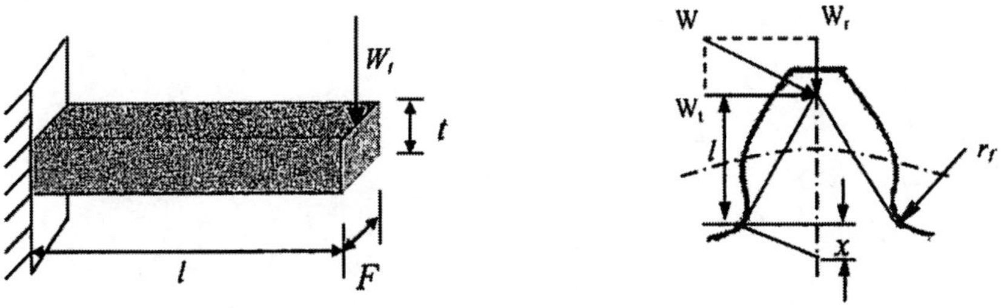

(a) (b)

Figure 1 (a) spur-gear modeled as cantilever beam (b) acting forces and dimensions of spur gears

Fig. 1 shows a spur gear tooth modeled as a cantilever beam to derive the Lewis equation. From the Lewis equation in reference [1, 3], the bending stress is

$$\sigma = \frac{W_t P}{K_V F Y} \tag{1}$$

where W_t = tangential force, lb
P = diametral pitch, in
K_V = velocity factor
F = face width, in
Y = Lewis form factor, table 14-2 reference [1]

Equation (1) is important because it forms the basis for the AGMA (American Gear Manufacturers Association) approach to determining the bending strength of gear teeth. It is most applicable for estimating the gear capacity when life and reliability are not important considerations. This equation is also useful in obtaining a preliminary estimate of gear sizes for various applications [1].
The gear tooth is modeled as a cantilever beam. Therefore, the mass of a gear tooth can be defined as

$$m_1 = \rho F t l \tag{2}$$

when ρ = density of gear materials,
F = Face width, in.
t = thickness, in.
l = length, in.
From eq. (1) and $\sigma = \sigma_e$, then

$$F = \frac{W_t P}{K_V \sigma_e Y} \tag{3}$$

, where σ_e is endurance limit of spur gear.

The Marin equation is employed to determine the fatigue-bending stress. The Marin equation [1] is

$$\sigma_e = k_a k_b k_c k_d k_e \sigma_e^{'} \qquad (4)$$

where $\sigma_e^{'}$ = endurance limit of test specimen

k_a = surface factor

k_b = size factor

k_c = load factor

k_d = temperature factor

k_e = miscellaneous-effect factor

In equation (4), the safety factors ($k_a k_b k_c k_d k_e$) do not influence the selection [2]. Therefore, they can be neglected. However, in this paper some safety factors will not be neglected because they do affect the selection process.

Surface Factor k_a

The modification factors depend on the quality of the surface finish and tensile strength of the materials. Surface factors can be expressed as

$$k_a = a\sigma_{ut}^{b}$$

where σ_{ut} = is the minimum tensile strength and **a** and **b** are constants. Now, **a** and **b** can be found from Reference [1] page 283. In the CMS software, there is no function for surface-finish so let

$$k_a = 1 \qquad (5)$$

Size Factor k_b

A small bar should have a higher fatigue life than a larger bar because there is less volume exposed to a given load [5]. Reference [1] presents the size factor for a round bar in bending as;

$$k_b = \left(\frac{d}{0.3}\right)^{-0.1133} \quad \text{in} \qquad 0.11 \le d \le 2 \quad \text{in} \qquad (6)$$

The spur gear model is assumed to be a rectangular cross-section. The effective diameter for a rectangular cross-section is shown in figure 2. This effective diameter is substituted for (d) in equation (6). The effective diameter is:

$$d_e = 0.808(Ft)^{1/2}$$

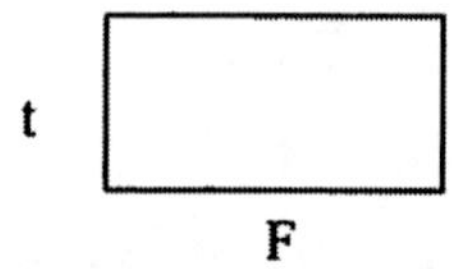

Figure 2 Effective diameter for rectangular cross-section

Substitute d_e in (6)

$$k_b = 0.8938(Ft)^{-0.05665} \qquad (7)$$

Load Factor k_c

Reference [1] suggested that specimens tested in bending show no change in fatigue life. For gear teeth in bending:

$$k_c = 1 \tag{8}$$

Temperature Factor k_d

Temperature changes affect the mechanical properties of gear materials. In low temperature conditions, brittle fracture is a strong possibility. In a high temperature environment, yielding is the primary cause of material failures [1]. Reference [1] presents the temperature factor as

$$k_d = \frac{S_T}{S_{RT}}$$

where S_T = tensile strength at operating temperature,

S_{RT} = tensile strength at room temperature.

For this design, the spur gears operate at room temperature. Therefore the temperature effect, is:

$$k_d = 1 \tag{9}$$

Miscellaneous-effects Factor k_e

In general, gear teeth are subjected to bending in only one direction. This is shown in figure 3 as repeated stress.

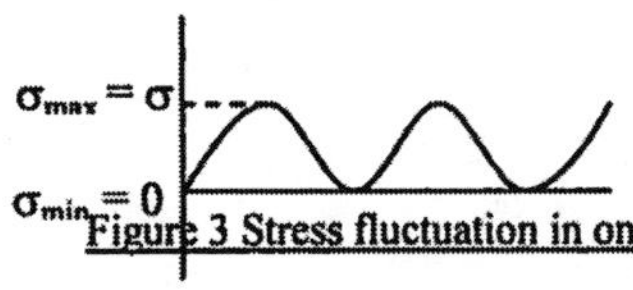

Figure 3 Stress fluctuation in one way bending

The mean stress (σ_m) and amplitude stress (σ_a) are expressed as:

$$\sigma_a = \frac{\sigma_{\max} - \sigma_{\min}}{2} = \frac{\sigma - 0}{2} = \frac{\sigma}{2}$$

$$\sigma_m = \frac{\sigma_{\max} + \sigma_{\min}}{2} = \frac{\sigma}{2}$$

The Goodman relation is employed to correct the endurance for repeated bending stress.

$$\frac{\sigma_a}{\sigma_e'} + \frac{\sigma_m}{\sigma_{ut}} = 1$$

, where σ_{ut} = ultimate tensile stress.

Replacing σ_a and σ_m with σ/2 and in dominator replacing σ_e' with $0.504\sigma_{ut}$ (for steel), it turns out that:

$$\sigma = 2\left(\frac{\sigma_e'\sigma_{ut}}{(\sigma_{ut} + 0.504\sigma_{ut})}\right) = 1.329\sigma_e'$$

This shows that a miscellaneous-effects factor is

$$k_e = 1.329 \tag{10}$$

The stress concentration at the root of the gear tooth is very important for determining fatigue failures. The fatigue stress concentration factor K_f is

$$K_f = 1 + q(K_t - 1)$$

,when K_t = stress concentration factor

q = notch sensitivity.

Fatigue stress concentration factor K_f is related to miscellaneous-effects factor k_e by

$$k_e = \frac{1}{K_f}$$

In preliminary design, there is usually no information available for q. Therefore, q should be 1 for safe design [6].

Now substitute (4) into (3)

$$F = \frac{W_t P}{K_V (k_a k_b k_c k_d k_e \sigma_e^{'}) Y} \tag{11}$$

Substitute (5), (7), (8), (9) and (10) into (11) then

$$F = \left(\frac{W_t P t^{0.05665}}{1.1878 K_V k_e \sigma_e^{'} Y} \right)^{1/0.94335} \tag{12}$$

Substitute (12) into (2)

$$m_1 = \left(\frac{W_t}{1.1878 K_V k_e} \right)^{1.06005} \left(\frac{Pt}{Y} \right)^{1.06005} l \left(\frac{\rho}{\sigma_e^{'1.06005}} \right) \tag{13}$$

From the bending constraint, a material index is

$$\boxed{M_1 = \frac{\sigma_e^{'1.06005}}{\rho}} \tag{14}$$

where $\sigma_e^{'}$ = endurance limit. M_1 should be maximized to minimize mass m_1.

Spur gear teeth have to resist pitting. Hertzian stress is useful for predicting pitting on the gear tooth surface. Contact stress or Hertzian stress [1, 3], can be expressed as

$$\sigma_c = \left\{ \frac{W_t}{C_V \pi F \cos\phi} \frac{\frac{1}{r_1} + \frac{1}{r_2}}{\left[\frac{(1-\upsilon_1^2)}{E_1} + \frac{(1-\upsilon_2^2)}{E_2} \right]} \right\}^{1/2} \tag{15}$$

where radius of curvature of the tooth profiles for pinion = $r_1 = \dfrac{d_P \sin\phi}{2}$,

radius of curvature of the tooth profiles for gear = $r_2 = \dfrac{d_G \sin\phi}{2}$,

velocity factor = $C_V = K_V$.

Young's modulus = E,
Poisson's ratio = ν,
face width = F,
tangential force = W_t,
pressure angle = φ.

Constraint (c) requires one material for spur gears. Therefore, the pinion and gear should be made from the same material.

$$E_1 = E_2 = E \text{ and } \nu_1 = \nu_2 = \nu \qquad (16)$$

Substitute (16) to (15)

$$\sigma_c = \left\{ \frac{W_t}{C_V \pi F \cos\phi} \frac{\frac{1}{r_1} + \frac{1}{r_2}}{\left[\frac{2(1-\upsilon^2)}{E}\right]} \right\}^{1/2} \qquad (17)$$

The term $(1-\nu^2) \approx 1$ in (17) because ν^2 is very small for all solids [2].
From (17)

$$F = \frac{W_t}{C_V \pi \sigma_c^2 \cos\phi} \frac{r_2 + r_1}{r_1 r_2} \frac{E}{2} \qquad (18)$$

The mass of a gear tooth is

$$m_2 = \rho F t l \qquad (19)$$

Substitute (18) to (19)

$$m_2 = \left(\frac{W_t}{2C_V \pi \cos\phi}\right)(lt)\left(\frac{r_2 + r_1}{r_1 r_2}\right)\left(\frac{E\rho}{\sigma_c^2}\right) \qquad (20)$$

From equation (20), a material index for pitting is:

$$\boxed{M_2 = \frac{\sigma_c^2}{E\rho}} \qquad (21)$$

Minimizing mass, $\mathbf{M_1}$ and $\mathbf{M_2}$ should be maximized.
Maximizing the material index will minimize mass $\mathbf{M_1}$ and $\mathbf{M_2}$.
A coupling equation can be derived from equation (13) and equation (20). Equation (13) is equal to equation (20) because the same spur gears should have the same mass,

$$m_1 = m_2$$

$$\boxed{\frac{M_2}{M_1} = \left(\frac{0.191 K_V^{0.06005} k_e^{1.06005}}{W_t^{0.06005} t^{0.06005} \cos\phi}\right)\left(\frac{Y}{P}\right)^{1.06005}\left(\frac{r_1 + r_2}{r_1 r_2}\right)} \qquad (22)$$

Equation (22) is the coupling equation between bending fatigue stresses and surface fatigue stresses. The coupling equation can determine the optimum selections for spur gear materials.

Results in Selection:

Specifications in Table II are applied to select spur-gear materials in preliminary design. A method to select gear materials is developed from material indices and the coupling equation from section 2.

Table II: Spur-gear specifications

Diametral pitch, P= 8 teeth/in
Number of pinion teeth, $N_p = 16$
Pressure angle, $\phi = 20°$
Horse power out put, H = 6.93 hp
Angular velocity of pinion, $n_p = 1200$ *rev/min*
Number of gear teeth, $N_G = 50$

From coupling equation (22), unknown variables are K_V, W_t, k_e, t, ϕ, Y, r_1 and r_2 which should be determined to construct a slope of equation (22) in CMS software. Samples of calculations of unknown variables are presented in Appendix1. The result of calculating equation (22) is

$$\frac{M_2}{M_1} = 0.0105 \qquad (23)$$

Plotting coupling equation (23) in CMS (Cambridge Materials Selector) version 2.01 [4], spur-gear materials can be chosen from Fig. 4. Material index M_1 is equation (14) and Material index M_2 is equation (21). Maximizing M_1 and M_2 is to minimize gear mass. The materials that are close to the slope line, 0.0102, are optimum materials for spur gears.

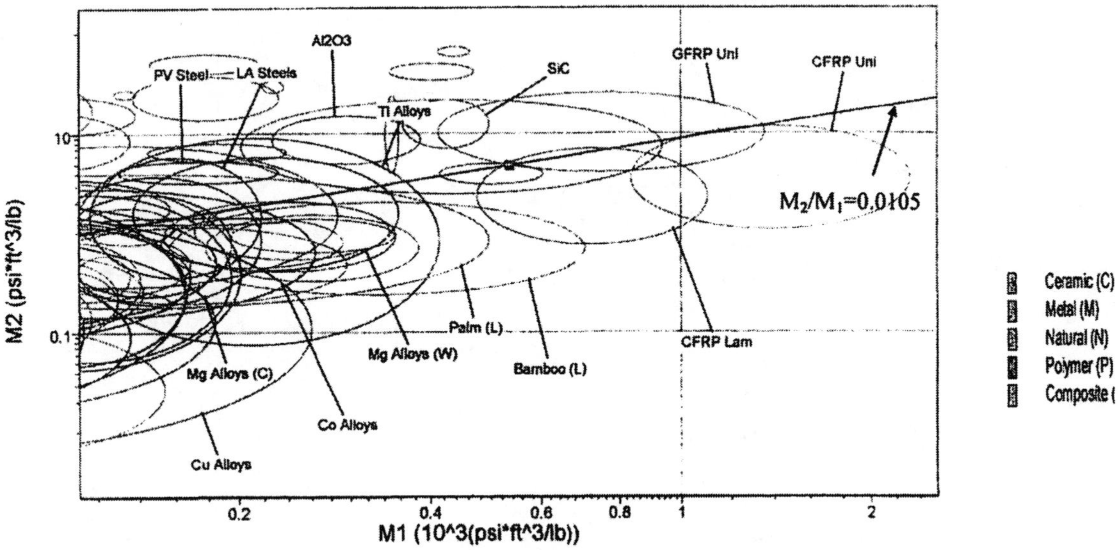

Figure 4 materials for spur gears from CMS database

Ceramic materials are eliminated from selection because ceramics cannot resist shock loads when misaligned installation of spur gears occurs. Composites and woods are also eliminated because it is difficult to form gear shapes. Five candidate materials and corresponding material processes are listed in table III.

Table III: Candidate materials for spur gears and materials processing

Materials	Material Processing
1. Ti-alloys	Cast, Die cast, Powder, Extrude, Machine
2. Mg-alloys(W)	Machine, Powder, Machine
3. LA-steels	Cast, Bar, Forge, Grind, Machine, Powder
4. Co-alloys	Cast, Machine, Powder
5. Cu-alloys	Cast, Die cast, Machine, Powder

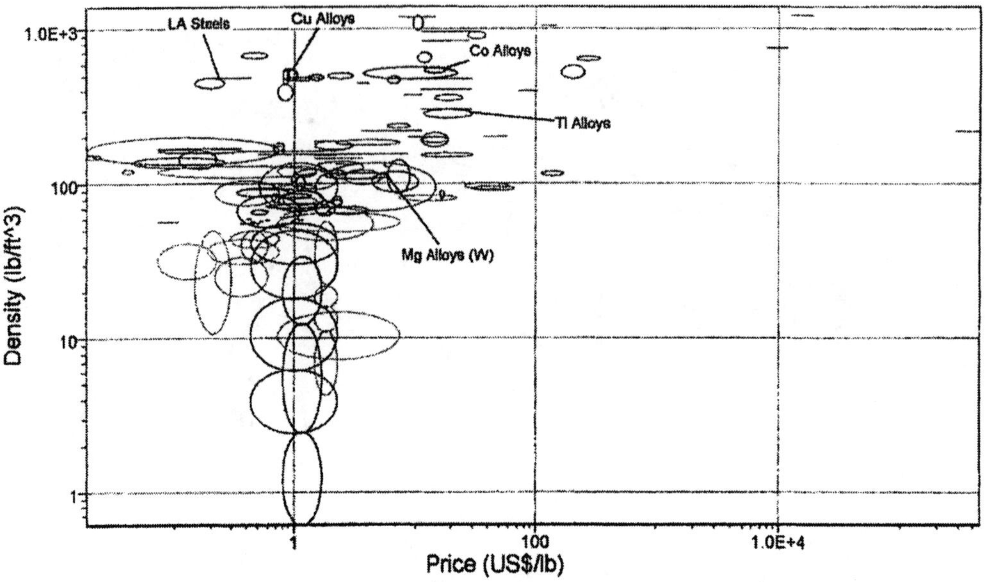

Figure 5: Density and Price from CMS for candidate materials

CMS plots material prices vs. density for candidate materials in Fig. 5. Ti-alloys are the best materials for light, strong, durable spur gears. However, Ti-alloys are not economical for producing spur gears. Mg-alloys are good but expensive. LA steels (low alloy steels) are optimum for producing spur gears because of strength and economy. Alloy steel castings are employed when design considerations combined with cost of manufacturing, show an economic advantage over other methods. LA steel castings have good resistance to shock load in the hardened and tempered condition [8].

Conclusion:

Screening material data is very significant for preliminary design of the spur gear. Material indices and CMS allow gear designers to screen tremendous material data by graphical methods. The advantages of graphical methods are to facilitate designers to interpret data without arbitrary decisions. Other advantages of graphical methods are shortened times for designers to interpret material data with optimum selection choices.

Material indices and the coupling equation are very effective to select suitable spur-gear materials. Light, strong, and durable spur gear materials are obtained from the coupling equation and the material indices plotted by CMS. Maximizing both material indices, M_1 and M_2, can reduce gear mass. Materials that are close to the line plotted from the coupling equation are the optimum materials for spur gears.

Acknowledgements:

The authors wish to thank Ms. Maria Guerrero for the preparation of the manuscript.

References:

[1] Joseph Edward Shigley and Charles R. Mischke. Mechanical Engineering Design. 5th International ed. Singapore: McGraw Hill; 1989. p. 282-288, p. 587-595.

[2] Michael F. Ashby. Material Selection in Mechanical Design. 2nd ed. Great Britain: Butterworth Heinemann; 1999. p. 65-71, p.123, p.217-218, p.229.

[3] Andrew D. Dimarogonas. Computer Aided Machine Design. UK: Prentice Hall; 1988. p.607-613.

[4] Joseph Edward Shigley and John Joseph Uicker Jr. Theory of Machines and Mechanisms. 2nd inernational ed. McGraw-hill; 1995. p.11, p.250-255.

[5] Franklin E. Fisher and Joy R. Fisher. Probability Application in Mechanical design, Marcel Dekker Inc.; 2000. p.71.

[6] Walter D. Pilkey. Peterson's Stress Concentration Factor. 2nd ed. USA: John Wiley & Sons; 1997. p.39.

[7] Cambridge Material Selector (CMS) Version 2.01, Granta Design Ltd., 1992.

[8] H J Watson. Modern Gear Production. 1st ed. Pergamon Press; 1970, p.52, p.75.

Appendix 1:

Reference 1 provides the methods to calculate unknown variables in equation (22). Following procedures are examples to calculate unknown variable (K_V, W_t, k_e, t, ϕ, Y, r_1 and r_2).

a) First compute thickness t

$$t = (4lx)^{1/2} \tag{24}$$

$$l = \frac{1}{P} + \frac{1.25}{P} = \frac{1}{8} + \frac{1.25}{8} = 0.281 \text{ in} \tag{25}$$

Find Y from P and Table14-2 in reference [1]. When P is 8, Y is 0.296. x in (24) can be calculated from

$$x = \frac{3Y}{2P} = \frac{3(0.296)}{2(8)} = 0.0555 \tag{26}$$

Substitute (25), (26) to (24), then

$$t = (4 \times 0.281 \times 0.0555)^{1/2} = 0.25 \text{ in} \qquad (27)$$

b) Next compute curvature of tooth profile r_1 and r_2

$$r_1 = \frac{d_p \sin\phi}{2}$$
$$r_2 = \frac{d_G \sin\phi}{2} \qquad (28)$$

d_p and d_G are diameters of pinion and gear. d_p and d_G can be calculated from following equation.

$$d_p = \frac{N_p}{P} = \frac{16}{8} = 2 \text{ in}$$
$$d_G = \frac{N_G}{P} = \frac{50}{8} = 6.25 \text{ in}$$

Substitute d_p, d_G and ϕ to (28), then

$$r_1 = 0.342$$
$$r_2 = 1.0688 \qquad (29)$$

c) Compute k_e from stress concentration. Stress concentration can occur at fillet radius (r_f) showed in Fig. 1.

$$r_f = \frac{0.3}{P} = \frac{0.3}{8} = 0.0375$$

From graph in reference [1], $\frac{r_f}{t} = \frac{0.0375}{0.25} = 0.15$ and $\frac{D}{d} = \infty$, then $k_t = 1.68$

Fatigue stress concentration factors can express as

$$k_f = 1 + q(k_t - 1)$$

In primary design, reference [6] suggests that Notch sensitivity $q = 1$ for safe design.

$$k_f = 1 + 1(1.68 - 1) = 1.68$$

$$k_e = \frac{1}{k_f} = \frac{1}{1.68} = 0.5952$$

d) Compute dynamic factor K_V from tangential velocity

Tangential velocity of pinion can be calculated from following equation.

$$V = \frac{\pi d_p n_p}{12} = \frac{\pi(2)(1200)}{12} = 628.31 ft/\min$$

Then, dynamic factor can find from Table 14-7 in reference1. When Q_v (AGMA transmission accuracy-level number) is 5 and V is 628.31 ft/min, $K_V = 0.65$.

e) Compute load W_t from horsepower

W_t can be calculated from following equation

$$H = \frac{W_t V}{33000} = 6.93hp$$

$$W_t = \frac{(6.93)(3300)}{628.31} = 363.97lb$$

f) Compute a slope from equation (22)

$$\frac{M_2}{M_1} = \left(\frac{0.191 \times 0.65^{0.06005} \times 0.5952^{1.06005}}{363.97^{0.06005} \times 0.25^{0.06005} \cos 20°}\right)\left(\frac{0.296}{8}\right)^{1.06005}\left(\frac{0.342 + 1.0688}{0.342 \times 1.0688}\right) \quad (30)$$

$$\frac{M_2}{M_1} = 0.0105$$

PROGNOSTICS IV

Chair: Mr. Mark Hollins
Naval Air Warfare Center/AD

INCIPIENT FAULT DETECTION AND REMAINING USEFUL LIFE PREDICTION FOR AVIONICS SYSTEM POWER SUPPLIES

Rolf Orsagh, Douglas Brown,
& Michael Roemer
Impact Technologies, LLC
200 Canal View Boulevard
Rochester, NY 14623
(585) 424-1990
rolf.orsagh@impact-tek.com

Thomas Dabney, & Andrew Hess
Joint Strike Fighter Program Office
Suite 600, Crystal Gateway Four
1213 Jefferson Davis Highway
Arlington, VA 22202-4316
(703) 601-5619
tom.dabney@jsf.mil

Abstract: This paper presents an integrated approach to switching mode power supply health management that implements techniques from engineering disciplines including statistical reliability modeling, damage accumulation models, physics of failure modeling, and sensor-based condition monitoring using automated reasoning algorithms. Novel features extracted from sensed parameters such as temperature, power quality, and efficiency was analyzed using advanced fault detection and damage accumulation algorithms. Using model-based assessments in the absence of fault indications, and updating the model-based assessments with sensed information when it becomes available provides health state awareness at any point in time. Intelligent fusion of this diagnostic information with historical component reliability statistics provides a robust health state awareness as the basis for accurate prognostic predictions. Complementary prognostic techniques including analysis of projected operating conditions by physics-based component aging models, empirical (trending) models, and system level failure progression models will be used to develop verifiable prognostic models. The diagnostic techniques and prognostic models have been demonstrated through accelerated failure testing of switching mode power supplies.

Key Words: Prognostics, Diagnostics, Avionics, Failure Progression, Switch Mode Power Supplies, Accelerated Failure Testing

Introduction: Electronic systems such as electronic controls, on-board computers, communications, navigation and radar perform many critical functions on-board military and commercial aircraft. All of these systems depend on electrical power supplies for direct current (DC) power at a constant (regulated) voltage to drive solid-state electronics. With these power supplies playing an important role in the operation of aircraft systems and subsystems, flight and ground crews need health state awareness and prediction tools that diagnose faults accurately, predict failures, and project life remaining of these components. A specific goal of the Joint Strike Fighter (JSF) technology maturation programs is to develop health state awareness and prediction capabilities through Prognostics and Health Management (PHM) for all JSF systems including avionics. Substantial safety and cost benefits are projected if the optimal application of LRU diagnostic and prognostic techniques is realized.

Switch-mode power supplies (SMPS's) are commonly used aboard aircraft where their weight, size, and efficiency make them preferable to conventional transformer-based power supplies. In addition to regulating the voltage of direct current (DC) power, these novel circuits can also serve as DC to DC converters that can step down ("bucking" design) voltage like conventional supplies or step up ("boost" or "fly back" design) voltage. However, SMPS's are no panacea. Many early SMPS designs suffered from sudden and catastrophic failures or generated excessive electromagnetic interference (EMI). More recent SMPS designs employ protective circuits to isolate sensitive components from damaging events.

The DC-DC converter at the heart of SMPS's uses a switching element, along with capacitors and inductors, to step up or step down voltage and current accordingly. High speed switching enables the transfer of energy packets from the input filter capacitor to the output filter capacitor. The last stage filters out any high frequency components from the DC output. Finally, the output is feedback into a control circuit that stabilizes the DC/DC converter. Figure 1 shows a simplified diagram of a SMPS.

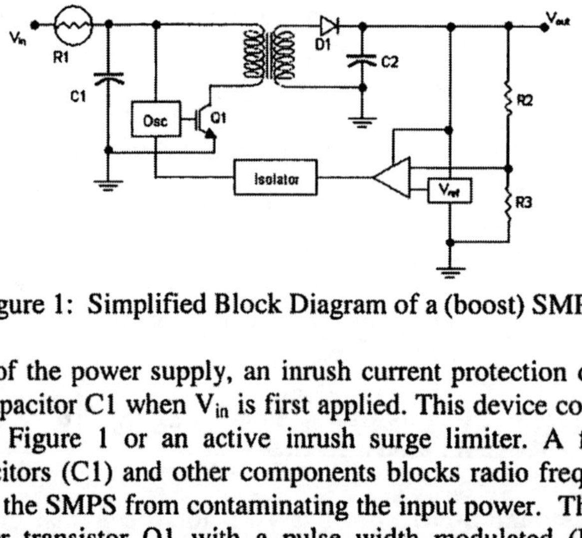

Figure 1: Simplified Block Diagram of a (boost) SMPS

On the input side of the power supply, an inrush current protection device controls the current charging capacitor C1 when V_{in} is first applied. This device could be a thermistor (R1) as shown in Figure 1 or an active inrush surge limiter. A filter on the input consisting of capacitors (C1) and other components blocks radio frequency interference (RFI) generated by the SMPS from contaminating the input power. The oscillator (OSC) switches the power transistor Q1 with a pulse width modulated (PWM) signal at a frequency around 100 kHz. During each PWM cycle the current through the primary side of the transformer ramps up when Q1 turns on. When Q1 turns off, the voltage applied to C2 rises as the inductor attempts to maintain current on its secondary side and charges C2. The duty cycle of the PWM signal determines the voltage applied to C2, and is set based on feedback (control) from comparison of a fraction of V_{out} with V_{ref}.

SMPS Failure Modes: Reliability studies of switching mode power supplies have shown that the majority of failures may be attributed to small number of components (Pareto effect) [i, ii, iii, iv]. The relative frequency of specific component failures may vary based on SMPS topology, type of component used, derating factors, and location in the system. The components that commonly fail are classified into three different categories:

- Switching Transistors
- Filtering Capacitors
- Rectifying Diodes

Three types of switching transistors are commonly used in SMPS applications; bipolar junction transistors (BJTs), metal-oxide semiconductor field-effect transistors (MOSFETs), and insulated-gate biplolar transistors (IGBTs). Each type of transistor exhibits unique failure modes and rates, but switching transistors are generally the leading cause of SMPS failures. Bipolar junction transistors are widely used in SMPS designs where low priorities for weight and efficiency allow lower switching frequencies. MOSFETs and IGBTs are more common in applications where weight and efficiency requirements mandate higher switching frequencies.

The primary purpose of output filter capacitors in a SMPS is to suppress high frequency noise generated by switching in the DC-DC converter. As a consequence, the output filter capacitor is subject to continuous voltage ripples. The magnitude of the voltage ripple is dependent on ESR, ambient temperature, output current, and the input voltage of the converter. [v] Stress is also applied to the capacitor when a load is removed from the power supply. If the control circuit cannot respond fast enough to the event, the energy stored in the inductor transfers to the capacitor, accumulating internal damage [[vi]].

The two types of filtering capacitors are commonly used in SMPS applications, tantalum capacitors and electrolytic capacitors are both susceptible to failure by dielectric breakdown, though each exhibit type of capacitor has unique root causes of failure. Filtering capacitors are the second largest source of SMPS failures. These capacitors are required to have large capacitance, high voltage ratings, and low equivalent series resistance (ESR). Tantalum capacitors are replacing electrolytic capacitors for their small form factor and higher reliability.

The diodes that are used to rectify the input and output voltage in switching mode power supplies are also a significant cause of failure. Shockley diodes are commonly used in rectifiers. The leading failure modes affecting switching mode power supplies are summarized in

Table 1 vii.

Table 1 SMPS Failure Modes

Failure Modes	Related Components	Cause of Failure	Possible Effects	Probability of Occurrence	Criticality	Possible Action to Reduce Failure Rate of Effect
Contact Migration	- BJTs - Diodes	- Electrical overstress - Temperature - High current density	-Decreased efficiency - Increased turn-on resistance	High	Catastrophic	-Improve heat dissipation -Choose a transistor with a higher current rating
Corrosion	- BJTs - Capacitors - Diodes - MOSFETs	- Moisture	- Internal corrosion -Bonding pad corrosion	Low	Marginal	-Operate in an environment with minimal moisture
Dielectric Breakdown	- Capacitors	- Electrical overstress - High operating voltage - Temperature	- Short circuit - Open circuit	High	Catastrophic	- Increase voltage rating - Add a small series resistor
Electromigration	- BJTs - Capacitors - Diodes - MOSFETs	- High current densities - Electrical overstress - Temperature	-Metal interconnect failure	Low	Marginal	- Use a transistor with a higher current rating
Gate Oxide Breakdown	- MOSFETs	- Electrical overstress - Temperature	-Increased threshold voltage - Increased turn-on resistance - Decreased efficiency	High	Catastrophic	- Choose a transistor with a higher current rating
Hot Carrier Effect	- MOSFETs	- Large electric field in the channel region	- Short circuit	Low	Marginal	- Reduce drain-to-source voltage
Thermal Runaway	- BJTs - Diodes - MOSFETs	- Increased operating temperature - Gate Oxide Breakdown	- Short circuit	High	Catastrophic	- Improve heat dissipation
Thermal Cycling	- BJTs - Capacitors - Diodes - MOSFETs	- Environmental temperature cycling - Power up/down cycle	- Thermal runaway	Medium / High	Critical	-Reduce exposure to rapid changes in environmental temperature

Criticality: ☐ Marginal ☐ Critical ☐ Catastrophic

Physics of Failure Models: Physics of failure models are used as the basis for incipient fault detection, fault to failure progression, and remaining useful life predictions. Critical transistor failure modes include thermal runaway, gate-oxide breakdown, contact migration and thermal fatigue. Thermal runaway and thermal fatigue can affect all types of transistors; while gate-oxide breakdown only affects MOSFETs and contact migration only affects BJTs. Thermal runaway, contact migration, and thermal fatigue also affect diodes. Critical capacitor failure modes include dielectric breakdown and thermal fatigue.

While thermal runaway affects all types of transistors, bipolar junction transistors are the most susceptible due to the positive feedback relationship between device temperature and heat generation. As the temperature of a BJT rises, so does its heat generation. When the rate of heat generation exceeds the capacity of the heat sink, the temperature of the device rises quickly until it fails. Temperature is therefore the primary indicator of thermal runaway. Thermal runaway may occur when a device cannot remove the additional power generated by electrical overstress events or an increase in internal resistance due to gradual device degradation. Thermal runaway eventually leads to a short circuit mode.

Gate oxide breakdown is one of the major concerns regarding metal-oxide semiconductor field effect transistors (MOSFET) devices. Damage to the gate oxide (Figure 2) can result in excessive leakage current, increased standby power, and a decrease in response time. Eventually, the damage will cause a MOSFET to short-circuit. The damage results from thinning of the oxide and an increase in the effective electric field. Two types of gate oxide breakdown are possible: A catastrophic breakdown can occur as the result of electrical overstress. Time-dependent dielectric breakdown (TDDB), takes place during operation within the rated conditions of voltage, temperature, and power dissipation. Time dependent dielectric breakdown refers to the damage accumulated in the gate oxide region of a MOSFET during use within its rated operating condition. The rated operating conditions include voltage, temperature, and the magnitude of the electric field between the drain and body. Both types of gate oxide breakdown ultimately result in a short circuit between the drain and source regions of a MOSFET.

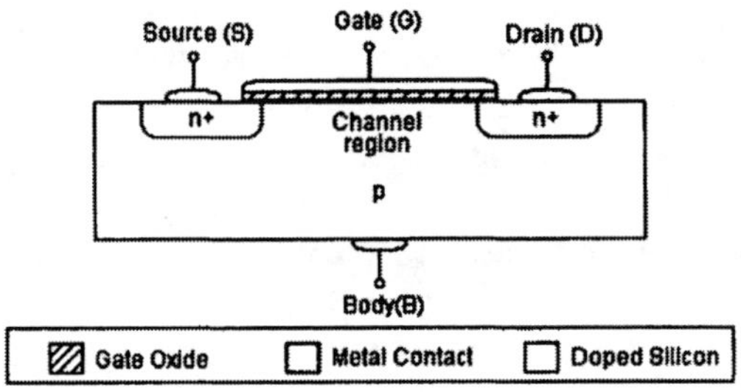

Figure 2: MOSFET Gate Oxide Channel

Modeling TDDB lifetime depends on the strength of the electric field so long as the oxide region is at least moderately thick. For relatively strong electric field, the expected life (τ_f) is proportional to the reciprocal of the applied electric field strength (Equation 1). In contrast, the expected life in a weak electric field is proportional to the electric field strength (Equation 2).

$$\ln(\tau_f) \propto \frac{1}{E_{db}} \tag{1}$$

$$\ln(\tau_f) \propto -E_{db} \tag{2}$$

Contact migration can affect rectifier diodes and BJT switching transistors in a SMPS. Contact migration occurs when material from aluminum contacts diffuses into the silicon region. The diffused aluminum forms "spikes" in the silicon region, which can eventually short out a pn junction as shown in Figure 3. Junction spiking occurs as a result of high current densities resulting from electrical overstress. A model for contact migration is given in Equation 3.

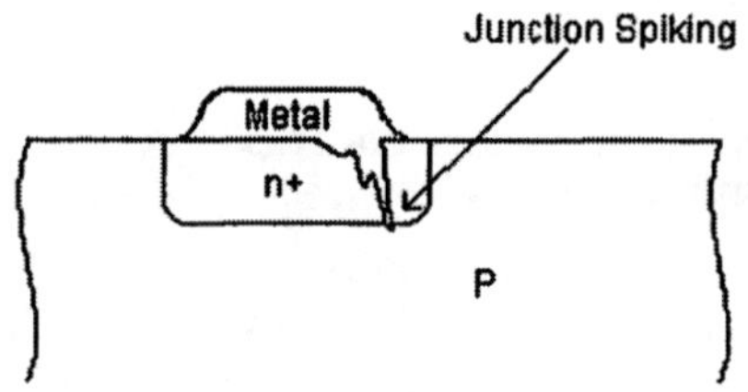

Figure 3: Contact Migration

$$\tau_f \propto \left(J - J_{crit}\right)^{-N} e^{(E_a/\,kT)} \tag{3}$$

J = Current Density
Ea = Thermal Activation Energy
Jcrit = Critical Current Density
N = Correction factor (1 – 2)

All components in a SMPS, particularly diodes, transistors, and capacitors, are susceptible to thermal fatigue failures. Fatigue damage due to thermal cycling accumulates every time a device experiences a power-up and a power-down cycle and every time a device experiences an environmental temperature cycle. Thermal fatigue eventually weakens metallic contacts and may cause: dielectric film cracking, lifted bonds, solder fatigue and cracked dies. The Coffin-Manson model is used to estimate the number of thermal cycles till failure is shown in Equation 4.

$$N_f = C_0[\Delta T - \Delta T_0]^{-q} \tag{4}$$

N_f = Number of cycles to failure
C_o = A material dependent constant
ΔT = entire temperature cycle-range for the device
ΔT_0 = elastic region temperature cycling
q = an empirically derived constant

Dielectric breakdown can occur in all types of capacitors. When dielectric breakdown occurs, current flows from one side of the capacitor through the insulating layer to the other side of the capacitor. Breakdown occurs differently in electrolytic and tantalum capacitors. Dielectric breakdown occurs gradually in electrolytic capacitors whereas dielectric breakdown occurs abruptly in tantalum capacitors. The temperature, operating voltage, electrical overstress (or current transients) are all parameters that accelerate dielectric breakdown. The results of dielectric breakdown are catastrophic, resulting in either a short circuit or open-circuit failure.

The effects of dielectric breakdown in electrolytic capacitors are catastrophic, but the symptoms develop gradually. A model of a real capacitor includes parasitic elements that cause a capacitor to take on very low resistive and inductive properties, shown in Figure 4. These non-ideal characteristics include some equivalent series resistance (ESR), as well as lead inductance and equivalent parallel resistance (EPR).

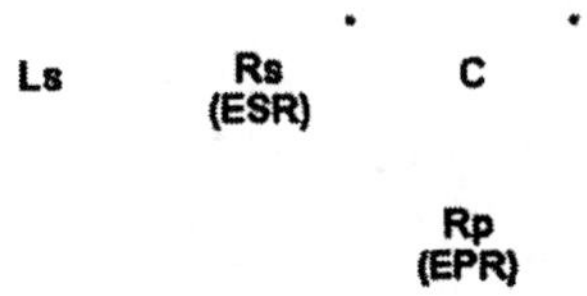

Figure 4: Model of a real capacitor.

As an electrolytic capacitor ages, the electrolytic fluid begins to evaporate drying out the capacitor and causing an increase in ESR. Since the magnitude of ESR is inversely proportional to the maximum amount of current a capacitor can handle, the internal temperature of the capacitor begins to increase. As the temperature and power dissipation begin to increase, the likelihood of dielectric breakdown occurring increases. As a result of dielectric breakdown, the capacitor enters a short circuit failure mode.

Tantalum capacitors possess a lower failure rate when compared to electrolytic capacitors when operated at the same rated conditions [[viii]]. The low failure rate of tantalum capacitors is due in part to the annealing process that enables 'self healing'. As faults develop in the dielectric, increased current flow generates heat that promotes the formation of a manganese trioxide patch as shown in figures 5, 6 and 7 respectively.

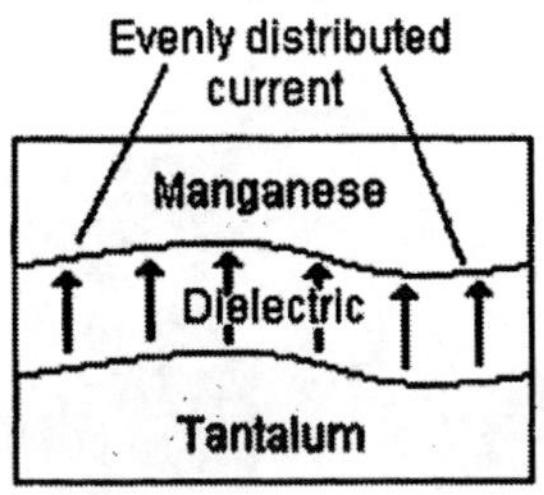

Figure 5: Cross section of a healthy capacitor

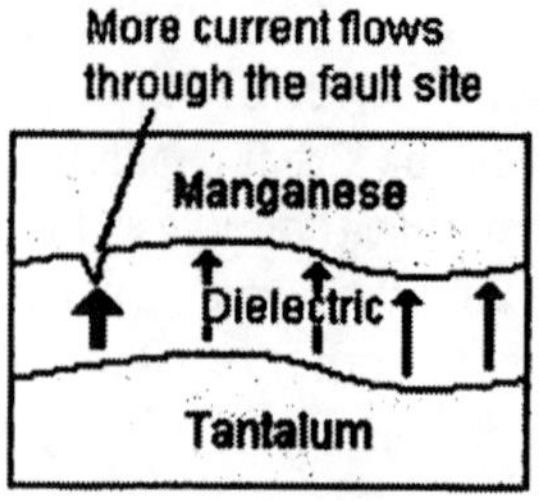

Figure 6: Cross Section of a Degraded Dielectric

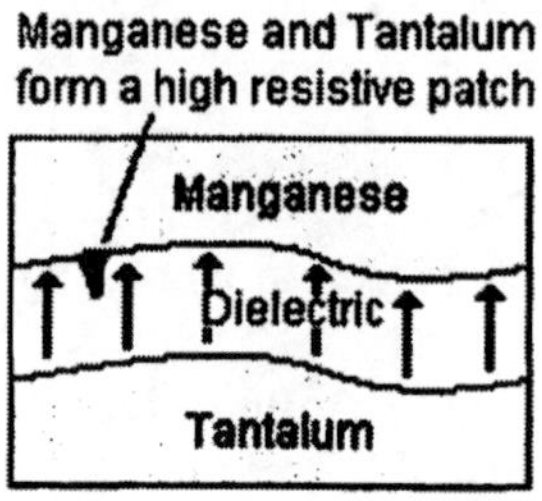

Figure 7: Cross section of degraded Manganese Dioxide

1. **Accelerated Failure Testing:** The SMPS fault to failure progression models and diagnostic features for incipient fault detection were verified through accelerated failure tests of commercially available computer power supplies. To generate failures quickly, a load emulator shown in

Figure 8 was designed and built subject the test specimens to extreme electrical and thermal stress.

The load emulator successfully generated multiple switching transistor and diode failures over the course of three months of testing. While tests designed to generate capacitor failures were performed, no catastrophic capacitor failures occurred.

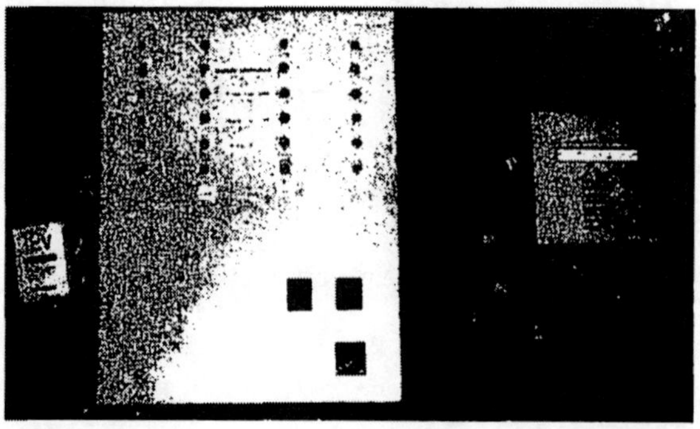

Figure 8: Outside View of the Load Emulator

The bipolar junction transistors in the test supplies exhibited both thermal runaway and contact migration induced failures. As thermal runaway occurs, the temperature of the switching transistor begins to increase exponentially until the semiconductor becomes permanently conductive and the transistor exhibits a "short circuit" failure mode. This failure mode was observed in the lab by attaching a thermocouple to a power transistor in a SMPS. According to the data that was obtained in the lab, thermal runaway began 67 minutes before failure.

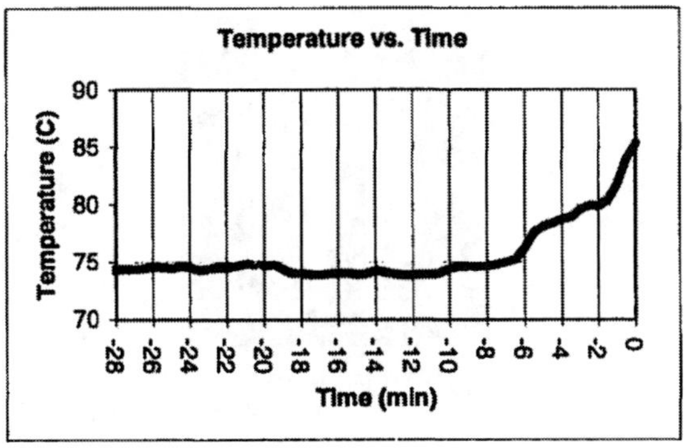

Figure 9: Transistor Temperature vs. Time During Thermal Runaway

The switching transistors also exhibited failures which are believed to be the result of contact migration. Electrical overstress applied by the load emulator causes metal from contacts to diffuse into the semiconductor. Eventually a junction spike occurs, shorting the two regions together causing a short circuit that allows a high current to flow through the device.

During these accelerated failure tests, the input and output power was monitored for transients that indicate the onset of damage. This is best illustrated in a plot the ratio of output power to input power or efficiency vs. time as shown in Figure 10. An electrical overstress event causes the efficiency to instantaneously decrease by almost 4% around 1240 minutes into the test. The transistor failed 160 minutes later.

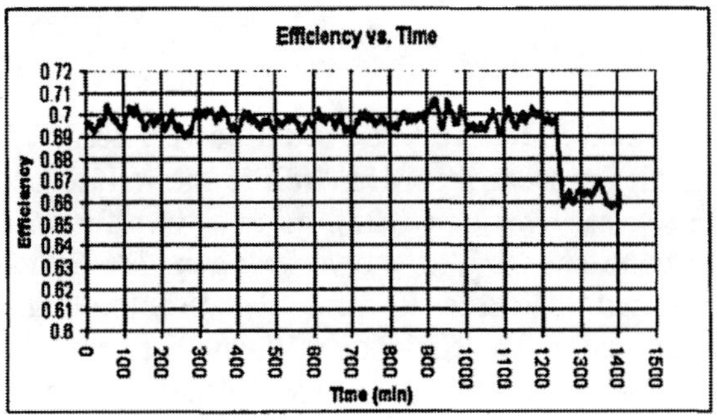

Figure 10: Efficiency vs. Time During BJT Failure

Destructive examination of the transistor revealed a bond wire failure as shown in Figure 11. These light gauge wires that connect the semiconductor to the external solder connections can fail when subjected to a large current.

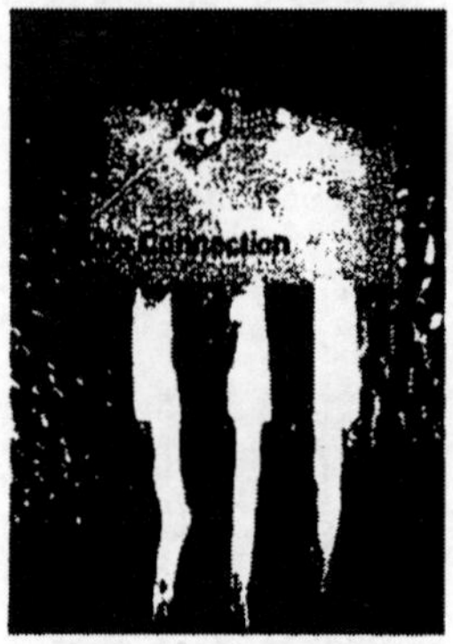

Figure 11: Failed bond wire

The electrical stress applied by the load emulator also induced a rectifier diode failure. The diode is believed to have failed due to from junction spiking as a result of electro migration. Figure 12 shows x-rays of a healthy (S10C40C) die and the damaged die. The circled area indicates damage to the silicon from a short circuit across a PN junction. This short was most likely caused by a electro migration induced spiking.

Healthy DIE Damaged DIE

Figure 12: Xray of a healthy and damaged DIE from a S10C40C Shockley Rectifier

During the same accelerated failure test, the input power, output power, output voltage, output current and efficiency were recorded. Before the failure occurred, the diode was subjected to severe electrical overstress 1250 minutes into the test. After the event, the regulation of the 5V output changed, as shown in Figure 13. Although the power supply was still functioning, it had difficulties keeping the 5V output constant, which is unique to a diode failure. Additionally, the efficiency of the power supply dropped by an average of 5% after the electrical overstress event occurred. The drop in efficiency is illustrated in Figure 14.

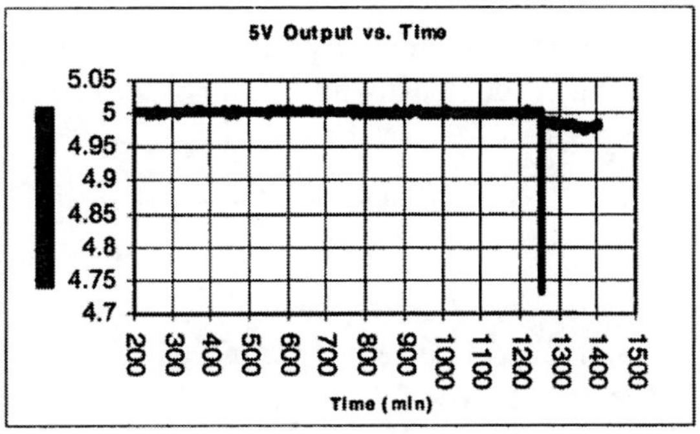

Figure 13: 5V Output Voltage vs. Time during the Accelerated Failure Test

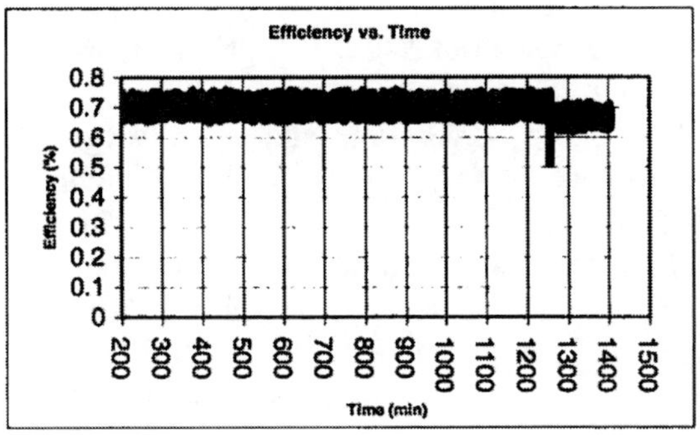

Figure 14: Efficiency vs. Time during the Accelerated Failure Test

Indications of diode degradation are similar to the symptoms of BJT transistor degradation with one important difference. During the late stages of diode degradation, the power supply has difficulty keeping the output voltage regulated at a constant DC level. In other words, when a rectifying diode becomes damaged, the output voltage deviates from its rated value, shown in Figure 13. Therefore, as the rectifying diode accumulates damage, the following symptoms occur:

Conclusion: An integrated approach to switching mode power supply health management that implements techniques from engineering disciplines including statistical reliability modeling, damage accumulation models, physics of failure modeling, and sensor-based condition monitoring using automated reasoning algorithms has been presented. These techniques are focused on the three types of switching mode power supply components (switching transistors, filter capacitors, and rectifier diodes) that reliability studies of have shown cause the majority of failures. Novel features extracted from sensed parameters such as temperature, power quality, and efficiency were analyzed using advanced fault detection and damage accumulation algorithms. Using model-based assessments in the absence of fault indications, and updating the model-based assessments with sensed information when it becomes available provides health state awareness at any point in time. Intelligent fusion of this diagnostic information with historical component reliability statistics provides a robust health state awareness as the basis for accurate prognostic predictions. The diagnostic techniques and prognostic models have been demonstrated through accelerated failure testing of switching mode power supplies.

References:

[1] A Lahyani, P. Venet, G. Grellet, and P. Vivegre, "Failure Prediction of Electrolytic Capacitors During Operation of a Switchmode Power Supply, " IEEE Trans on Power Electronics, vol. 13, NO. 6, pp. 1199-1200, Nov. 1998

[2] Tyco Electronics, Application Notes for Reliability Concepts for Electronic Power Supplies

[3] Born, Frank, Boenning, Robert, Marginal Checking – A Technique to Detect Incipient Failures, IEEE, 1989

[4] Texas Instruments, Application Note for Off-Line SMPS Failure Modes PWM Switchers and DC-DC Converters

[5] A. Lahyani, P. Venet, G. Grellet, and P. Viverge, "Failure Prediction of Electrolytic Capacitors During Operation of a Switchmode Power Supply", in EPE'98 Proc, 1998, vol. 13.

[6] Lambda Inc., "Linear vs Switch Mode Power Supplies"

[7] Amerasekera, A. E., Farid, N. N., "Failure Mechanisms in Semiconductor Devices", 2nd Ed., Chichester, 1997

[8] Kemet, Application Notes for Tantalum Capacitors, p. 80.

A DOMAIN INDEPENDENT DATA MINING METHODOLOGY FOR PROGNOSTICS

Sylvain Létourneau Chunsheng Yang Chris Drummond
Elizabeth Scarlett Julio Valdes Marvin Zaluski

Institute for Information Technology
National Research Council of Canada, Ottawa
{*firstname.lastname*}@nrc-cnrc.gc.ca

Abstract: Modern operation of complex systems such as trains and aircraft generates vast amounts of data. This data can be used to help predict component failures which may lead to considerable savings, reduce the number of delays, increase the overall throughput of the organization, and augment safety. Many data mining algorithms, such as neural networks, decision trees, and support vector machines, exist to learn models from vast amounts of data but their application to real world operational data from systems such as aircraft and trains is very challenging. For successful prognostics, several difficulties need to be carefully addressed including data selection, data fusion, data labeling, model integration, and model evaluation. This paper explains these issues and presents a methodology that we have developed to address them in a systematic manner. The paper discusses the application of the methodology to the rail and aerospace industries and highlights open problems.

Key words: Data mining; aircraft health monitoring; train wheel monitoring; component failure prediction.

1 Introduction

An adequate health management system for complex machinery requires diagnostic, prognostic, and repair management technologies. Although significant research is still going on in all three aspects, prognostic technologies lag behind the other two particularly in the maturity of the technologies and adoption by operators of complex machinery. Over the last few years, we have seen a number of researchers working on improving prognostics by incorporating techniques from Artificial Intelligence and Data Mining. In this paper, we will push this idea much further by proposing the construction of prognostic models using only data mining techniques, applied to historical data, and a minimal amount of engineering knowledge. The result is a domain independent methodology to construct prognostic models. These models are capable of predicting failures of specific components of complex systems such as aircraft and trains.

The goal of our methodology is to generate models to predict when to replace a component. These models must accurately recognize particular data patterns that indicate upcoming component problems. They must also be able to recognize problems within a reasonable period of time prior to the actual occurrence of the failure. For most components, a period of one day to three weeks in advance is appropriate. For components that are very expensive or difficult to obtain, an earlier warning would be preferable.

The proposed methodology is entirely based on data collected during the normal operation of the machinery. There are two kinds of data: maintenance data, describing repair actions carried out by the maintenance staff, and sensor data, recording values acquired from sensors monitoring system parameters. No particular domain knowledge is required. So as long as these two kinds of data are available, the methodology is applicable to any complex system. The information in the maintenance data is fairly consistent across applications. For each maintenance action, we have: the date of the action, the equipment identifier, the identifying numbers for parts installed and removed, the employee name, and typically a textual description of the work performed. The characteristics of the sensor data depend on the machinery considered. For instance, an Airbus A320 generates up to 19 types of sensor measurement reports corresponding to different stages of operation (e.g., engine start, takeoff, cruise) and each one contains between 20 and 150 data items (numeric and symbolic). On the other hand, the WILD (Wheel Impact Load Detector) system, which monitors the impact of train wheels on the rails, collects only a few measurements (e.g.; dynamic impact, train speed, train direction) for each wheel of the train.

There are two reasons to replace a component: as part of regular maintenance (imposed by regulations or operator policy) or because the part's condition has deteriorated and might fail. Only the latter is relevant for predictive models, as maintenance staff are already aware of regular maintenance requirements. Of course, a system cannot predict replacement for components without relevant data. Since the data available in our projects are related to aircraft engines and train wheels, our focus is on models that predict problems with components from these systems. It is also important to notice that we can only address components for which we have already observed a minimal number of failures.

Our methodology does not attempt to capture all possible failures. In particular, we cannot identify component failures due to maintenance or design problems. We do not have sufficient data for such problems; fortunately, they represent a small percentage of all component failures. The rest of the paper begins by describing the methodology, it then discussed two case studies, and then goes on to highlight open problems. Létourneau et al. [5] offers an alternative description of the methodology which focuses on predicting aircraft component replacement.

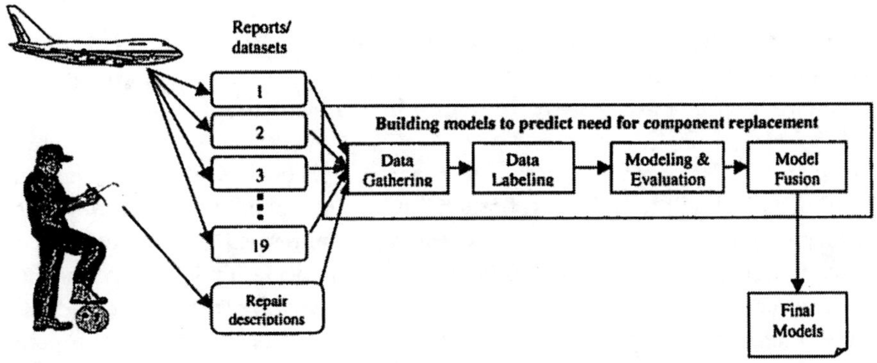

Figure 1: The 4 steps process to build models for component replacements.

2 The Methodology

Other research has used data mining to infer models from historical data. Our methodology builds on this by addressing issues not identified in earlier work. As illustrated in Fig 1, we stress four steps of particular importance in this type of application. They are data gathering, data labeling, model evaluation, and model fusion. The following sections explore these issues in turn and propose corresponding solutions.

2.1 Data Gathering

Most data mining algorithms require as input a data set containing examples consisting of vectors of attribute values. Modern machinery often generates many such data sets. The Airbus A320 generates up to 19 different data sets reporting the status of the aircraft in different phases of operation. The number of examples in each data set varies considerably. Our first problem is to select the data set(s) to use to build models for a particular component. Domain experts, if available, may select the most suitable data set(s) for the component of interest. Otherwise, we rely on data set descriptions and include all data set(s) that appears to be somehow related as to limit the risk of disregarding potentially useful information.

Not only must we select a data set, we must also select a subset of instances to use in the analysis. The data sets are typically so large, it is inefficient to build models using all instances. Simple solutions, such as random sampling, are also inappropriate. To build the desired predictive models, we must be much more focused. We base our analysis on data generated around each occurrence of component replacement. We first identify when replacement occurred and then retrieve instances around the time of this occurrence.

Retrieving replacement cases:

The maintenance databases contain reports on the repair and replacement of many different components. We might hope a simple query, matching all reports with the desired part number, would retrieve all relevant replacement cases. Unfortunately, retrieving occurences of failures may be turned out to be quite difficult. The difficulties come from errors in maintenance reports and the need to process free form texts.

We see many kinds of problems with maintenance reports. Some reports mention a component although it didn't fail. Some reports mention a component that failed but not in a way that we are trying to predict. Sometimes the part number entered is incorrect. Sometimes alternative part numbers used. Sometimes the right information is entered but in the wrong field. Sometimes the right part number is used but the wrong component is identified, there are often multiple components of the same type within the machinery. If a person studies the report it is usually possible to determine what has been replaced and why. But because the maintenance databases are typically very large, manual analysis is not possible. Since the types of error differ between applications, customized solutions are required. Section 3 presents the two solutions we have developed to retrieve failure information for train wheels and aircraft components.

Selecting instances: Once we have the date and the part identifier for each component replacement, we retrieve the relevant instances from the selected sensor data sets. For each data set and replacement, we retrieve the data obtained m days prior to the replacement and n days after. The numbers m and n depend on the data sets and the component; we usually select m so that we have at least 200 instances available for learning. For example, when we have an average of two reports per machine per day, m will be set to 100 or above. For n, we simply set it as $n = 0.15 * m$. The selected values for m and n influence how we set the k parameter (defined in Section 2.2).

Finally, we add two new attributes to the initial sensor data: the time between sensor measurement and the actual replacement ('time from failure'), and a tag associating each instance with a specific replacement case ('problem ID'). We combine all instances from a selected data set to create the data set that we'll use to build the predictive model. The number of output data sets is equal to the number of data sets selected as potentially useful for the target component.

2.2 Data Labeling

Predicting the need for component replacement can be viewed as a classification task with two classes: replace component or don't replace component. The learned model will classify each new report as one of these two classes. Supervised learning cannot be directly applied to the sensor data, however, because it does not contain a class attribute. Our solution is automatic data labeling: we add a preprocessing step that computes the class value of all instances used in the analysis.

We set the class to 1 for all instances in a target window, defined as k days before component replacement. We set the class attribute to 0 for all instances outside the window. Obviously, k has to be smaller than m. We have used days to define the window rather than usage information (e.g., mileage, hours of operation, or cycles - pairs of takeoffs and landings) mainly because such information is generally not provided with the sensor data. In some applications, users have specifically asked for predictions in days. From a data analysis viewpoint, this choice is not ideal as repairs are more related to usage than to time. However, we experimented with cycles in an aerospace application and the results were quite similar to those using days. This is probably because the commercial aircraft considered in this study all had similar usage patterns.

For each component, we perform several experiments to find the best value for k. When designing the experiments, we consider the following factors:

- The target window for predicting component replacement. Different components will have different sized windows.
- The proportion of positive and negative instances. Without a significant proportion of positive instances, data mining approaches often have difficulties with model learning. To ensure that at least 10% of the examples are positive, we use the constraint $k > 0.1 * (m + n)$
- Pattern characteristics. To facilitate the learning of the models, we try to synchronize labeling with qualitative changes we observe in the data.

2.3 Modeling and Evaluating Models

For building models, we use standard data mining algorithms but we propose a novel method to evaluate models since current ones are not adequate for this kind of application. We build models with techniques such as decision trees, nearest neighbor, naive Bayes, rough sets, regression, and neural networks. Depending on the algorithm, extra preprocessing steps may be needed: selecting the most suitable subset of attributes, normalizing the attribute values, creating new features from the original ones, and discretizing continuous attributes. Techniques to perform these tasks can be drawn from different areas of research such as machine learning, statistics, pattern recognition, and data mining.

In practical applications such as ours, results must be properly evaluated. The evaluations must fairly estimate the model's performance on future data and account for domain specific criteria. There are several approaches for computing the expected performance, including hold-out validation, cross-validation, and bootstrapping. Unfortunately, none of these approaches is suitable for our application as they rely on random sampling to select instances from the population. Random sampling assumes that the instances are independent. In our application, the assumption of instance independence is simply not viable. Because of variations in system construction, operation, and maintenance, any two instances from a given system have a greater dependence than any two instances from different systems. If we were to use data from the same system for training and for validation, success would be much more likely than if we used data from a different system for validation.

Similar reasoning applies to component replacement data: two instances from the same problem will be more related than two instances from different problems. Given that the number of component replacements is significantly less than the total number of instances, random sampling would likely generate training and validation sets containing data from the same problem. The end result would be an overly optimistic evaluation of the model's performance. We solve this by splitting the data so that training and validation instances come from different subgroups of component replacement events. This is achieved by using the 'Problem ID' attribute added during data gathering. We obtain a more robust estimate by adapting the cross-validation method to account for the subgroups as follows:

1. Split the data into batches using the 'Problem ID' (one for each failure case).
2. Keep one batch for validation and use all others for training. Repeat this step so that each batch is used once for testing.
3. Average the validation results from each of the batches.

Evaluation function: At the heart of our modified cross-validation method is a measure of classifier performance. In machine learning research, performance is often summarized by either error-rate or accuracy. The error-rate is defined as the expected probability of misclassification: the number of classification errors over the total number of test instances. The accuracy is 1 minus the error-rate.

Because some errors can be more costly than others, it's often desirable to minimize the misclassification cost rather than the error-rate. The ROC method[7] is very popular in practical applications since it allows the user to evaluate models independent of the particular class frequency. Other possible metrics are precision and recall commonly used in the information-retrieval community[6]. Recall is usually defined as the ratio of relevant documents retrieved for a given query over the number of relevant database documents; precision is the ratio of relevant documents retrieved over the total number of documents retrieved.

Unfortunately, these metrics fail to capture two important aspects of our application. The first aspect is that the usefulness of a prediction is function of the time between the prediction and the actual replacement. Warning too early about a potential failure leads to non-optimal component use; warning too late makes proper repair planning difficult. We need an evaluation method that takes alert timeliness into account. The second aspect relates to coverage of potential failures. Because the learned model classifies each report into one of two categories (replace component; don't replace component), a model might generate several alerts before the component is actually replaced. More alerts suggests a higher confidence in the prediction. However, we clearly prefer a model that generates at least one alert for most component failures over one that generates many alerts for just a few failures. That is, the model's coverage is very important to minimizing unexpected failures. Given this, we need an overall scoring metric that considers alert distribution over the various failure cases. The next section presents a reward function to take into account first aspect (timeliness of the alerts) while the following section introduces a new scoring metric that addresses the second aspect (coverage of failures).

Reward function: In our methodology, we define a reward function for predicting the correct instance outcome. For each target component, the reward for predicting a positive

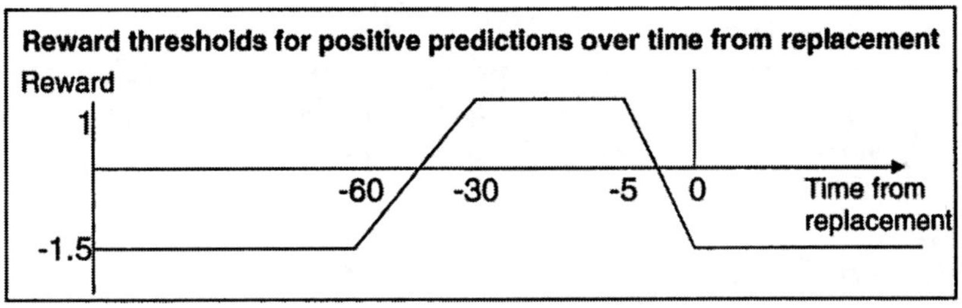

Figure 2: Example of a reward function for prediction of positive instances.

instance is based on the number of days between instance generation and the actual replacement. Figure 2 shows a graph of this function. The maximum gain is obtained when the model predicts the need for replacement five to thirty days prior to the component's replacement. Outside this target period, predicting a need for replacement can lead to a negative reward threshold, as such a prediction corresponds to misleading advice. Accordingly, false-positive predictions (predictions of a failure when there is no failure) are penalized by a reward of -1.5 in comparison to a 1.0 reward for true-positive predictions (predictions of failure when there is a failure).

Although the reward function's target period is not necessarily the same as the labeling period, the two are related. In general, we determine the reward function's target period according to the user requirements, and then fix the labeling period. For example, if the user says that predictions for a given component should be within one to three weeks in advance, we set the reward function's target period for between -7 to -21 days. We then experiment with various labeling periods around this target period and use the one that offers the best result.

The reward function in Figure 2 follows a piecewise, linear model. Such a model is convenient because it is comprehensible to the maintenance staff and is sufficiently complex to capture the relevant information. There are several ways to improve the reward function's precision. One possibility is to use higher-order polynomials instead of straight lines. Another possibility is to try to smooth the overall function. However, according to domain experts, increased complexity is rarely needed.

Scoring metric: Our reward function accounts for alert timeliness; to evaluate model coverage we must look at alert distribution over the different failure cases. The overall performance metric we propose to evaluate a model is:

$$score = (\sum_{i=1}^{p} score_i) * (NbrDetected/NbrOfCases)^{SignOfSumOfScores} \tag{1}$$

where p is the number of positive predictions made by the model on validation instances during application of the evaluation method, $score_i$ is the score from the reward function for the ith instance classified as positive, $NbrDectected$ is the number of replacement cases for which we have at least 1 positive prediction in the target interval (e.g., -30 to -5 days), $NbrOfCases$ is the total number of replacement occurrences we have for the given component, and $SignOfSumOfScores$ is the sign of the first term in the expression. The term

$(NbrDetected/NbrOfCases)^{SignOfSumOfScores}$ is introduced in the evaluation function to favor models that optimize the recall (The term $(NbrDetected/NbrOfCases)^{SignOfSumOfScores}$ is set to zero when $NbrDetected = 0$ and $SignOfSumOfScores$ is negative). The value of $\sum_{i=1}^{p} score_i$ makes the final score sensitive to precision of the model.

2.4 Model Fusion

We use model fusion for two reasons. First, model fusion can be used to combine models derived from different data sets. When more than one data set is relevant for a given component then we build models independently for each data set and use model fusion to combine predictions from the various models. Second, we can apply model fusion to help improve performance even if there is a single data set. We can learn various models using various techniques or parameter settings and combine them to obtain better performance than using any single model.

Bagging and boosting are two popular techniques to combine models but they are only applicable when there is a single data set and one kind of model (a single learning algorithm). For heterogeneous models or multiple data sets, we apply methods based on voting or stacking strategy. As we explain in the discussion of the WILDMiner project (next section), we successfully applied meta-learning as a way to combine heterogeneous models.

3 Case Studies

In the following we discuss our experience in applying the proposed methodology to two maintenance applications for the railways and for the airlines. Our description focuses on data gathering since customized solutions where required. In the case of the train wheel failure predictions, we also elaborate on model fusion because of its positive impact on results. All other steps were applied as described above.

3.1 The WILDMiner project

Wheel failures on trains represent a significant proportion of the operational cost for railways. For instance, Canadian railroads spend about $65M/year to repair and replace freight car wheels. Moreover, when not detected and fixed in a timely fashion, wheel failures may cause rail breakage and even train derailment. To avoid such events, many railroads have invested in the WILD technology. In Canada alone, there are more than twenty WILD systems installed at strategic locations on main line tracks.

The WILD systems measure the vertical force of each passing wheel over the site. Problem wheels typically produce higher WILD measurements due to an increase in the impact on the rail. The measurements are sent in real-time to a central site where fleet supervisors decide whether or not to replace/repair a given wheel. Depending on the actual WILD measurement, recommended actions range from scheduling a wheel inspection to stopping the train until the wheel is repaired. Often, the train is allowed to continue at a reduced speed to the nearest shop or siding where the offending car is uncoupled from the train. The current strategy is reactive and leads to many changes in the route or speed of the train during operation. This is problematic because it introduces delays, disrupts the overall schedule, degrades customer service, and reduces the railway's throughput capacity.

To minimize the actual number of disruptions during operation, we need a pro-active maintenance strategy based on a system that monitors the data in real-time and identifies the deteriorating wheels before they force speed reduction or train stoppage. This strategy requires models to predict wheel failures, which we are developing using the methodology introduced above.

We retrieve information about past failures from the maintenance data and then gather relevant sensor data. In this application, the maintenance database is relatively clean because the staff use a menu driven system to enter most of the information. The options selected through the menus are translated into codes in the maintenance database. To retrieve occurrences of wheel replacements, we simply search for specific codes. The only difficulty is in identifying the exact location of the wheel that has been replaced. The cars we worked on have 12 axles for a total of 24 wheels. To identify a particular wheel location, the maintenance staff enters in free form text the number of the axle (1 to 12) and the side of the faulty wheel (right or left). Unfortunately, we found many inconsistencies and errors in this text. For instance, technicians use many symbols to refer to the axle number 10 such as: 'a', '10', 'x', '0x'. The same problem also exists for other axles above 10. Moreover, we found that the staff sometime enters the wrong axle number (e.g., 6 or 8 instead of 7) or wrong side (left instead of right). These errors can negatively influence the modeling process by making it construct failure models that fit non-failing wheel data. We try to deal with errors in wheel location through extensive manual validation. When the ambiguity is too high, we generally ignore the case and ensure that potentially related sensor data is not included in the modeling process.

The sensor data for this application is as follow. Each time a train goes over a WILD system, sensors installed along the tracks measure the dynamic impact of each wheel on the rail. The system then sends these measurements to a central system along with contextual information such as the ID of the car, the speed of the train, and its direction. These information are then stored in a single table. We create the data set to build the predictive models by extracting from this potentially huge table all the data around occurrences of wheel failure replacements. We did most of our experiments with $m = 100$ and $n = 15$.

To build the models, we experimented with various learning and feature extraction techniques. The scores obtained from various learning algorithms were all around 290 (for the dataset considered, score values vary from -9500 to 1200) with a decent coverage but also a relatively high rate of false alerts. We applied meta-learning to perform model fusion. This process involves learning a meta-model that predicts failures based on predictions from a number of low-level models directly relying on the sensor data. We obtained our best meta-model using the J48 algorithm implemented in WEKA[8]. In addition to a global score of 650, this model has both a good coverage and very low false positive errors. The four heterogeneous low-level models used in this experiment were learned with the J48 and

Naive Bayes classifiers (two versions for each classifier). We consider that this model could be deployed with significant financial benefits. However, as we discussed in Section 4, a number of issues could be addressed to further improve the performance.

3.2 The ADAM Project

Our methodology was developed for ADAM[1], a collaborative project with a Canadian airline, and then generalized to other domains. The aim of this project is to develop a monitoring system that predicts aircraft component failures and identifies abnormal system behavior. The models are learned from operational and maintenance data for more than 75 aircraft: about half of these are Airbus A320 and the other half are Airbus A319. For the A320, we have accumulated more than 10 years of historical data.

The data gathering stage of the methodology was particularly difficult for this project. Most of the data stored in the maintenance database was input by hand resulting in great deal of variability in the quality of the reports. Often the field for the part-removed identifier is empty or contains erroneous information. The correct part identifier may be written in the textual description instead of the part-removed field. In fact, the free text portion of the reports were the often most reliable indicators of a part being replaced.

Therefore, we must process the textual descriptions to find all occurrences of component replacement. Automatically processing such descriptions is very difficult, as the maintenance staff uses informal language, which often includes abbreviations and acronyms, major grammatical and syntactical problems, typing mistakes, and inconsistent use of abbreviations among reports[3]. We extract key phrases by analyzing the textual descriptions retrieved. We tried to automate the keyword extraction process using a keyword-generation system but we later realized that it is often more efficient to perform this step manually.

We use an information retrieval based approach to solve this problem, called query reformulation. We start with the part number in part-removed field as the basic query. We then explore the reports that match this query and identify new values in different fields that seem associated with this part being replaced. Text fields are often the source of these new values. For example, if we find the key phrases 'starter motor', 'starter', and '49400126', then the new query will search for any maintenance report that contains the part identifier in the appropriate field or any of the key phrases in the textual descriptions. We typically find new terms in these new reports and they are added to the query. New terms may also be added to make existing terms more specific and thus reduce the number of records they match. We continue reformulating the query until we have most of the relevant reports returned. To make sure of this we over-generalize the query and then manually filter reports that do not deal with part replacement.

Our approach provides no guarantee that all related reports will be selected. Its success depends on the completeness of the final queries we use to retrieve the reports. Specialists may be able to estimate the number of failures for the component of interest. In such case, we can evaluate our coverage by comparing specialists' estimate to the number of replacements found. Although this process still requires much human intervention, the number of reports that require manual validation is very small in comparison with the overall number in the database. So our approach can be viewed as reducing manual analysis by automatically preselecting potentially relevant reports. We have also looked at ways to semi-automate this process to further reduce the work required by the human [2].

In Létourneau et al. [5], we report results from a fully automated large scale experiment involving components from the APU (auxiliary power unit) and the main engines. Since then, we performed extensive evaluations and studies focusing on components of interest to the airline involved in the project. The results are highly positive on a few components such as the APU starter motors. For instance, the models developed for the A320 APU starter can predict around 95% of failures between 1 to 21 days in advance without producing false alerts. The models for the A319 APU starter are also very good but with a small ratio of false alerts, which we explain by the lack of data for that fleet. For other components such as the various valves used to fine tune engine performance, we have been unable to build accurate predictive models. The next section elaborates on issues that we believe need to be addressed to overcome current difficulties.

4 Open Issues

The methodology presented in this paper is based on the steps needed to better apply data mining to predicting failures in complex equipment but many open problems remain. This section discusses some of these open problems and links them to the methodology and applications discussed above.

Improvements to quality of maintenance data: To gather the right data for the analysis, we need to be able to identify all relevant failure events. Retrieving these events is quite difficult in practice due to data consistency and reliability issues. This could be improved by enhancing the tools available for data entry. For instance, in our aerospace application, the introduction of tools to validate the information on parts removed and installed would be of great help. Additionally, personal should understand that our methodology is dependent on the quality of the maintenance practices. If components are replaced without good reasons, the models are likely to recommend replacement of healthy components. Making maintenance personnel aware of the detrimental effects of inaccurate maintenance data on prognostics and health management should help motivate them to increase data quality.

Improvements to quality of sensor data: Most of the sensors in current commercial aircraft have been installed for diagnostic purposes. Prognostics may require different or additional information. An issue that needs to be accounted for in the design of future complex equipment. Current projects like the Joint Strike Fighter (JSF) which clearly includes prognostics and health management as a core functionality are particularly encouraging in that respect. In addition to integrating the right sensors, we also need to ensure their reliability over the life of the equipment. In practice, we observe that organizations invest considerably in data collection and storing technologies but they rarely implement mechanisms to ensure that the data they get is reliable. For instance, in the WILDMiner project we have seen WILD systems that appear highly uncalibrated. Similarly, some of the aircraft sensor values we get are questionable. A complete prognostics and health management system should definitely include data validation techniques.

Data normalization to control variance: Inevitably, when we mix sensor data from complex equipments operating in heterogeneous environments we obtain very high variances. Such deviations may prevent learning techniques from discovering patterns of interest in the data. This is particularly true when there is limited data (few examples of failures) or when a failing component has only marginal effects on sensor measurements. We believe that this explains why we have difficulties building accurate predictive models for some aircraft engine components.

To resolve this problem, we need to integrate a number of complementary techniques. Domain engineering models can help to account for part of the variance associated with equipment characteristics and operational conditions. Configuration management systems can help keeping track of physical differences between individuals in a given fleet. Finally, techniques for contextual normalization [4] and data reduction may be used to further reduce the variance in the data that cannot be explained by component failures.

Methods to perform cost/benefit analysis: The evaluation method presented in this paper accounts for important technical constraints such as failure coverage and alert timeliness. However, this is not sufficient to assess the potential business value of models to predict failures of a given component. Ideally, we would like an extension of the current method that takes into accounts the various costs (undetected failures, early replacement, false alerts, implementation of the models) and returns the expected gain in dollars. However, this may not be practical as organizations may be unable to adequately evaluate these costs. We should therefore also investigate other alternatives.

5 Conclusion

This paper generalizes a methodology we have developed to build predictive models for complex equipment using data mining and historical data. The methodology focuses on the use of classification techniques to build the predictive models. We describe how to select relevant historical data and a method for automatic labeling. The methodology includes a novel method to evaluate models. After discussing the application of the methodology to predict train wheel failures and aircraft component failures, the paper describes a number of open problems. The case studies demonstrate the applicability of the proposed methodology to various domains. In particular, we observe that most of the steps are applied without any customization. Our current work in this area tries to address the open problems discussed.

References

[1] ADAM project at the National Research Council of Canada. Website. `http://iit-iti.nrc-cnrc.gc.ca/projects-projets/adam_e.html`.

[2] Drummond, C. (2004). Iterative semi-supervised learning: Helping the user to find the right records. In *Proceedings of the International Conference on Industrial & Engineering Applications of Artificial Intelligence & Expert Systems*, (pp. 1239–1240).

[3] Farley, B. (1999). From free-text repair action messages to automated case generation. In *Proceedings of the AAAI 1999 Spring Symposium on AI in Equipment Maintenance and Support*, (pp. 109–118)., Stanford University, Stanford, California, U.S.A.

[4] Létourneau, S., Famili, A. F., & Matwin, S. (1998). A normalization method for contextual data: Experience from a large-scale application. In *Proceedings of the 10th European Conference on Machine Learning (ECML-98)*, (pp. 49–54).

[5] Létourneau, S., Famili, A. F., & Matwin, S. (1999). Data mining for prediction of aircraft component replacement. *IEEE Intelligent Systems and their Applications*, *14*(6), 59–66.

[6] Makhoul, J., Kubala, F., Schwartz, R., & Weischedel, R. Performance measures for information extraction. In *Proceedings of DARPA Broadcast News Workshop, (Herndon, VA), February 1999.*

[7] Provost, F. & Fawcett, T. (1997). Analysis and visualization of classifier performance: Comparison under imprecise class and cost distributions. In Heckerman, D., Mannila, H., Pregibon, D., & Uthurusamy, R. (Eds.), *Proceedings of the Third International Conference on Knowledge Discovery and Data Mining (KDD-97)*, (pp. 43–48). AAAI Press.

[8] Witten, I. H. & Frank, E. (1999). *Data Mining: Practical Machine Learning Tools and Techniques with Java Implementations.* Morgan Kaufmann.

AN AUTOMATED GOAL SEEKING APPROACH TO MACHINERY PROGNOSTICS AND LIFE EXTENSION

Kenneth Blemel and Carl Stern
Management Sciences, Inc.
Albuquerque, New Mexico, USA
ken_blemel@mgtsciences.com
carl_stern@mgtsciences.com
Eric C. Hansen
Autonomous Maritime Navigation
Combatant Craft Department
Naval Surface Warfare Center Carderock Division
hansenec@nswccd.navy.mil

Abstract:
In this paper we discuss ongoing work in development of an intelligent platform for automated health management and machinery life extension. This work exploits the rich opportunities that are arising from the integration of a new generation of onboard intelligent technologies. These technologies include smart monitoring, prognostic modeling, and intelligent control. We describe two new technologies that our research group is bringing to onboard health management: *teleo-adaptive control* provides adaptive goal-seeking behavior that optimizes mission performance while minimizing stress and mitigating risk. *Dynamic Bayesian models* provide continuous online estimates for equipment condition, performance and stress levels, and evolving fault trajectories. We show how these two technologies can be fruitfully integrated into a single proactive framework called *Teleo Health Management* (THM). THM is a total systems approach to machinery health management and life extension, providing a range of functionality beginning at the level of conventional CBM and extending to machinery life extension through *in situ* stress reduction and risk mitigation.

Keywords: Goal-seeking; health management; intelligent; prognostics; stress reduction

Introduction:
We discuss ongoing work in development of an intelligent platform for automated health management and machinery life extension. This work exploits the rich opportunities arising from the integration of a new generation of onboard intelligent technologies. These technologies include smart monitoring, prognostic modeling, and intelligent control. We describe two new technologies that our research group is bringing to onboard health management: *Teleo-adaptive control* provides adaptive goal-seeking behavior that optimizes mission performance while minimizing stress to equipment and mitigating risk to successful mission execution. *Dynamic Bayesian models* provide continuous online estimates for equipment condition, performance and stress levels, and evolving fault trajectories. We show how these two technologies can be fruitfully integrated into a

single proactive framework called *Teleo Health Management* (THM). The integration of these two technologies into a single platform for automated health management and machinery life extension has resulted in some very important synergies that will be described here.

We believe that systems for automated self-maintenance and life extension represent the future state of the art in health management, a proper successor to CBM and PHM approaches. We describe our own first cut at such a system, the *Teleo Health Management* (THM) framework. THM is based on a total systems approach to machinery health management and life extension including both ground-based maintenance operations, onboard self-maintenance operations, and *in situ* stress and risk reduction. We have identified four levels of potential THM functionality:

Level 1. Condition-based management of the schedule and scope of human maintenance operations;
Level 2. Onboard condition-based management of autonomous self-maintenance operations;
Level 3. Onboard real time control for machinery stress reduction and life extension;
Level 4. Onboard prognostic operations management for machinery life extension.

Current and emerging CBM/PHM technologies for health monitoring and prognostic health management provide support for **Level 1** functionality. Current CBM approaches can be further enhanced by exploiting intelligent technologies for onboard monitoring and situation awareness. Monitoring of indicator parameters (e.g., vibration) and performance parameters (e.g., engine efficiency) can be correlated with situation components such as operational state, stress level, and environmental conditions (temperature, air currents, turbulence), leading to more accurate systems for state-aware health monitoring.

Our research interest in capabilities at **Levels 2 – 4** is partly motivated by the requirements of health management for the current and emerging generations of unmanned vehicles, particularly vehicles that are required to operate without human maintenance for extended periods of time. In the case of unmanned boats, for example, these requirements might include the need for self-maintenance functions such as autonomous oil filter replacement, cleaning of water filters, oil processing, etc. This form of self-maintenance clearly benefits both in efficiency and effectiveness from an autonomous CBM-style condition-based approach.

A natural next step for a system capable of state-aware monitoring, a system capable of detecting, for example, that oil temperature or engine rotation speed is rising rapidly and will soon exceed the limits of normal non-destructive operation, is to anticipate and prevent damage or accelerated wearout before it occurs. This has motivated our interest in **Level 3** technology, technology for intelligent reactive control that is capable of reducing the destructive effects of equipment stress while robustly pursuing mission goals, strategies, and requirements. We describe in the next section a technology that is capable of performing this simultaneous management of performance goals and stress reduction goals, a technology for adaptive task-execution called *teleo-adaptive control.*

Teleo-adaptive control provides rapid reactive stress management based on short time horizons. If prognostic models are also available that allow estimation of remaining useful life relative to projected equipment utilization and stress levels, a **Level 4** approach becomes feasible. A **Level 4** approach involves performance and stress management over longer time horizons. It uses a planning and scheduling algorithm to find a performance schedule that optimizes expected utility with respect to mission goals while managing the risk of potential equipment failure. A schedule that optimizes expected utility in this case is one that handles the equipment's expected remaining useful life as a resource to be allocated amongst mission goals so as to minimize the risk of premature equipment failure. Implementing a **Level 4** long term stress management approach that is effective and that is capable of rapidly adapting to changes in equipment state, environment, or mission goals, presents a set of interesting and formidable challenges falling outside the scope of the current paper.

We turn now to a description of the key enabling technologies of the *Teleo Health Management* framework.

Teleo-adaptive Control:

Teleo-adaptive control is based on the *Teleo-Reactive Control framework* originally developed by Nils Nilsson and his students in the Stanford robotics lab [1]. The term *teleo-reactive* comes from the Greek work *telos* meaning goal. Teleo-Reactive programs are reactive goal-seeking programs that implement linear plans or plans with conditional branching by iteratively sequencing the actions in those plans on a moment to moment basis in accordance with current state of environment and the current state of success in achieving the plan's subgoals. The Teleo-Reactive Control framework is designed to provide robust goal-pursuit in a dynamic partially unpredictable environment where the ability of actions to achieve their goals within a specified window of time is itself uncertain. Teleo-Reactive Control is intended to be used as a middle level of executive control between an agent's top level planning and scheduling layer and its bottom level of primitive behaviors implemented in hardware or with conventional fast closed loop control.

More formally, a teleo-reactive control program consists of a sequence of condition-action pairs:

$$C_o \rightarrow \alpha_0$$
$$C_1 \rightarrow \alpha_1$$
$$\ldots$$
$$C_n \rightarrow \alpha_n$$

At each cycle, the action associated with the highest true condition executes. C_o corresponds to top level goal and α_0 is typically the null action (i.e., do nothing when the highest level goal is achieved). Actions are expected to (eventually) make some condition higher in the tree true. Under this assumption, it is mathematically provable that a TR program will achieve its top level goal.

Insofar as a teleo-reactive program typically contains alternative paths to the same goal, it is also called a *Teleo-Reactive Tree*. In the TR tree in **Figure 1**, the circles, labeled C_i, represent conditions or states and the arrow labels α_i represent actions. When the program starts, execution typically begins near the bottom of the tree, selecting branches based on whether a condition on the left or right is true. As progress is made, execution tends to move up the tree, although some unexpected event may undo this progress and push execution back down the tree. In the case of such an unexpected event, the program will relentlessly re-execute the lower level actions that push performance back up the tree. At the same time, if the environment is already in a favorable state, the program will start opportunistically from a higher level of the tree and skip all the unnecessary actions at the lower levels.

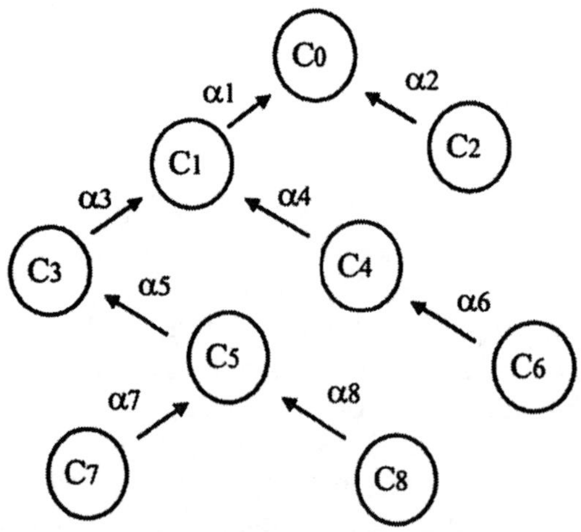

Figure 1. A typical Teleo-Reactive Tree with conditional branching.

Stern, Klein, and others [2, 3] have extended and enhanced Nilsson's original *Teleo-Reactive Control* framework in a number of ways. While Nilsson's original formalism clearly supports symbolic condition variables (as opposed to numeric values from sensor data) in a TR program, our implementation includes a managed framework for *condition monitors*, small real-time classification programs that implement decision rules for determining whether a symbolic condition is true.

While Nilsson's original formalism also supports hierarchically structured TR Programs, programs in which an entire TR program can is treated as a primitive action of a higher level TR program, we have developed a formal approach for aggregating simple behaviors into larger more complex or sophisticated behaviors. This approach is based on the idea that a teleo-reactive program is intended to organize and sequence behaviors rather than simple actions. The intelligent organization of behaviors means that behaviors become active or dormant based on the prioritized challenges or opportunities presented by a situation. For example, the set of behaviors for collision avoidance becomes the dominant behavior in a situation where a collision is imminent whereas vigorous waypoint navigation or other behaviors dominate when no collision threats are imminent.

The most important and critical extensions to the TR framework that we have developed address the need to anticipate emerging situations and to react in a timely fashion, taking into account the intrinsic time delays in many types of control actions. For example, in navigation control, it is often the case that collision avoidance behavior must be initiated well ahead of time based on the anticipation of a collision. This presents a significant challenge for developers of teleo-reactive programs in that a critical element for the success of these programs will be the ability to define effective trigger conditions for transitioning between behaviors, i.e., for transitioning into collision avoidance behavior far enough in advance. Our solution has been to develop an algorithm for deriving effective behavior transition triggers using prognostic Bayesian models that model condition and state trajectories. Using these models, together with Monte Carlo and other simulation frameworks, we have implemented a method for automatically generating effective behavior transition triggers in the form of the condition monitoring programs and decision rules described above.

We now turn to a brief description of our work in prognostic Bayesian modeling.

Prognostics with Dynamic Bayesian Models:
A key enabling technology for our work in prognostic modeling is a new probabilistic modeling framework called *Loopy Logic* originally invented at the University of New Mexico (UNM) [4] and later refined and extended in joint work with Management Sciences (MSI). *Loopy Logic* extends the first order probabilistic logic, characterized by Kersting and de Raedt, [5] with recursion, allowing it to represent recursive time series models such as Hidden Markov Models (HMMs) and Dynamic Bayesian Networks (DBNs). The *Loopy Logic* framework combines this recursive probabilistic logic with an approximate inference algorithm called *Loopy Belief Propagation* (LBP) [6].

The Loopy Belief Propagation algorithm (also called Iterative Belief Propagation) for inference and learning constitute a significant step towards real-time performance of probabilistic models in computationally resource-limited environments. The UNM/MSI implementation of probabilistic learning, based on Loopy Expectation Maximization, has provided the ability to train probabilistic models to recognize dynamic behavioral signatures rapidly enough so that the trained model can be immediately used in the same situation for object re-identification and/or object classification. It has allowed MSI to build applications that interleave inference and learning, so that a rough initial model can be quickly induced and then continuously refined for ongoing improvement in pattern recognition and classification accuracy.

We have used the framework of Loopy Logic to experiment with a large range of probabilistic models. Hidden Markov Models are already widely used in domains such as speech recognition, protein sequence analysis, etc., providing linear time pattern recogni-

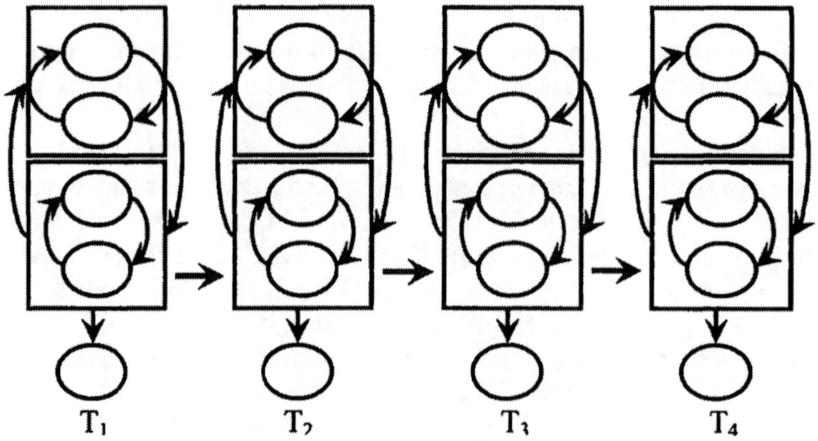

Figure 2. A Loopy Dynamic Bayesian Model

tion. However the simple state representation of HMMs sometimes precludes their application to complex pattern recognition and prediction problems, particularly problems requiring multi-scale pattern recognition and prediction. For this reason we have exploited the recursive representation of Loopy Logic to implement a number of other model types, including Hierarchical Hidden Markov Models, Segmented Semi-Markov Models, and Loopy Dynamic Bayesian Networks. These models give us the capability of analyzing complex patterns with multiple levels of granularity and multiple time-scales (Figure 2). We are able to employ these more powerful and sophisticated probabilistic models without losing near real-time performance because of the linear time scaling of the LBP algorithm for many problems of interest. We have found that Hierarchical Hidden Markov Models, in particular, have worked particularly well in pattern recognition and analysis of vibration data coming from misaligned or cracked rotors, because of the presence of elements in these vibration patterns occurring at two different time scales.

Example Applications:

In a Department of Defense project we are developing use of THM for monitoring and diagnostics during mission execution with rules for goal based self-maintenance. Dynamic Bayesian models are used to anticipate emerging fault conditions and extend survival of propulsion systems and fighting capability. In flight THM will perform stress limiting and self-maintenance control functions. Stress limiting functions include anticipatory vibration damping and pre-emptive thermal control to avoid thermal stress. Self-maintenance functions include engine lubrication analysis, thermal control, and multi-engine vibration management. Self preservation functions include control of thrust and RPM limits to avoid consequential damage. We are preparing for engine test cell demonstrations that will demonstrate THM that anticipates, prepares, reacts, and adapts to accomplish mission objectives while maintaining the sustained health of the engine propulsion system.

In another example, we are developing an extended form of THM that deals with machinery failure prevention for the propulsion systems of unmanned military boats.

Teleo-reactive programs produce significant behavior transitions depending on the context of surveillance, training and warfighting. Learning how to cope with debris is common to all and knowing health status to work the edges of the envelope of performance is important to survival. Requirements for completing the mission and preventing failures often conflict, and priorities change as the battle progresses. Much knowledge engineering work remains to be done to address the complexity of these battlefield situations.

Summary:

We have discussed ongoing work in development of an intelligent platform for automated health management and machinery life extension. This work exploits the rich opportunities that are arising from the integration of a new generation of onboard intelligent technologies. We describe two new technologies that our research group is bringing to onboard health management: *teleo-adaptive control* provides adaptive goal-seeking behavior that optimizes mission performance while minimizing stress and mitigating risk. *Dynamic Bayesian models* provide continuous online estimates for equipment condition, performance and stress levels, and evolving fault trajectories. We show how these two technologies can be fruitfully integrated into a single proactive framework called *Teleo Health Management* (THM). THM is a total systems approach to machinery health management and life extension, providing a range of functionality beginning at the level of conventional CBM and extending to machinery life extension through *in situ* stress reduction and risk mitigation.

References:

[1] Nilsson, N.J., Teleo-Reactive Programs for Agent Control. Journal of Artificial Intelligence Research, 1, 1994, 139-158.

[2] Stern, C., Olsson, E., Kroupa, M., Westervelt, R., Luger, G., Klein, W. A Control System for Accelerator Tuning Combining Adaptive Plan Execution with Online Learning, Proceedings of ICALEPCS'97, 1997.

[3] Klein, W., Stern, C., Luger, G., Olsson, E.. An Intelligent Control Architecture for Accelerator Beamline Tuning, Proceedings of IAAI 97, 1997.

[4] Pless, Dan and Luger, George F. "Toward General Analysis of Recursive Probability Models", Published in the *Proceedings of the Uncertainty in Artificial Intelligence Conference-2001*, San Francisco: Morgan Kaufmann, 2001.

[5] Kersting, K., De Raedt, Luc, "Bayesian Logic Programs", Proceedings of the Work-in-Progress Track at the 10th *International Conference on Inductive Logic Programming*, 2000.

[6] Murphy, K., Weiss, Y., and Jordan, M., Loopy Belief Propagation for Approximate Inference: An Empirical Study, *Proceedings of Uncertainty in AI*, 1999.

PREVENTING FAILURES BY MINING MAINTENANCE LOGS WITH CASE-BASED REASONING

Mark Devaney, Ashwin Ram
Enkia Corporation
[markd@enkia.com, ashwin@enkia.com]
Hai Qiu, Jay Lee
Center for Intelligent Maintenance Systems
University of Wisconsin-Milwaukee
[haiqiu@uwm.edu, jaylee@uwm.edu]

***Abstract**: The project integrates work in natural language processing, machine learning, and the semantic web, bringing together these diverse disciplines in a novel way to address a real problem. The objective is to extract and categorize machine components and subsystems and their associated failures using a novel approach that combines text analysis, unsupervised text clustering, and domain models. Through industrial partnerships, this project will demonstrate effectiveness of the proposed approach with actual industry data.*

Key Words: Data Mining; Semantic Web; Maintenance log analysis; Natural Language Processing; Machine Learning, Case-Based Reasoning

1. Introduction:

This research addresses the problem of discovering knowledge required for critical business performance improvements by mining the equipment maintenance logs collected by the owners of complex machinery. These logs typically consist of terse free-text input manually entered by maintenance personnel. Implicit in these logs is information about equipment components and associated failures and repairs. If this information can be made explicit, it provides valuable knowledge resulting in improvements in operating and maintenance processes.

The approach described here is an inter-disciplinary combination of sophisticated text analytics, a novel bootstrapping algorithm for unsupervised text clustering, and an extensible semantic representational framework for the construction of domain models of equipment used in the manufacturing industry. This representational framework builds on the standards of OWL (Web Ontology Language) and the Semantic Web concepts of RDF (Resource Description Framework) in order to provide a powerful extensible representational component to the work.

Our approach extracts from the free text logs a canonical set of physical machine components and subsystems and their associated failures. The extraction of such information can provide the data necessary to characterize operating cycles, maintenance schedules, periodic breakdowns, and most importantly, to identify and address abnormal failure rates before critical problems arise. This ability has long been a goal of the owners of this expensive equipment, and some manufacturing companies have been retaining maintenance log records for over ten years with this objective in mind. However, due to the nature of the data, extracting useful data from these maintenance logs has proven to be a difficult endeavor. A data-mining tool to extract this valuable information can provide owners and maintainers of complex equipment with the knowledge necessary to improve their operating efficiencies, reduce critical downtimes, and determine optimal maintenance schedules.

2. Background:

Maintenance logs describe every problem, repair, adjustment, or alteration made to complex, expensive machinery over their entire lifespan. Each log record corresponds to some physical action taken on some physical component or subsystem of a particular machine. This data is collected for every machine in use and retain the historic logs for many years with the intent of extracting the operating information that is critical to improving efficiency and reducing downtimes and unexpected failures. However, because the log entries are made by humans and entered in free-text natural language without descriptive or other standards, the variation in the input is extremely large, exhibiting all of the expressive freedoms possible in natural language with the additional freedoms of not having to adhere to the usual constraints of spelling, grammar, or vocabulary.

The information that must be extracted from this multitude of free text representations is the canonical set of actions and the associated components that were made to a particular machine or class of machines. From this information it is possible to identify repair trends, spot deviations from expected failure rates, construct lifecycle models, and a number of other activities that increase efficiency and provide opportunities for process improvement. There are three specific objectives:

- To identify categories of components at a suitable "natural" level of description (e.g., "clamps")
- To identify categories of problems and/or repair actions associated with these components (e.g., "hydraulic oil leak", "adjust pressure")
- To learn distributions of the problems and/or repair actions (e.g., "hydraulic oil leak" accounts for 20% of "clamp" problems)

We formulate this task as a *clustering* problem (Hanson, 1990) in which the system must learn the natural categories of the data as well as learn a model for distributing new input into those categories. More specifically, it is a *text clustering* problem, because the maintenance data is represented solely as free-text. However, there are a number of characteristics of the maintenance log data that make the application of traditional approaches unviable and there are real-world constraints that require the innovative approach described here.

Due to the realities of the operating environment and personnel, maintenance logs are often recorded in fairly haphazard and non-standard ways. These logs may be written by hand and then transcribed or entered directly from the operating environment via limited input devices such as PDAs or handheld computers. The subsequent input data, while nominally natural language, exhibit many characteristics that make standard approaches to natural language processing and text classification difficult:

Vocabulary used in descriptions is inconsistent. There is typically no standard set of terms used for the names of mechanical parts or the activities performed on them. For example, one record might contain "change the oil" whereas another might contain "replace the fluid", both referring to the same action. Additionally, unlike this example, many of the descriptions in the real-world data are not amenable to traditional dictionary or thesaurus-based approaches as the terms used may be highly specific to the particular machine and/or laced with jargon.

Vocabulary may not correspond directly to systems or components of interest. Repair personnel may refer to a component or part of a system without explicitly mentioning the large system that the company is interested in monitoring. For example, "broken hose" may actually be "broken coolant hose" which needs to be categorized as a coolant system fault and not a hose fault *per se.*

Input is not well-formed. Because of character length limitations and probable treatment of the log entries as secondary task, the text descriptions are limited to short phrases or sentence fragments. Additionally, attention is not typically paid to spelling and other language rules, and the data therefore exhibits very high degrees of grammatical and spelling errors. For example, "add oil to leakey[sic] fixture" contains a typo whereas "add oil to fixture" neglects to mention that the fixture is leaky.

Jargon and extremely terse abbreviations are common. Due to the limited input space and time pressures of maintenance engineers, log entries are usually laced with creative abbreviations and contain large amounts of jargon. For example, "HD-7" might be a kind of "hydraulic oil". The jargon may be range in use from generally-known to very local terms for particular machines or tools.

Large amounts of data are not available. We work with some of the largest manufacturing companies in the United States. These companies, some of whom have been collecting data for almost ten years, have maintenance log databases of at most 5-10,000 records for a single class of machines. Thus, a new type of machine put into service will have on the order of 50-100 maintenance actions performed per year. While this is too large for manual analysis, it is too sparse for purely data-intensive approaches.

The following are examples of actual maintenance log records that exhibit some of the properties described above:

128272	"HIGH PRESSURE COOLANT COMES ON AT WRONG
83618	"HYD LEAK ON CLAMPS, POSSIBLE BLOWN GUAGE"

414181	"Hyd. oil leak, B-Axis table."
89666	"HYDRAULIC PUMP MAKING NOISE, LOCATED IN
181353	adjust clamping pressure
241772	adjust clamps
409507	ADD COOLANT HOSE TO 4264
674894	ADJUST CLAMPS
476269	ADD OIL TO LEAKEY FIXTURE
594549	add HD-7
398880	add hydraulic oil
124594	adjust clamps from hitting heads
644999	ADJUST CLAMPS OVER DATUM POINTS
199925	BROKEN HOSE

The very limited amount of data available and its sparseness of content make purely data-intensive statistical text clustering approaches impractical. Conversely, knowledge-intensive approaches, such as supervised learning models, require a degree of knowledge engineering (e.g., to construct a labeled training set) that is impractical for a real-world application, as companies typically do not have the personnel or other resources to devote to this task.

3. Solution:

Our solution is based on four key ideas:

- Develop a domain ontology through a knowledge engineering effort
- Represent the domain in OWL/RDF
- Use text analytics to sanitize the data (spelling, stemming, stop words) and reduce problem dimensionality
- Apply a conceptual clustering approach to learn categories from ontology-based data
- Employ case-based reasoning (Kolodner, 1993) to identify patterns of maintenance activities or component malfunctions that lead to systemic failure.

This approach has its roots in Clerkin, Cunningham & Hayes (2001), who propose a method based on the concept formation system COBWEB (Fisher, 1987) to learn ontologies represented in RDF schema (Brickley & Guha, 2000). Their goal was to cluster documents in order to generate class hierarchies for the semantic web. Although our data set shares some of the characteristics of theirs, our problem is somewhat different. Instead of learning ontologies, our goal is to use existing ontologies to learn natural categories of objects that occur in the data set.

In order to accomplish this, we need to develop a method of evaluating similarity of ontological objects that might be described at different levels in the ontological hierarchy. Maedche & Zacharias (2002) discuss methods for similarity assessment for clustering ontology-based meta-data, again for the semantic web. Our approach also requires clustering of ontology-based data, but our data is somewhat different from the meta-data in their approach.

Building on these approaches, our unsupervised conceptual clustering approach uses the OWL ontologies to classify and assess similarities between objects, problems, and repair actions in the data set. The output is a set of clusters representing families of <object, repair> pairs that occur in the data. Our approach uses automated techniques to augment the manual creation of ontologies. Specifically, we extend Blum & Mitchell's (1988) co-training approach, which allows a system to learn over a large set of unlabeled data using a small subset that is manually labeled. Finally, the learned categories are used to construct a *case library* database of maintenance patterns leading to systemic failure. These patterns are used by a Case-based Reasoning engine (CBR) to predict future failures based on observed patterns of maintenance activity as well as allow more efficient troubleshooting and diagnosis in the event of a failure.

The component overview of this architecture is shown in Figure 1 below.

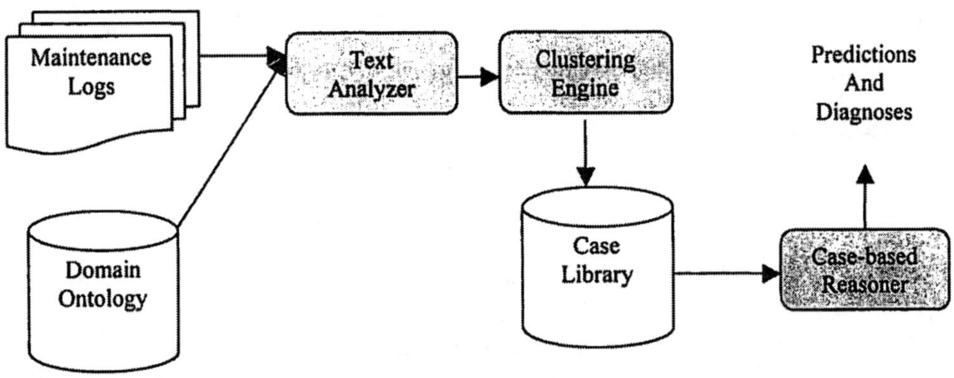

Figure 1: System Architecture

3.1. Generic Representations:

There is a great deal of effort underway to construct ontologies and other meta-data representations in order to help make use of the vast number of information repositories available. The "Semantic Web" (Berners-Lee, 2001) is the notion that in addition to low-level information, there will be machine-readable databases of meta-data and other background knowledge that will enable computers to more effectively fuse different sources of data and make much more powerful inferences from that data. Similarly, in the manufacturing industry (among others), standards organizations (e.g., ISO) have been creating taxonomies, ontologies, and vocabularies to describe the equipment and processes used in those businesses. For example, The International Electrotechnical Commission (IEC) standard 61499-1 describes architectures for industrial-process measurement and control. However, these standards are more typically focused on processes rather than physical objects and are extremely complicated and detailed.

Our approach is to fuse the ideas of the Semantic Web and industrial standards organizations and develop ontologies for the domain of machine tool maintenance. These ontologies are represented using OWL (Web Ontology Language), an XML-based extension of RDF (Resource Description Framework) that is an official W3C standard (World Wide Web Consortium). Our ontologies are developed to be generally-applicable across a number of application domains and types of machinery. For example, all tools share high-level systems and subsystems such as hydraulics, electronics, pneumatics, etc. and each of these systems has certain characteristics that are independent of the particular machine in which they reside. All hydraulic systems have hoses, pumps, and fluid, all of which have specific functions and interconnections. And so on.

OWL is a hierarchical representational framework, so that in addition to a generic representation, more specific ontologies or models can be constructed within the OWL framework. Thus, for example, an ontology of a specific machine such as an *"HPC 500/630 TS Multi-Spindle CNC"* can be written to define components and systems specific to that machine while inheriting general properties from a higher representational class. By using an open and widely accepted representational framework we can leverage the contributions of others outside of this particular effort. For example, more general ontologies and models of physical systems that have been created elsewhere may be incorporated into our application to increase its capabilities.

3.2. Text Analytics:

Advanced text analytics algorithms including the use of standard English dictionaries and thesauri are used for parsing and analyzing records in the maintenance database in order to disambiguate jargon, misspellings, and abbreviations to the degree possible. Higher-level terms are identified through the use of part of speech tagging, *n*-gramming, and correlation techniques such as Latent Semantic Analysis. These techniques are used to reduce the representational dimensionality of the problem, i.e., to increase the number of apparent similarities between records. For example, in the sample data above, the two records "add HD-7" and "add hydraulic oil" actually refer to the same thing because HD-7 is the local name for the hydraulic fluid used. A thesaurus-based analysis allows the system to correctly equate those two records.

While text analytics represent an important component of this project, our primary focus is on the definition and construction of a powerful representational framework as described above and the development of a clustering algorithm to produce natural categories from the raw log data as described below.

3.3. Bootstrapping Clustering:

Even after applying text analytics to reduce discrepancies between records that describe the same physical action on the same underlying component (e.g., replacing a coolant hose), there is still a great deal of representational variety—that is, the number of unique log records is much larger than the number of unique <component, repair> pairs. The heart of the solution to this problem of discovery natural categories that can be used for business-process analysis is the use of *clustering* techniques to group the disparate log records into their functional groupings.

Domain models are used in a bootstrapping clustering algorithm by providing the *categories* represented by the data as a whole but no category labels for any individual records as needed by supervised classification methods. This approach provides the opportunity for improved performance over strictly unsupervised methods and presents new technical challenges. An additional component of our approach is the inclusion of advanced text analytics algorithms that also take advantage of the generic domain models in order to disambiguate text as much as possible, reducing the dimensionality of the input to the greatest degree.

Building on the work of Clerkin, Cunningham & Hayes (2001) and Maedche & Zacharias (2002) discussed earlier, our unsupervised conceptual clustering system uses the bootstrapping approach to discover natural groupings of components and repair actions in the text-sanitized repair logs. Domain ontologies from text analytics are used to assess similarities between objects, problems, and repair actions in the data set, and to classify these items into appropriate categories.

Taken together, these techniques enable an AI system to identify types of components, types of problems and repair actions, and their distribution in an automated manner.

3.4. Case-based Reasoning Engine

In case-based reasoning (CBR), solutions to new problems are obtained by finding similar past problems and analyzing the solutions to those problems for the answer to the current one. CBR is a mature technology and has been used successfully in a number of commercial and industrial applications. It has proven successful in a number of diagnostic and maintenance applications, for example, 3Com Corporation's on-line troubleshooting system (http://knowledgebase.3com.com/) and Iomega's "i-man" online support system (http://iomega.com). The case-based reasoning process consists of a number of steps as shown in Figure 2 below.

Retrieval : In the first stage of CBR, a new maintenance log (or subset) is provided to the system. The indexer module retrieves a set of related maintenance histories ("cases") from the case library. For a case to be retrieved, a mathematical combination of its similarity to the new maintenance history and its historical behavior must meet a performance threshold relative to the other cases in the library – e.g., each case must "compete" against all others in order to be chosen for retrieval. Similarity is computed by the co-sine between the two feature vectors:

$$s^{(C)}(\mathbf{x}_a, \mathbf{x}_b) = \frac{\mathbf{x}_a^{\dagger}\mathbf{x}_b}{\|\mathbf{x}_a\|_2 \cdot \|\mathbf{x}_b\|_2}$$

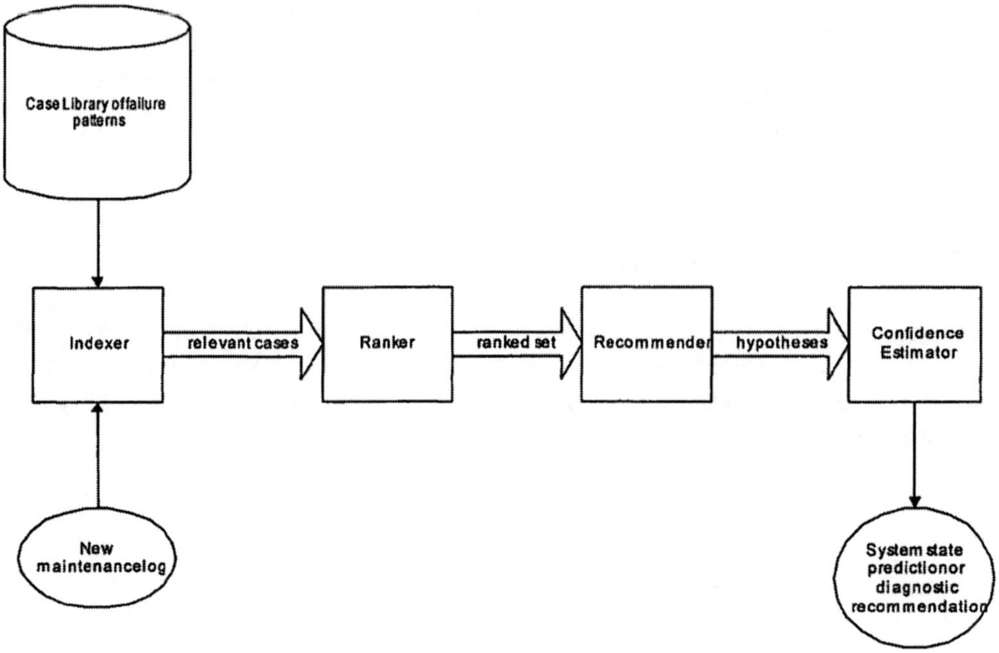

Figure 1: Case-Based Reasoning Process

Ranking: The set of related prior cases from the indexer is then ordered or prioritized by the ranker module. This ordering is based on similarity vs. quality metric computed during the retrieval phase and results in a creating a distribution of the items in the retrieval set against that smaller population.

Recommendation: In the recommendation stage of CBR, the ranked set of most similar past cases are combined to produce a set of hypothesized corrective actions for the new discrepancy. This is done by computing a weighted normalized histogram of cluster patterns based on all of the cases in the ranked set.

Confidence estimation: In the final stage of the CBR process, each hypothesis generated by the recommendation process is given a confidence value (0..1), which helps a human operator to decide whether to pursue the recommendation of the automated system. Confidence values are derived from internal properties of the system such as the relative cohesiveness of the ranked set of more similar past log histories and the degree to which one or more of the predictions "stands out" from the others.

4. Evaluation Plan

A comprehensive evaluation of the approach described above is in progress and we are anticipating significant results based on our related experience in applying text analytics techniques to complex maintenance problems. For example, the chart in Figure 3 below shows the application of text analytics and user feedback to identifying appropriate troubleshooting recommendations based on free-text problem descriptions (Ram & Devaney, 2005).

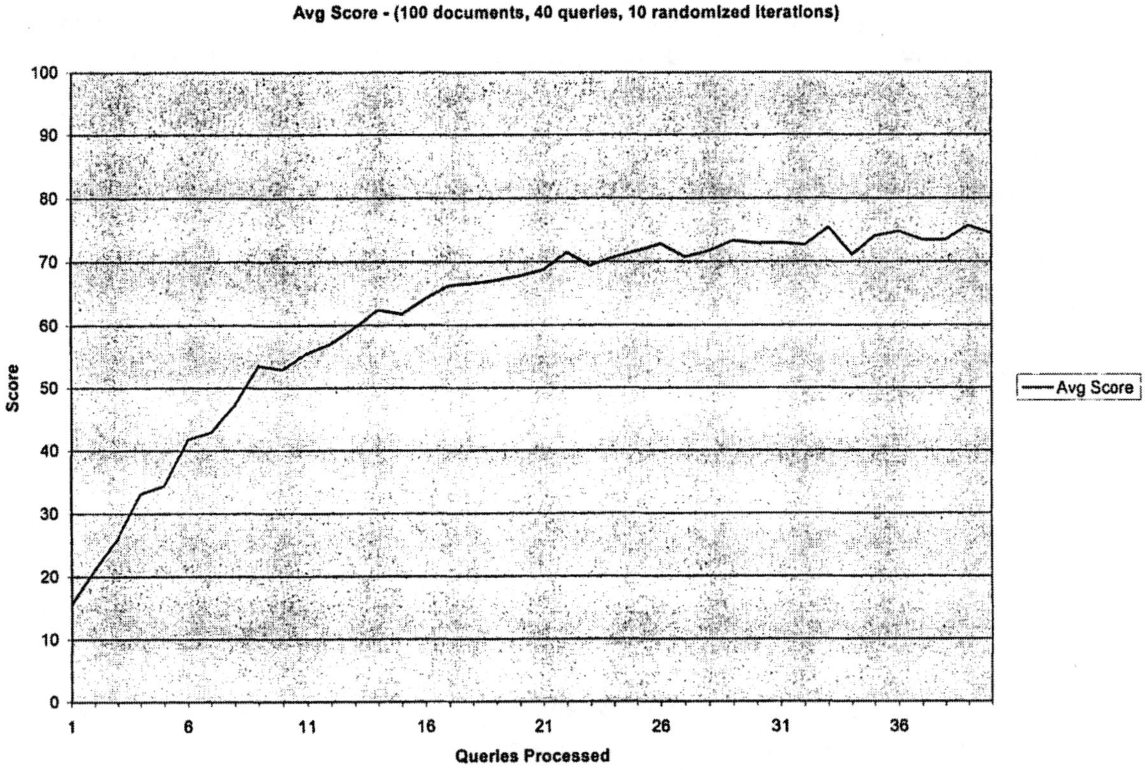

Figure 2: Troubleshooting problem identification with user feedback

Our industry partners have committed to providing us with real-world data with which to evaluate the effectiveness of the proposed approach. Our evaluation plan will focus on quantitatively measuring the quality of the classification of log records into natural categories through the use of domain experts as well as our industry partner. These evaluations will be primarily useful for benchmarking against the results obtained using naïve clustering algorithms as described in section 2.1 as well as measuring the improvements that are made as the domain ontology and model evolves over time. Additionally, evaluations will be conducted to measure the business impact of the approach by running the system and generating a taxonomy and frequency of repair actions. These results will be provided to experienced maintenance engineers and managers who will assess the utility of the results and provide rough characterizations of the usefulness of the systems performance. Ideally, we also hope to obtain quantitative estimates of business metrics that would be improved through use of the system, such as cost and time savings, failures avoided, etc.

5. Conclusions

Maintenance logs contain a potential wealth of information that can be used to improve the maintenance of complicated machinery, reduce downtimes, and prevent failures. Advanced text analytics techniques show great promise in extracting this useful information from the complicated, diverse, and inconsistent entries typically found in these maintenance logs. This paper has described our approach in employing these techniques along with advanced knowledge representations in order to construct a powerful Data Mining software architecture. Our next step is to conduct extensive evaluations of the system using real-world data and collecting both quantitative and qualitative metrics.

6. References

Berners-Lee, T. (2001). Semantic Web. http://www.w3.org/2001/sw/

Blum, A., & Mitchell, T. (1988). Combining Labeled and Unlabeled Data with Co-Training. In *Proceedings of the Eleventh Annual Conference on Computational Learning Theory*, 92-100.

Brickley, D., & Guha, R.V. (2000)., 2000] Dan Brickley und R.V. Guha. *Resource Description Framework (RDF) - Schema Specification 1.0.* http://www.w3.org/TR/2000/CR-rdf-schema-20000327.

Clerkin, P., Cunningham, P., & Hayes, C. (2001). Ontology Discovery For The Semantic Web Using Hierarchical Clustering. In *Proceedings of the Workshop in Semantic Web Mining, Twelfth European Conference on Machine Learning*, Freiberg, Germany.

Fisher, D.H. (1987). Knowledge Acquisition with Incremental Conceptual Clustering. *Machine Learning*, 2:139-172.

Hanson, S.J. (1990). Conceptual clustering and categorization: Bridging the gap between induction and causal models. In R. S. Michalski and Y. Kodratoff, editors, *Machine Learning: An Artificial Ingelligence Appproach (Volume III)*, chapter 9, pages 235-268. Morgan Kauffmann, San Mateo, CA, 1990.

Kolodner, J. (1993). *Case-Based Reasoning*. Morgan Kaufmann, San Mateo, CA.

Maedche, A., & Zacharias, V. (2002). Clustering Ontology-Based Metadata in the Semantic Web. *Principles of Data Mining and Knowledge Discovery*, 348-360.

Ram, A., & Devaney, M. (2005 in progress). Interactive CBR for Precise Information Retrieval. In *Case-Based Reasoning in Knowledge Discovery and Data Mining*, S. Pal & D. Aha (Eds.). To be published.

CORROSION

Chair: Mr. John V. Kelley
U.S. Army Research Laboratory

PREDICTIVE ANALYSIS OF CORROSION

John Fassino and Nancey Maegerlein
Naval Surface Warfare Center, Crane Division
Explosive Sciences Branch, Code 4052, Bldg 2707
300 Highway 361
Crane, IN 47522

Abstract: One of the biggest problems faced by the U.S. Navy is corrosion. From the harsh environments experienced shipboard to erosion corrosion of gun barrels, there are continuing efforts to extend the life of products, reduce costs, and maintain superior quality. One way to achieve all these goals is to accurately predict corrosive environments and corrosion rates, and effectively minimize such atmospheres. The corrosive environments can be improved in several ways, innovative protective coatings, new materials, better monitoring, and new technologies to better predict corrosion and how it affects a given material. Predicting corrosion allows for service lives to be extended and maintenance costs to be reduced by knowing ahead of time what to expect and at what rate changes will occur. Peltier sensors and heat flow microcalorimetry (HFMC) allow for a determination if an environment is corrosive, the corrosion rate, and the predicted life of the test coupon. HFMC allows for service life predictions as opposed to the current salt flog testing which gives little meaningful data other than an item's condition after 48 hour exposure. The following study details how HFMC was able to determine the corrosive effects of several propellant formulations on gun barrel steel to aid in a product improvement plan. The HFMC analysis was coupled with standard chemical and metallurgical tests to validate the data.

Keywords: Barrel; corrosion; microcalorimetry; predictive analysis; propellant

Background. Code 4022, the In-Service Engineering Branch from Crane Division Naval Surface Warfare Center (NSWC Crane) contacted the Explosives Sciences Branch (ESB) about performing analyses on medium caliber propellant samples as part of a Product Improvement Plan (PIP). Included in the analyses were chemical, thermal, and metallurgical tests. The details and results of a corrosion study to determine the corrosive effects of the different propellant types on gun barrel steel coupons are described below.

Introduction. The purpose of this corrosion evaluation is to determine what propellants or propellant constituents are the most detrimental to the life of the gun barrels, as related to corrosion. The corrosion study was set up as follows. First, the different propellant types were burned in a PARR bomb and the by-products were contained to simulate the by-products actually experienced inside a gun barrel. Next, coupons were obtained from a new gun barrel so that actual gun barrel material would be tested with actual propellant by-products. The barrel coupons were combined with the by-product solutions and analyzed using heat flow microcalorimetry (HFMC) to determine the rate, and relative amount of corrosion on the barrel coupons caused by the propellant by-products. Prior to the HFMC analysis, ion exchange chromatography was performed on the propellant by-product solutions to determine what inorganic by-products were present and the amount of each. After the HFMC analysis, the solutions were tested for pH (acidity), and the barrel coupons underwent a metallurgical investigation to determine how the grains were attacked.

Analyses Completed. The details of the various analyses included in the propellant corrosion evaluation are included in the subparagraphs below.

Bomb Calorimetry. A PARR bomb was used to burn six different propellant types. The propellants were burned in an inert environment (argon) so that the only oxygen present would be from the propellants. The gases from the explosion were bubbled through deionized water to capture their by-products. The inside of the bomb was then rinsed with deionized water to capture the propellant residue. The rinse residue solution and solution from the gases were combined so that all byproducts from a propellant type were contained in one solution to be used in additional analyses. For further reference the final solution will be referred to as the propellant by-product solution. Bomb calorimetry and the solution formation were completed on six propellant types, summarized in Table 1.

Table 1: Propellant Sample Identification

Propellant Type	ESB Lab Number
X6162	6734
X6163	6735
X6164	6736
WC867	6737
WC868	6738
WC872	6739

Ion Exchange Chromatography (IEC). IEC was performed on the different propellant by-product solutions to determine the anions present. Determining what anions are present, and their respective amounts gives an indication of the acidity (corrosiveness) of each propellant type.

Barrel Sectioning and Coupon Preparation. Seven total barrel coupons were obtained by sectioning remaining strips of a new gun barrel used in a previous metallurgical analysis. The sectioning was performed utilizing a diamond coated wire saw, which would allow for little or no heat generation into the barrel material. The barrel coupons were approximately 1.5" long, 0.25" wide, and 0.125" thick. After the sectioning was complete the coupons underwent five grinding stages is succession, 60 grit, 240 grit, 320 grit, 400 grit, and 600 grit. Once the grinding was complete on all coupons, they underwent three polishing stages, 9 micron diamond, 3 micron diamond, and 1 micron diamond, in succession. The purpose for such extensive surface preparation was to rid all surfaces of any flaws that would act as stress points to accelerate the corrosion process.

Heat Flow Microcalorimetry (HFMC). HFMC is used to monitor the rate of physical processes and/or chemical reactions. It uses twin heat conduction calorimeters, i.e. it consists of two calorimeters (one for the sample and one for a reference). HFMC is sensitive to all physical and chemical processes associated with a heat flow. Thus, the monitored heat flow may contain contributions from several processes. Individual contributions may be distinguished by varying the experimental conditions. In this investigation, the HFMC was used to determine the rate of corrosion caused to the barrel coupons as a result of the propellant by-product solutions. One coupon was combined with deionized water only to act as a blank for comparison with the other solutions, for a total of seven samples.

pH. After the HFMC analysis, pH tests were performed on the propellant by-product solutions to determine the acidity of each. The acidity of the solutions should correlate with the corrosiveness of a given solution.

Metallurgical Microscopy. The barrel coupons were examined under a metallurgical microscope (metallograph) upon completion of the HFMC analysis. The purpose of the metallograph examination was to determine the extent of corrosion on the grain structure of the barrel steel, and to correlate the results from the HFMC with a historical approach for determining corrosion severity. Two metallurgical samples were obtained from each barrel coupon tested. The original coupon was cut in half so that one sample could be examined transversely, and the other sample could be examined longitudinally. The ESB lab numbers originally assigned to each sample were modified with an "A" for the longitudinal sample and a "B" for the transverse sample. For example, ESB sample 6734A would designate the longitudinal sample of barrel steel that underwent HFMC analysis with the X6162 propellant by-product solution.

Results. The comprehensive results from all tests are described below. Following the results, the ESB details the conclusions and plans for future analyses.

Bomb Calorimetry. The six propellant types were successfully burned in the PARR bomb. The gases and residues from each sample were combined with deionized water and retained for further analyses.

Ion Exchange Chromatography (IEC). Each propellant by-product solution was analyzed for anion content using IEC. Special emphasis is given to the presence of nitrates (NO3) and nitrites (NO2) because they represent the corrosive nitrous and nitric acidic gases that attack the gun barrels in service. The results from the IEC analysis are summarized below in Table 2.

Table 2: IEC Results

ANION Content [mg/kg]				
TYPE	**Cl**	**NO2**	**NO3**	**SO4**
WC867	51.32	11.77	48.60	59.85
WC868	33.36	16.88	164.92	3894.80
WC872	48.99	18.32	70.60	73.20
X6162	59.86	24.11	114.65	430.90
X6163	49.27	14.86	100.42	1568.80
X6164	42.81	28.01	140.04	2714.83

Table 2 above clearly illustrates the following results for the IEC analysis for free anion content in the various propellant by-product solutions.

First, the X6162 and X6164 by-product solutions appear the most probable corrosive solutions based on their high nitrate and nitrite content. The X6162 by-product solution also has high chloride content giving further evidence to its acidity. Special attention should also be given to the WC868 by-product solution for its extremely high nitrate content.

<u>Barrel Sectioning</u>. Seven gun barrel coupons were successfully cut using a diamond coated wire saw. A digital picture can be seen below, figure 1, showing a barrel coupon and a glass ampoule used for the HFMC analysis.

Figure 1: 4cc Glass Ampoule, Gun Barrel Coupon

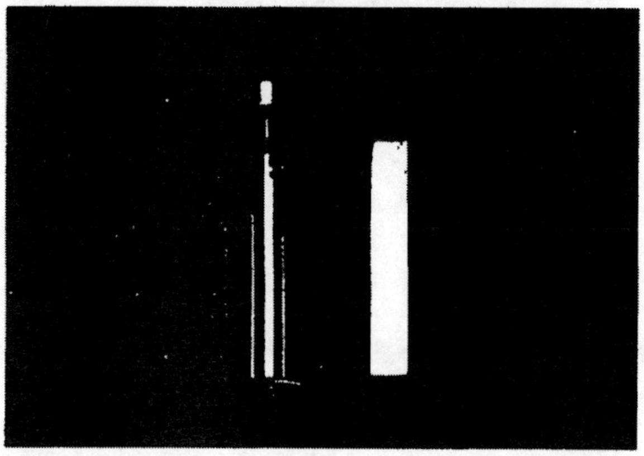

<u>Heat Flow Microcalorimetry (HFMC)</u>. The barrel coupons, immersed in the propellant by-product solutions were analyzed with HFMC at 35°C for 72 hours to monitor the rate of heat produced from the corrosion process. Figure 2, below, shows a graph of total corrosion energy from the barrel coupons by the propellant by-product solutions. From figure 2, the plot suggests that the addition of Calcium Carbonate to the formulation may be a contributing factor to the corrosion of the barrels. Calcium carbonate is not added to WC867 and yet it shows a very similar initial response as the X6162, which has Calcium Carbonate, added as pellets. The overall amount of corrosion, however, is quite small. This is apparent by comparing the energy produced by the propellant by-product solutions with that of the deionized water, which is a measure of the relative reactivity. Samples X6163, X6164, and WC868 are less corrosive than the water. WC867 and WC872 are more corrosive than water initially. Only the X6162 sample is more corrosive than water after 72 hours, although the WC868 sample is still significantly increasing in energy and the X6163 sample is still slightly increasing.

Figure 2: HFMC Results

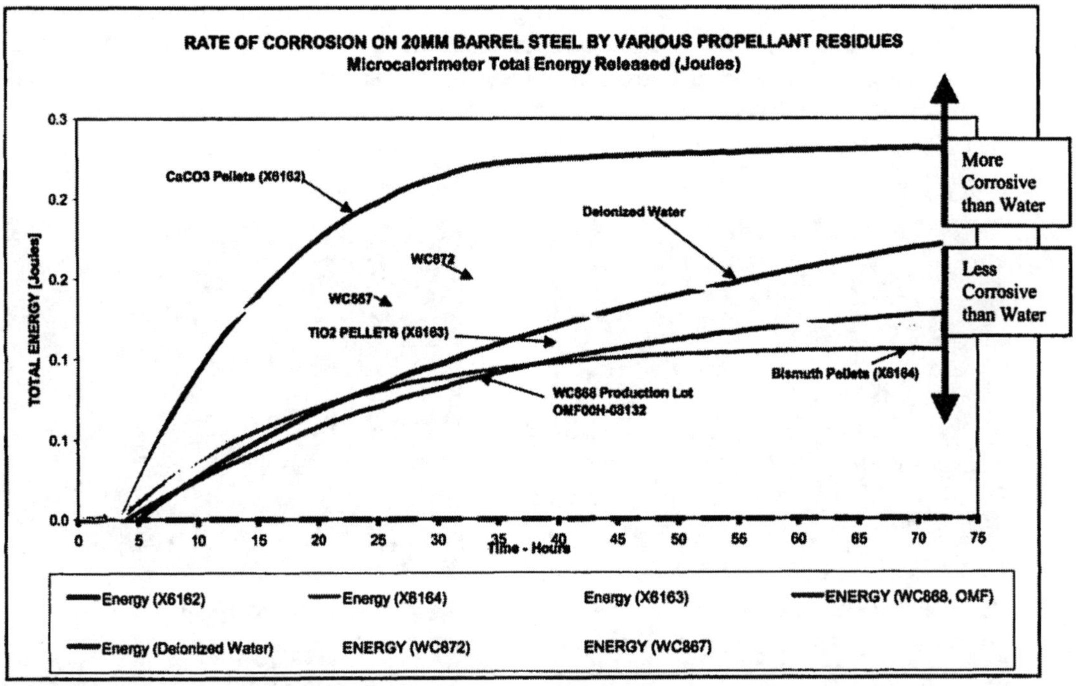

pH. The results of the pH analysis are tabulated below in table 3. The total energy after 72 hours from the HFMC analysis is also included in table 3 for comparison. The lower the pH, the more acidic the solution is considered. More acidity should correspond to more corrosion, which, in turn, should show more total energy from the HFMC analysis. The pH results correlate well with the HFMC results.

Table 3: pH and HFMC Results

ESB Lab Number	PH	HFMC – Total Energy [Joules]
6734 (X6162)	8.12	0.231
6735 (X6163)	8.35	0.121
6736 (X6164)	8.38	0.106
6737 (WC867)	8.54	0.118
6738 (WC868)	8.28	0.128
6739 (WC872)	8.62	0.138
6740 (Deionized H_2O)	8.24	0.172

Metallurgical Microscopy. The overall small amount of corrosion that took place in the HFMC analysis caused very little to be determined during the metallograph investigation. The majority of the samples showed mostly normal grain structure (Figure 3). In fact, nine of the fourteen samples examined had no evidence of corrosion at all. Several of the cross-section or transverse view samples showed slight areas of corrosion (spots) in the center of the samples, see figure 4. Five of the seven cross section samples had the slight center corrosion (spots). Five of the seven cross section samples also showed edge corrosion. An example of the edge corrosion can be seen in figure 5. The most severe case of edge corrosion can be seen in figure 6, which was found on sample number 6736B, which is the cross section sample tested with the X6164 propellant solution. Nonetheless, the amount of corrosion evident on the samples was overall quite small. Due to the limited overall amount of corrosion noted, strong correlations between the HFMC results and microstructure examinations could not be made.

Figure 3: Typical grain structure

Figure 5: Edge corrosion

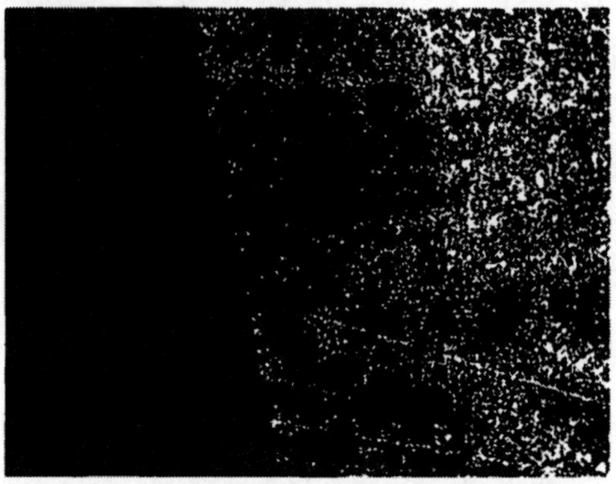

Figure 4: Center corrosion

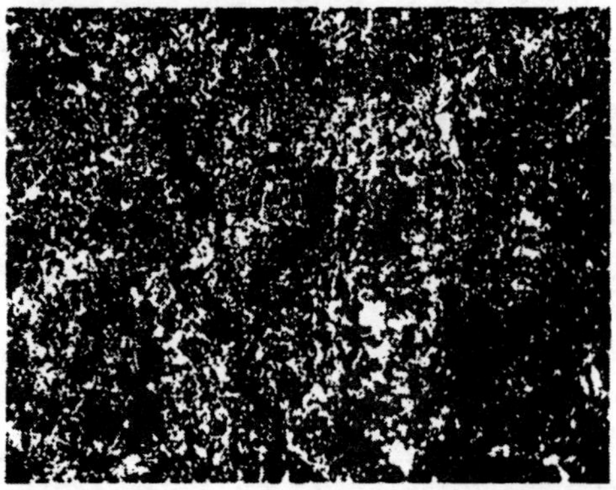

Figure 6: Corner corrosion

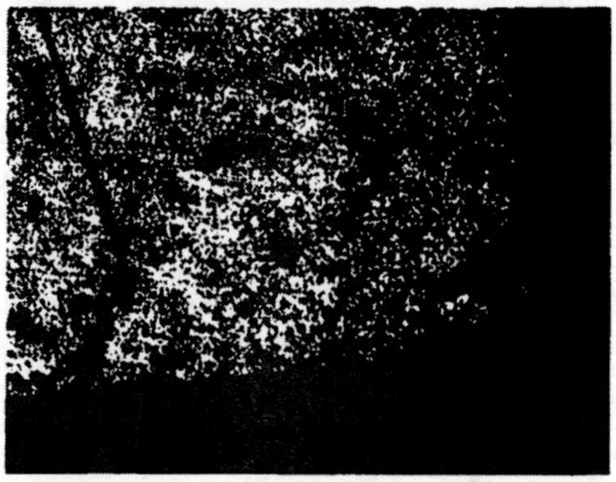

Conclusions. As determined from previous analyses, the by-products from the propellants are corrosive to the gun barrel steel. The intent of the testing described above was to determine the severity of corrosion caused by each propellant's by-products. The HFMC analysis, which measured the rate of corrosion, and total corrosive energy for the 72-hour test, gave the best relationship for how each propellant type affects the gun barrel steel. Initially, the rate of corrosion for the WC867, WC872, and X6162 propellant by-products solutions were much greater than that of the WC868, X6163, and X6164 types. However, over time, only the X6162 propellant by-product solution continued to corrode more severely than water. The X6162 propellant contains calcium carbonate pellets, which are believed to be a contributor to the increased amount of corrosion. The other PIP samples, X6163 and X6164, did not contain the calcium carbonate pellets and were less corrosive to the barrel steel. The additives to the X6163 and X6164 propellants were titanium dioxide pellets and bismuth pellets, respectively. At the end of the 72-hour HFMC analysis, the heat flow for most of the propellant by-product solutions had leveled off. The sample of barrel steel in water, and the barrel steel in the WC868 solution were still significantly increasing in energy. The barrel steel in the X6163 solution was slightly increasing in total energy as well.

As a result of the corrosion study to determine the effects of the different propellant by-products on barrel steel, the ESB determines the X6164 propellant (WC868 + 1.5% bismuth pellets) to be the least corrosive. It is believed that the addition of the bismuth pellets could help to reduce the corrosion experienced by the gun barrel steel. Extensive testing is required to confidently determine the effects of the bismuth pellets on corrosion.

Further Evaluations. The results from this investigation were very promising and gave many insights for how the different propellant types affect gun barrel steel. However, to gain more definitive results, the following tests will be performed.

Barrel coupons will be sectioned from remaining segments of a new gun barrel so that they can fit inside the cup of the PARR bomb. This will allow the barrel steel to experience the heat, gases, and residues that it would experience in an actual environment. The heat of combustions for each propellant type will also be measured and compared to later results. Two sets of coupons will be obtained for each propellant type.

The first set of coupons, after bomb calorimetry is performed, will be placed in a room temperature desicator maintained at 50% relative humidity for 7 days. This will allow complete reaction of the propellant residues with the barrel steel. After seven days, the barrel coupons will be cleaned with deionized water and dried in a nitrogen desicator for one day. After the drying process, the coupons will be immediately potted in a room temperature cure epoxy for a metallurgical investigation. The metallurgical investigation will consist of a metallurgical microscope examination to determine the severity of, and how the grains were attacked. If cracking is noted during the metallurgical microscope examination, they will be measured utilizing an optical comparator.

The second set of barrel coupons will immediately be placed inside glass ampoules for microcalorimic analysis. The microcalorimic analysis will be very similar to that performed in this investigation, only the samples will be tested until no further reaction is seen in the data (i.e. the heat flow returns to zero). Upon completion of the microcalorimetry testing, the barrel coupons will be subjected to a metallurgical examination as described in the previous subparagraph.

Analysis almost identical to that described in this enclosure will again be performed. However, there will be several key differences. One, the propellants will be burned in an oxygen environment because that will more closely resemble the actual environment the propellant will experience. The gases and residues will again be contained in a deionized water solution for IEC and pH analysis to verify previous results. Barrel coupons will be combined with the solutions for microcalorimetry to again determine the rate of corrosion for each. There will also be a few differences in the microcalorimetry procedure. First, the ampoules containing the barrel coupons will only be filled 50% with the propellant by-product solutions. This will allow the steel to face both the propellant by-products and an atmospheric environment. The samples will also be tested at a higher temperature (approximately 55°C), and tested for a longer amount of time. The coupons will again undergo metallurgical testing after the microcalorimetry analysis.

Depending on the amount of funding and time available, the ESB will perform an identical analysis to that described herein, with a couple of differences, to directly compare the results from burning the propellant in an oxygen environment with burning in an inert environment. The changes from the analysis detailed in this report will mimic those described above. One, in the microcalorimic analysis, the ampoules will be filled 50% with the propellant by-product solutions. Two, the test will be performed at a higher temperature (~55°C). Three, the microcalorimic testing will be conducted for a longer period of time, until the heat flow returns to zero.

Purpose Of Future Testing. The purpose of the additional analyses described above tests is two-fold. First, and most importantly, is to determine the affects of the different propellants on the gun barrel steel. Second, by comparing and correlating all the results, the process of determining the rate of corrosion utilizing microcalorimetry will be proofed, thus validating the technique. The different types of testing, and amount of data that will be generated, should allow the ESB to draw definitive conclusions pertaining to the affects of the different propellant types on gun barrel steel, relating to corrosion. Assuming the performance characteristics of all propellant types are similar, using the less corrosive propellant should extend the life of the gun barrels.

MEASURING THE THICKNESS OF COATINGS USING ULTRASONICS

James A. Kidd, Jr.
US Army Research Laboratory
AMSRD-ARL-WM-MC
APG, MD 21085-5069

Abstract

The purpose of this report is to illustrate how the thickness of several different primers and topcoats can accurately be determined with ultrasonic instrumentation. The coating systems involve various combinations of pretreatment, primer (chromate and nonchromate), and topcoat. The coatings were applied to one composite and four different metallic substrates that are found in aviation. The theoretical background of the methodology is explained quantitatively.

Key Words: Acoustic contrast; corrosion; density; elastic modulus; laminas; primer; sonic velocity; topcoat.

1. Introduction

Quantifying the thickness of coatings is necessary for quality control and consistency of coatings for commercial and military applications. Thickness measurements can be employed as part of a production process to provide feedback for adjustments to the application of coatings.

This report describes the use of a Pulse Echo Layer Thickness (PELT) system to quantitatively evaluate several different combinations of pretreatment, primer (chromate and nonchromate), and topcoat on one composite and several different metallic substrates. The coatings are candidates for application on US Army rotary-wing aircraft such as the Black Hawk helicopter. The test matrix that was prepared by the US Army Aviation and Missile Command (AMCOM) and the US Army Research Laboratory contains five different substrate materials. The panels are described in Table I.

Table I. Description of Substrates

Material	Specification	Size, inch
Composite	G-11, E-Glass	3 X 5 X 0.062
Aluminum 2024-T351	T3 => Solution heat treated and cold worked 51 => Stress relieved by elongation, for plate stock: 1.5 to 3.0% Rolling direction is longitudinal, i.e. parallel to 5 inch direction	3 X 5 X 0.25
Aluminum 7075-T651	T6 => Solution heat treated then artificially aged 51 => Stress relieved by elongation, for plate stock: 1.5 to 3.0% Rolling direction is longitudinal, i.e. parallel to 5 inch direction	3 X 5 X 0.25
High Strength Steel	American Iron and Steel Institute, AISI 4340	3 X 5 X 0.25
Magnesium ZE41A-Temper 5	Alloyed with aluminum, manganese, zinc, silicon, copper, and nickel	3 X 5 X 0.25

Results of the measurements that were taken for the Department of Defense project are included. Some representative thickness specifications for commercial applications are also provided.

2. Theory of Operation

The Pulse Echo Layer Thickness (PELT) system is a nondestructive measurement device that applies an acoustic pulse to the sample that is being measured. The pulse echoes from interfacial boundaries between coatings and off the substrate. The amplitude of an echo is proportional to the acoustic contrast (the difference in acoustic properties of the laminas). To gauge an individual layer, it must be of sufficient thickness and be discernable (adequate acoustic contrast) from adjacent layers. The minimal thickness for resolution is typically 0.4 mil (10 micron). [1]

Sonic velocity through a coating is dependent on the elastic modulus and the density of the coating. The PELT system is calibrated with sonic velocity in units of inch/second. The following calculation demonstrates how a representative sonic velocity is dependent on the bulk modulus and the density. Modulus and density are inversely proportional to temperature; assume that the coating is at room temperature, 68°F or 20°C. For a topcoat such as MIL-DTL-64159, a water dispersible aliphatic polyurethane Chemical Agent Resistant Coating (CARC), the modulus is approximately $2.5(10)^{10}$ dyne/cm^2 [2]. For this example, the units of dyne/cm^2 must be converted to lb$_f$/in^2:

$$E = 2.5(10)^{10}\,\frac{\text{dyne}}{\text{cm}^2}\left(\frac{\dfrac{\text{lb}_f}{\text{in}^2}}{68{,}947.59087\dfrac{\text{dyne}}{\text{cm}^2}}\right) = 362{,}594.25\,\frac{\text{lb}_f}{\text{in}^2}. \qquad (1)$$

A representative density for the topcoat is: $\rho = 43.657\dfrac{\text{lb}_m}{\text{ft}^3}$

Assume that $a = 32.174\dfrac{\text{ft}}{\text{sec}^2}$ so that $\text{lb}_f = \text{lb}_m$

$$\text{Sonic velocity, } v_s = \sqrt{\frac{Eg}{\rho}} \qquad (2)$$

$$v_s = \sqrt{\frac{\left(362{,}594.25\dfrac{\text{lb}_f}{\text{in}^2}\right)\left(144\dfrac{\text{in}^2}{\text{ft}^2}\right)\left(32.174\dfrac{\text{ft}}{\text{sec}^2}\right)}{43.657\dfrac{\text{lb}_m}{\text{ft}^3}}}$$

$$v_s = 6{,}203.22\,\frac{\text{ft}}{\text{sec}}\left(\frac{12\text{ in}}{\text{ft}}\right) = 74{,}438.64\,\frac{\text{inch}}{\text{second}}.$$

The PELT system measures elapsed time for a pulsed sound wave to traverse the coating and return to the transducer; the thickness measurement is inferred from correlating the transit time with the sonic velocity. The following calculation illustrates this methodology. For this example, assume that the elapsed time for the echoed sound wave to return to the transducer is

$t = 54.0$ nanosecond $= 54.0(10)^{-9}$ second.

Assume that the sonic velocity is constant for each individual measurement.

$$s = v_s t = \left(74{,}438.64 \frac{\text{inch}}{\text{second}}\right)\left[54(10)^{-9} \text{ second}\right] = 0.004 \text{ inch} \tag{3}$$

Total distance traversed, $s = 2t_c$ where t_c is the thickness of the measured coating, solve for t_c

$$t_c = \frac{s}{2} = \frac{0.004 \text{ inch}}{2} = 2.0 \text{ mil} \tag{4}$$

Therefore, for this illustration the thickness of the CARC topcoat is two mil.

The thicknesses of each primer and topcoat were measured with a PELT system that was manufactured by JSR Ultrasonics. The system is based on a calibrated sonic velocity for a specific combination of primer and topcoat. One panel from a control set was sent to the manufacturer so that calibration velocities could be determined. The actual calibration velocities were obtained by cross-sectioning representative coated panels, using an optical microscope to measure the thickness of the primer and the topcoat layers, and using the PELT to quantify the transit time for an echoed sonic wave. From the distance (twice the thickness of the coating) and the time, the calibration velocity can be determined. Each of the four calibration velocities shown in Table II is the result of weight averaging four data points. The composite panels did not provide adequate acoustic contrast to provide a significant ultrasonic echo for determining a calibration velocity. Each coating system required specific calibration velocities. The calibration velocities were determined for the metallic substrates and extrapolated for the composite panels.

Table II. Calibration Velocities

Calibration	Calibration Velocity, $\frac{\text{inch}}{\text{second}}$	
	Topcoat	**Primer**
I	88,860	99,320
II	95,197	81,881
III	78,189	102,525
IV	74,439	101,024

Contact between the probe and the painted surface was improved by using a drop of glycerin at the interface.

3. Results

The results cover two test series: Series A, ASTM D 3359, "Standard Test Method for Measuring Adhesion" and Series C, ASTM D 4541, "Standard Test Method for Pull-Off Strength of Coatings Using Portable Adhesion Testers." The PELT system was not able to reliably measure the coating thickness on the cast magnesium panels because of roughness of the surface.

For Test Series A and C, each individual value represents the average of three measurements.

3.1 Results for Thickness of Coatings, Test Series A

Tables III, IV, V, and VI contain the results for the four indicated substrates. The thickness in thousandths of an inch for the primer, topcoat, and total coating system is included; the mean of the primer, topcoat, and total is also provided.

Table III. Thickness in Mil of Primer and Topcoat on Composite Panels

Panel	Primer	Topcoat	Total	Panel	Primer	Topcoat	Total	Panel	Primer	Topcoat	Total
1A1	2.41	1.71	4.12	2A1	1.28	1.95	3.23	3A1	1.31	2.24	3.55
1A2	2.32	1.50	3.82	2A2	1.54	1.93	3.47	3A2	1.02	2.35	3.37
Average	2.37	1.61	3.97	Average	1.41	1.94	3.35	Average	1.17	2.30	3.46

Table IV. Thickness in Mil of Primer and Topcoat on Aluminum 2024-T3 Panels

Panel	Primer	Topcoat	Total	Panel	Primer	Topcoat	Total	Panel	Primer	Topcoat	Total
4A1	1.28	1.85	3.13	5A1	0.79	2.08	2.87	6A1	0.80	2.21	3.01
4A2	1.69	1.55	3.24	5A2	0.89	2.10	2.99	6A2	0.82	2.09	2.91
Average	1.49	1.70	3.19	Average	0.84	2.09	2.93	Average	0.81	2.15	2.96

Panel	Primer	Topcoat	Total	Panel	Primer	Topcoat	Total	Panel	Primer	Topcoat	Total
8A1	0.70	1.86	2.56	9A1	0.83	2.15	2.98	10A1	0.75	1.85	2.60
8A2	0.67	1.66	2.33	9A2	0.71	1.91	2.62	10A2	0.81	1.81	2.62
Average	0.69	1.76	2.45	Average	0.77	2.03	2.80	Average	0.78	1.83	2.61

Panel	Primer	Topcoat	Total	Panel	Primer	Topcoat	Total	Panel	Primer	Topcoat	Total
11A1	0.70	1.66	2.36	12A1	0.81	1.91	2.72	13A1	0.74	2.21	2.95
11A2	0.78	1.84	2.62	12A2	0.91	2.22	3.13	13A2	0.72	2.23	2.95
Average	0.74	1.75	2.49	Average	0.86	2.07	2.93	Average	0.73	2.22	2.95

Panel	Primer	Topcoat	Total	Panel	Primer	Topcoat	Total
14A1	1.60	1.84	3.44	15A1	1.66	1.80	3.46
14A2	1.71	1.84	3.55	15A2	1.25	1.67	2.92
Average	1.66	1.84	3.50	Average	1.46	1.74	3.19

Table V. Thickness in Mil of Primer and Topcoat on Aluminum 7075-T6 Panels

Panel	Primer	Topcoat	Total	Panel	Primer	Topcoat	Total	Panel	Primer	Topcoat	Total
16A1	1.45	1.53	2.98	17A1	0.90	2.01	2.91	18A1	0.81	2.01	2.82
16A2	1.39	1.67	3.06	17A2	0.84	1.97	2.81	18A2	0.78	2.04	2.82
Average	1.42	1.60	3.02	Average	0.87	1.99	2.86	Average	0.80	2.03	2.83

Panel	Primer	Topcoat	Total	Panel	Primer	Topcoat	Total	Panel	Primer	Topcoat	Total
19A1	0.72	1.76	2.48	20A1	0.99	1.98	2.97	21A1	0.79	1.92	2.71
19A2	0.60	1.96	2.56	20A2	1.04	1.94	2.98	21A2	0.74	1.85	2.59
Average	0.66	1.86	2.52	Average	1.02	1.96	2.98	Average	0.77	1.89	2.66

Panel	Primer	Topcoat	Total
22A1	0.62	1.68	2.30
22A2	0.79	1.71	2.50
Average	0.71	1.70	2.41

Table VI. Thickness in Mil of Primer and Topcoat on High Strength Steel AISI 4340 Panels

Panel	Primer	Topcoat	Total	Panel	Primer	Topcoat	Total
23A1	1.30	1.54	**2.84**	**24A1**	0.59	1.73	**2.32**
23A2	1.77	1.18	**2.95**	**24A2**	0.69	1.65	**2.34**
Average	**1.54**	**1.36**	**2.90**	**Average**	**0.64**	**1.69**	**2.33**

Panel	Primer	Topcoat	Total	Panel	Primer	Topcoat	Total
25A1	0.71	1.70	**2.41**	**26A1**	0.86	1.09	**1.95**
25A2	0.79	1.47	**2.26**	**26A2**	0.90	1.14	**2.04**
Average	**0.75**	**1.59**	**2.34**	**Average**	**0.88**	**1.12**	**2.00**

Table VII correlates the types of coatings with a summary of the thickness results for test Series A.

Table VII. Consolidated Results for Mean Thickness of Coatings, Test Series A

Substrate	Panel	Calibration Velocity	Pretreatment	Primer C=> Chromate N=> Nonchromate; Mil	Topcoat; Mil	Total Mil
Composite	1A	I	None	MIL-P-2377C 2.37	MIL-C-46168, IV 1.61	3.97
	2A	II		MIL-PRF-85582N 1.41	MIL-C-53039 1.94	3.35
	3A	III		85582N 1.17	MIL-DTL-64159, II 2.30	3.46
Aluminum 2024-T3	4A	I	MIL-C-5541	23377C 1.49	MIL-C-46168, IV 1.70	3.19
	5A	II	5541	85582N 0.84	53039 2.09	2.93
	6A	III	5541	85582N 0.81	64159, II 2.15	2.96
	8A*	IV	Trivalent Chrome Process	85582C 0.69	64159, II 1.76	2.45
	9A	II	TCP	85582C 0.77	53039 2.03	2.80
	10A	III	TCP	85582N 0.78	64159, II 1.83	2.61
	11A	IV	Alodine 5200	85582C 0.74	64159, II 1.75	2.49
	12A	II	Alodine 5200	85582N 0.86	53039 2.07	2.93
	13A	III	Alodine 5200	85582N 0.73	64159, II 2.22	2.95
	14A	IV	X-It Precoat	85582C 1.66	64159, II 1.84	3.50
	15A	III	X-It Precoat	85582N 1.46	64159, II 1.74	3.19
Aluminum 7075-T6	16A	I	5541	23377C 1.42	46168, IV 1.60	3.02
	17A	II	5541	85582N 0.87	53039 1.99	2.86
	18A	III	5541	85582N 0.80	64159, II 2.03	2.83
	19A	II	TCP	85582C 0.66	53039 1.86	2.52
	20A	II	TCP	85582N 1.02	53039 1.96	2.98
	21A	III	TCP	85582N 0.77	64159, II 1.89	2.66
	22A	IV	5541	85582C 0.71	64159, II 1.70	2.41
High Strength Steel AISI 4340	23A	I	SAE-AMS-QQ-P416	23377C 1.54	46168, IV 1.36	2.90
	24A	II	QQ-P416	85582N 0.64	53039 1.69	2.33
	25A	III	QQ-P416	85582N 0.75	64159, II 1.59	2.34
	26A	IV	QQ-P416	85582C 0.88	64159, II 1.12	2.00

* => Panel 7 was not tested.

3.2 Results for Thickness of Coatings, Test Series C

Tables VIII, IX, X, and XI contain the results for the four indicated substrates. The thickness in thousandths of an inch for the primer, topcoat, and total coating system is included; the mean of the primer, topcoat, and total is also provided.

Table VIII. Thickness in Mil of Primer and Topcoat on Composite Panels

Panel	Primer	Topcoat	Total	Panel	Primer	Topcoat	Total	Panel	Primer	Topcoat	Total
1C1	2.44	1.54	3.98	2C1	1.54	1.94	3.48	3C1	1.72	2.05	3.77
1C2	2.23	1.86	4.09	2C2	1.45	2.09	3.54	3C2	1.39	2.36	3.74
1C3	2.56	1.62	4.18	2C3	1.53	2.37	3.90	3C3	1.40	2.03	3.43
1C4	2.34	1.66	4.00	2C4	1.41	2.00	3.41	3C4	1.34	2.22	3.56
1C5	2.23	1.81	4.04	2C5	1.41	1.89	3.30	3C5	1.36	2.09	3.45
Average	2.36	1.70	4.06	Average	1.47	2.06	3.53	Average	1.44	2.15	3.59

Table IX. Thickness in Mil of Primer and Topcoat on Aluminum 2024-T3 Panels

Panel	Primer	Topcoat	Total	Panel	Primer	Topcoat	Total	Panel	Primer	Topcoat	Total
4C1	1.63	1.46	3.09	5C1	0.98	2.26	3.24	6C1	0.61	1.95	2.56
4C2	1.56	1.44	2.99	5C2	0.89	1.83	2.72	6C2	0.68	2.18	2.85
4C3	1.42	1.39	6.09	5C3	0.86	1.79	2.65	6C3	0.65	1.93	2.58
4C4	1.39	1.43	2.81	5C4	0.80	1.89	2.69	6C4	0.68	1.75	2.43
4C5	1.68	1.34	3.02	5C5	0.69	1.77	2.46	6C5	0.61	1.61	2.22
Average	1.53	1.41	2.94	Average	0.85	1.91	2.76	Average	0.64	1.89	2.53

Panel	Primer	Topcoat	Total	Panel	Primer	Topcoat	Total	Panel	Primer	Topcoat	Total
8C1	0.54	1.72	2.26	9C1	0.65	1.83	2.48	10C1	0.70	1.70	2.40
8C2	0.67	1.69	2.35	9C2	0.87	1.73	2.61	10C2	0.63	1.94	2.57
8C3	0.59	1.63	2.22	9C3	0.74	1.83	2.56	10C3	0.73	1.83	2.56
8C4	0.53	1.51	2.04	9C4	0.80	1.70	2.50	10C4	0.83	1.86	2.69
8C5	0.58	1.64	2.22	9C5	0.77	1.74	2.51	10C5	0.69	1.73	2.42
Average	0.58	1.64	2.22	Average	0.77	1.77	2.54	Average	0.71	1.81	2.52

Panel	Primer	Topcoat	Total	Panel	Primer	Topcoat	Total	Panel	Primer	Topcoat	Total
11C1	0.63	1.66	2.28	12C1	1.07	1.85	2.92	13C1	0.64	2.18	2.82
11C2	0.73	1.85	2.58	12C2	0.88	2.15	3.04	13C2	0.66	1.92	2.58
11C3	0.63	1.84	2.46	12C3	0.84	2.00	2.83	13C3	0.65	1.95	2.60
11C4	0.73	1.71	2.43	12C4	0.88	1.83	2.71	13C4	0.66	1.79	2.45
11C5	0.64	1.48	2.11	12C5	0.92	1.91	2.83	13C5	0.67	2.05	2.71
Average	0.67	1.71	2.38	Average	0.92	1.95	2.87	Average	0.65	1.98	2.63

Panel	Primer	Topcoat	Total	Panel	Primer	Topcoat	Total
14C1	1.66	1.55	3.21	15C1	1.55	1.78	3.33
14C2	1.46	1.72	3.18	15C2	1.29	2.00	3.29
14C3	1.51	1.63	3.14	15C3	1.38	1.80	3.18
14C4	1.41	2.02	3.44	15C4	1.38	1.67	3.05
14C5	1.50	1.60	3.10	15C5	1.27	1.86	3.13
Average	1.51	1.71	3.22	Average	1.37	1.82	3.19

Table X. Thickness in Mil of Primer and Topcoat on Aluminum 7075-T6 Panels

Panel	Primer	Topcoat	Total	Panel	Primer	Topcoat	Total	Panel	Primer	Topcoat	Total
16C1	1.58	1.38	2.96	17C1	0.86	1.87	2.73	18C1	0.62	1.84	2.46
16C2	1.39	1.40	2.79	17C2	0.85	1.96	2.80	18C2	0.62	1.76	2.38
16C3	1.56	1.40	2.96	17C3	0.80	1.74	2.54	18C3	0.57	2.00	2.57
16C4	1.70	1.34	3.04	17C4	0.90	1.96	2.86	18C4	0.79	1.84	2.63
16C5	1.68	1.42	3.10	17C5	0.93	1.75	2.68	18C5	0.76	1.70	2.46
Average	1.58	1.39	2.97	Average	0.87	1.86	2.73	Average	0.67	1.83	2.50

Panel	Primer	Topcoat	Total	Panel	Primer	Topcoat	Total	Panel	Primer	Topcoat	Total
19C1	0.66	1.80	2.46	20C1	0.78	1.87	2.65	21C1	0.78	1.78	2.56
19C2	0.70	1.82	2.51	20C2	0.96	1.79	2.75	21C2	0.79	1.87	2.66
19C3	0.55	1.72	2.27	20C3	1.04	1.75	2.79	21C3	0.72	1.71	2.42
19C4	0.59	1.69	2.28	20C4	0.92	1.82	2.73	21C4	0.72	1.93	2.65
19C5	0.62	1.70	2.31	20C5	0.85	1.82	2.66	21C5	0.72	2.02	2.74
Average	0.62	1.74	2.36	Average	0.91	1.81	2.72	Average	0.74	1.87	2.61

Panel	Primer	Topcoat	Total
22C1	0.59	1.80	2.39
22C2	0.53	1.50	2.03
22C3	0.70	1.66	2.36
22C4	0.58	1.60	2.18
22C5	0.63	1.63	2.26
Average	0.60	1.64	2.24

Table XI. Thickness in Mil of Primer and Topcoat on High Strength Steel AISI 4340 Panels

Panel	Primer	Topcoat	Total	Panel	Primer	Topcoat	Total
23C1	0.94	1.69	2.63	24C1	0.77	1.67	2.44
23C2	1.30	1.64	2.94	24C2	0.91	1.70	2.60
23C3	1.24	1.35	2.60	24C3	0.69	1.94	2.63
23C4	0.97	1.51	2.49	24C4	0.63	1.89	2.52
23C5	1.27	1.40	2.67	24C5	0.99	1.70	2.69
Average	1.14	1.52	2.66	Average	0.80	1.78	2.58

Panel	Primer	Topcoat	Total	Panel	Primer	Topcoat	Total
25C1	0.76	1.45	2.21	26C1	1.00	1.05	2.05
25C2	0.70	1.60	2.29	26C2	1.07	1.11	2.18
25C3	0.64	1.39	2.02	26C3	0.97	1.05	2.02
25C4	0.74	1.79	2.53	26C4	0.84	1.21	2.06
25C5	0.72	1.55	2.27	26C5	0.61	1.12	1.73
Average	0.71	1.56	2.27	Average	0.90	1.11	2.01

Table XII correlates the types of coatings with a summary of the thickness results for Test Series C.

Table XII. Consolidated Results for Mean Thickness of Coatings, Test Series C

Substrate	Panel	Calibration Velocity	Pretreatment	Primer C=> Chromate N=> Nonchromate; Mil	Topcoat; Mil	Total Mil
Composite	1C	I	None	MIL-P-2377C 2.36	MIL-C-46168, IV 1.70	4.06
	2C	II		MIL-PRF-85582N 1.47	MIL-C-53039 2.06	3.53
	3C	III		85582N 1.44	MIL-DTL-64159, II 2.15	3.59
Aluminum 2024-T3	4C	I	MIL-C-5541	23377C 1.53	MIL-C-46168, IV 1.41	2.94
	5C	II	5541	85582N 0.85	53039 1.91	2.76
	6C	III	5541	85582N 0.64	64159, II 1.89	2.53
	8C*	IV	Trivalent Chrome Process	85582C 0.58	64159, II 1.64	2.22
	9C	II	TCP	85582C 0.77	53039 1.77	2.54
	10C	III	TCP	85582N 0.71	64159, II 1.81	2.52
	11C	IV	Alodine 5200	85582C 0.67	64159, II 1.71	2.38
	12C	II	Alodine 5200	85582N 0.92	53039 1.95	2.87
	13C	III	Alodine 5200	85582N 0.65	64159, II 1.98	2.63
	14C	IV	X-It Precoat	85582C 1.51	64159, II 1.71	3.22
	15C	III	X-It Precoat	85582N 1.37	64159, II 1.82	3.19
Aluminum 7075-T6	16C	I	5541	23377C 1.58	46168, IV 1.39	2.97
	17C	II	5541	85582N 0.87	53039 1.86	2.73
	18C	III	5541	85582N 0.67	64159, II 1.83	2.50
	19C	II	TCP	85582C 0.62	53039 1.74	2.36
	20C	II	TCP	85582N 0.91	53039 1.81	2.72
	21C	III	TCP	85582N 0.74	64159, II 1.87	2.61
	22C	IV	5541	85582C 0.60	64159, II 1.64	2.24
High Strength Steel AISI 4340	23C	I	SAE-AMS-QQ-P416	23377C 1.14	46168, IV 1.52	2.66
	24C	II	QQ-P416	85582N 0.80	53039 1.78	2.58
	25C	III	QQ-P416	85582N 0.71	64159, II 1.56	2.27
	26C	IV	QQ-P416	85582C 0.90	64159, II 1.11	2.01

* => Panel 7 was not tested.

4.0 Discussion

The results of the PELT measurements provide feedback to the Department of Defense, the vendors that manufactured the coatings, and the personnel that applied the coatings. The thickness information is used in conjunction with the data obtained from performing Series A, ASTM D 3359, "Standard Test Method for Measuring Adhesion" and Series C, ASTM D 4541, "Standard Test Method for Pull-Off Strength of Coatings Using Portable Adhesion Testers" for selection of the coating system to be used on US Army aircraft. The barrier properties of a coating are proportional to the lack of porosity and the thickness of the specific coating. The specified thickness of coatings is dependent on the type of coating, the intended use, and the environmental exposure. Table XIII provides typical requirements for coating thickness for several commercial applications in various environments. Note that coating systems on offshore platforms have a combined barrier thickness in excess of 1/32 inch and a total cumulative thickness of nearly 3/64 inch.

Table XIII. Typical Protective Coating Thickness Specifications [3]

Application	Coat	Material	Thickness Mil
Steel Bridge	Primer	Epoxy Zinc Phosphate	2.95
	Barrier	Epoxy Micaceous Iron Oxide	4.92
	Finish	Polyurethane	1.97
	Finish	Polyurethane	1.97
External Steel on Buildings	Primer	Zinc Rich Epoxy	2.95
	Barrier	Epoxy Micaceous Iron Oxide	4.92
	Finish	Polyurethane	1.97
Swimming Pool Environment	Primer	Epoxy Zinc Phosphate	2.95
	Barrier	Epoxy Micaceous Iron Oxide	3.94
	Finish	Polyurethane	1.97
Offshore Platforms	Primer	Epoxy Zinc Phosphate	0.98
	Barrier	Glass Flake Epoxy	19.68
	Barrier	Glass Flake Epoxy	19.68
	Finish	Polyurethane	1.97

5.0 Conclusions

Although using ultrasonics is not the only method of determining the thickness of individual layers of a coating system, it is a highly effective nondestructive technique for determining the consistency of application. The barrier properties and corrosive protection of a coating are proportional to the lack of porosity and the thickness of the specific coating. However, applying an excessively thick layer of a coating is not necessarily cost-effective or beneficial. Provided that there is a substrate that is discernibly harder than the coatings, ultrasonic technology can measure the thickness of several layers, typically up to four lamina, with a cumulative thickness of approximately 4.0 mil (100 micron).[4] The technique can be used for both chromate and nonchromate primer as well as various topcoats. Ultrasonic testing can provide an accurate way of measuring coating layers during periodic inspections of US Army materiel.

6.0 References

[1] JSR Ultrasonics PELT 500 Series Owner's Manual, Page 9, September 2003.

[2] Crawford, D. M. and Escarsega, J. A. "Dynamic Mechanical Analysis of Novel Polyurethane Coating for Military Applications" in Thermochimica Acta, page 164, 2000.

[3] www. Corrosionsource.com; Graham Gedge, 10 November 2000

[4] JSR Ultrasonics PELT Explorer User Manual, Page 11, 2003.

7.0 Acknowledgment

Jay Luckner with JSR Ultrasonics; Pittsford, New York generously provided a Pulse/Echo Layer Thickness (PELT) instrument, appropriate calibration velocities, and technical assistance.

METALLURGICAL TESTING OF GUN BARRELS

John Fassino, Nancey Maegerlein, and David Southern
Naval Surface Warfare Center, Crane Division (NSWC Crane)
Explosive Sciences Branch, Code 4052, Bldg 2707
300 Highway 361
Crane, IN 47522

Abstract: The demands on today's military for lighter, cheaper, longer lasting products only increase the risk for catastrophic failure of items already facing adverse service conditions. After a medium caliber barrel failure, the Explosive Sciences Branch (ESB) of Crane Division, Naval Surface Warfare Center (NSWC Crane) performed metallurgical investigations on one damaged gun barrel, six additional barrels pulled from service and one new gun barrel. to determine if any inherent flaws or service conditions could have caused the failure. The metallurgical analysis included detailed microhardness profiling on several areas of each barrel, complete microstructure examination with a metallurgical microscope, crack depth determination via optical comparator, optical microscopy and SEM analyses on down-bore sections of the barrels, and geometrical analysis on the rifling of the barrels. Hardness differences were found between the outside edge and/or inside edge of the barrels versus the center or core of the barrels. Additionally, areas of no chrome coating, cracked coating, base metal cracking, and stress corrosion cracking were evident in the breech end of the barrel, while land damage/deformation was prominent in down-bore barrel sections. The ESB determined coating cracking and slight interface corrosion between the base metal and coating would lead to accelerated degradation (erosion, corrosion, cracking, etc.) of barrels in service. These findings show that many barrels still in use should be pulled from service and that better erosion/corrosion monitoring techniques need to be in place to better assess the condition and risk of barrels in service.

Keywords: Barrel; corrosion; erosion; failure; gun; metallurgy

Background. The Explosive Sciences Branch (ESB) was contacted by Code 4022, the Medium Caliber Ammunition Branch, about an incident that ultimately resulted in catastrophic failure of a gun barrel. The ESB was initially tasked to perform chemical analyses on the different lots of propellant used in the gun ammunition, and decided to additionally complete metallurgical testing on one damaged gun barrel to determine if a defect in the barrel caused the catastrophic failure.

The results of the metallurgical testing determined that the sample gun barrel exhibited the following potentially problematic characteristics.

1. A significant hardness difference between the inside/outside edge and the center of the barrel.

2. The sample, taken from a heavily damaged portion of the barrel, exhibited significant loss of the chrome plating.

3. Considerable cracking of the base metal. Several of the cracks noted penetrated the base metal and branched off, which is characteristic of stress corrosion cracking.

4. The grain structure of the base metal immediately adjacent to the chrome plating demonstrated characteristics consistent with hydrogen embrittlement.

As a result of the above findings, the ESB suggested that a complete metallurgical investigation of additional gun barrels be performed to definitively determine if the gun barrel contributed significantly to the catastrophic failure.

Introduction. The ESB was tasked to complete full-scale metallurgical testing on seven separate gun barrels. Six barrels were selected for testing based on erosion level, type of ammunition, and number of rounds fired. The seventh barrel is new and was used as the control sample. The paragraphs that follow will detail each of the parts included in the metallurgical testing of the gun barrels.

1. Barrel identification and specific characteristics of each barrel.

2. Gun barrel cutting to accommodate completion of testing.

3. Sample identification for tracking all samples during the course of the testing.

4. Tests completed on each individual sample groups.

5. A description of each test performed, which particular samples were subjected to each test, how the test was performed and what the purpose of the test was.

6. ESB conclusions, based on the data and information generated from completion of all testing.

7. ESB recommendations as a result of test results and conclusions.

Gun Barrel Identification. The gun barrels selected for testing and provided to the ESB are identified in table 1.

Table 1 – Barrel Identification

Barrel ID	Propellant Lots	Total Number of Rounds Fired	Erosion Level Air Force/Navy
A	None – NEW	None – NEW	None – NEW
B	PGU27B	20263	0.030/0.028
C	PGU27A/B	16392	0.035/0.032
D	PGU27A/B	16392	0.039/0.034
E	PGU27A/B	19000	0.040/0.034
F	PGU28A/B	16606	0.043/NA
G	PGU28A/B	21780	0.053/NA

Gun Barrel Cutting. All transverse cuts were completed using a metal band saw with water-soluble lubricant. The longitudinal cuts were completed utilizing an end mill in conjunction with a specially designed fixture, and additional water-soluble lubricants. The step-by-step procedure used to obtain the necessary samples is detailed in the steps which follow.

<u>Step #1</u> – Figure 1 depicts a barrel as they appeared when initially received for testing.

Figure 1: Initial Sample

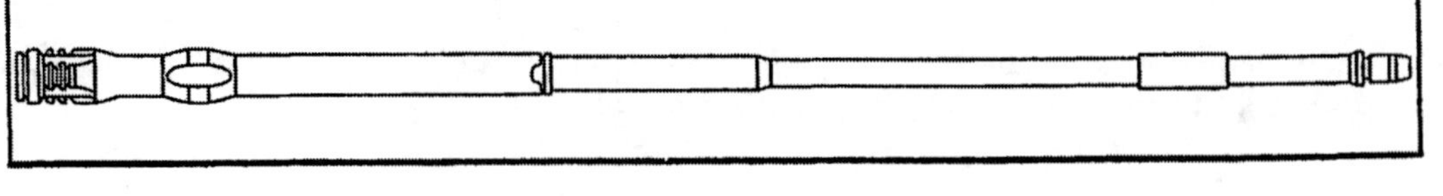

Step #2 – Each barrel was transversely cut at a point approximately eighteen inches from the breech end. This cut is depicted in Figure #2 below.

Figure 2: 18 inch Transverse Cut

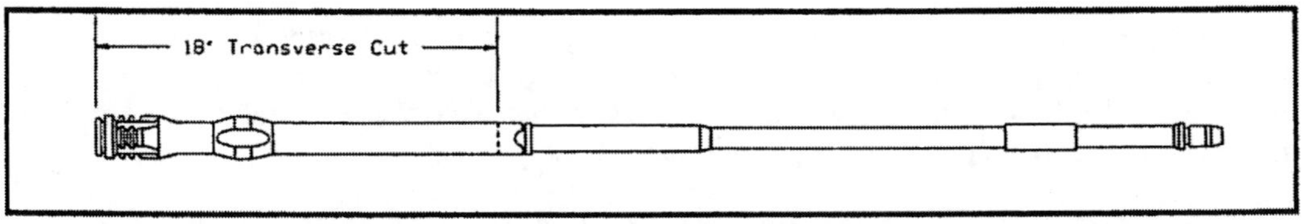

Step #3 – Each eighteen inch section of each barrel obtained in step #2 was cut in half longitudinally. One half of the eighteen inch section was transversely cut at the area of the Initiation of Lands and at the area of the Initiation of Erosion of Lands. The other half of the eighteen-inch section from each gun was longitudinally cut into six equally spaced pieces at approximate 30 degrees per section. Figure #3 depicts the eighteen-inch longitudinal section. Figure #4 depicts a transverse view of the entire barrel and illustrates the six pieces on the one half versus the complete barrel half on the other.

Figure 3: 18 inch Section, Transverse and Longitudinally Cut

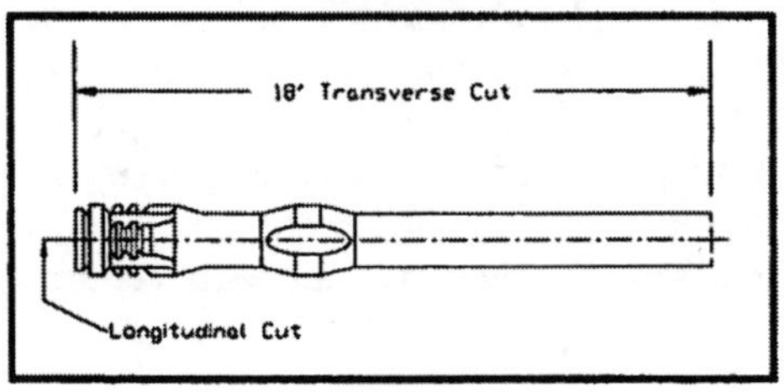

Figure 4: Cutting of 18 inch Section

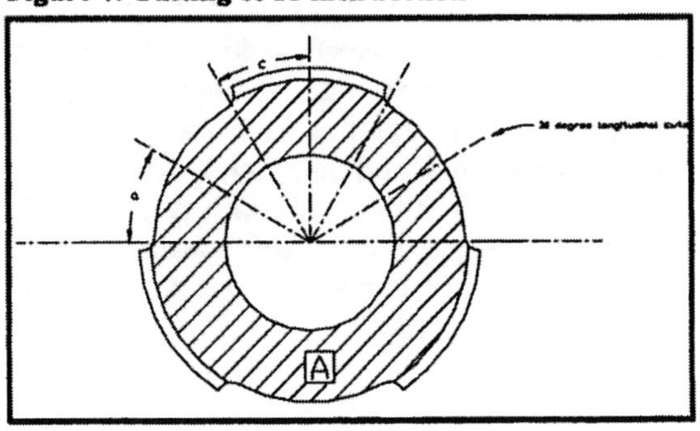

Step #4 – The ESB selected longitudinal strips "a" and "c" for cross-sectioning into eighteen one inch pieces, for a total of 36 one inch pieces from each gun barrel.

Sample Identification and Tracking. Due to the large volume of samples being tested, the ESB found it necessary to implement a sample identification and tracking system. The identification and tracking system consisted of the following parameters.

1. A CAPITAL LETTER was assigned to each of the seven gun barrels to be tested.

2. Each of the 36 individual pieces from each barrel were identified as to whether they originated from strip "a" or strip "c". Each of the 36 pieces from each barrel were also identified as to the exact inch of their origination on that strip. These identifications were 1 through 18, with number 1 meaning the first inch beginning from the breech end of the barrel.

3. Further, before mounting the samples for metallographic preparation, a "potting number" was assigned to each, a tag prepared bearing that number, and the tag bearing the number attached directly to the specific sample to which it was assigned.

The above steps taken by the ESB allowed for easy identification, tracking and cross-checking of all the micro-analysis samples. The ultimate result can be summarized as follows, Sample Number Aa1 = Barrel A, strip a, inch 1. Potting Number 1001 was assigned to it, recorded in a permanent branch record and attached to it.

Sample Preparation. All 36 one-inch samples for metallurgical microscope inspection were mounted, ground and polished before micro-hardness analysis was performed. After completion of the micro-hardness profiling, the samples were etched with 5% Nital solution (5% nitric acid/95% methanol) prior to microscopic examination with the metallurgical microscope.

Analytical Tests. The following paragraphs detail the various testing completed on the metallurgical test samples from each of the gun barrels. Each paragraph includes a description of the test in general, the general purpose of the test and which samples were subjected to the particular test.

Micro-hardness Testing. The samples subjected to micro-hardness testing were the 36 one-inch samples from each of the seven gun barrels. The samples were first prepared via mounting, grinding and polishing of each individual sample.

The test consisted of indenting the base metal of each of the samples utilizing a Knoop micro-hardness indenter, at a test load of 500gf, dwell time of approximately 10 seconds and indentions magnified for measurement at 40X. The first indention was completed on each sample at a depth of 0.002 inch from the interface with the chrome plating. This point is referred to throughout the remainder of this report as the inside edge. The second indention on each sample was completed at 0.009 inch, the third indention at 0.017 inch and remaining indentions at succeeding intervals of 0.015 inch from the 0.017 inch point until the last indention was completed at a point 0.001 inch to 0.005 inch from the outside edge of the sample. These particular increments were chosen, primarily, because of results obtained from the only historical sample of gun barrels subjected to micro-hardness testing.

The historical sample results demonstrated substantial hardness differences within the sample. Such difference was noted in hardness at the inside edge versus the middle or core of that sample. Similar differences were noted at the inside edge versus the hardness noted at a depth of 0.017 inch. In order to ascertain differences between the two depths, the current samples were also indented at a depth of 0.009 inch.

The purpose of the succeeding indentions being at 0.015 inch increments was to ensure detection of a dramatic change, as nearly as possible, at its point of inception.

Results of the micro-hardness testing revealed the following:

(a) The softest point on Barrel A, the new barrel, was determined to be the inside edge, on average.

(b) Barrel A inside edge hardness was generally less than five Rockwell hardness points softer than the hardness determined at 0.009 inch and 0.017 inch depths.

(c) Barrel A inside edge hardness was generally less than six Rockwell hardness points softer than the hardness determined at the core or middle of the samples.

(d) The 36 samples from each of the other six barrels tested demonstrated similar results.

(e) Approximately 92 percent of all samples tested fit the above generalizations.

The only obvious trend elucidated by the combined data of the test is that the softest point of each sample is predominant at a depth of 0.002 inch, the inside edge. This indicates the likelihood that the plating process has affected, at least to a degree, the hardness at the inside edge.

The hardest point on the samples is generally and consistently at a minimum depth of 0.317 inch. The depth of the hardest point is far enough away from the inside edge, combined with the difference being generally less than six Rockwell points, that the hardness difference is considered to be negligible as relates to being a direct or ultimate cause of barrel catastrophic failure via rupture.

The fired or used barrels all had a hardness average that was harder than that determined for the new barrel.

Barrel G, which had the greatest amount of erosion and the greatest amount of rounds fired through it of the two 28A/B barrels was determined to have the highest hardness value of all the fired barrels. Barrel B, which had seen only 27B rounds fired through it, was determined to have the second highest hardness values. Both of these barrels had seen over 21,000 rounds.

Metallurgical Microscope Examination. Each of the 36 one-inch samples from each of the seven barrels, a total of 252 one-inch samples, was examined utilizing the metallurgical microscope. The examination centered on the following areas: (a) Chrome Plating Cracking; (b) Chrome Plating Loss; (c) Chrome Plating/Base Metal Interface Corrosion; (d) Base Metal Erosion; (e) Base Metal Cracking; (f) Micro-structure/Grain Structure Anomalies.

The results of the metallurgical microscope examination of the samples are listed in the paragraphs that follow.

(a) Chrome Cracking. Each of the 252 total samples was examined for the presence of and extent of cracking within the chrome plating. The extent of cracking within the chrome plating for each sample was given a rating based on the scale in table 2.

Table 2 – Definition of Chrome Plating Cracking Levels

Level	Definition of Each Chrome Cracking Level
1	No cracking of chrome detected
2	**MINIMAL** chrome cracking – 1 crack in 100X photograph frame
3	**SLIGHT** chrome cracking – 2 cracks in 100X photograph frame
4	**MODERATE** chrome cracking – 3 or 4 cracks in 100X photograph frame
5	**SIGNIFICANT** chrome cracking – 5 or more cracks in 100X photograph frame

The results of the examination are summarized in the following paragraphs.

The cracking is evident on the new barrel that had no rounds fired through it. Although the cracking was evident on only half the samples from the barrel and the cracking seen was primarily minimal to slight, it is significant in that it exists at all.

Barrel B, exposed to 20,263 27B rounds, has at least level 3 cracking (slight) on each of the individual 36 one-inch samples from the barrel.

All of the other barrels also have at least level 3 cracking (slight) on each of the individual 36 one-inch samples from each barrel.

Significant and moderate cracking, levels 5 and 4, are most predominant in barrel C. Barrel D has the second highest incidence of significant level cracks in the chrome plating. Barrels F and G have similar levels of cracking of the chrome and Barrel E has the least moderate and significant levels of all the used barrels tested.

(b) Chrome Loss. Each of the 252 one-inch samples from the barrels was examined for loss of chrome plating. Chrome plating loss is defined as areas where the thickness of the plating decreases from the normal or original thickness at any given point along the one-inch sample. The extent of chrome plating loss on each sample was rated in accordance with table 3.

Table 3 – Definition of Chrome Loss Levels

LEVEL	DEFINITION OF LEVEL OF CHROME PLATING LOSS
1	No chrome plating loss
2	**MIMINAL** – thin amount of plating lost
3	**SLIGHT** – Up to ½ of plating gone
4	**MODERATE** – More than ½ of the plating is gone
5	**SIGNIFICANT** – All plating is gone

The results of the examination are summarized below.

Barrel A, the new barrel, exhibits chrome loss, primarily in the minimal to slight levels, on approximately one-third of the 36 samples tested from that barrel. This is considered significant because the barrel exhibits this characteristic with no rounds having ever been fired through it. If rounds were to be fired through it the chrome loss would be exasperated. A new barrel ideally should exhibit no noticeable chrome loss.

Barrel B, the barrel having seen only 27B rounds, exhibits the most comprehensive chrome loss. Some level of chrome loss was detected on 35 of the 36 samples tested from it. Fifteen of the samples exhibited chrome loss at the moderate to significant levels and 20 more at the slight level.

All of the remaining barrels exhibited widespread chrome loss throughout the 36 samples tested from each. Barrel C, exposed to 16,392 27A/B rounds, was determined to have moderate to significant chrome loss in 21 of the 36 samples tested from it. Barrel G, exposed to 21,780 28A/B rounds, was determined to have moderate to significant chrome loss in 28 of the 36 samples tested from it. Barrel F, exposed to 16,606 28A/B rounds, had only six of the 36 samples exhibiting significant level of chrome loss. However, barrel F was also determined to have moderate levels of plating loss in 14 of the 36 samples tested from it. Barrels D and E, having seen 16,392 and 19,000 rounds of 27A/B respectively, exhibit less significant and moderate levels of chrome loss. Barrel D and barrel E exhibited the majority of chrome loss in the slight level.

(c) Interface Corrosion. Each of the 252 one-inch samples from each of the tested barrels were examined utilizing the metallurgical microscope for evidence of corrosion or other anomalous conditions that disrupt the integrity of the chrome plating/base metal interface. The presence of corrosion at the interface of the plating and base metal greatly increase the possibility of the base metal being weakened and increases the possibility of cracking occurring and propagating. Crack propagation would, in turn, increase the propensity of the metal to eventually break or rupture in service, when under intense heat and pressure. The extent of existence of corrosion at the plating/base metal interface on each of the 252 samples was rated in accordance with table 4.

Table 4 – Definition of Plating/Base Metal Interface Corrosion levels

LEVEL	DEFINITION OF LEVELS OF CHROME PLATING/BASE METAL INTERFACE CORROSION
1	No areas of corrosion at the interface note
2	**MINIMAL** – One or two areas noted on the sample
3	**SLIGHT** – 3 to 5 areas noted on the sample
4	**MODERATE** – 5 to 10 areas noted on the sample
5	**SIGNIFICANT** – Greater than 10 areas noted on the sample

The results of the analysis on the samples for levels of corrosion at the plating/base metal interface on the samples from each of the seven barrels are summarized below.

Barrel A, the new barrel, has corrosion at the interface on 29 of the 36 samples tested. The corrosion levels are primarily minimal to slight, with 4 of the 36 samples demonstrating corrosion presence at the interface at the moderate level. The fact that corrosion already exists at the interface, without exposure to fire and the accompanying heat, pressure and gases, increases the potential for problems relating to cracking and erosion in service. The existence of such corrosion in the new barrel at the interface hints that such was possibly the case in other, if not all, barrels in use and in storage. The relationship between the formation and propagation of base metal cracking to the presence of such corrosion is unmistakable. Therefore, the level of such corrosion on any of the barrels is important to their eventual service life.

Barrel G clearly has the most instances of moderate to significant corrosion at the interface. Twenty-five of the 36 samples examined from barrel G demonstrate interface corrosion at these levels. Additionally, the remaining eleven samples have such corrosion at the slight level. This fact, in combination with the high incidences of chrome loss and chrome cracking seen in barrel G, leads the ESB to conclude its service life, beyond that already experienced, would have been extremely limited. The same general conclusions may be reached concerning both barrel B and barrel C. Although these two particular barrels demonstrate a lower incidence frequency at the significant level, all but one of the 36 samples from each of these barrels demonstrate corrosion at the plating/base metal interface to be between slight and significant.

Barrel F did not demonstrate as high a level of incidence of interface corrosion as did barrels B, G and C. Eight of the 36 samples from barrel F were determined to have no evidence of, or minimal corrosion. Another 13 samples showed only slight levels. This barrel would likely have a little more service life left. The same is true of barrel D. Barrel D demonstrated no incidences of significant corrosion at the interface and 27 of the 36 samples from it show corrosion at the slight or lower levels. Barrel E showed the least evidence of corrosion at the interface, with only 3 of the 36 samples demonstrating corrosion at the moderate level and none at the significant level. The ESB concluded barrel E would likely have the greatest amount of service life remaining of the barrels tested.

(d) Base Metal Erosion. Each of the 252 samples from the seven barrels tested was examined, utilizing the metallurgical microscope, for evidence of existing erosion of the base metal. The presence of such erosion has the same significance to the probability of base metal cracking and eventual service life of the barrel as the presence of corrosion at the plating/base metal interface discussed above. In short, the greater the incidence of such erosion of the base metal the greater the propensity for formation and propagation of cracks and eventual catastrophic failure of the barrel via rupture. The extent of erosion determined to be present on each of the 252 samples tested was rated in accordance with the definitions contained in table 5.

Table 5 – Definition of Levels of Erosion of Base Metal Material

LEVEL	DEFINTION OF LEVELS OF BASE METAL EROSION
1	Discoloration of the base metal at the point of corrosion
2	Erosion of the base metal is noticeable, some of the metal is gone
3	Erosion of the base metal is significant, a clear void in the base metal is present

The results of the metallurgical microscope examination for the presence and level of erosion on the samples from each of the seven barrels are summarized in the following paragraphs.

Barrel C shows the greatest severity of base metal erosion. Eighteen samples tested from barrel C were determined to have erosion of the base metal at the noticeable or significant levels, levels 2 and 3 respectively. Level 1, discoloration of the base metal, was noted on another four of the 36 samples. Barrel G also demonstrates levels 2 and 3 cracking at a propensity level that would be unacceptable for any substantial service life to have remained for that barrel. Barrel F would be somewhat suspect, with eight of the 36 samples demonstrating base metal erosion at the 2 and 3 levels. Neither barrel D nor barrel E demonstrate serious incidence of base metal erosion.

The most interesting findings emanating from this examination is that barrel B, which has seen only 27B rounds, shows an incidence of base metal erosion on only eight of the 36 samples from it, with all the instances being at the significant level, level 3. The severity of the incidences that do exist in barrel B leads the ESB to conclude its useful service life to be suspect as well.

Barrel A, the new barrel, shows no incidence of base metal erosion, but does have noticeable discoloration at many interface corrosion points.

(e) Base Metal Cracking. Each of the 252 individual samples from the seven barrels was examined for the presence and extent of cracking within the base metal. The examinations were completed utilizing the metallurgical microscope. A high incidence of base metal cracking or presence of significant to moderate base metal cracking within the samples from a barrel or the barrel in general, make catastrophic failure of the that barrel exponentially more likely to occur. The extent of base metal cracking observed on the 252 samples from the barrels was rated in accordance with the definitions in the following table (table 6).

Table 6 – Definition of Base Metal Cracking Levels

LEVEL	DEFINITION FOR RESPECTIVE LEVEL OF BASE METAL CRACKING
1	**SLIGHT** – Crack depth of 0.0001 to 0.001 inch
2	**MODERATE** – Crack depth of 0.0011 to 0.0075 inch
3	**SIGNIFICANT** – Crack depth of 0.0076 inch or greater

The results of the metallurgical microscope examination on each of the samples from the seven barrels for the presence and severity of base metal cracking are summarized in the following paragraphs.

Barrel G and Barrel C demonstrated the most numerous incidences of samples with base metal cracking. Barrel G examination resulted in 28 of the 36 samples demonstrating such incidence.

Ten of the incidences were of the moderate and significant level. Barrel C demonstrated incidence of base metal cracking in 18 of the 36 samples examined. Four of the 18 incidences were at the moderate and significant levels.

Barrel B and Barrel D had incidences of base metal cracking on 11 of the 36 samples examined from each barrel. 5 samples from Barrel B indicated moderate or significant levels of cracking. Barrel D was determined to have moderate and significant levels of base metal cracking on 3 of the 36 samples.

Barrel F showed a significant number of incidences of base metal cracking, 16 of the 36 samples examined. However, fifteen of the sixteen incidences were at the slight level.

Samples from barrel E demonstrated fewer incidences of base metal cracking, only 8 of the 36 samples examined. Two of the eight incidences were at the significant level.

Barrel A, the new barrel, was determined to have incidences of base metal cracking on three of the 36 samples from that barrel. Two of the three incidences were at the slight level. One incidence was at the moderate level, and, was within about 0.003 inch of significant.

The ESB concludes that all samples show enough evidence of incidence and severity of cracking to make their life expectancy suspect as to longevity. Barrels D and E would likely last for a while. Barrels G, C and F would likely need to be pulled from service at this point due to a combination of the number of incidences and severity of base metal cracking within them. Barrel B should also be pulled due to the severity of the incidences occurring in it.

(f) Microstructure/Grain Structure Anomalies. Each of the 252 samples from the seven barrels was etched with a solution consisting of 5 percent nitric acid and 95 percent methanol. Each sample was examined using the metallurgical microscope for anomalies within the grain structure of each of the samples. With the exception of strip "a" from barrel E, all the strips from all barrels examined had at least one incidence of an anomalous area within their respective microstructures. Various types of anomalies noted within the entire group of 252 samples examined. One type was porous structure indicative of hydrogen embrittlement, usually existing near the inside edge next to the plating. Some samples exhibited a finer or smaller grain within the structure at the edge next to the plating when compared to the interior or core of the same sample. Another common anomalous characteristic noted on several samples were a layered look to the microstructure. The layered look would be evidence by an area of darker and then lighter structure, alternating, giving an appearance of stripes. There were also some samples that had corrosion and the microstructure had been obviously changed near the corrosion area.

Final Conclusions. The ESB concludes the following, based on the combination of results from all analyses performed, on all samples and sample types from each barrel.

(a) An inherent problem exists with the new barrels. This conclusion is predicated on the postulation that other new barrels likely exhibit characteristics similar to what ESB observed on Barrel A, the new barrel analyzed in this examination. Barrel A exhibited no unusual hardness, but, did demonstrate evidence of chrome cracking, chrome loss, corrosion at the chrome plating/base metal interface, very small amount of base metal discoloration, and base metal cracking. These characteristics indicate further cracking and susceptibility to catastrophic failure via rupture when exposed to live fire and rounds.
(b) All of the barrels tested that had seen fire have a high probability of catastrophic failure via rupture. Each barrel, B through C, exhibited some evidence of chrome cracking, chrome

loss, corrosion at the chrome plating/base metal interface, base metal erosion and base metal cracking. The ESB, based upon the results from all analyses performed on the barrels, would rank the likelihood of such catastrophic failure for each barrel as follows, from most to least likely.

(1) Barrel G – Should be removed from service now.
(2) Barrel C – Should be removed from service now.
(3) Barrel F – Should be removed from service now.
(4) Barrel B – Should be removed from service now.
(5) Barrel D – SHORT amount of service remaining.
(6) Barrel E – SHORT, but longer, service left.

(c) Future catastrophic failure via rupture is likely, predicated on the results of the analyses of these barrels, unless a preventative program is instituted to monitor new barrels and barrels already in service.

Recommendations. The ESB recommends that a monitoring program be instituted to guard against the occurrence of additional catastrophic failures of the gun barrels in the future. Monitoring programs should be instituted for new barrels, and, a separate one for barrels in use. There is evidence to suggest that new barrels have inherent problems. The evidence from the analyses also indicate at least a loose connection between erosion levels of the barrel and the amount and/or severity of adverse characteristics analyzed for. Therefore, the ESB recommends the following preventative measures for each type barrel respectively.

(a) New Barrels. The ESB recommends that a representative sample of new barrels be taken from storage, or upon receipt from the manufacturer, and subjected to testing.

The barrels should be cut transversely at nine different spots, equally spaced, beginning about two inches from the breech end and ending about two inches from the muzzle end. This would amount to nine separate pieces per barrel. Each of these pieces from each barrel should be subjected to inspection of the land areas to determine the presence of corrosion at the interface of the plating and base metal, and other locations as well as for cracking and crack propagation.

Also, one one-inch piece should be cut from each of the nine pieces on each barrel and subjected to full micro-analysis as was completed herein, i.e., micro-hardness and metallurgical microscope examination.

(b) Used Barrels. Barrels in use or having been in use should be subjected to surveillance testing. The testing should consist of pulling them out of service long enough to test their erosion levels. If the erosion level is 0.040 inch or greater they should be removed from service. If their erosion level is determined to 0.030 inch or greater, a representative number of such barrels should be subjected to the same analyses as noted above for new barrels.

EXAMINATION OF ALTERNATIVE PRETREATMENTS TO HEXAVALENT CHROMIUM BASED DoD-P-15328 WASH PRIMER FOR MIL-A-46100 HIGH HARD STEEL ARMOR

Brian E. Placzankis, Chris E. Miller and John V. Kelley
U.S. Army Research Laboratory, Weapons and Materials Research Directorate
Aberdeen Proving Ground, Maryland 21005-5066

ABSTRACT

Hexavalent Chromium based compounds with their associated health and environmental risks have necessitated reductions and even eliminations of their usage in U.S Army weapon systems.[1] These compounds are primarily associated with pretreatments and conversion coatings applied prior to primers and topcoats that make up the Chemical Agent Resistant Coating (CARC) coating system. Despite these hazards, for armor steels such as MIL-A-46100[2] high hard alloy, hexavalent chromium based DoD-P-15328[3] wash primer is still commonly used. This study explores the effectiveness of wash primer alternatives with respect to corrosion, adhesion, and susceptibility to Environmentally Assisted Cracking (EAC) issues. The alternatives studied were: spray based zinc phosphate in accordance with TT-C-490[4], grit blast only with no other pretreatment, and grit blast with a rust inhibiting sealer.

Keywords: Armor Steel; Chromium; Hexavalent; Phosphate; MIL-A-46100, Wash Primer

INTRODUCTION

Historically, DoD-P-15328 wash primer has extensively been used as one possible pretreatment portion of the CARC coating system for steel and aluminum substrates. The hexavalent chromium component present in the wash primer promotes adhesion of the primer to the substrate and inhibits the nucleation and growth of corrosion sites. In order to comply with current Army policies and DoD directives, all new Army acquisitions must be hexavalent chromium free. Three hexavalent chromate free solutions were selected for study based upon feedback from the Brigade Combat Team (BCT) Stryker Interim Armored Vehicle (IAV) Environmental Management Team and process engineers at the General Dynamics Land Systems (GDLS) London, Ontario, Canada assembly facility for the Stryker family of armored vehicles (Figure 1). These three pretreatments were: Abrasive blast only, Abrasive Blast + PPG CHEMINHIB 420[5], Abrasive Blast + Henkel Surface Technologies BONDERITE 952 spray based zinc phosphate. Performance criteria applied to replacement candidates for hexavalent chromate based DoD-P-15328 wash primer included corrosion resistance and adhesion, as well as process controls and ease of application during manufacturing.

A particular concern for this study is the armor steel itself. The yield strength for MIL-A-46100 High Hard armor steel used in Stryker and other weapon systems is at or beyond the 1400 MPa[6] level commonly noted for high risk to cracking via hydrogen exposure near the surface. This form of hydrogen attack usually referred to as EAC, can occur as a result of corrosion in a natural environment and can also be introduced via aqueous surface processing procedures such as cleaning or plating. Zinc phosphate as a pretreatment for steels is common in the automotive industry, however, it is normally used on much lower strength steels. The phosphating process is aqueous by nature and includes acidic steps. Due to the high strength level of the MIL-A-46100 armor steel, the potential for environmentally assisted cracking from potential hydrogen sources such as a spray zinc phosphate remains a critical concern and cannot be underestimated. Funding for this study was provided by the U.S. Army Tank and Automotive Command (TACOM) via PM BCT.

EXPERIMENTAL PROCEDURE

High hard armor steel test panels (120 each nominally 10.16 cm x 15.24 cm x 0.635 cm) of MIL-A-46100 were prepared. Prior to pretreatment and testing, all panels were clearly designated for their respective pretreatments. Due to the extreme hardness of the material, panel identification was executed via removal of panel corners through surface grinding. The three panel processing sets and their quantites prepared were identified as follows:

(1) All 4 corners intact - Abrasive Blast (30)
(2) One corner removed – Abrasive Blast with CHEMINHIB 420 (30)
(3) Two corners removed – Abrasive Blast with spray zinc phosphate (30)

Each of the 3 sets of test panels were further subdivided and coated with 2 different primer/topcoat systems:

(1) MIL-P-53022[7] (type II) primer with MIL-C-53039[8] topcoat (15)
(2) MIL-P-53022 (type II) primer with MIL-PRF-64159[9] (type II) topcoat (15)

An additional control set of 30 panels representative of legacy processing methods was prepared with hexavalent chromium containing DoD-P-15328 and was coated with MIL-P-53022 (type I) primer and MIL-C-46168[10] topcoat. All of the test panels' respective primer coats were applied within 24 hours after the surface preparation and pretreatment. The abrasive blasting step was performed on all panels and consisted of SiC media applied to a SSPC-SP 10[11] near white metal finish with a 1.5 mil profile. Applications of spray zinc phospate and CHEMINHIB 420 surface pretreatments were performed on the blasted panels following manufacturer's instructions and procedures used at the GDLS London facility. To avoid confusion, it should be noted that KLEENINHIB 486[12] product used in cleaning and shipping phases for bare Stryker hulls is also known by the trade name CHEMKLEEN 340 in the United States. Both of these products are chemically identical. The CHEMINHIB 420 was prepared from PPG Industries RECOOL® ADDITIVE 15[13] concentrate. Additional spray zinc phosphated panels were set aside and sampled for coating weight using a heated chromic acid stripping solution. The spray zinc phosphate applied on the 46100 corrosion test panels fell within the range for zinc phosphate coating weight specified in TT-C-490 and averaged 320mg/ft^2. The application procedures for the different pretreatment scenarios as performed in production and in laboratory conditions are described in Figures 2 - 4.[14-17]

Two corrosion test chambers, a Harshaw Model 22 standard neutral salt fog and an Attotech Model CCT-NC-30 cyclic corrosion test chamber were used to evaluate the armor steel test panels. Immediately prior to exposure, the panels were all "X" scribed using a standard carbide tipped hardened steel scribe. The scribed panels were placed into the respective chambers and exposed to ASTM B 117[18] and General Motors 9540P[19] accelerated corrosion test environments. The ASTM B 117 neutral salt fog conditions are 95°F with saturated humidity and atomized fog of 5% NaCl solution. The GM 9540P test consists of 18 separate stages that include the following: saltwater spray, humidity, drying, ambient, and heated drying. The environmental conditions and duration of each stage for one complete 9540P cycle are provided in Table 1. The standard 0.9% NaCl, 0.1% $CaCl_2$, 0.25% $NaHCO_3$ test solution was used. In addition the cyclic chamber was calibrated with standard steel mass loss calibration coupons as described in the GM 9540P test specification.

In order to quantify the corrosion, ASTM B 117 salt fog panels were numerically rated for scribe corrosion creepback damage at weekly intervals up to 1512 hours using method ASTM D 1654.[20] The GM 9540P panels were also rated using ASTM D 1654 for scribe creepback corrosion at 10, 20, 40 60, 70, and 80 cycles. In addition to the numeric ratings, representative panels chosen from the first exposure interval were digitally scanned at each rating interval and saved as high quality graphics files at 600 dpi resolution to visually chronicle the corrosion damage. To facilitate easier viewing of the inevitable large quantities of raw data from this test matrix, color codes were assigned based upon ranges of ASTM D 1654 ratings. Table 2 depicts the ASTM D 1654 rating parameters and also defines the colors and their respective rating ranges. For panels where coating blistering from corrosion occurred away from the scribe, a diagonal cross hatched pattern was added to the data cells in addition to the color coding.

For evaluating coating adhesion to the different surface treatments, pull-off adhesion was chosen as the basis for adhesion evaluation over scribe and tape based methods such as ASTM D 3359[21]. Tape based measurements were attempted, however, the coarse, irregular surface topography and extreme substrate hardness inherent to 46100 armor steel led to rapid dulling and even breaking of scribing tools as well as inconsistent and unusable results. The pull-off adhesion measurements assessing the effects of pretreatment or surface preparation at the substrate were performed in accordance with ASTM D 4541[22]. An Elcometer Model 108 Hydraulic Adhesion Test Equipment (HATE) was used for this procedure. In addition to being a more quantitative test method, pull-off adhesion is also less prone to human elements in testing such as variations in pressure applied during scribing as well as interpretation and perception of results. For the pull-off adhesion test, a loading fixture commonly referred to as a "dolly" is secured normal to the coating surface using an adhesive. The adhesive used was cyanoacrylate. After allowing the adhesive to cure for 24 hours at 40 °C in dry conditions (Table 3), the attached dolly was inserted into the test apparatus. The load applied by the apparatus was gradually increased and monitored on the gauge until a plug of coating was detached. The failure value (in psi) was recorded and the failure mode was characterized. The pull-off test apparatus and dolly configuration are illustrated in Figure 5. For pull-off data to be valid, the specimen substrate must be of sufficient thickness to ensure that the coaxial load applied during the removal stage does not distort the substrate material and cause a bulging or "trampoline effect." When a thin specimen is used, the resultant bulge causes the coating to radially peel away outwards from the center instead of being uniformly pulled away in pure tension and thus results in significantly lower readings than for identically prepared specimens at greater substrate thickness. At 0.25 inches, all of the armor panels evaluated in the test matrix had adequate thickness for valid pull-off test results. Measurements were obtained by utilizing the 5 remaining specimens from each pretreatment coated with MIL-P-53022 (type II) primer with MIL-PRF-64159 (type II) topcoat except for the controls which used DoD-P-15328 and were coated with MIL-P-53022 (type I) primer and MIL-C-46168 topcoat. Pull-off adhesion measurements attempted with the MIL-C-53039 topcoated specimens all resulted in coating interlayer adhesive failures at the primer/topcoat interface and thus could not be used to assess the relative effect on coating adhesion imparted by the pretreatment or surface preparation at the substrate. Any other failure measurements due to coating separation between the primer and topcoat layers or at the topcoat surface with the cyanoacrylate adhesive were also rejected.

In order to assess the relative resistance of pretreatment alternatives to flash rusting in a production environment, humidity testing was conducted on test panels prepared with only the surface preparation or pretreatment alternatives. Surplus panels from the adhesion study were abrasive blasted and prepared in four different configurations with 5 replicate panels for:

1.) Abrasive blast only
2.) Abrasive blast with KLEENINHIB 486, sealed with CHEMINHIB 420
3.) Abrasive blast and spray zinc phosphate (Bonderite 952 w/Henkel clean stages)
4.) Abrasive blast with KLEENINHIB 486 and spray zinc phosphate (Bonderite 952)

The humidity exposure was conducted in accordance with ASTM D 1735[23] at saturated levels for 16 hours. Additional panels were also set aside in ambient laboratory conditions with relative humidity ranging from 60 - 70 % for one week. For both exposures, the panels were digitally scanned prior to as well as after exposure.

Finally, resistance to environmentally assisted cracking was assessed using the rising step load method for determination of K_{IEAC}. For this procedure, CV2 Charpy specimens of MIL-A-46100 and VAR 4340 were machined in Longitudinal - Transverse (L-T) and Transverse - Longitudinal (T-L) orientations in accordance with ASTM E 399[24,25]. Table 4 contains mechanical properties typical for these alloys. The pretreatments evaluated for these specimens included: DoD-P-15328 wash primer, and spray zinc phosphate for the MIL-A-46100 steel and only spray zinc phosphate for the 4340 steel. Unlike the 46100 test panels, the charpy specimens were not abrasive blasted prior to pretreatment. The 4340 steel specimens were included for cross referencing of MIL-A-46100 to known materials. Specimen fatigue precracking was carried out using three stages, each consisting of decreasing loading levels. In the first precracking stage, the load was maintained to keep stress intensity values below 80% of the estimated experimental critical stress intensity and the stress ratio ($\sigma_{max}/\sigma_{min}$) was kept between –1 and +0.1. In the intermediate stage, the cycling load was reduced to maintain the stress intensity value as crack growth occurred and the intact cross section was reduced. For the final stage of precracking, the load was further reduced so the final value of K_{max} was unlikely to exceed 60 % of the estimated value for K_I during experimentation. Additionally, the final value for K_{max}/E did not exceed 0.0032 $m^{1/2}$, where E is Young's Modulus. Precrack length, represented by the dimensionless expression a/W (crack length over specimen width), was maintained near 0.5.

Specimens were then fastened into a double cantilever array test fixture under aqueous conditions with 3.5% NaCl solution at open circuit potential conditions. Specimens were loaded by incremental steps in accordance with ASTM F 1624[26] using an Instron model 4301 hydraulic load frame apparatus. The specimen load values versus time were recorded via Labview interface to a computer data file. The calculation for the onset of environmentally assisted cracking, or K_{IEAC}, was derived as follows for cantilever bending from the four point bending expression.

$$K_{IEAC} = \left(\frac{6M_{IEAC}\sqrt{\pi a}}{BW^2} \right) \times f(a/W) \qquad [1]$$

where:

$$f(a/W) = 1.122 - 1.40(a/W) + 7.33(a/W)^2 - 13.08(a/W)^3 + 14.00(a/W)^4 \qquad [2]$$

Where: M_{IEAC} = the applied moment necessary to induced cracking,
B = the specimen thickness,
W = the specimen depth, and
a = the initial crack length.

For double cantilever bending conditions, the term M_{IEAC} becomes P_{IEAC}, where P_{IEAC} is the load necessary to induce critical cracking and S is the length of the moment arm. This yields the utilized expression for K_{IEAC}[27,28]

$$K_{IEAC} = \left(\frac{6P_{IEAC}S\sqrt{\pi a}}{BW^2} \right) \times f(a/W) \qquad [3]$$

Where f(a/W) varies between 2.3 and 3.2 for a/W values between 0.45 and 0.55, respectively. The initial crack length was determined by averaging the measured values visible on each side of the charpy specimen that ran perpendicular to the crack front. The K_{IEAC} values calculated were compared among the pretreatments and against historic values measured for untreated high hard steel to determine whether any pretreatment degraded the material and exacerbated environmentally assisted cracking.[29,30] The dimensions for the CV2 specimens, their orientations with respect to the major axis' of the rolled armor plate, and the configuration of the rising step load equipment are shown in Figures 6 - 9.

RESULTS

Pull-Off Adhesion:

With the exception of specimens pretreated with DoD-P-15328 wash primer, all of the surface pretreatments showed consistent repeatable readings for adhesion data collected and plotted in Figures 10 - 13 and listed in Table 5. As mentioned in the procedure, the readings for abrasive blast, abrasive blast with CHEMINHIB 420, and spray zinc phosphate, were all collected from panels coated with MIL-P-53022 (Type 2) primer and MIL-DTL-64159 waterborne topcoat. Pull-off readings were attempted on the MIL-C-53039 coated specimens however the pull-off separations all occurred consistently in the 2200 – 2500 psi range as adhesive separations *between* the MIL-P-53022 (Type 2) primer and the MIL-C-53039 topcoat and were thus not applicable for evaluation of pretreatment effects on adhesion. Abrasive blast and abrasive blast prepared with CHEMINHIB 420 water soluble flash rust inhibiting coating, the two procedures currently used at GDLS for Stryker production in particular had very tightly grouped data and both averaged within 100 psi of each other at 3411 psi and 3494 psi respectively. For spray zinc phosphate pretreated panels the average pull-off adhesion reading was 3304 psi, down roughly 150 psi from the average readings from panels prepared with abrasive blast and abrasive blast with CHEMINHIB 420. All of the current and proposed pretreatments had significantly better adhesion performance than the legacy system consisting of DoD 1538 wash primer with MIL-P-53022 (Type 1) primer with MIL-C-46168 2 component solventborne topcoat. The increases ranged from 16% for spray zinc phosphate to 21% for abrasive blast with CHEMINHIB 420 over the wash primer baseline average pull-off adhesion value of 2775 psi.

Accelerated Corrosion:

For all of the panels exposed in ASTM B 117 neutral salt fog and GM 9540P, the dominant failure mode was creepback from the scribe via blistering as rated under ASTM D 1654. In addition to the creepback rating, severity color coding described in Table 2, additional corrosion characterization such as blistering away from the scribe was illustrated through application of *diagonal* cross-hatching texture to the color coded cells.

For neutral salt fog testing, many of the panels, especially the abrasive blast and the abrasive blast with CHEMINHIB 420 panels, had secondary blistering away from the scribed areas. For these panels, onset exposures for the nonscribe blistering occurred at 500 and 1176 hours, respectively, for the waterborne and solventborne topcoated specimens. Interestingly, despite poorer performance for nonscribe areas, scribe creepback ratings for waterborne topcoated abrasive blast, and abrasive blast with CHEMINHIB 420 were slightly improved versus their solventborne counterparts. The spray zinc phosphate and wash primer treated panels had much better performance, only 2 of the waterborne topcoated zinc phosphate had any blistering away from the scribe, the solventborne topcoated zinc phosphated panels and all of the wash primer treated panels remained blister free in the nonscribe areas through 1500 hours. For blistering at the scribe, the wash primer panels fared no better than abrasive blast or abrasive blast with CHEMINHIB 420. When evaluating the neutral salt fog panels for overall performance, in scribed and nonscribed areas, the vast majority of spray zinc phosphated panels performed better than or equal to the other pretreatments across all replicates. Table 6 displays the ratings at various exposures and characterizes the corrosion using the color coding and patterns described above. Digital photographs comparing the pretreatments with waterborne and solvent-borne topcoats and also the legacy system with DoD-P-15328 and MIL-C-46168 at 1500 hours are displayed in Figures 14 - 16.

The performance gains noted in neutral salt fog exposure for spray zinc phosphate were even more pronounced in GM 9540P cyclic exposure. Spray zinc phosphate scribe creepback ratings showed roughly a 2 to 1 improvement in performance ratio throughout the 80 cycle duration vs. abrasive blast and abrasive blast with CHEMINHIB 420. None of the panels tested in GM 9540P exhibited blistering away from the scribe. In contrast to salt fog, no significant differences were observed between waterborne and solventborne topcoated panels. Cyclic corrosion performance at 80 cycles for DoD-P-15328 wash primer was inconsistent with some of the replicate panels performing very well while others rated low or failed completely due to excessive blister growth from the scribed region. The improvement in scribe creepback corrosion performance gained using the spray zinc phosphating is visible in Table 7 and is apparent across all of the exposure intervals. As in neutral salt fog, digital photographs comparing the pretreatments with waterborne and solventborne topcoats and also the legacy system with DoD-P-15328 and MIL-C-46168 at 80 cycles are displayed in Figures 17 - 19.

Resistance to Flash Rusting:

In order to assess pretreatments from a production perspective, the three main test pretreatments: abrasive blast, abrasive blast with CHEMINHIB 420, spray zinc phosphate were applied to the 46100 steel and exposed without primer and topcoat to ambient laboratory conditions at 27°C with relative humidity ranging from 60 - 70% and to ASTM D 1735 at 35°C with 100% relative humidity. One additional hybrid set of panels was prepared using the KLEENINHIB 486 process which is the surface cleaning step used for CHEMINHIB 420 but was then spray zinc phosphated using the BONDERITE 952 with the PARCOLENE 99 rinse thus effectively substituting the existing production cleaning step for the Henkel surface prep stages. Before and after exposure scans of the treatments are pictured in Figures 20 - 23. The initial surfaces showed almost no difference in appearance between the abrasive blast and abrasive blast with CHEMINHIB 420 prepared panels. The spray zinc phosphate and spray zinc phosphate cleaned with KLEENINHIB 486 both showed effective deposition of the zinc phosphate crystals however, there was also noticeable flash rusting present on the KLEENINHIB 486 cleaned spray zinc phosphate panels. Panels exposed in 16 hours at 100% RH show rusting on all of the pretreated panels. Of these panels, the CHEMINHIB 420 treated abrasive blast panels had the least amount of rusting among the four surface pretreatments with most of the rusting occurring within the surface pores commonly found on 46100 steel. Surprisingly, despite the initial head start in flash rust, there was less rusting present on the KLEENINHIB 486 prepared spray zinc phosphate panels after 16 hours versus the standard Henkel cleaned spray zinc phosphate panels. For the laboratory exposed panels, there was no significant visible damage to any panels except for the panels prepared with only abrasive blast which only showed localized flash rusting.

Rising Step Load (K_{IEAC}):

Figure 24 shows the average K_{IEAC} value for each alloy/pretreatment configuration plotted with respect to notch orientation. The complete listing of all K_{IEAC} values, details, and the crack initiation loads are listed in Table 8. It can be seen that experimental results fall within the historic ranges of the respective alloys without a protective pretreatment. As expected, MIL-A-46100 performed better than the VAR 4340. It was also expected that, for all samples tested, long-transverse (LT) orientations would be more resistant to environmentally assisted cracking than transverse-longitudal (TL) orientations. While the LT wash primer 46100 samples had a highest average K_{IEAC}, it was only 5% higher than its TL counterpart. Moreover, the reverse was true of the phosphated 46100 where the K_{IEAC} for the TL orientation was 4% higher than the LT. These results show no clear trend to confirm that either orientation is more resistant to EAC than the other. In addition, there is no evidence to suggest that either spray zinc phosphate or DoD-P-15328 wash primer had an adverse influence on the K_{IEAC} of either 46100 or 4340.

DISCUSSION

The goal of this study was to investigate alternative pretreatments for hexavalent chromium based DoD-P-15328 wash primer. Due to the less aggressive nature of the current GDLS surface preparations, most of the discussion will concern the proposed spray zinc phosphate process. It is the aqueous nature of the spray zinc phosphate application process that provides questions as to the possible risk for EAC. During early production of the U.S. Marine Corps Light Armored Vehicle (LAV) at the London, Ontario facility (formerly GM Defense) that is the current site for Stryker production, there were numerous problems with EAC in the 46100 armor plate, notably the hatch covers and doors. The typical areas of crack nucleation were near the edges where cuts had been made. A previous U.S. Army Materials Technology Laboratory (MTL) investigation revealed that most of this damage from cracking could be avoided by changing the method for cutting the armor plate. MTL recommended a change from underwater plasma arc cutting to laser cutting techniques. By using laser cutting, the thicknesses of the remaining untempered martensite layer and the heat affected zone at the cut were significantly reduced.[31] After implementing the MTL recommendations, the cracking problems were greatly reduced. In addition to these improvements, the enhanced precision inherent with the laser cutting improved the overall fit of individual armor plate components and significantly reduced the preloading induced at subsequent joining stages. Despite these significant production improvements the risk for EAC remains a concern for Stryker fabrication, especially for any process or situation that when introduced could cause an influx of hydrogen.

Among the alternative pretreatments investigated, the hydrogen ions in the phosphoric acid component, present in the Henkel spray zinc phosphate process remains the key area of concern. Before the procedures that investigated sensitivity to EAC on phosphated high hard are discussed, it should be noted that the DoD-P-15328 wash primer pretreatment coatings, used extensively on LAV's and on early production Stryker vehicles also contained phosphoric acid. In addition, the CHEMINHIB 420 and KLEENINHIB 486 surface preparation stages currently used in conjunction with abrasive blast at the GDLS London facility and at the Militex subcontracted facility both are water soluble solutions for cleaning and inhibitor purposes. Any of these "wet" preparation stages listed represent varying degrees of vulnerability to hydrogen infiltration. The proposed zinc phosphating method, like the wash primer is a spray process so the risks for hydrogen influx in spray application versus immersion methods for phosphating are likely to be inherently diminished. Another decrease in spray zinc phosphate risk comes from the recently revised Federal Specification TT-C-490D allowing the use of nonchromate final rinses; this eliminates hexavalent chromium based chromic acid, another potential source of hydrogen.

For the spray zinc phosphate sprayed rising step load EAC specimens there was no significant drop in K_{IEAC} versus the DoD-P-15328 wash primer treated specimens for either the L-T or T-L orientations. All values were within the K_{IEAC} historic range for 46100 high hard steels. Spray zinc phosphate that was applied to precracked samples even appeared to impart more EAC resistance than when the spray zinc phosphate was applied prior to precracking. As seen in Table 8, regardless of orientation, the phosphate applied to precracked samples had on average, approximately 16% higher K_{IEAC} values. One possible explanation is that during the application process, phosphate may have penetrated into the precrack and thus provided the additional EAC resistance. Despite these interesting results, more tests with a larger statistical sampling would be necessary to validate this observation.

Results for coating adhesion for spray zinc phosphate as well as the abrasive blast and abrasive blast with CHEMINHIB 420 were significantly higher as well as more consistent than results for DoD-P-15328. Coating adhesion performance should improve regardless of which of the wash primer alternatives is selected.

Corrosion resistance in neutral salt fog was greater than or equal to DoD-P-15328 for all pretreatment alternatives. For GM 9540P, wash primer fared better in corrosion resistance versus the abrasive blast and abrasive blast with CHEMINHIB 420 treated panels. Performance for the wash primer samples was mixed when compared to spray zinc phosphate with some panels performing significantly better and others performing significantly worse. Some of the creepback failures of the wash primer exhibited very large blistering yet with little red rust staining. This rust free blistering, and inconsistent data in GM 9540P for the DoD-P-15328 wash primer correlated with the lower adhesion readings. This relationship suggests that while wash primer effectively inhibits corrosion (rusting) of the substrate, its promotion of adhesion to the substrate is less reliable. The spray zinc phosphate provided improved corrosion performance versus the abrasive blast and abrasive blast with CHEMINHIB 420 prepared panels and nearly matched the adhesion performance of abrasive blast prepared panels.

A key area of interest for surface pretreatments is the processing in the production environment. Currently abrasive blasting for the GDLS London plant is done off-facility at Militex, a local subcontractor responsible for cleaning and mechanical surface preparation and also a portion of the MIL-P-53022 primer coating application. The vehicles that are not primer coated at Militex receive an additional KLEENINHIB 486 water soluble cleaner/inhibitor process step in preparation for the interim transit period before the primer application at the GDLS facility. The processes as they stand currently are quite simple, Figures 2 - 4 highlight differences between the surface preparation and pretreatment stages conditional to the location of primer application as well as include the Henkel Bonderite spray zinc phosphate steps. In addition to introducing several more surface preparation stages, the pot temperatures on three of the Bonderite chemical spray steps must be heated to within a specified range. Additional care must be taken by the process personnel to control the Bonderite 952 stage with respect to "free acid" and "total acid" through a series of quality control inspections via titrations as specified in the Henkel Technical Process Bulletin. When the free acid levels were elevated, more orange-brown staining occurred on the zinc phosphated panels at the PARCOLENE 99 rinse stage. This staining was present in varying degrees for all of the spray zinc phosphated test panels that were prepared. Specifications TT-C-490 and MIL-DTL-16232[32] for phosphating both mention the acceptability of these stains. Despite these stains, the zinc phosphate treated panels performed well in corrosion resistance and adhesion suggesting that zinc phosphating is somewhat forgiving from a process control standpoint and that the specification allowance for some staining appears to be reasonable. Through optimizations to free acid and total acid levels and periodic inspections to check the zinc phosphate coating weight versus the specifications, the quality and appearance of the phosphate coating can be optimized as seen in Figure 25.

If spray zinc phosphating is implemented in production as the pretreatment step before primer application, additional steps can be taken to reduce the risk for EAC. Phosphating specifications TT-C-490, MIL-DTL-16232, and a previous phosphating study[33] all include instructions for hydrogen embrittlement heat treatment. Heat treatment schedules found in these publications consist of ageing at room temperature and should be executed before any loading of the material is permitted. The room temperature dwell times recommended differ between the specifications at 120 hours and 240 hours respectively for MIL-DTL-16232 and TT-C-490. Although the ageing period in MIL-C-16232 is shorter, this specification also includes a quality control test to determine the adequacy of the hydrogen embrittlement relief. The specification recommends periodic checks at intervals of 90 to 120 days in a production environment to ensure that the hydrogen embrittlement relief remains adequate.

Regardless of what pretreatment method is implemented, every attempt should be made throughout all stages of production to limit and relieve possible residual tensile stresses in the 46100 at locations such as welds or where cold working and/or machining has occurred. The manufacturing improvements made during previous LAV production cycles have done much to limit these residual stresses. Additional guidelines for residual stress relief as it applies to phosphating are also available in MIL-DTL-16232.

CONCLUSIONS

- Spray Zinc Phosphate pretreatment improved the creepback resistance for CARC coated MIL-A-46100 High Hard armor steel for both ASTM B 117 neutral salt fog and GM 9540P cyclic corrosion.
- Coating adhesion for Spray Zinc Phosphate Pretreatment vs. abrasive blasting alone was marginally lower but was still significantly improved when compared to DoD-P-15328 wash primer.
- PPG CHEMINHIB 420 water soluble flash rust inhibiting coating did not degrade adhesion of the MIL-P-53022 (Type II) primer to the substrate versus substrates that were prepared with only abrasive blasting.
- PPG CHEMINHIB 420 water soluble flash rust inhibiting coating greatly increased resistance to flash rusting when applied to abrasive blasted panels.
- If spray zinc phosphating is not used as a pretreatment after abrasive blasting, the continued use of PPG CHEMINHIB 420 over abrasive blasted surfaces is recommended.
- The results of the rising step load tests, indicate that spray zinc phosphate had no adverse effect on the K_{IEAC} for MIL-A-46100 High Hard armor.
- When spray zinc phosphate is used, quality control and inspection procedures in accordance with MIL-DTL-16232 for hydrogen embrittlement relief should be followed.

ACKNOWLEDGEMENTS

The authors thank the following people for their knowledge and support during the performance of this study. Carmen Uffen, Keith Clavett, and Lilliana Pereira of GDLS, London, Ontario for information and assistance on current coating processes. Janet Byrne of PPG Industries, Euclid, OH, who cross referenced the Canadian and U.S. product names, supplied PPG product samples for evaluation, and information for product mixing/application procedures. John Beatty and Jonathan Montgomery of the U.S. Army Research Laboratory for their additional knowledge in high strength armor steels and assistance in troubleshooting the tensile test load frame. Finally, the authors wish to recognize Terry Dean and Geoff Hoerauf of PM BCT for their funding, guidance, and most of all their patience with respect to the unanticipated additional time needed for completion of this study.

REFERENCES

1. H. Gibb, P. Lees, P. Pinsky, B. Rooney, "Lung Cancer Among Workers in Chromium Chemical Production," American Journal of Industrial Medicine, pp. 115-126 Volume 38, Issue 2, Published Online: 7 Jul 2000

2. MIL-A-46100D, (w/Amendment 1) "Armor Plate, Steel, Wrought, High-Hardness," U.S. Department of Defense, Washington DC, June 1990

3. DoD-P-15328D (Note 1), Primer (Wash), Pretreatment (Formula No. 117 for Metals)," U.S. Department of Defense, Washington DC, March 1992

4. TT-C-490D Federal Specification, "Chemical Conversion Coatings and Pretreatments for Ferrous Surfaces (Base for Organic Coatings)," General Services Administration, Washington DC, July 2002

5. CHEMINHIB® 420 Water Soluble Short Term Corrosion Protective Coating - Product Data Sheet, PPG Industries, Inc., Euclid, OH, May 1996

6. Bruce, Craig, "Hydrogen Damage," Metals Handbook Ninth Edition, Vol. 13, p 168

7. MIL-P-53022-10, "Primer, Epoxy Coating, Corrosion Inhibiting, Lead and Chromate Free," U.S. Department of Defense, Washington DC, February 1992

8. MIL-C-53039A, "Coating, Aliphatic Polyurethane, Single Component, Chemical Agent Resistant," U.S. Department of Defense, Washington DC, May 1993

9. MIL-DTL-64159, "Coating, Water Dispersible Aliphatic Polyurethane, Chemical Agent Resistant," U.S. Department of Defense, Washington DC, January 2002

10. MIL-C-46168, "Coating, Aliphatic Polyurethane, Chemical Agent Resistant," Revision D, Amendment 3, U.S. Department of Defense, Washington DC, May 1993

11. SSPC-SP 10, "Near White Blast Cleaning," Steel Structures Painting Council, Pittsburgh, PA, November 1982

12. KLEENIHIB® 486 (CHEMKLEEN® 340 – U.S. Trade Name)) Cleaning Solution - Product Data Sheet, PPG Industries, Inc., Euclid, OH, May 1996

13. RECOOL® ADDITIVE 15 Concentrated Rust Inhibiting Solution Coating - Product Data Sheet, PPG Industries, Inc., Euclid, OH, March 2003

14. PARCO® CLEANER 319 Alkaline Cleaner - Technical Process Bulletin No. 0993, Henkel Surface Technologies, Madison Heights, MI, April 1990

15. FIXODEEN® ZL Conditioning Pretreatment - Technical Process Bulletin No. 1569, Henkel Surface Technologies, Madison Heights, MI, November 1993

16. BONDERITE® 952 Polycrystalline Conversion Coating Process - Technical Process Bulletin No. 1046, Henkel Surface Technologies, Madison Heights, MI, April 1997

17. PARCOLENE® 99A Chromium Free Post Treatment - Technical Process Bulletin No. 1695, Henkel Surface Technologies, Madison Heights, MI, July 1997

18. ASTM B117 – 90, "Standard Method of Salt Spray (Fog) Testing," ASTM, West Conshohocken, PA, 1990

19. GM 9540P, "Accelerated Corrosion Test; GM 9540P," General Motors Engineering Standards, 1997

20. ASTM D 1654-79A, "Standard Method for Evaluation of Painted or Coated Specimens Subjected to Corrosive Environments," ASTM, West Conshohocken, PA, 1984

21. ASTM D 3359, "Standard Test Methods for Measuring Adhesion by Tape Test," American Society for Testing and Materials, West Conshohocken, PA, 1987

22. ASTM D 4541, "Standard Test Method for Pull-Off Strength of Coated Specimens Subjected to Corrosive Environments," American Society for Testing and Materials, West Conshohocken, PA, 1989

23. ASTM D 1735, "Standard Practice for Testing Water Resistance of Coatings Using Water Fog Apparatus," American Society for Testing and Materials, Philadelphia, PA, 1987

24. ASTM E 399-97, "Standard Test Method for Plane Strain Fracture Toughness of Metallic Materials," ASTM, West Conshohocken, PA, 1998

25. R. Pasternak, M. Pepi, G. Pelletier, C. Amos, M. Valavanis, "Test Specimens for Mechanical Property Determination," U.S. Army Materials Technology Laboratory Special Publication, MTL SP 87-5, Watertown, MA, October 1987

26. ASTM F 1624-95, "Standard Test Method for Measurement of Hydrogen Embrittlement in Steel by the Incremental Loading Technique," ASTM, West Conshohocken, PA, 1995

27. P.S. Tyler, M. Levy, and L. Raymond, "Investigation of the Conditions for Crack Propagation and Arrest Under Cathodic Polarization by Rising Step Load Bend Testing," Proceedings from the 1992 Tri-Service Conference on Corrosion, Plymouth, MA, May 1992

28. S. Grendahl, R. Kilbane, "Environmentally Assisted Cracking Concerns for Cadmium Replacement," pp. 227-245, Proceedings from the 2002 Tri-Service Corrosion Conference, San Antonio, TX, January 2002

29. W. A. Herman, G. M. Campbell "Environmental Assisted Cracking in High Hardness Armor Steel," U.S. Army Materials and Mechanics Research Center Technical Report, TR 85-28, Watertown, MA, 1985

30. D. Dawson, "Development of a Stress Corrosion Cracking Test for Armor Alloys," U.S. Army Materials and Mechanics Research Center Technical Note, TN 72-5, Watertown, MA, 1972

31. M. G. Wells, R. Weiss, J. Montgomery, T. Melvin, "LAV Armor Plate Study," U.S. Army Materials Technology Laboratory Technical Report, MTL TR 92-26, Watertown, MA, April 1992

32. MIL-DTL-16232G, "Phosphate Coating, Heavy, Manganese or Zinc Base," U.S. Department of Defense, Washington DC, November 1978

33. P. Austin, S. Chek, W. Lurie, "Investigation of the Effect of Hydrogen Embrittlement and its Relief in Phosphate Coated AISI 4340 and 1095 Steels," U.S. Naval Weapons Plant, NAVORD Report 6509, NWPW-T-20-59, Washington DC, August 1959

TABLES

Table 1: GM 9540P Cyclic Corrosion Test Details

Interval	Description	Time (min)	Temperature (±3C)
1	Ramp to Salt Mist	15	25
2	Salt Mist Cycle	1	25
3	Dry Cycle	15	30
4	Ramp to Salt Mist	70	25
5	Salt Mist Cycle	1	25
6	Dry Cycle	15	30
7	Ramp to Salt Mist	70	25
8	Salt Mist Cycle	1	25
9	Dry Cycle	15	30
10	Ramp to Salt Mist	70	25
11	Salt Mist Cycle	1	25
12	Dry Cycle	15	30
13	Ramp to Humidity	15	49
14	Humidity Cycle	480	49
15	Ramp to Dry	15	60
16	Dry Cycle	480	60
17	Ramp to Ambient	15	25
18	Ambient Cycle	480	25

Table 2: Evaluation of Scribed Coated Specimens Subjected to Corrosive Environments - ASTM D 1654[20]

Rating of Failure at Scribe (Procedure A)		
Representative Mean Creepage From Scribe		Rating Number
(Millimeters)	(Inches)	
[illegible]	[illegible]	[illegible]
Over 1.0 to 2.0	1/32 to 1/16	7
Over 2.0 to 3.0	1/16 to 1/8	6
[illegible]	1/8 to 3/16	5
[illegible]	3/16 to 1/4	4
[illegible]	[illegible]	[illegible]
Over 16.0 to more	5/8 to more	0

Table 3: Laboratory Conditions for Pull-Off Adhesion ASTM D 4541

Adhesive Type	Cyanoacrylate
Cure time (hours)	24
Temperature (C)	40
Percent Relative Humidity	~0
Substrate Material	MIL-A-46100
Substrate Thickness (in)	0.25
Substrate Surface Profile	Abrasive Blast (1.5 mil)
Pretreatment Types	DOD-P-15328 NONE CHEMINHIB 420 Spray Zinc Phosphate
Coating Types	CARC 53022(I)/46168 CARC 53022(II)/64159
Coating Thickness (mils)	~5.5 (Including Profile)

Table 4: Typical Mechanical Properties of MIL-A-46100 High Hard Armor and Other High Strength Steels[29,30]

Mechanical Property	46100 HRc 48-51	VAR 4340 HRc 52-54	AISI 1095 HRc 48-50
UTS (MPa)	1825	1916	1696
0.2% YS (MPa)	1440	1572	1393
Elongation (%)	10	12.5	7.0
RA (%)	40	37.9	19.6

Table 5: Readings (psi) for ASTM D 4541 Pull-Off Adhesion*

Abrasive Blast (None) 53022(II)/64159	CHEMINHIB 420 53022(II)/64159	Spray Zinc Phosphate 53022(II)/64159	DoD-P-15328 (Wash Primer) 53022(I)/46168
3260	3500	3250	2400
3380	3500	3400	2400
3460	3460	3450	2150
3610	3410	3400	2770
3520	3570	3260	2980
3410	3510	3300	2720
3300	3600	3200	3040
3200	3520	3310	3000
3470	3410	3460	2300
3460	3520	3610	3200
3440	3600	3090	2860
3410	3680	3400	2800
3400	3620	3540	2970
3370	3510	3390	2990
3380	3660	3240	2810
3550	3580	3380	3100
3570	3450	3410	2720
3420	3490	3400	2780
3400	3480	3000	2890
3290	3610	3240	2720
3170	3520	3420	2340
3460	3620	3160	2250
3480	3490	3260	3110
3610	3400	3240	2310
3240	3410	3310	2300
3050	3400	3410	3060
3490	3220	3320	2960
3520	3190	3200	2800
3410	3300	3260	2690
3370	3490	3050	3030
3690	3390	3300	3260
	3710	3560	2860
		3140	3100
		3230	2690
		3270	
		3830	
		3200	
		3300	
		3510	
		3480	
		3280	
		3420	
		3190	
		3100	
		3290	
		3340	
		3190	
		3260	
		2870	
		3320	

*All pull-off separation modes were adhesive at the substrate

Table 6: ASTM D 1654 Scribe Creepback Ratings for 1500 Hour ASTM B 117 Salt fog Exposure*

Panel #	Pretreatment	Primer/Topcoat	144 Hours	336 Hours	500 Hours	672 Hours	840 Hours	1008 Hours	1176 Hours	1344 Hours	1512 Hours
GB1s	Abrasive Blast	53022(II)/53039			7						
GB2s	Abrasive Blast	53022(II)/53039			7						
GB3s	Abrasive Blast	53022(II)/53039			6						
GB4s	Abrasive Blast	53022(II)/53039									
GB5s	Abrasive Blast	53022(II)/53039			8						
GB1w	Abrasive Blast	53022(II)/64159		7	6	8					
GB2w	Abrasive Blast	53022(II)/64159			7	6					
GB3w	Abrasive Blast	53022(II)/64159			8	6	6				
GB4w	Abrasive Blast	53022(II)/64159		7	7						
GB5w	Abrasive Blast	53022(II)/64159			6						
P1s	Spray Zn Phosphate	53022(II)/53039			6	6					
P2s	Spray Zn Phosphate	53022(II)/53039			7	6					
P3s	Spray Zn Phosphate	53022(II)/53039			7	8	8				
P4s	Spray Zn Phosphate	53022(II)/53039			6						
P5s	Spray Zn Phosphate	53022(II)/53039			7	6					
P1w	Spray Zn Phosphate	53022(II)/64159		7	7	6	6				
P2w	Spray Zn Phosphate	53022(II)/64159			6						
P3w	Spray Zn Phosphate	53022(II)/64159		6	5	5	6				
P4w	Spray Zn Phosphate	53022(II)/64159			6	6					
P5w	Spray Zn Phosphate	53022(II)/64159			8	6					
C1s	CHEMINHIB 420	53022(II)/53039			6						
C2s	CHEMINHIB 420	53022(II)/53039									
C3s	CHEMINHIB 420	53022(II)/53039			7						
C4s	CHEMINHIB 420	53022(II)/53039		7	7						
C5s	CHEMINHIB 420	53022(II)/53039			7						
C1w	CHEMINHIB 420	53022(II)/64159			7						
C2w	CHEMINHIB 420	53022(II)/64159			7						
C3w	CHEMINHIB 420	53022(II)/64159			7	5					
C4w	CHEMINHIB 420	53022(II)/64159			6						
C5w	CHEMINHIB 420	53022(II)/64159			7						
WP1	DoD-P-15328	53022(I)/46168	No Data	No Data	No Data	No Data	No Data	No Data	No Data	No Data	
WP2	DoD-P-15328	53022(I)/46168	No Data	[illegible]							
WP3	DoD-P-15328	53022(I)/46168	No Data	No Data	No Data	No Data	No Data	No Data	No Data	No Data	
WP4	DoD-P-15328	53022(I)/46168	No Data	No Data	No Data	No Data	No Data	No Data	No Data	No Data	
WP5	DoD-P-15328	53022(I)/46168	No Data	No Data	No Data	No Data	No Data	No Data	No Data	No Data	

* Diagonal Cross-hatched Pattern Denotes Blistering Away from Scribed Areas

Table 7: ASTM D 1654 Scribe Creepback Ratings for 80 Cycle GM 9540P Cyclic Corrosion Exposure*

Panel #	Pretreatment	Primer/Topcoat	10 Cycles	20 Cycles	40 Cycles	60 Cycles	70 Cycles	80 Cycles
GB1s	Abrasive Blast	53022(II)/53039	7	[illegible]	[illegible]	0	0	0
GB2s	Abrasive Blast	53022(II)/53039	7	[illegible]	[illegible]	[illegible]	[illegible]	[illegible]
GB3s	Abrasive Blast	53022(II)/53039	7	[illegible]	[illegible]	[illegible]	[illegible]	0
GB4s	Abrasive Blast	53022(II)/53039	6	[illegible]	[illegible]	0	0	0
GB5s	Abrasive Blast	53022(II)/53039	6	[illegible]	[illegible]	[illegible]	0	0
GB1w	Abrasive Blast	53022(II)/64159	6	[illegible]	[illegible]	[illegible]	[illegible]	[illegible]
GB2w	Abrasive Blast	53022(II)/64159	6	[illegible]	[illegible]	0	0	0
GB3w	Abrasive Blast	53022(II)/64159	6	[illegible]	[illegible]	0	0	0
GB4w	Abrasive Blast	53022(II)/64159	7	[illegible]	[illegible]	[illegible]	0	0
GB5w	Abrasive Blast	53022(II)/64159	[illegible]	[illegible]	[illegible]	[illegible]	[illegible]	0
P1s	Spray Zn Phosphate	53022(II)/53039	[illegible]	6	[illegible]	[illegible]	[illegible]	[illegible]
P2s	Spray Zn Phosphate	53022(II)/53039	[illegible]	6	[illegible]	[illegible]	[illegible]	[illegible]
P3s	Spray Zn Phosphate	53022(II)/53039	[illegible]	[illegible]	[illegible]	[illegible]	[illegible]	[illegible]
P4s	Spray Zn Phosphate	53022(II)/53039	[illegible]	[illegible]	[illegible]	[illegible]	[illegible]	[illegible]
P5s	Spray Zn Phosphate	53022(II)/53039	[illegible]	6	[illegible]	[illegible]	[illegible]	[illegible]
P1w	Spray Zn Phosphate	53022(II)/64159	[illegible]	[illegible]	[illegible]	[illegible]	[illegible]	[illegible]
P2w	Spray Zn Phosphate	53022(II)/64159	[illegible]	[illegible]	[illegible]	[illegible]	[illegible]	[illegible]
P3w	Spray Zn Phosphate	53022(II)/64159	[illegible]	6	[illegible]	[illegible]	[illegible]	[illegible]
P4w	Spray Zn Phosphate	53022(II)/64159	[illegible]	6	[illegible]	[illegible]	[illegible]	[illegible]
P5w	Spray Zn Phosphate	53022(II)/64159	[illegible]	7	[illegible]	[illegible]	[illegible]	[illegible]
C1s	CHEMINHIB 420	53022(II)/53039	7	[illegible]	[illegible]	[illegible]	[illegible]	0
C2s	CHEMINHIB 420	53022(II)/53039	7	[illegible]	[illegible]	[illegible]	[illegible]	0
C3s	CHEMINHIB 420	53022(II)/53039	6	[illegible]	[illegible]	[illegible]	0	0
C4s	CHEMINHIB 420	53022(II)/53039	[illegible]	[illegible]	[illegible]	[illegible]	[illegible]	[illegible]
C5s	CHEMINHIB 420	53022(II)/53039	6	[illegible]	[illegible]	0	0	0
C1w	CHEMINHIB 420	53022(II)/64159	6	[illegible]	0	0	0	0
C2w	CHEMINHIB 420	53022(II)/64159	[illegible]	[illegible]	[illegible]	[illegible]	[illegible]	0
C3w	CHEMINHIB 420	53022(II)/64159	6	[illegible]	[illegible]	[illegible]	0	0
C4w	CHEMINHIB 420	53022(II)/64159	6	[illegible]	[illegible]	[illegible]	0	0
C5w	CHEMINHIB 420	53022(II)/64159	6	[illegible]	[illegible]	[illegible]	[illegible]	0
WP1	DoD-P-15328	53022(I)/46168	No Data	No Data	No Data	No Data	No Data	[illegible]
WP2	DoD-P-15328	53022(I)/46168	No Data	No Data	No Data	No Data	No Data	[illegible]
WP3	DoD-P-15328	53022(I)/46168	No Data	No Data	No Data	No Data	No Data	[illegible]
WP4	DoD-P-15328	53022(I)/46168	7	6	[illegible]	[illegible]	No Data	0
WP5	DoD-P-15328	53022(I)/46168	No Data	No Data	No Data	No Data	No Data	[illegible]

* Diagonal Cross-hatched Pattern Denotes Blistering Away from Scribed Areas

Table 8: Experimental Conditions and K_{IEAC} Values for Rising Step Load Specimens

Sample ID	Sample Material	Sample Orientation	Hardness Rc	Surface Pretreatment	Crack Initiation Load P_{II} (lbs.)	K_{IEAC} (ksi√in)	Pretreatment Applied Before/After Precracking
715064	4340	TL	50.8	Spray Zinc Phosphate	52.4	15.1	After
715057	4340	LT	50.9	Spray Zinc Phosphate	54.2	15.7	After
715038	46100	LT	50.7	Spray Zinc Phosphate	70	19.5	After
715037	46100	LT	50.4	Spray Zinc Phosphate	68	19.1	After
715042	46100	LT	51	Spray Zinc Phosphate	59	17.6	Before
715048	46100	TL	51.4	Spray Zinc Phosphate	79.5	22.5	After
715047	46100	TL	51.1	Spray Zinc Phosphate	62.9	18.0	After
715051	46100	TL	51.1	Spray Zinc Phosphate	58	16.6	Before
715043	46100	LT	51.4	DoD-P-15328	71.4	20.4	After
715053	46100	TL	51.2	DoD-P-15328	68.2	19.1	After
715054	46100	TL	51.2	DoD-P-15328	68.9	19.9	After

Note: Rockwell C hardness values were averaged from 3 readings per specimen

FIGURES

Figure 1: Brigade Combat Team (BCT) Stryker Interim Armored Vehicle (IAV) in Standard Infantry Carrier Vehicle (ICV) Configuration

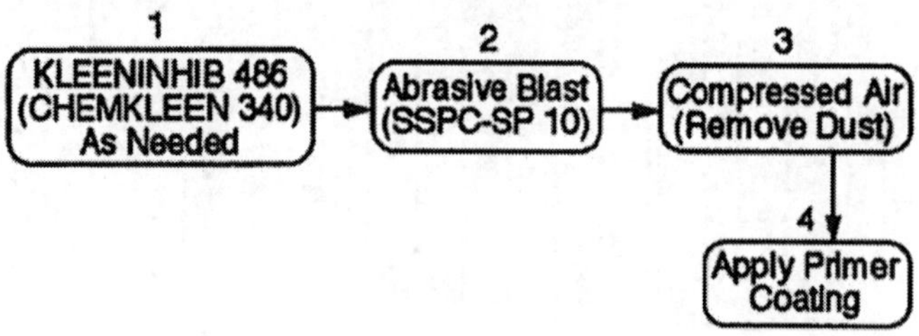

Figure 2: Process Steps for Abrasive Blast With Primer Coating Applied at MILITEX and Followed for the Laboratory Procedure

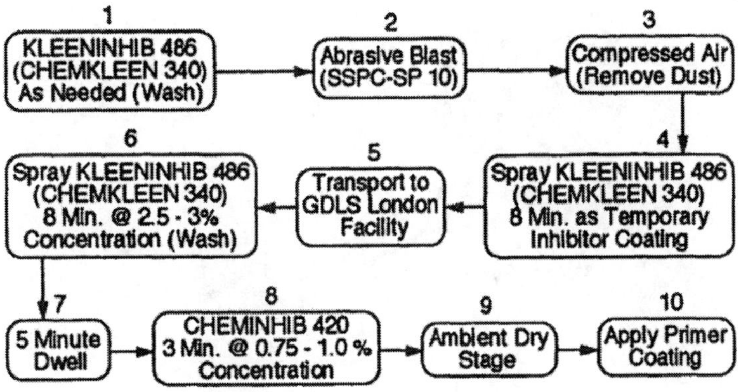

Figure 3: Process Steps for Abrasive Blast With CHEMINHIB 420 Abrasive Blast at MILITEX with Primer Applied at GDLS, London Facility and Followed for the Laboratory Procedure

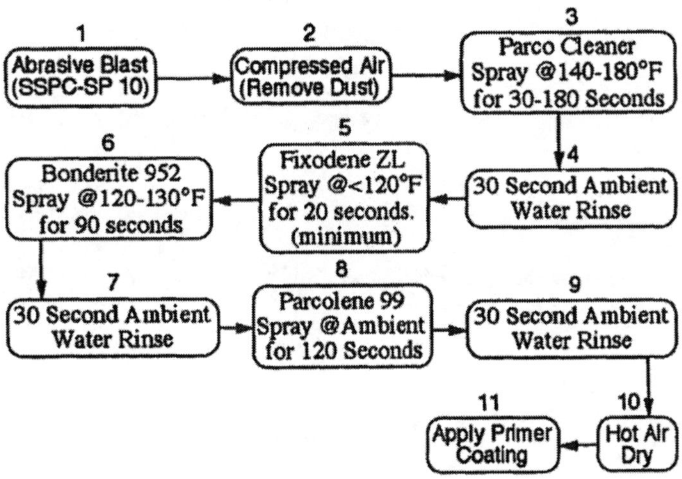

Figure 4: Process Steps for the Proposed Spray Zinc Phosphate Followed in the Laboratory Procedure *(Henkel)*[14-17]

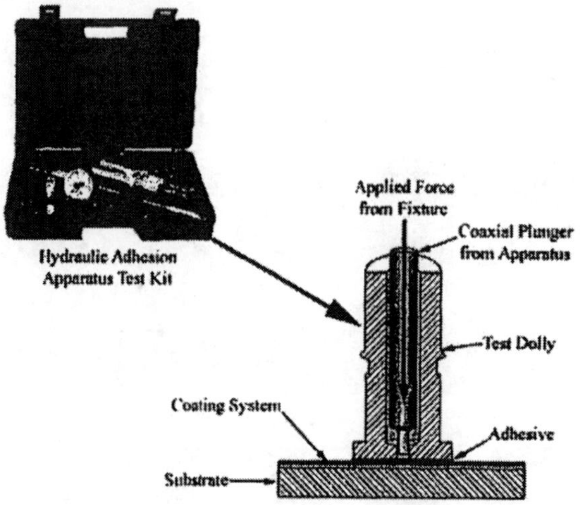

Figure 5: Pull-Off Hydraulic Adhesion Test (ASTM D 4541)

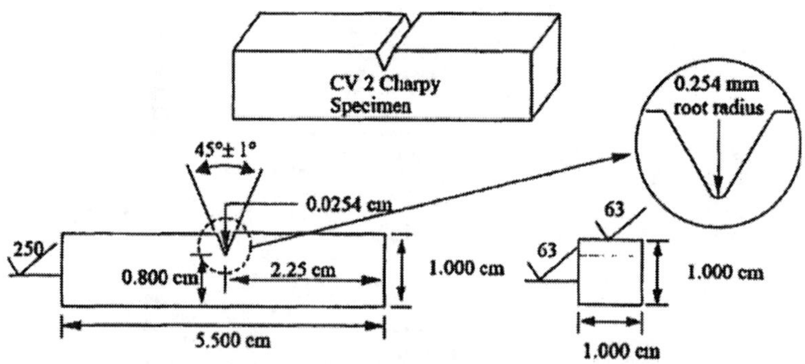

Notes:

1. Tolerances, unless noted.

Squareness	⊥ perpendicular ± 0.001 TIR
	‖ parallel ± 0.001 TIR
Decimal	.XX ± 0.127 mm
	.XXX ± 0.0254 mm
	.XXXX ± 0.0127 mm

Figure6: K_{IEAC} Test Specimen Configuration[25]

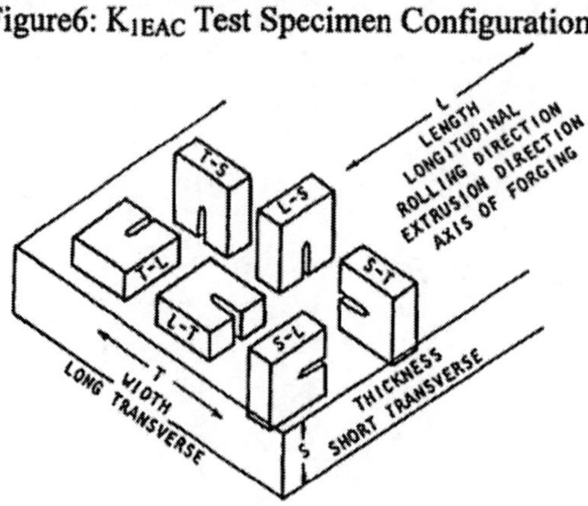

Figure 7: Crack Plane Orientation Identification Codes for Rolled Plate[24]

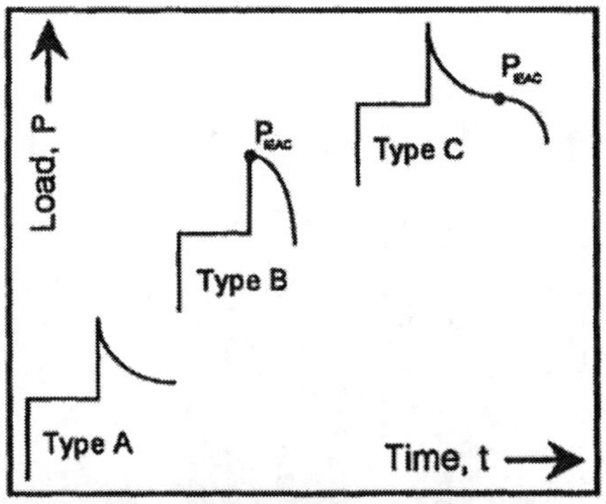

Figure 8: Identification of Crack Initiation Load P_{IEAC}

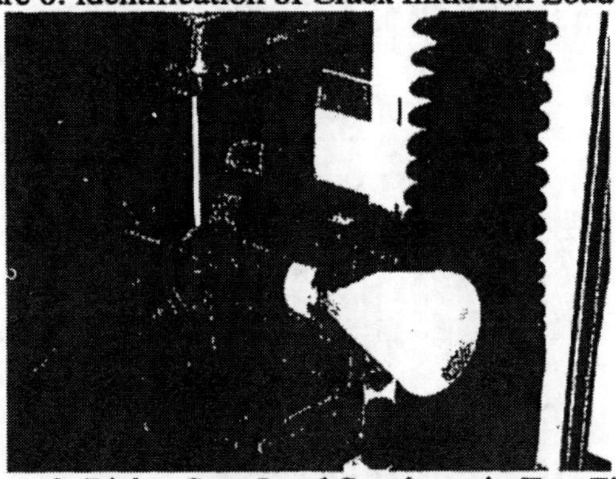

Figure 9: Rising Step Load Specimen in Test Fixture

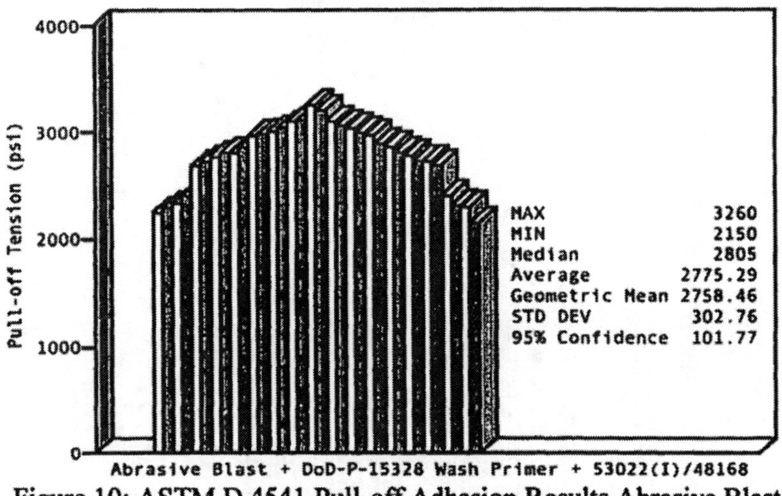

Figure 10: ASTM D 4541 Pull-off Adhesion Results Abrasive Blast with DoD-P-15328 (Wash Primer) Cr^{+6} Based Legacy Pretreatment

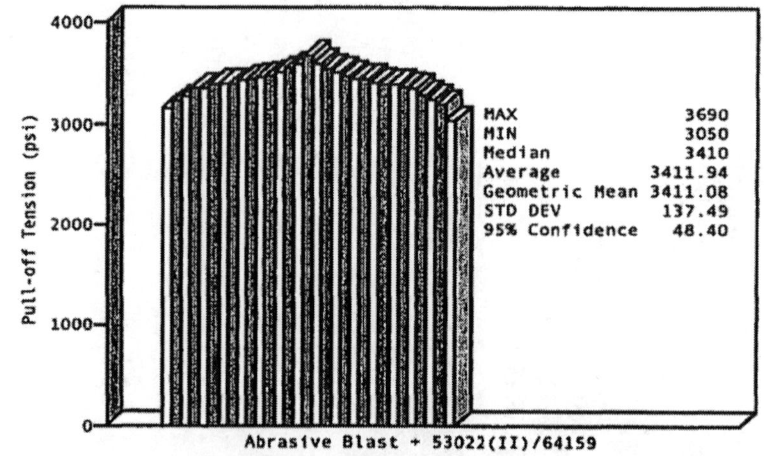

Figure 11: ASTM D 4541 Pull-off Adhesion Results for Abrasive Blast Only

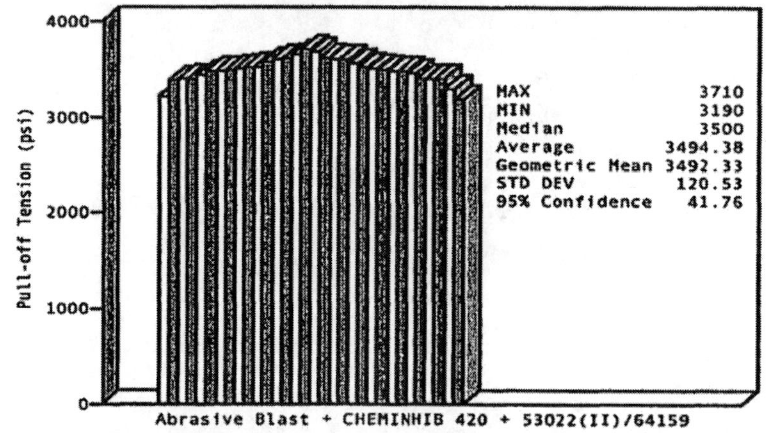

Figure 12: ASTM D 4541 Pull-off Adhesion Results for Abrasive Blast with CHEMINHIB 420 Water Soluble Inhibitor Coating

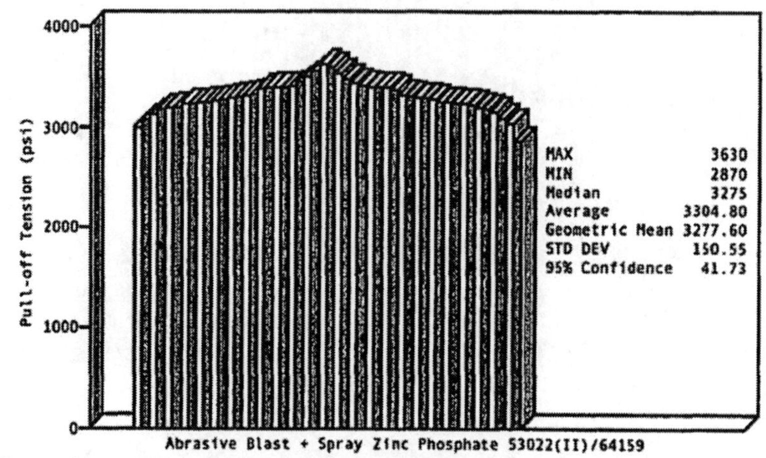

Figure 13: ASTM D 4541 Pull-off Adhesion Results for Abrasive Blast with Spray Zinc Phosphate Pretreatment

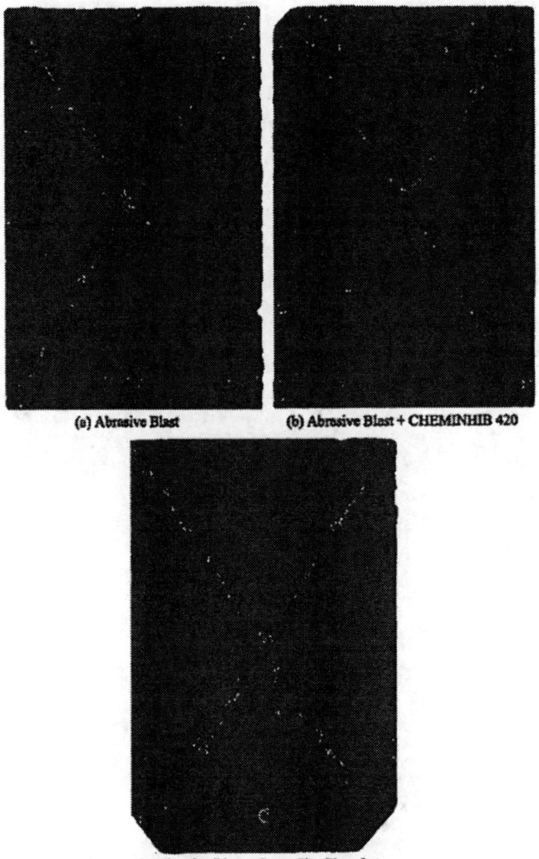

(a) Abrasive Blast (b) Abrasive Blast + CHEMINHIB 420

(c) Abrasive Blast + Spray Zinc Phosphate

Figure 14: 53022(II)/53039 @ 1500 hrs ASTM B117

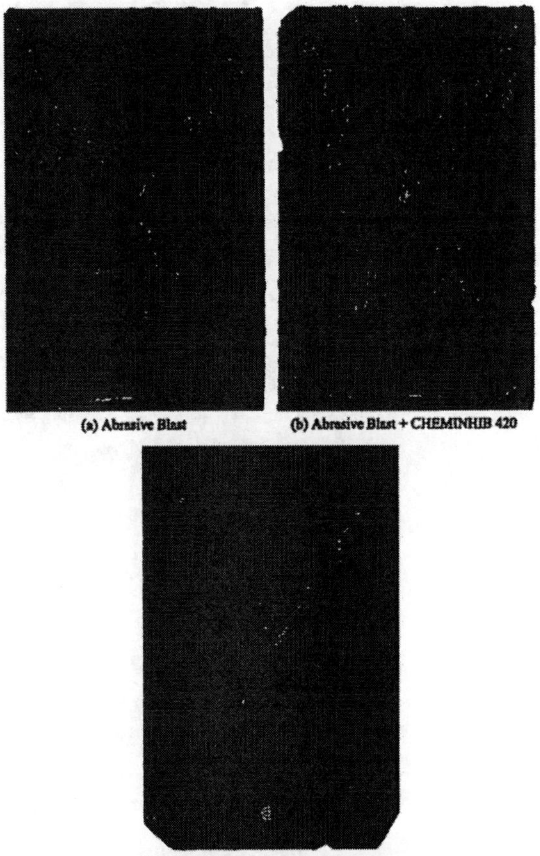

(a) Abrasive Blast (b) Abrasive Blast + CHEMINHIB 420

(c) Abrasive Blast + Spray Zinc Phosphate

Figure 15: 53022(II)/64159 @ 1500 hrs ASTM B117

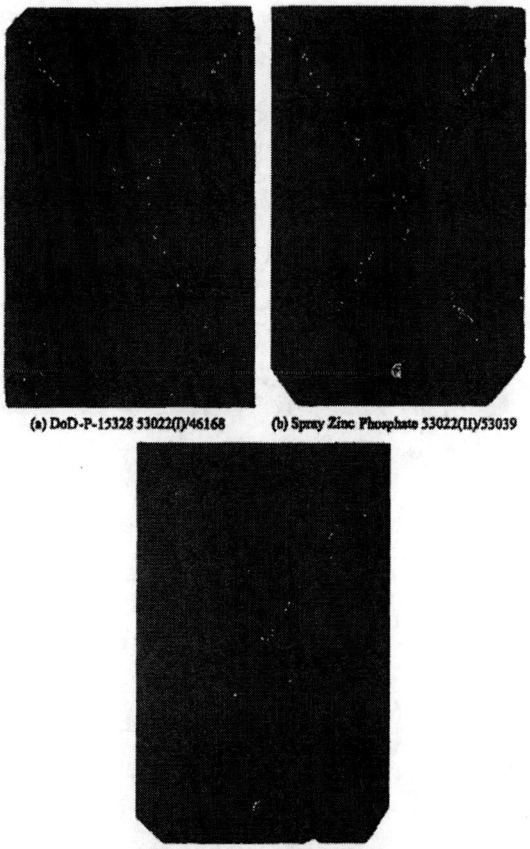

(a) DoD-P-15328 53022(I)/46168 (b) Spray Zinc Phosphate 53022(II)/53039

(c) Spray Zinc Phosphate 53022(II)/64159

Figure 16: Legacy coating system (a) vs. current systems (b&c) preteated with spray phosphate @ 1500hrs ASTM B117

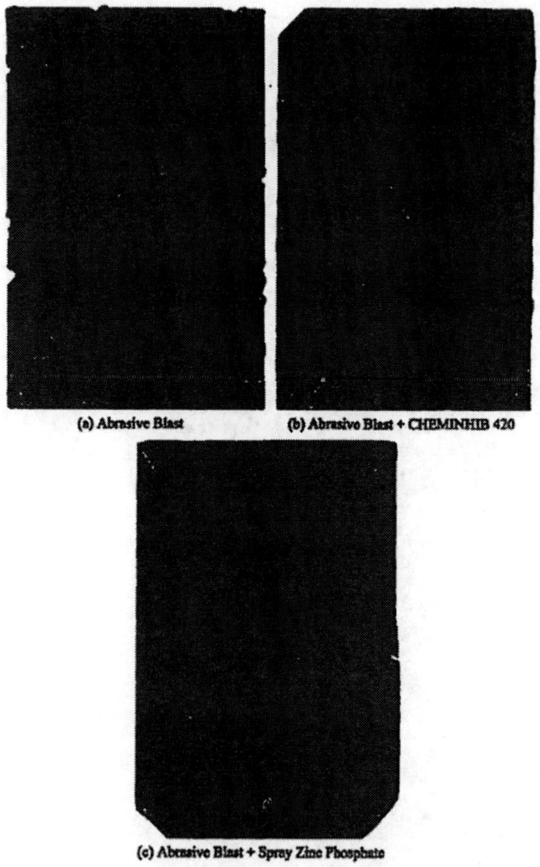

(a) Abrasive Blast (b) Abrasive Blast + CHEMINHIB 420

(c) Abrasive Blast + Spray Zinc Phosphate

Figure 17: 53022(II)/53039 @ 80 cycles GM9540P

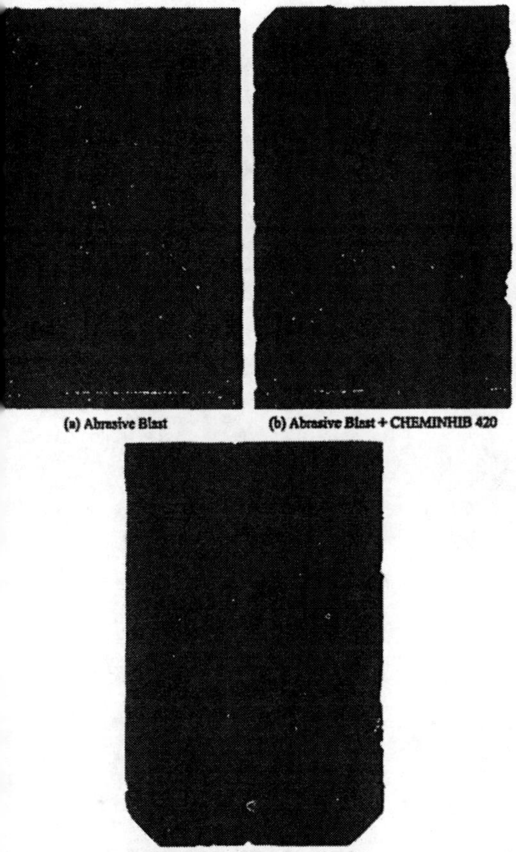

(a) Abrasive Blast (b) Abrasive Blast + CHEMINHIB 420

(c) Abrasive Blast + Spray Zinc Phosphate

gure 18: 53022(II)/64159 @ 80 cycles GM9540P

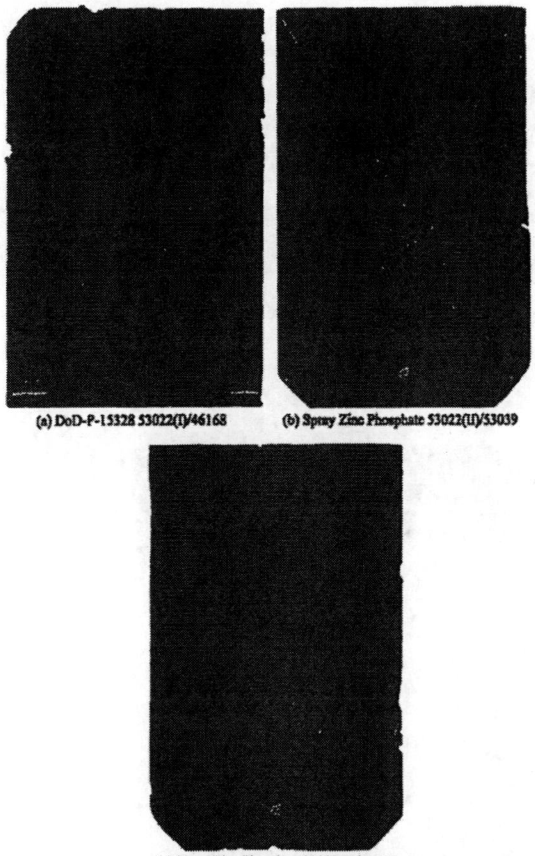

(a) DoD-P-15328 53022(I)/46168 (b) Spray Zinc Phosphate 53022(II)/53039

(c) Spray Zinc Phosphate 53022(II)/64159

Figure 19: Legacy coating system (a) vs. current systems (b&c) preteated with spray phosphate @ 80 cycle GM 9540P

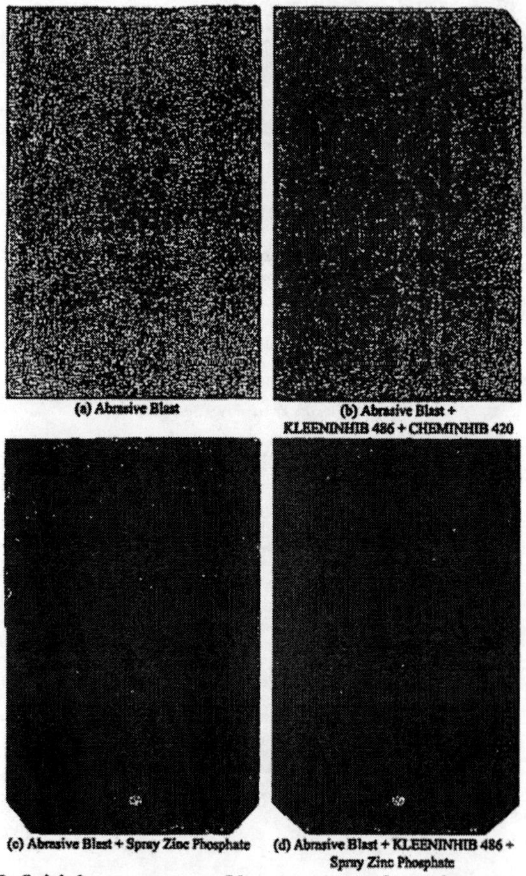

Figure 20: Initial appearance of bare pretreated specimens prior to 100% humidity exposure

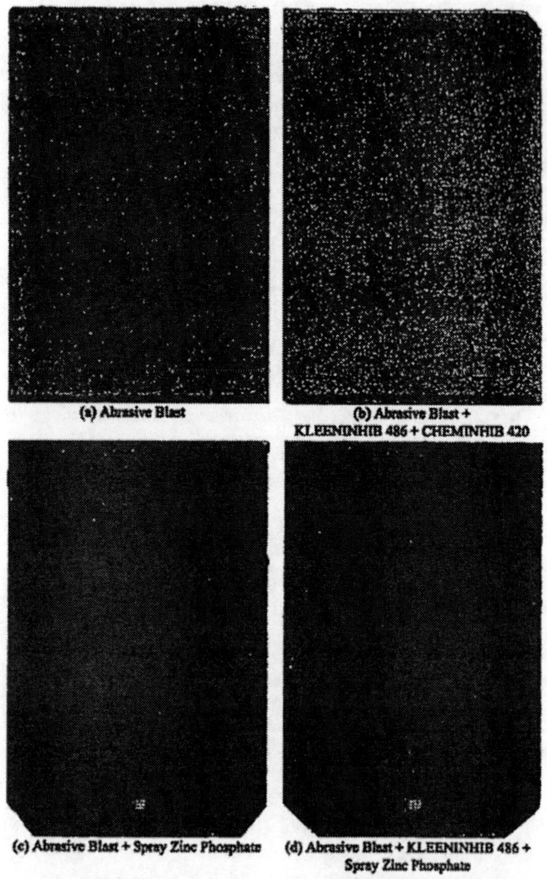

Figure 21: Appearance of bare pretreated specimens after 16 hrs exposure to ASTM D 1735 100% humidity exposure

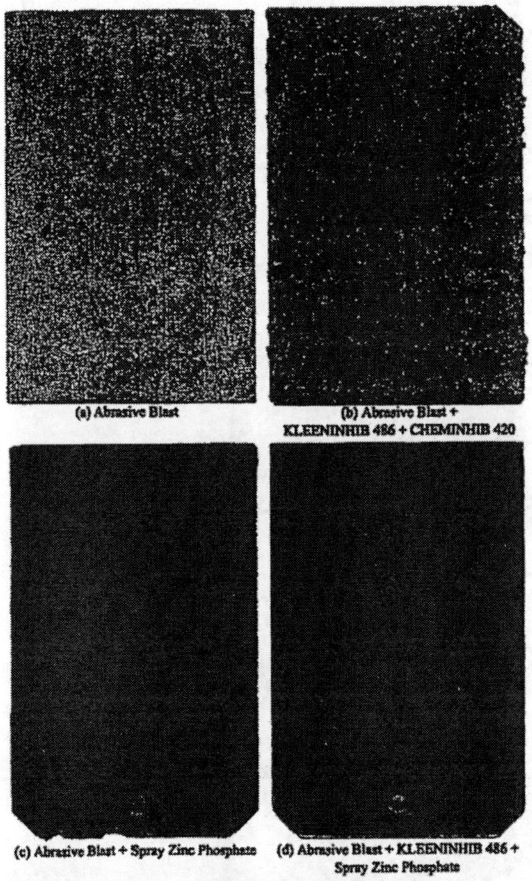

gure 22: Initial appearance of bare pretreated specimens ior to ambient lab exposure (69%RH, 27°C)

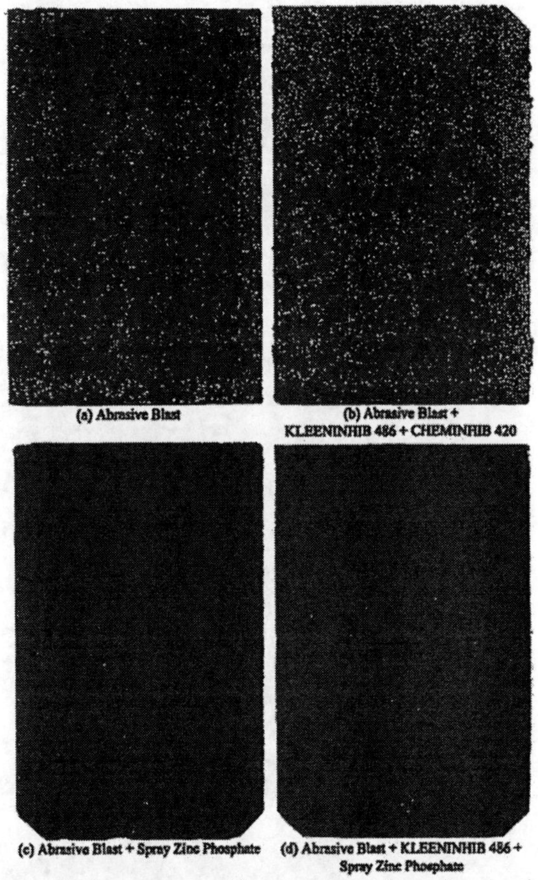

Figure 23: Appearance of bare pretreated specimens after 1 week ambient lab exposure (69%RH, 27°C)

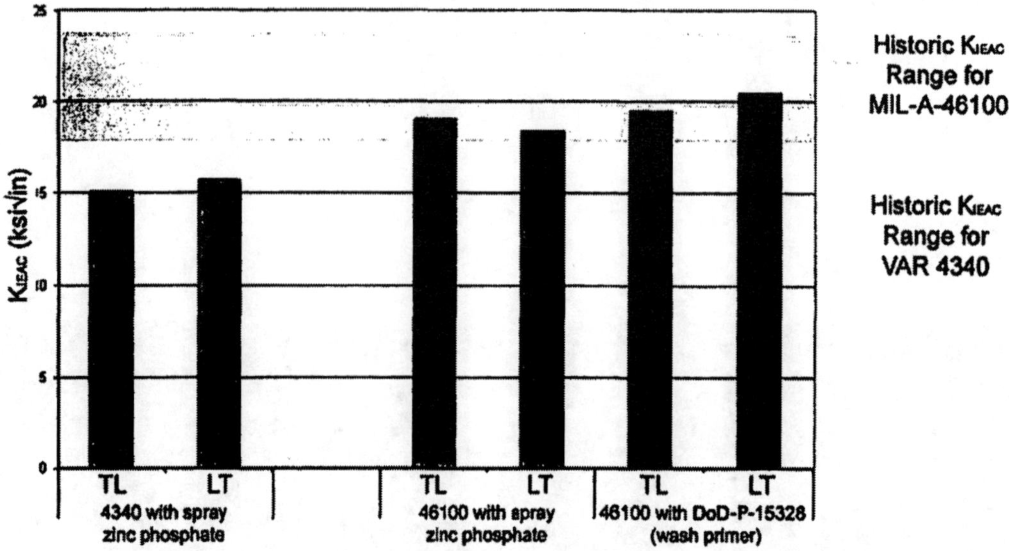

Figure 24: Average K_{IEAC} for pretreated 4340 and 46100 by orientation

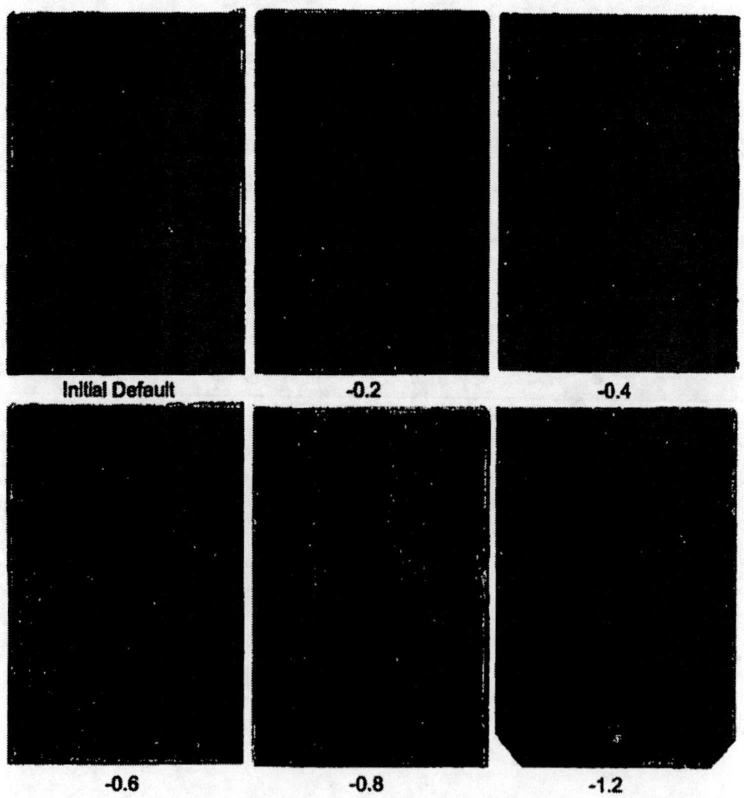

Figure 25: Changes in Phosphate Quality/Appearance from Adjustments to Free Acid Concentration Point Values via Increases of Henkel Primer 40 to the BONDERITE 952 Spray Zinc Phosphate Solution

SUPERSONIC PARTICLE DEPOSITION FOR MILITARY APPLICATIONS

V. K. Champagne and W. Ziegler

US Army Research Laboratory
Weapons and Materials Research Directorate
Building 4600
Aberdeen Proving Ground, MD 21005-5069

Abstract: This paper will discuss an emerging technology 'Supersonic Particle Deposition (SPD)" commonly referred to as 'cold spray' and its possible use as a metal deposition technique to form wear and corrosion resistant coatings for DOD applications. Cold spray is a material deposition process whereby metal powder particles are utilized to form a coating or a free-standing structure by means of ballistic impingement upon a suitable substrate. Conventional methods of surface preparation of the substrate, such as grit blasting, which are necessary for thermal spray processes, are often not required for cold spray. In the cold spray process, metal powder particles are accelerated by a supersonic stream of compressed gas. The spray nozzle design incorporates the use of a converging-diverging throat through which the preheated gas stream is expanded to attain supersonic velocities, with an accompanying decrease in temperature. The informal term "cold spray" arose due to the relatively low temperatures of the expanded gas and particle stream that exits the nozzle. The advantages of cold spray over current and emerging technologies are to be realized in terms of economics, environmental compliance and performance. Cold spray coatings offer certain advantages over those applied by conventional thermal spray techniques in that they can contain less oxidation since they are applied below the melting temperature and do not experience the effects of thermal stress as a result of resolidification. In addition, some cold spray coatings have been shown to adhere to the substrate by a solid state bonding mechanism having much higher bond strengths.

Keywords: Cold spray; coatings; metal deposition.

Background: The cold gas-dynamic spray method or "cold spray" was originally developed in the mid-1980's by Dr. Anatoli Papyrin and colleagues at the Institute of Theoretical and Applied Mechanics (Russian Academy of Sciences).[1,2] Cold spray is a material deposition process whereby metal powder particles are utilized to form a coating or a free-standing structure by means of ballistic impingement upon a suitable substrate. Since, in some instances, research has indicated that the deposited material appears to adhere to the substrate as a result of a metallurgical bond, conventional methods of surface preparation, such as grit blasting, which are necessary for thermal spray processes, are often not required for cold spray. Additionally, in the initial stages of the cold spray process, before the individual particles begin to adhere to the surface, the substrate is in fact being grit blasted, which may facilitate in bonding. The metal powders range in particle size from <5–50 μm and are accelerated by a supersonic (300-

1200 m/s) stream of compressed gas. The spray nozzle design incorporates the use of a converging-diverging throat through which the preheated gas stream is expanded to attain supersonic velocities, with an accompanying decrease in temperature. The informal term "cold spray" arose due to the relatively low temperatures (-17°C to 704°C) of the expanded gas and particle stream that exits the nozzle. It is important to note that the temperature of the gas stream is always below the melting point of the particulate material and the resultant coating and/or freestanding structure is formed in the solid state. Currently, the types of carrier gases most often used in the cold spray process are nitrogen and helium. Atmospheric air may also be used but then oxidation of the powder particles may become an issue and higher velocities that are required to deposit some of the harder materials cannot be achieved. Since adhesion of the metal powder to the substrate, as well as the cohesion of the deposited material, is accomplished in the solid state, the characteristics of the cold spray deposit are quite unique in many regards. There are a variety of metals that have been successfully deposited by the cold spray process including iron, stainless steel, nickel, copper, aluminum, zinc, cobalt, titanium, molybdenum, silver, tin, cadmium, bronze, Inconel 718, nickel based superalloys and many other alloys.

Advantages and Applications: The cold spray process does not use thermal energy to melt the particles to be deposited, but instead relies upon the ability of the particles, which are accelerated by a supersonic jet of compressed gas (300-1200 m/s)[3], to plastically deform upon impact with a substrate and cause cratering (Figure 1).[4] The bonding mechanism of cold spray has been compared to that of explosive welding[5,6] whereas the formation of a solid-state jet of metal occurs at the impact point between the particle and substrate.[4] The same phenomena is believed to occur between particles, such that residual stresses in cold sprayed samples have been reported to be predominately compressive rather than tensile.[7] Theoretically, such a condition can be substantiated by the fact that the subsequent solid particles impacting previously deposited material at these high velocities would be continuously peening the surface as the coating is built up.[8] Recrystallization and evidence of severe cold work have been reported, to support this claim. There has been no indication that melting occurred during impact but significant plastic deformation has been observed of the particle and the substrate. Figure 2 shows the high velocity impact of a single copper particle against a stainless steel substrate. It is well understood that shot-peened metallic surfaces result in the formation of compressive stress layers. Since the bonding mechanism appears to be directly related to the plasticity of the powder used, it stands to reason that the cold spray process is limited to powders that are predominately metallic. However, research has shown that a metallic powder can be doped with a significant amount of non-metallic powder and be successfully deposited. Amateau and Eden have demonstrated this by applying a coating of Al-9-Ce5Cr-2.8Co + 15 volume percent SiC onto an aluminum substrate.[9] Other hard materials such as chromium-carbide (Cr3C2) have also been applied after being mixed with 25 percent metallic powder serving as a binder.[9]

The most significant advantage of cold spray is the ability to apply metal and metal alloy particles at temperatures much lower than the melting points of either the coating or the substrate. This characteristic of the cold spray process alone overcomes a multitude of problems associated with coatings applied by conventional thermal spray techniques such as phase transformation, recrystallization, oxidation and residual stress. Figure 3 shows the operating temperatures and particle velocities of the cold spray method as compared with other thermal coating methods. The exceptional adhesion of cold spray coatings is, in part, due to the low temperatures at which the coatings are deposited. One of the most deleterious effects of depositing coatings at high temperatures is the residual stresses that develop, especially at the coating-substrate interface, upon solidification from the molten state. This problem is compounded when the substrate material has a significantly different thermal expansion coefficient from the coating material. This variation often causes the metal coating to debond from the substrate upon cooling. Sometimes the residual stresses may not be enough to cause the coating to separate from the substrate but may weaken it and reduce its effective service life. This problem can be minimized when cold spray is used.

Figure 1. Cross section of the impact site between a copper particle and a stainless steel substrate. [4]

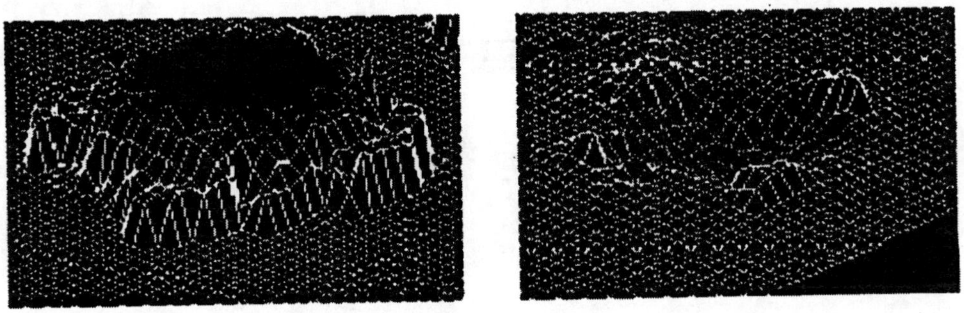

Figure 2. Profilometry results before and after removal of copper splat from a 600 m/s impact. [4]

Another important attribute of cold spray is the high coating density that can be achieved with minimal oxide content, which is a distinct advantage over that of thermal spray coatings. Surface oxides are broken up during impact, exposing a clean metal surface, which promotes good bonding and low oxide content. The resultant cold spray coating retains the properties of the original metal powder, such as thermal and electrical conductivity, making it a good candidate for applications in the electronics industry. High deposition rates are another significant advantage. Deposition rates exceeding 15 kg/min have been achieved [9] compared to approximately 0.23 kg/min for high velocity oxygen fuel (HVOF)[10]. The high deposition rates, combined with low oxide and residual stress levels, allow for increased adhesion and extremely thick coatings (>2.54 cm in a matter of minutes), as well as the formation of freestanding structures. This characteristic makes cold spray a possible alternative to laser rapid prototyping.

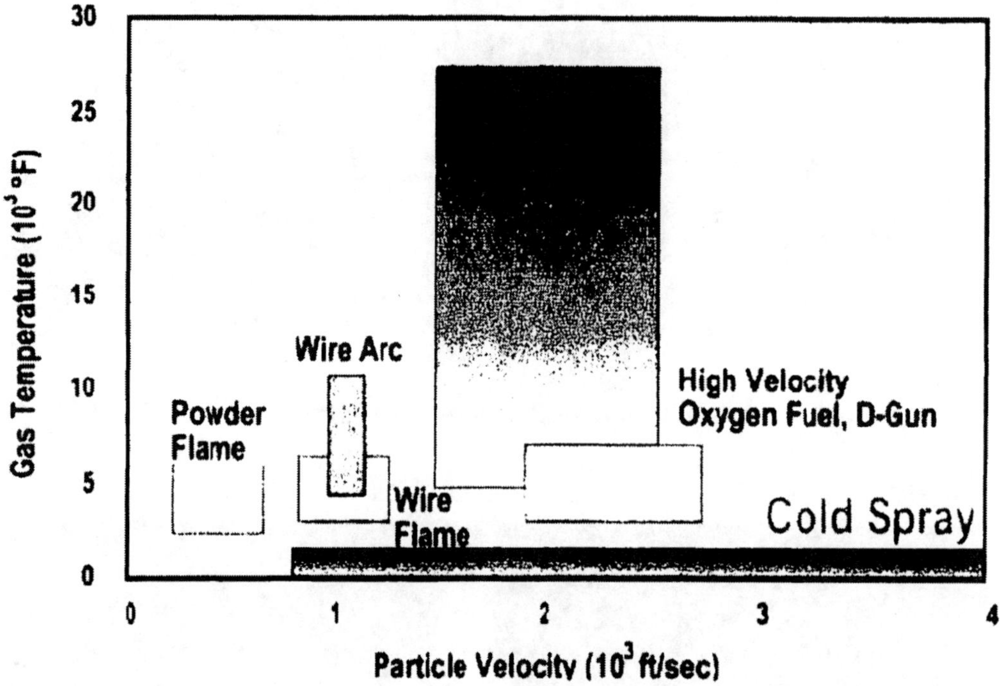

Figure 3. Comparison of thermal spray and cold spray particle velocity and gas temperature. [11]

The ARL Cold Spray System: The ARL cold spray facility consists of a liquid nitrogen source and compressor; a gas control module that regulates the high-pressure gas; a gas heater; a powder feeder; the supersonic nozzle; and a data acquisition and control system. A photograph of the ARL system is shown in Figure 4.

Figure 4. ARL SPD System.

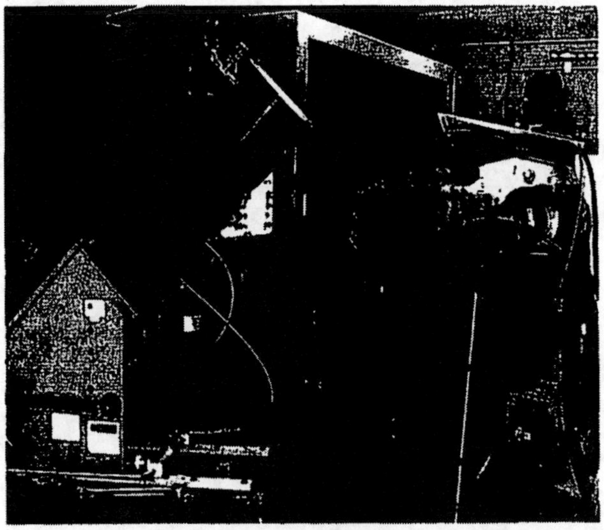

Liquid nitrogen is evaporated and then compressed to a working pressure of 500 psi. A 400 gallon storage tank serves as a buffer between the compressor and the Gas Control Module (GCM). The GCM ties the control inputs from the operator to the hardware operation. Instructions to the GCM are conveyed from the operator via the Data Acquisition System (DACS). The GCM splits the nitrogen flow into two streams – the main gas flow to the gas heater and nozzle and the feeder flow to the powder feeder. The pressures are controlled and the flow rates are measured for each stream.

The gas heater heats the main gas flow to the selected operating temperature. This is accomplished by means of a 25 KW heating coil. Current to the coil is controlled by the DACS. Bulk powder feedstock is contained in the powder feeder. Slots in a rotating wheel within the powder feeder fill with powder, and the wheel's rotation transfers powder to the exit port. Varying the rotation rate through the DACS controls the powder feedrate. The flow of nitrogen from the GCM transports the powder from the powder feeder to the nozzle. Precautions should be taken when attempting to cold spray pyrophoric materials such as titanium and aluminum, in order to mitigate the risk of an ignition or an explosion.

The heated main gas flow from the gas heater and the powder flow from the powder feeder combine at the nozzle prechamber. The prechamber mixes the powder with the heated main gas for subsequent acceleration into and through the supersonic nozzle. The nozzle can be stainless steel or silicon carbide and has a 0.079" throat diameter. Depending on the gas pressure and temperature as set by the DACS, the gas exit velocity can reach up to Mach 4 (Table 1). Figure 5 shows the major components and the arrangement of the cold spray system at ARL.

Table 1. The ranges of gas and powder conditions of the cold spray system.

Parameter	Value
Main Gas Stagnation Pressure	100 to 500 psi
Main Gas Stagnation Temperature	0 to 1300 degree F
Main Gas Flow Rate	30 to 100 CFM
Powder Feeder Gas Flow	40 to 110 CFM
Powder Feed Rate	10 to 30 pounds/hour
Gas Heater Power	0 to 25 KW

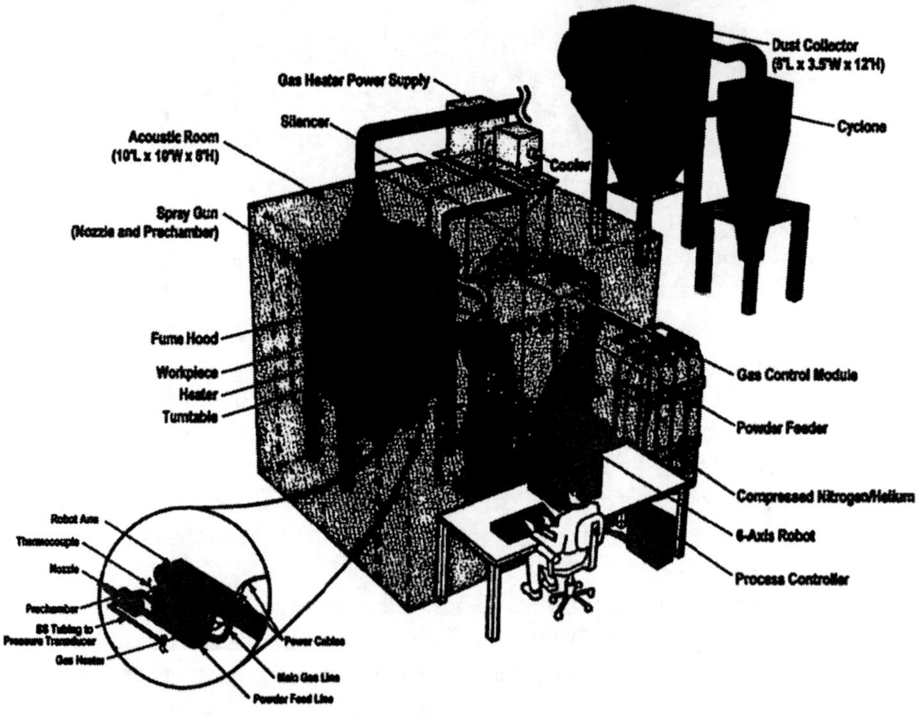

Figure 5. Schematic of the cold spray system.[12]

Military Applications for Cold Spray: The cold spray process shows immediate promise in the electronics since the deposited material has nearly the same properties as the original powder, such as electrical and thermal conductivity. This is a result of low oxide contamination and high density. Potential applications include electrical conductors for integrated circuit boards, electrical contacts in bomb fuzes, heating elements and electrical insulation coatings for metals, ceramics and polymeric materials.

Sacrificial coatings for use on steel and aluminum are very good candidates for cold spray. Amateau and Eden have produced data indicating that the corrosion rate of aluminum, zinc and 85zinc-15aluminum coatings sprayed onto steel were lower than those produced by thermal spray (Figure 6). The corrosion tests were conducted at the La Que Center for Corrosion Technology in Wrightsville Beach, North Carolina and at PennState. Sections of the appliqué steel armor of the AAV (Amphibious Assault Vehicle) were coated by the cold spray process and also thermal spray. The tests performed at the La Que Center were accomplished by immersing the panels in filtered seawater for 1 hour followed by a 23 hour drying cycle. This procedure war repeated over the course of 6 months to represent an accelerated corrosion test. PennState conducted similar tests where the immersion cycle was 10 minutes in the seawatater and 50 minutes in air. The corrosion rates were derived from polarization measurements taken over time.

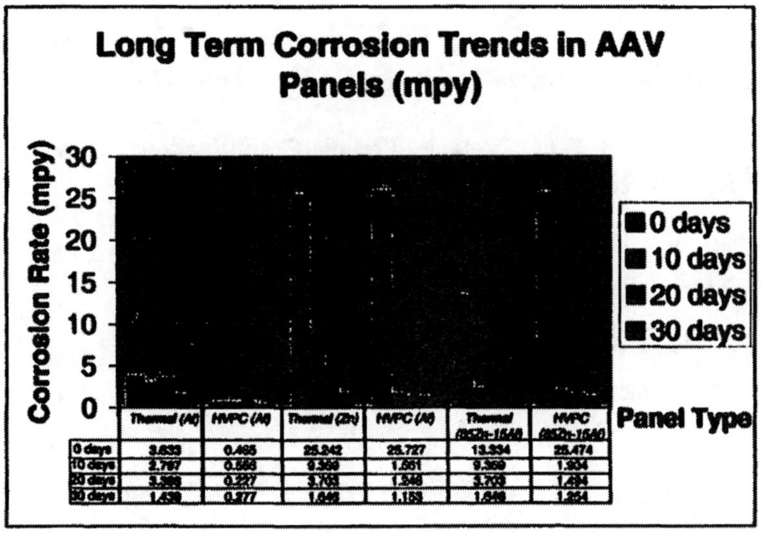

Figure 6. Corrosion rates for AAV appliqué armor panels coated with Al, Zn and 85Zn-15Al. [9]

References:

1. A.P. Alkhimov, A.N. Papyrin, V.F. Dosarev, N.I. Nesterovich, and M.M. Shuspanov, "Gas Dynamic Spraying Method for Applying a Coating," U.S. patent 5,302,414,12 April 1994.

2. A. O. Tokarev, Structure of Aluminum Powder Coatings Prepared by Cold Gas Dynamic Spraying, Met. Sci. Heat Treat; Vol. 38 (No. 34), 1996, p. 136-139.

3. R. Dykhuisen and M. Smith, "Gas Dynamic Principles of Cold Spray," Journal of Spray Technology, Volume 7, No. 2, 1998.

4. R. Dykhuisen, et al, "Impact of High Velocity Cold Spray Particles," Journal of Spray Technology, Vol. 8, No. 4, p. 564, Dec. 1999.

5. B. Crossland, "Explosive Welding of Metals and its Application," Clarendon Press, Oxford, 1982.

6. H. El-Sobky, "Mechanics of Explosive Welding," Explosive Welding Forming and Compaction, T. Blazynski, Ed., Applied Science Publishers, London, 1983.

7. R. McCune, et al, "Thermal Spray: Practical Solutions for Engineering Problems," C. Berndt, Ed., ASM International Proceedings, 1996.

8. M. Smith, et al, "Cold Spray Direct Fabrication-High Rate, Solid State, Material Consolidation," Materials Research Society Proceedings, Nov. 1998.

9. M. Amateau and T. Eden, "High-Velocity Particle Consolidation Technology," IMAST Quarterly, p.4, No.2, 2000.

10. M. Thorpe, "Try Thermal Spray for Specialty Metal and Carbide Coatings," Chemical Engineering Progress, p. 55, November, 1991.

11. R. McCune, et al, "An Exploration of the Cold Gas-Dynamic Spray Method for Several Materials Systems," Proceedings of the 8th National Thermal Spray Conference, September, 1995.

12. Ktech Cold Spray Operator's Manual, Revision 0, Oct. 10, 2001; pg ix.

FEASIBILITY OF ANODIC COATINGS ON AL_2O_3 FIBER REINFORCED ALUMINUM METAL MATRIX COMPOSITES FOR WEAR AND CORROSION RESISTANCE

By

John V. Kelley
U.S. Army Research Laboratory
jkelley@arl.army.mil

ABSTRACT

Aluminum and its alloys exposed to air are protected by a native oxide film that is heterogeneous in nature and does not provide adequate corrosion resistance. For this reason, anodized coatings are often used for corrosion as well as wear protection on aluminum alloys. While there is a wealth of information available about the effectiveness of anodized coatings on aluminum and aluminum alloys, little can be found about the coatings effectiveness on aluminum matrix composites. Metal matrix composites (MMC) in general, offer new challenges in corrosion protection. Depending on the fiber material, galvanic reactions can occur at the fiber/matrix interface that could accelerate the corrosion process. Fiber "pull-out" can also affect the surface by leaving behind crevices that can also lead to preferential corrosion sites. The object of this study is to investigate the effectiveness of a type III anodic hardcoat on a 3M NEXTEL (Al_2O_3) fiber reinforced aluminum matrix composite. The corrosion and wear properties will be presented and compared with an anodized monolithic baseline.

Key Words: Aluminum matrix composite, anodizing, anodic coatings, MMC

INTRODUCTION

The U.S. Army is continually seeking to lighten the force in an effort to increase its mobility, deployability and lethality. For this reason, the potential use of metal matrix composites, specifically aluminum matrix fiber reinforced composites is coming more to the forefront. Fiber reinforcement of aluminum alloys offer higher strength-to-weight ratio, stiffness and modulus [1] making them ideal candidates for many Army applications. One such application is for use in advanced artillery projectiles. MMCs are being looked at for shell bodies that are 50% lighter than steel shells allowing more payload to be carried for a given weight and volume [2]. But MMC's can introduce new challenges in field applications. Corrosion protection is a prominent issue with many of the Army's systems and replacing monolithic materials with MMC's can complicate matters further. For example, depending on the fiber material, galvanic reactions can occur at the fiber/matrix interface which would accelerate the corrosion process. Also, fiber "pull-out" at the surface will leave behind a void or (crevice) that also creates a region of preferential corrosion. Therefore, a coating is usually necessary for corrosion protection. Currently, the Army requires a type III hard coat anodizing per MIL-A-8625F [3] on all aluminum ordinance shells.

The two primary objects of this study are: 1) To investigate the effectiveness of type III anodic hardcoat on a 3M NEXTEL fiber reinforced aluminum matrix composite relative to the baseline type III anodic hardcoat on monolithic aluminum 6061; and 2) To determine if an aluminum rich outer layer is necessary to form an effective anodic coating. Aluminum and its alloys exposed to air are already protected by a native oxide film. However, this layer is heterogeneous in nature and does not provide adequate corrosion resistance [4]. Anodizing can be viewed as a deliberate controlled corrosion of an aluminum surface to create (grow) a uniform, continuous protective oxide film [5]. This can be achieved using several types of electrolytes baths including chromic and sulfuric acid. Type III coatings are generally thought of as wear resistant coatings. However, the corrosion resistance can be enhanced by adding various coatings and sealers. While there is an abundance of available information about the effectiveness of anodic coatings on aluminum and aluminum alloys, little can be found about the process and effectiveness of anodic coatings on aluminum matrix composites.

What is known is that, on monolithic aluminum, an anodic coating consists of a thin continuous barrier layer beneath a thicker porous columnar structured layer. This porous layer has been characterized elsewhere, [5, 6] as a closed packed array of columnar hexagonal cells that contain a central pore. The columns are oriented normal to the substrate surface (Figure 1). The concern is whether an anodic coating, as described can adequately be formed on and cover the surface of an aluminum matrix composite. And if not, will the anodization of the exposed matrix material be enough to provide adequate corrosion and wear protection to satisfy the specification.

EXPERIMENTAL PROCEDURE

<u>Specimen Preparation:</u>

All substrate material was supplied by 3M Corporation. MMC coupons were manufactured using 3M proprietary Nextel fibers. Because of the cost of the MMC material, it was purchased in limited supply. Thus to maximize the amount of samples available for testing, they were machined by 3M to the geometry shown in Figure 2. Five samples of each were prepared as listed below in Table 1. Monolithic aluminum 6061 matrix material samples were also supplied by 3M and included in the test matrix as the anodized baseline comparison.

Some MMC coupons were coated with a thin layer of pure aluminum using an ion vapor deposition process (IVD). The class I, IVD was deposited at 1/4 amp and 600-700 volts. 4 passes were made for a final thickness of approximately 0.0005 inch. Anodized samples were prepared according to MIL-A-8625F using a sulfuric acid anodizing process and type III hardcoat anodized to an average thickness of approximately 0.002 inch. The anodizing is a four step process that includes: 1) degreasing 2) deoxidizing, 3) anodizing, 4) sealing. First, all coupons were degreased using acetone. Then to remove the native oxide layer on the coupons, deoxidizing was performed at room temperature using an ISO McDermot solution [7]. The anodizing was done as a type III in sulfuric acid starting at 0 volts ramping up to 24-26 volts over three minutes. The deinonized hot water seal (HWS) was performed at 95 degrees F for 15 minutes.

Electrochemical Evaluation:

Electrochemical Impedance Spectroscopy was employed to measure coating integrity over time. The apparatus used consisted of a Princeton Applied Research PAR Model 283 potentiostat and a Schlumberger Model 1255 frequency response analyzer (FRA), and a Panasonic Model CF-71 laptop computer. Periodic measurements were taken at 0, 24, 48, 72, 144, 168, 336, 504, 720, and 1008 hours of continuous exposure using the cell configuration illustrated in Figure 3. The cell was filled with 25 mL of 0.5 N NaCl solution and allowed to equilibrate for at least 3 hours prior to taking measurements. The software package Zplot was used for data acquisition, and Zview for data analysis. Measurements were taken at the corrosion potential of the sample over the frequency range of 100 KHz-0.01Hz. The single sine technique was utilized, with an applied amplitude of 5mv.

The EIS data was plotted and evaluated in Bode and Nyquist formats. The Bode formats display the magnitude (log | Z |) and phase angle (θ) of the impedance as a function of applied frequency (log f). The total impedance of the specimen, defined as the log | Z | value at 10 mHz in the Bode magnitude plot, was also plotted as a function of exposure time for comparison.

Abrasion Resistance:

The relative wear resistance of each specimen was measured using the Ball-on-Disc technique. The system consisted of an Implant Sciences Corp ISC-200 tribometer and a computer interface data acquisition unit, PC-stripchart, which is a computer based chart recorder used to display and store data in real time.

The samples were mounted on the top of a rotating platform of the tribometer (Figure 4). A 0.5 inch stainless steel ball is attached to a precision balanced lever arm and is used to both apply vertical loads to the ball and to read the friction force on the pin. The ball is put in contact with the surface of the sample and a load is applied. To measure friction coefficient of the sample the load was set to 100 grams, (a load of about 1 N). The sample is then rotated and the total distance the steel ball travels on the sample is set. In this case the total distance traveled was 1000 m. All the friction measurements were conducted dry (without lubrication). Finally, the groove depth was measured using a Taylor-Hobson Laser Profilometer.

Accelerated Corrosion Testing:

Salt fog testing in accordance with ASTM B117 [8] was used to evaluate the corrosion resistance of the anodized coatings. The samples were placed into the salt fog chamber for a total duration of 1100 hours. The samples were periodically removed for observation and were photographed at 24, 48, 144, 216, 336, 504, 811, 1100 hours intervals.

RESULTS and DISCUSSION

EIS was used to measure the dielectric response of the coating as it degraded over the 30 days of immersion in 0.5 N sodium chloride (NaCl) solution. A total of 8 low frequency impedance measurements were recorded for each sample over the 30 days and plotted as a function of exposure time in Figure 8. As expected, both the sealed and unsealed anodized monolithic 6061 maintained the highest impedance values throughout the 30 days exposure. In fact, the plots for all samples with the exception of bare aluminum (Al NT) remained relatively flat with no significant change in impedance. Moreover, the impedance values for all the MMC samples shown in Figure 8 were within the same range indicating that the sealed anodized coating did not improve the wet electrical resistance over the unsealed on MMC. This is likely because of the surface roughness and protruding fibers present at the surface of the samples that cause pathways for the electrolyte to penetrate the coating.

Evidence of the corrosion performance can be seen in the results of the ASTM B117 salt fog testing (Figure 5-7). The minimum required exposure to ASTM B117 salt fog chamber, according to MIL-A-8625, is 336 hours. As expected, it is evident in Figure 5 that the baseline, Al 6061 Type III HWS outperformed all others beyond the 336 hour minimum. However, the MMC Type III HWS had adequate corrosion resistance up to 336 hours of B117 exposure. Beyond that significant degradation was observed. Figure 5 shows the results at 1000 hours. Figure 6 illustrates the effects of 3 different scenarios on each of the substrate materials at 336 hours of B117 salt fog exposure. Here it can be seen that the anodic coating alone provided some corrosion protection to the MMC. By including a HWS, the corrosion resistance of the anodic coating is enhanced further.

As mentioned earlier, a thin coating of aluminum, though porous, provided some corrosion protection for the MMC samples. Figure 7 illustrates the level of corrosion protection provided by the IVD aluminum, anodized and seal coated. When compared to the anodized and sealed MMC sample in Figure 5, the IVD coated, anodized and sealed sample performed comparably at 336 hours and considerably better at 1000 hours.

Scanning electron microscopy was used in order to ascertain the quality of the coatings as well as measure the coating thicknesses. A comparison of the coated monolithic samples and the MMC samples is made in Figures 9 and 10. As can be seen in Figures 9a, a relatively dense coating is achieved on the anodized monolithic sample. No gaps or voids were observed at the coating / substrate interface indicating that good adhesion was also probable on this sample. Although the thickness of the coating achieved, 0.0014 inches (35 μm) was somewhat less than the target thickness of 0.002 inches (50.8 μm), it is significantly thicker than what was realized on the MMC sample (Figure 9b). It's likely that the Nextel fibers at or near the surface of the sample hindered the growth of the anodic coating on the MMC. The thickness of the anodic coating on bare MMC was on approximately 0.0002 – 0.0003 inches) (5-8 μm) thick, nearly an order of magnitude less than the target thickness of 0.002 inches. When the MMC is precoated with a thin layer of aluminum, it becomes a slightly better candidate for anodization. The process of ion vapor deposition leaves behind a rather porous coating. This can clearly be seen in Figure 10 where the porosity appears to be continuous in some places even after anodizing. The anodizing of the porous coating appears to penetrate beyond the aluminum coating and into parts of the aluminum matrix. When sealed, the anodized IVD becomes an adequately corrosion resistant coating.

The relative abrasion resistance was measured using a ball-on-disc tribometer. The depth of groove for each coating scenario was recorded and presented in Figure 11. Using the monolithic samples as a baseline, it can be seen that the bare sample suffered the most wear in this set. When anodized, a significant increase in wear resistance is achieved. However, sealing the anodic coating reduced the wear properties slightly. This was expected as type III coatings are typically specified unsealed when the main function of the application is to obtain the maximum degree of abrasion or wear resistance [3]. Thus, sealing to improve the corrosion resistance will diminish the wear resistance slightly.

There is no clear trend was observed for wear resistance of anodized MMC. Figure 11 shows that all MMC's subjected to the ball-on-disc test performed poorly regardless of whether it was deoxidized prior to anodizing and/or sealed.

The IVD aluminum coated MMC samples showed incremental improvements in wear resistance when anodized and sealed. Contrary to the monolithic samples, the IVD coated, anodized and sealed samples displayed the best wear resistance of the group. This is likely because the ion vapor deposited aluminum is porous. When anodized, the coating is hardened but the porosity still exists. By eliminating some of the porosity when the coating is sealed, the wear properties are slightly improved.

SUMMARY AND CONCLUSIONS

1. The anodic coating formed more uniformly on the surface of the monolithic 6061 substrate than on MMC's.
2. A sealed anodic coating on MMC provided little corrosion protection beyond the minimum 336 hours of ASTM B117 Salt Fog exposure.
3. The anodic coating did not provide improved abrasion resistance to the MMC. However an aluminum rich layer applied to the MMC prior to anodizing allowed a more continuous anodic coating to form, thus providing better abrasion protection.
4. Although porous, the anodized IVD aluminum interlayer provided improved corrosion resistance for the MMC samples which demonstrates that an aluminum rich outer layer is recommended for MMC to achieve performance comparable to an anodic coating on monolithic material

ACKNOWLEDGEMENTS

The author would like to thank several people for their technical contribution. Dr. Constantine Fountzoulas who painstakingly performed the ball-on-disc tests, Mr. Paul Huang who performed profilometry measurements, and Mr. Cliff Hubbard who helped with the scanning electron microscopy. Consultations with Dr. Ernest Chin, aided in the overall understanding of metal matrix composite structures. Finally we would like to thank Dr. Christopher Hoppel for the opportunity to contribute in this way.

TABLES

Table 1: List of data set identification along with the treatment conditions tested

Description of Data Set Identifications		
Data Set ID	**Substrate Material**	**Anodization Treatment**
Al DT3s	Monolithic aluminum 6061	Deoxidized / Type III / HWS
Al NT	Monolithic aluminum 6061	No Treatment
Al DT3	Monolithic aluminum 6061	Deoxidized / Type III / no seal
MMC DT3s	Metal Matrix Composite	Deoxidized / Type III / HWS
MMC IDT3	Metal Matrix Composite	IVD coated / Deoxidized / Type III / no seal
MMC IDT3s	Metal Matrix Composite	IVD coated / Deoxidized / Type III / HWS
MMC T3s	Metal Matrix Composite	Type III / HWS
MMC IVD	Metal Matrix Composite	IVD only
MMC DT3	Metal Matrix Composite	Deoxidized / Type III / no seal
MMC T3	Metal Matrix Composite	Type III / no seal
MMC NT	Metal Matrix Composite	No Treatment

FIGURES

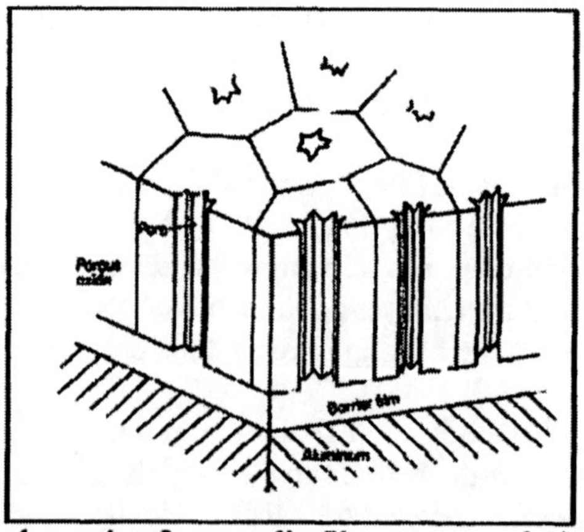

Figure 1: An accepted schematic of an anodic film structure derived from metallographic investigation [5]

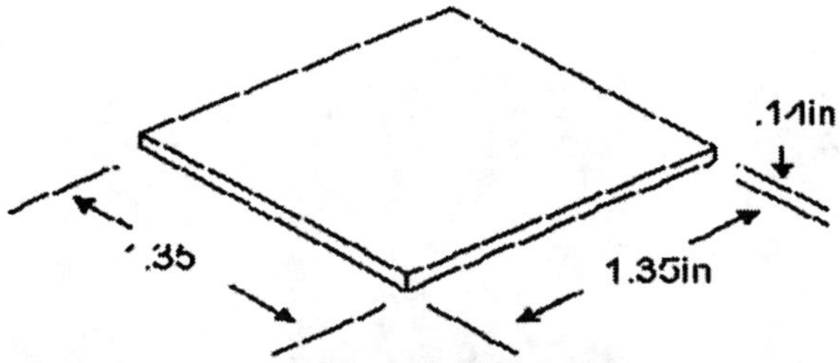

Figure 2: Specimen geometry of both monolithic and Nextel metal matrix composite coupons

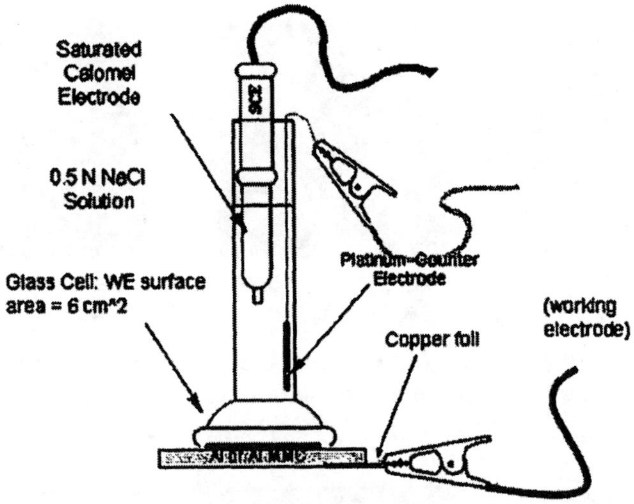

Figure 3: Electrochemical corrosion cell configuration used for AC impedance measurements.

Figure 4: Ball-on-Disc tribometer test apparatus which also includes a PC for data acquisition.

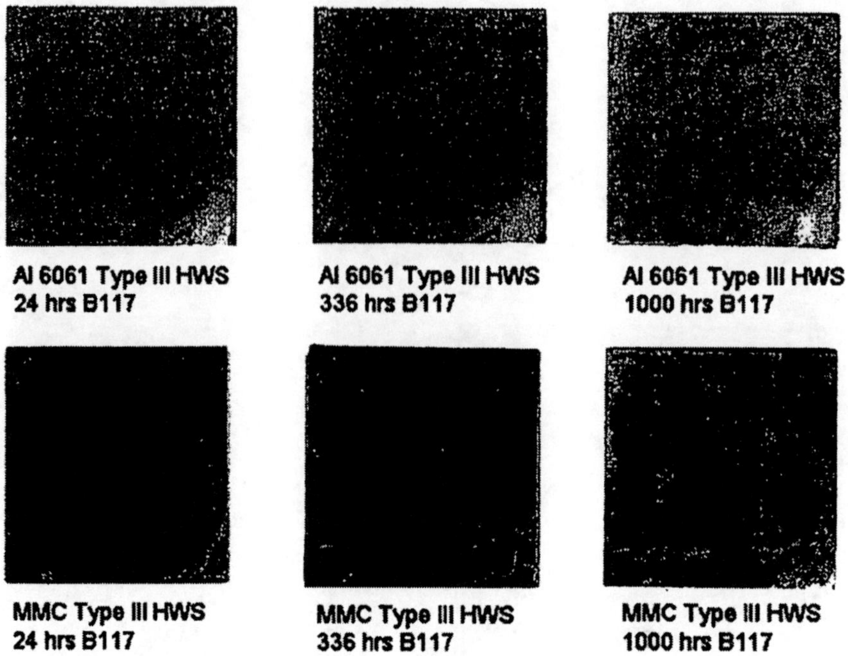

Figure 5: Comparison of aluminum 6061 vs. MMC over a range of salt fog exposure times. The 336 hour duration represents the minimum requirement of MIL-A-8625F.

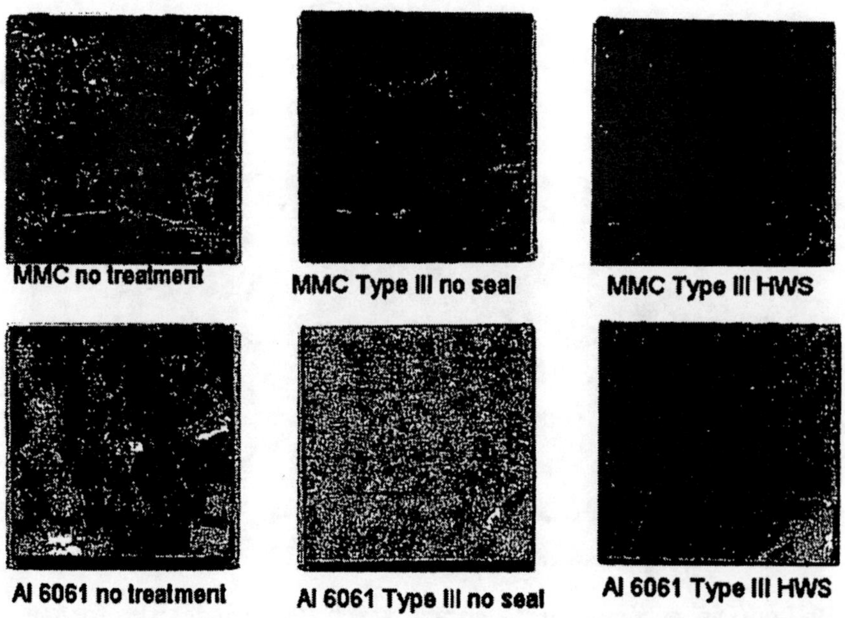

Figure 6: A comparison of 3 treatment scenarios for each substrate material after 336 hours of B117 salt fog exposure.

24 Hrs B117 336 Hrs B117 1000 Hrs B117

Figure 7: B117 exposure results of the IVD aluminum coated, anodized and sealed MMC samples.

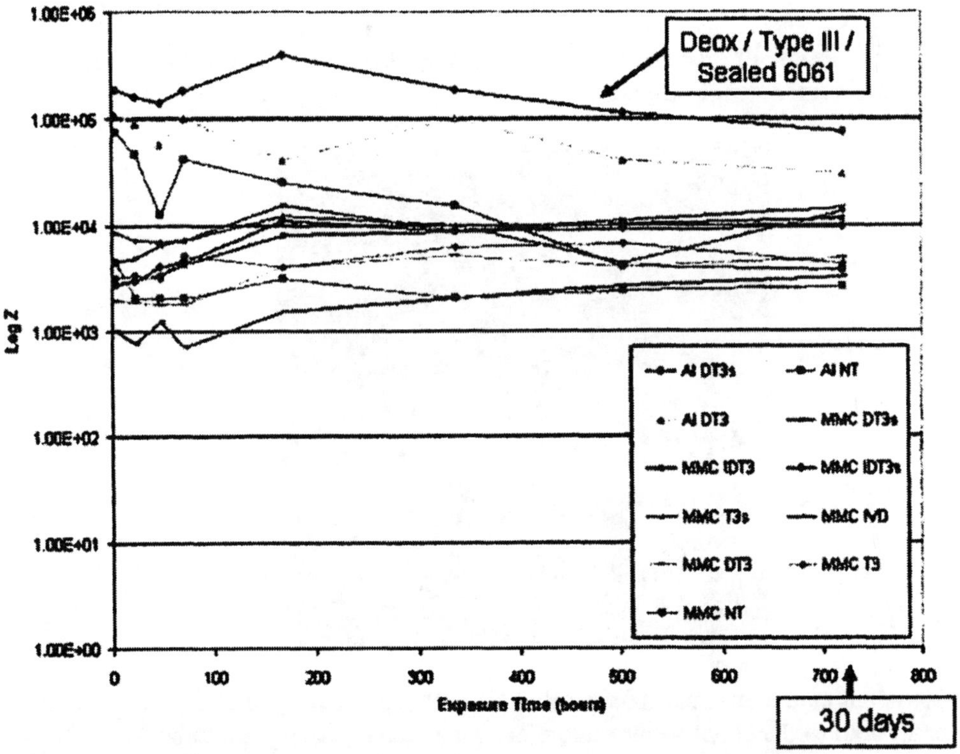

Figure 8: Low frequency impedance versus exposure time in 0.5N NaCl

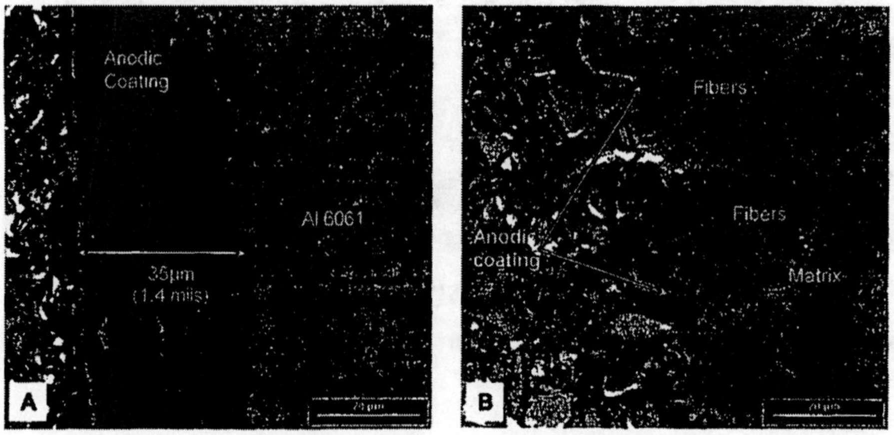

Figure 9: Cross-section of type III anodized and HWS; A) monolithic 6061 aluminum alloy; B) Aluminum Matrix Composite substrate

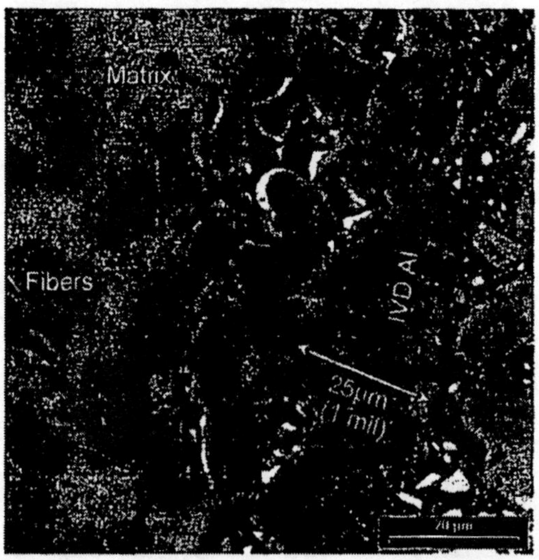

Figure 10: Aluminum matrix composite substrate with a precoating of IVD Aluminum then type III anodized and hot water sealed. The anodized layer appears to be 1 mil thick in some places.

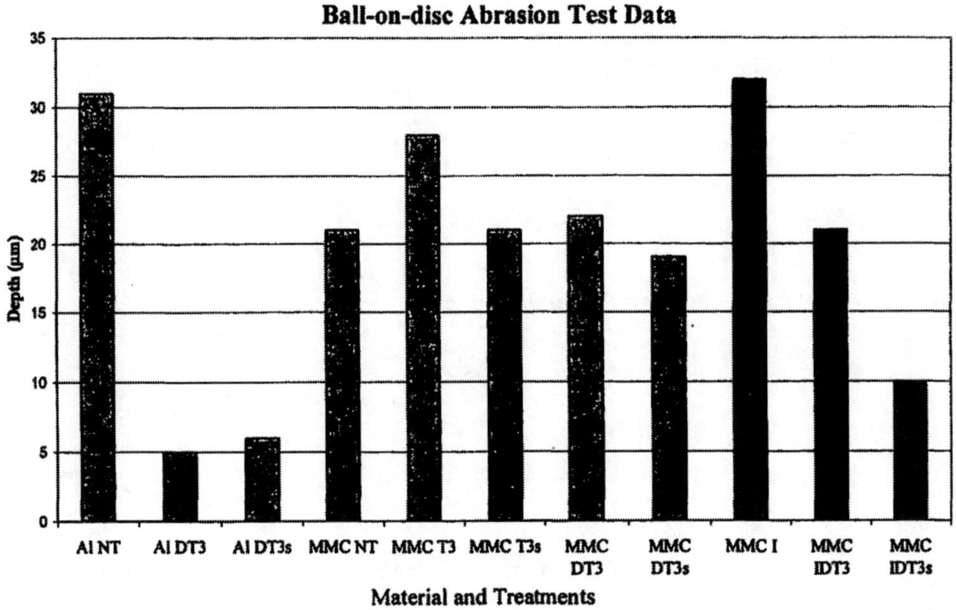

Figure 11: Depth of groove measurements from 1000m distance of ball-on-disc wear testing. Refer to table 1 for sample preparation and identification.

REFERENCES

1. Corrosion Behavior and Protection Efficiency of 2024Al and SiC /2024Al Metal Matrix Composite, Chunlin HE, Changsheng LIU, Fengqin LI and Qingkui CAI, School of Materials Engineering, Northeastern University, Shenyang 110004, China, Journal of Material Science Technology, Vol 18, pp351-353, No.4, 2002

2. Metal Matrix Composites for Ordinance Applications, C.P.R. Hoppel, J.H. Beatty, J.S. Montgomery, J.M. Bender, and T.A. Bogetti, U.S. Army Research Laboratory, May 2001, TR# ARL-RP-22

3. U. S. Department of Defense. "Anodic Coatings for Aluminum Alloys", MIL-A-8625F, Washington DC, September 1993

4. Characterization of Sealed Anodic Layers on 7050 T74 and 2214 T6 Aluminum Alloys, F. Snogan, C. Blanc, G. Mankowski, and N. Pebere, Centere Interuniversitaire De Recherche et d'Intgenierie des Materiaux, UMR CNRS 5085-ENSIACET, 118 route de Narbonne, 31077 Toulouse Cedex 4, France.

5. Anodic Oxide Film Formation Relating Mechanism to Composition and Structure, Jude M. Rudge, and Aaron J. Pomis, The C.J. Plating Company, Cicero, IL, USA

6. Corrosion Protection of Al/SiC Metal Matrix Composites by Anodizing, Lin S, Greene H, Shih H, and Mansfield F, Corrosion, Vol. 48, pp61-67, 1992

7. Teleconference and Correspondence with Mr. Jesse Garman of the Naval Air Warfare Center, Cherry Point SC, April, 2003

8. American Society for Testing Materials, "Standard Method of Salt Spray (Fog) Testing" ASTM B-117, Philadelphia, PA 1990.

This report has been approved for public release

PROGNOSTICS V

Chair: Mr. Andrew J. Hess
Joint Strike Fighter Program Office

REAL PROGNOSTICS – CHALLENGES, ISSUES, AND LESSONS LEARNED
CHASING THE BIG "P"

Mr. Andrew Hess
Joint Strike Fighter Program Office
200 12'th Street South
Arlington, VA 22202-4304
703-601-5551
Andrew.Hess@jsf.mil

Ing. Giulio Calvello
Joint Strike Fighter Program Office
200 12'th Street South
Arlington, VA 22202-4304
703-604-3852
Giulio.Calvello.it@jsf.mil

Dr. Peter Frith
Joint Strike Fighter Program Office
200 12'th Street South
Arlington, VA 22202-4304
703-601-5545
Peter.Frith@jsf.mil

Abstract: The desire and need for real predictive prognostic capabilities have been around for as long as man has operated complex and expensive machinery. This has been true for both mechanical and electronic systems. There has been a long history of trying to develop and implement various degrees of prognostic and useful life remaining capabilities. Stringent Diagnostic, Prognostic, and Health Management (PHM) capability requirements are being placed on new applications, like the Joint Strike Fighter (JSF), in order to enable and reap the benefits of revolutionary Autonomic Logistic support concepts. While fault detection and fault isolation effectiveness with very low false alarm rates continue to improve on these new applications; prognostics requirements are even more ambitious and present very significant challenges to the system design teams. These prognostic challenges have been aggressively addressed for mechanical systems for some time; but are only recently being fully explored for electronics systems. This paper will explore some of these design challenges and issues; discuss the various degrees of prognostic capabilities; address potential development methodologies; and draw heavily on lessons learned from previous prognostic development efforts.

Key Words: Autonomic Logistics; Diagnostics; Logistics; Prognostics; Support; Health Management.

1. Introduction

This paper will explore background, benefit impacts, and the current playing field for predictive prognostics; it will highlight some specific design challenges and issues, discuss the various degrees of prognostic capabilities and draw heavily on lessons learned from previous and current prognostic development efforts.

Real predictive prognostic capabilities are just one element among many interrelated and complementary functions in the playing field of PHM. The challenges involved with developing and implementing prognostic capabilities are several; and this paper will focus on those impacting and enabling benefits associated with new performance based logistic support concepts, global sustainment strategies, and new business practices. Similarly, the technical issues involved with developing real predictive prognostic capabilities are large and many, and here some of the issues associated with uncertainty will be discussed.

The challenges, issues, and the lessons learned are many and varied and these will certainly continue to grow, as this discipline matures and it is more widely applied. The readers should not expect to find here all the answers to the complex new world of "prediction", but they will discover new horizons to start their travels in the winding and deceptive forest of Prognosis.

2. Background, Needs, and Impacts

If you were to ask a fleet user or maintainer what their most pressing needs were you most probably would get a mixture of many answers. They would identify many needs including some of the following responses: maximize sortie generation rates and aircraft availability; very low Can Not Duplicate (CND) and Returned Tested OK (RTOK); minimum or no periodic inspections; low number of spares required and small logistics footprint; quick turn around time; maximized life usage and accurate parts life tracking; no false alarms; etc. Figure 2.1 depicts some of these and other fleet user needs. What the maintainer may actually want the most of all is no surprises. They would like to be able to have the ability to accurately predict future health status and to anticipate problems and required maintenance actions ahead of being surprised by "hard downing" events. Capabilities that would let the maintainers see ahead of both unplanned and even some necessary maintenance events would enable a very aggressive and beneficial opportunistic maintenance strategy.

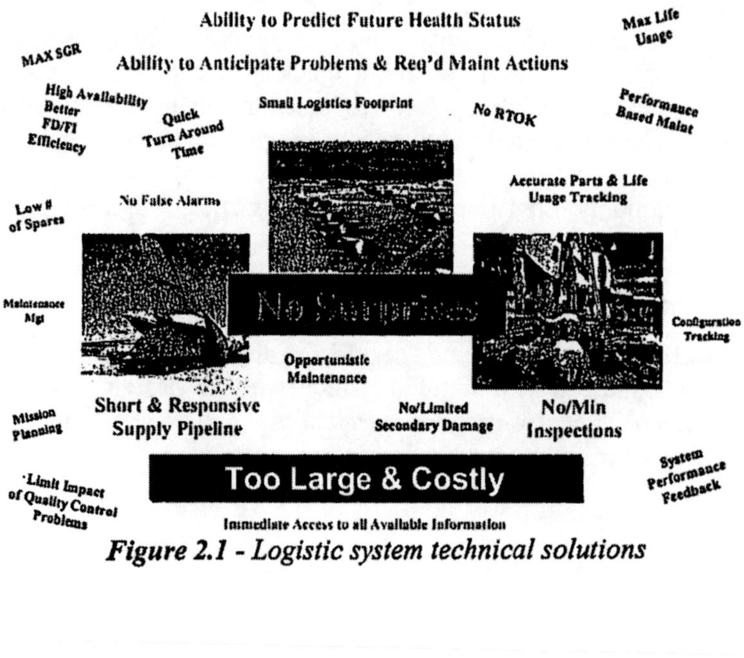

Figure 2.1 - Logistic system technical solutions

These combined abilities to both anticipate future maintenance problems and required maintenance actions coupled with an ability to predict future health status, are key enablers and attributes to any Condition Based Maintenance (CBM) concept and to "enlightened" opportunistic maintenance decisions. These predictive abilities and prognostic capabilities are also key enablers and very necessary attributes to both the evolving Performance Based Logistics (PBL) concepts and paradigm changing New Business Case approaches.

Substantial improvements in both reliability and maintainability over legacy aircraft are essential if the logistics footprint, sortie generation rate, mission reliability and anticipated O&S cost reductions are to be achieved. A modern and comprehensive Prognostics and Health Management system needs to interface with the Logistic Information System to trigger: (1) the activities of the Supply Chain Management System promptly to provide required replacement parts; (2) planning needed to perform the required maintenance; and (3) training of the maintainer. Typical Turn-Around-Time (TAT) to perform any required depot level maintenance action is significantly reduced over that of legacy aircraft as a result of the advanced notification of maintenance requirements, and readily available tools and replacement parts. Within security limitations, failure data should be available to all levels, including sustaining engineering personnel, OEMs, major command staff and unit level warfighter personnel. Some of the benefits of such a PHM system are listed below.

<u>Change of Maintenance Philosophy</u>
·On-condition
·Opportunistic
·Not "on-failure" nor "per schedule"
·Less interruption of mission schedule
Reduction in Test Equipment
·Less intermediate and flight line TE
·35% less peculiar support equipment during SDD
·Eliminated O-level TE

<u>Benefits to the Maintainer</u>
·Unprecedented insight into vehicle/squadron/fleet health
·Less time spent on inspections
·Better ability to plan maintenance
·Simplified training
·Improved fault detection

3. Prognostic Definitions

Prognostics and Health Management (PHM) is the name given to the capability being developed by JSF to enable the vision of Autonomic Logistics and so meet the overall affordability and supportability goals. In PHM, the term Prognostics takes on a much more global meaning covering the broader functions of fault / failure detection, fault / failure isolation, enhanced diagnostics, assess material condition, performance monitoring, life tracking and prognostics, rather than just the prognostic function alone.

The JSF's top-level definitions for enhanced diagnostics, prognostics, and health management are given in Figure 3.1. These definitions are specific to JSF and are more a reflection of the requirements being used to drive the design and the overall architecture selected to meet them. In particular, the separation of the Failure Detection and Isolation requirements from the Assess Material Condition requirements, the latter being used to drive both the prognostic and condition based maintenance capabilities. However, such definitions are subject to further refinement and evolution as the System Design and Development program proceeds and as the various PHM system capabilities are better defined and developed.

Prognostics and Health Management

What is it?

- **Enhanced Diagnostics – the process of determining the state of a component to perform its function(s), high degree of fault detection and fault isolation capability with very low false alarm rate**
- **Prognostics – actual material condition assessment which includes predicting and determining the useful life and performance life remaining of components by modeling fault progression**
- **Health Management – is the capability to make intelligent, informed, appropriate decisions about maintenance and logistics actions based on diagnostics/prognostics information, available resources and operational demand.**

***Figure 3.1** - JSF PHM Definitions*

What then is the "Big P" – real predictive prognostics. We are familiar with the dictionary definitions of prognostics, where such words as foretell, foresee, forecast, omen, prophesy, advance indication, prediction, and forecast of course of disease (medical) are used. Whilst such words all involve predicting future events or outcomes, they do not mention how or imply the method of doing so. Historically, such predictions would have been more of a black art than a science. Essentially, prognostics provides the predictive part of a comprehensive health management system and so complements the diagnostic capabilities, that detect, isolate and quantify the fault, and prognostics in turn is dependent on the quality of the diagnostics system.

From an operators, maintainers or logisticians - the user's view point- what distinguishes prognostics from diagnostics is the provision of a lead-time or warning time to the useful life or failure, and that this time window is far enough ahead for the appropriate action to be taken. Naturally, what constitutes an appropriate time-window and action is dependent on the particular user of the information and the overall system of system design one is trying to optimize. So the "Big P" is that part of the overall PHM capability that provides a prediction of the lead-time to a fault / failure event in sufficient time for it to be acted on.

What then constitutes a "Big P" method – essentially any method that provides a sufficient answer (time window) to meet the user's needs. This is an inclusive approach and one that is seen as necessary to get the system of system wide coverage and benefits of prognostics. Some would argue that the only true prognostics methods are those based on the physics of failure and one's ability to model the progression of failure. Whilst this proposition has scientific merit, such an unnecessarily exclusive approach is inappropriate given the current state of the art in prognostic technologies and the most likely outcome of insufficient coverage to provide system wide prognostics.

So "Big P" methods can range from the very simple to the very complex. Examples of the many approaches are: simple performance degradation trending; more traditional life usage counting (cycle counting with damage assessment models and useful life remaining prediction models; the physics of failure based, sometime sensor driven and probabilistic enhanced, incipient fault (crack, etc) propagation models; and others.

4. The Playing Field

Modern and comprehensive PHM Systems include many types of functional capabilities. Figure 4.1 lists most of these constituent functions, capabilities, and many associated processes. Many of these functional capabilities, though much advanced and more accurate, are an evolution of capabilities developed by legacy diagnostic systems. Through the philosophy and concept of health management, the modern PHM system uses these functional capabilities to complement each other and provide a much larger impact and a boarder set of maintenance oriented benefits than any single function by itself.

Though most of these functional capabilities will be developed together and can complement each other during PHM system development for a particular platform; real predictive prognostics, the "Big P", represents the new, hard, and often "risky" technology capability. The rest of this paper will address a PHM playing field focused only on this new predictive prognostic capability.

PHM Constituent Functions and Processes

Fault Detection
Fault Isolation
Advanced Diagnostics
Predictive Prognostics
Useful Life Remaining, Time-to-Failure Predictions
Component Life Usage Tracking
Performance Degradation Trending
False Alarm Mitigation
Warranty Guarantee Tracking,Tools and Information for Enabling New Business Practices
Health Reporting
–Only tells pilot what NEEDS to be known immediately
–Informs Maintenance of the rest
Aids in Decision Making & Resource Management
Fault Accommodation
Information Fusion and Reasoners
Information Management
–Right Info to Right People at Right Time

***Figure 4.1** – Some PHM System Functional Capabilities*

To understand the role of predictive prognostics; one has to understand the relationship between diagnostics and prognostics capabilities. Envisioning an initial fault to failure progression timeline is one way of exploring this relationship. Figure 4.2 represents such a failure progression timeline. This timeline starts with a new component in proper working order, indicates a time where an early incipient fault develops, and depicts how under continuing usage the component reaches a component or system failure state and eventually, if under further operation, reaching states of secondary system damage and complete catastrophic failure.

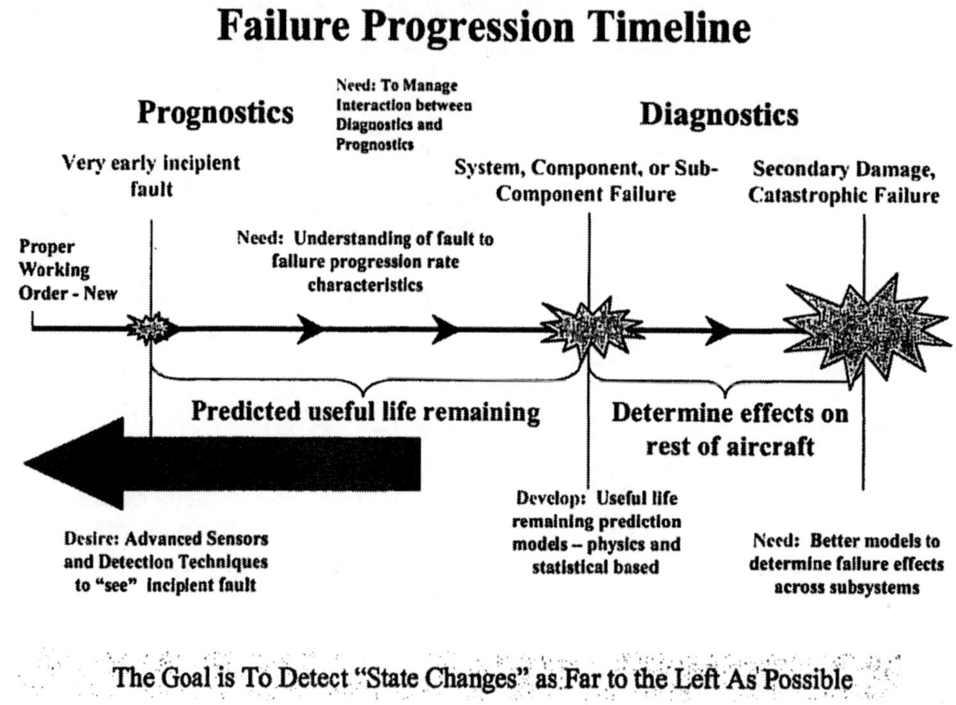

***Figure 4.2** - Failure progression timeline*

Diagnostic capabilities have traditionally been applied at or between the initial detection of a system, component, or subcomponent failure and complete system catastrophic failure. More recent diagnostics technologies are enabling detections to be made at much earlier incipient fault stages. In order to maximize the benefits of continued operational life of a system or subsystem component; maintenance will often be delayed until the early incipient fault progresses to a more severe state but before an actual failure event. This area between very early detection of incipient faults and progression to actual system or component failure states is the realm of prognostics.

If an operator has the will to continue to operate a system and/or component with a known detected incipient fault present; he will want to ensure that this can be done safely and will want to know how much useful life remains at any point along this particular failure progression timeline. This is the specific domain of real predictive prognostics, the "Big P", being able to accurately predict useful life remaining along a specific failure progression timeline for a particular system or component.

To actually accomplish these accurate useful life remaining prediction capabilities requires many tools in your prognostic tool kit. Sometimes available sensors currently used for diagnostics provide adequate prognostic state awareness inputs; and some times advanced sensors or additional incipient fault detection techniques are required. This brings up the questions: how early or small of an incipient fault do you want to detect and how small of a material state change to want or need to "see"? Other needed tools in your prognostic tool kit include: a model or set of models that represent the understanding of a particular fault to failure progression rate; material physics of failure models; statistical and/or probabilistic based models; models to represent failure effects across interconnected subsystems; and models to account for and address future operational mission usage.

As true prognostic capabilities evolve and are developed for particular applications many difficult questions need to be addressed. These questions can be very application specific. Fig. 4.3 provides a depiction of some of these questions from a prognostic perspective using a bus traveling around a globe as an analogy. This depiction can be used to address the concept of predictive prognostic horizon; or in this case, how early do you have to "see" or detect the bus in order to do something useful with this information.

Prognostic Horizon Level Targets

How Far Do You Want to See Into the Future?

Prognostics: What's Your Perspective?

- **Needs and Benefits**
- **Capabilities: Available and Desired**
- **Technology "Holes" to be Filled**
- **Philosophy and Strategy**
- **Integration and Implementation**
- **Questions:**
  - **Is It Possible?**
  - **How are you going to use It?**
  - **What's Good Enough?**

Choose One

- **Detect Bus Just Before It Hits You,**
  or
- **Detect Bus Far Enough In Advance to Take The "Right" Evasive Action**

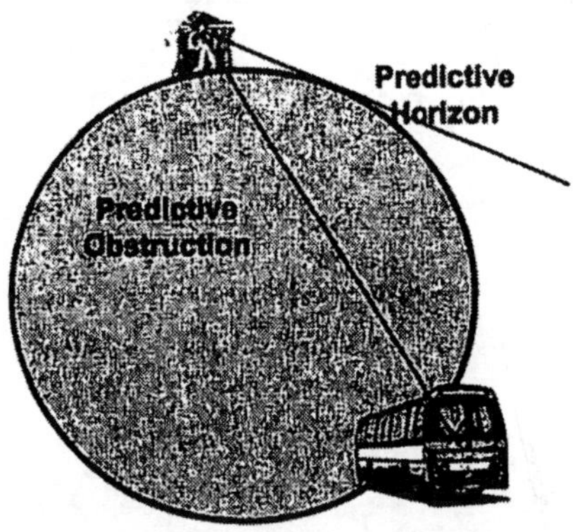

***Figure 4.3** - Prognostic Perspectives and Questions*

Perhaps the first basic question to ask is; how far does your specific application want or need to "see" into the future? The answer to this question will very much depend on your application specific prognostic perspectives. From this perspective, this first question generates many additional considerations. Some of these include: specific needs vs. recognized benefits; what is possible and doable; capabilities available vs those desired or high valued; technology shortfalls to be filled; integration, implementation, and usage strategies; and what is good enough.

5. CHALLENGES

The future vision for Logistic Support to military aircraft is a structured organization responsible for the provision of Global Technical Support to the "warfighters" during peace operations as well as in crisis phases. The above-mentioned organization would be managed by a joint Government and Contractor team. Most of the "legacy" activities, performed under Government responsibility, will be reallocated under the contractors management, but the Government will maintain full visibility. Work is underway in developing the details of this future Defense acquisition / sustainment strategy and there is urgent need to properly identify the roles and responsibilities of PHM / predictive prognostics in this strategy.

In order to realize the above vision, nine major operational groups has been identified to address the main Logistics Support functions (Fig.5.1):

1. Logistic Field Operations elements focus on providing effective on-time support to warfighter, providing support in the customer's environment, when and where it is needed.

2. Fleet Management provides non-routine support planning and execution, vehicles allocation, management of modifications and special programs, storage facilities and aircraft maintenance, configuration management.

3. Sustaining Engineering provides technical problem resolution, defect reports, damage recovery, maintenance planning, technical documentation, and infrastructure harmonization.

4. Supply Chain Management provides all aspect of supply support, component procurement, transportation, packaging, handling and storage.

5. Support Equipment includes ground equipments required to support and operate the aircraft.

6. Information System Support is the information system that links together all the elements of support and operations.

7. Training Systems Operations provides all aspects of training of pilots and maintainers.

8. Business Operations gives elements that integrate military priorities with commercial support under a management contract to guarantee aircraft availability, Life Cycle Cost (LCC) affordability business metrics, management contracts, procedures for operations, contracts, strategic partnerships/supplier.

9. Logistic Support Development would be a team, designing and developing major Engineering Change Proposals (ECP) solutions.

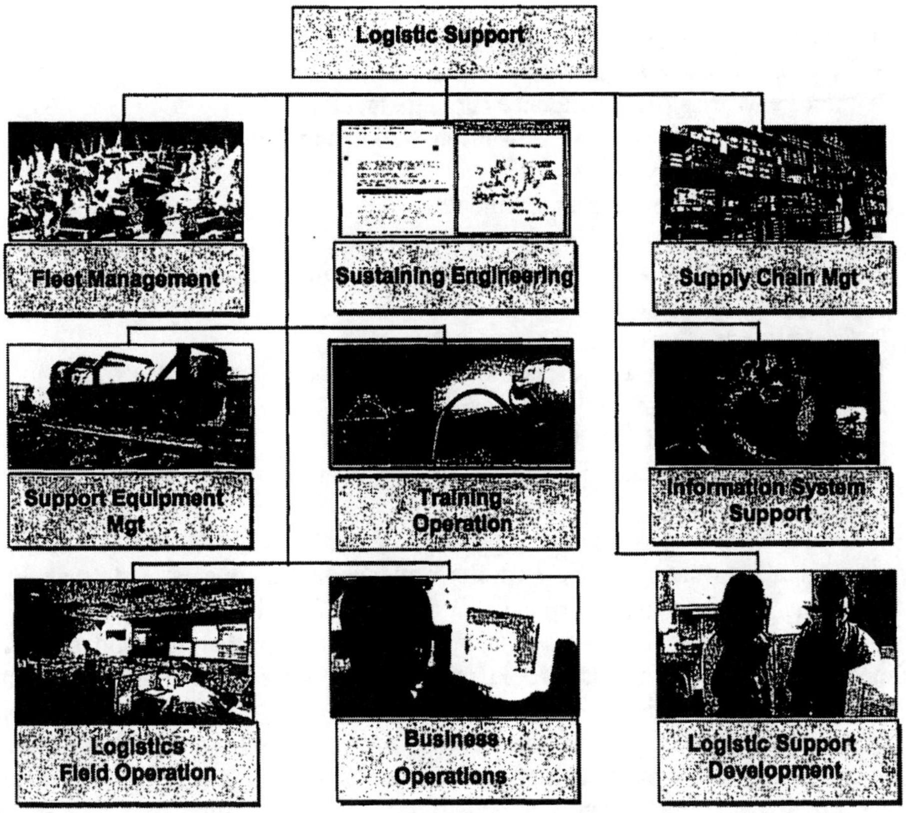

Figure 5.1 *- Logistic Operations*

A key target of this new Logistic Strategy is the move to long-term Performance-based Contracting, an approach for buying set levels of performance, such as a payment per flight hour approach. Development of well-defined performance, cost, and schedule metrics, financial and other incentives, and award fee and award term plans are prerequisites to establishing successful long-term performance-based arrangements. Performance-based contracting is envisioned as a mechanism to incentivize the designing of products that enhance reliability & maintainability and reduce cost throughout its service life.

Predictive Prognostics is one of the fundamental factors that influence the decision to shift to this new business approach. With this concept, Original Equipment Manufacture (OEM) and key Suppliers should be incentivized to make capital investments in new innovative "P" technology to improve logistic support capability, knowing that their involvement (and related profits) will extend well into the production and post-production phases of modern military aircraft. Additional long-term performance-based contractual arrangements would be established between the prime contractors, their sub-suppliers and government organic Depots.

A performance-based approach is intended to:

- Reduce Total Ownership Costs.
- Increase Warfighter Confidence and Satisfaction.
- Facilitate Contractor-Government Integration and Communication.
- Reduce the demand for logistics.
- Incentivize Reliability Enhancements.
- Decrease the Resource Requirements for Support.
- Encourage early & continuing emphasis on diminishing manufacturing sources and obsolescence planning.
- Centralize Management
- Create a Real Time Problem Response (24/7)
- Optimize the Technology Insertion
- Utilize a consistent Life-Cycle Cost Analysis
- Optimize infrastructure harmonization and asset utilization.

The vision is that legacy renegotiable price based contracting will abandoned over time shifting gradually to the new long-term Performance-based concept. For example, cost-type contracts could be used during LRIP (low rate initial production) phase. As the design matures, defined enablers are achieved, sufficient reliability and cost data are obtained; therefore the contract would transition to a fixed-price type contract, with performance incentives linked to metrics identified during LRIP. Through joint co-operative forums, client (i.e. Government) and contractor would establish the criteria and performance metrics for the transition in accordance with substantiated data, maturity of process, pricing, and performance responsibility. Total transition to a fixed-price type contract would be tied to obtaining sufficient reliability and cost data. However, the actual transition point would depend on the system and air vehicle peculiarities and maturity. The final decisions for transition sequencing will be determined during LRIP using factors such as achieved and stable reliability, business case analysis impact, incentives, and the maturity of depot level repair.

In order to implement the Performance-based approach, the strategy will revolve around how core capabilities for dealing with OEMs/suppliers are established, refined, and executed in relationship to delivering performance to the user community. This is a challenge in itself and it is an additional challenge figuring out how "P" capabilities and their associated data product will assimilate it.

Recognizing the need of the above-mentioned contractual approach, to ensure that system design is optimally balanced between total ownership cost and system/equipment performance requirements, it is mandatory to describe how to develop a best value business arrangement (which contains affordability initiatives targeted at continuous cost reduction and technology refresh/insertion). The establishment of a business model is the framework for assessing the correlation of business elements and attributes, that are the functions and tasks required to perform the nine logistics operations and that will allow one to assess the best value solution.

The figure 5.2 depicts an example of business model cost categories with the business elements included in the lower part of the left column. The options for five (5) management categories are displayed across the horizontal continuum for each element and attribute, ranging from full Government management responsibility (Category I) on the left to full Prime Contractor management responsibility (Category V) on the right. Selection of the level of management accountability for individual elements and attributes should be examined through a process to determine the different solutions.

	Category				
	I	II	III	IV	V
Management – Government	Total	Program Management	Approval	Oversight On Limited Tasks	Insight/Oversight by Exception
Management – Prime	None	As Directed	Program Management	Shared Approval	Total Approval [illegible]
Performance	Government	Government with Prime Support	Prime with Gov't Support	Prime Only or Shared	[illegible]
Business Model Elements					
Sustainment Integration	Government	⟵	⟷	⟶	Total Contractor
Joint Fleet Management	Government	⟵	⟷	⟶	Total Contractor
Sustaining Engineering	Government	⟵	⟷	⟶	Total Contractor
Logistic Field Operations	Government	⟵	⟷	⟶	Total Contractor
Logistic Information Systems	Government	⟵	⟷	⟶	Total Contractor
Supply Chain Management	Government	⟵	⟷	⟶	Total Contractor
Support Equipment	Government	⟵	⟷	⟶	Total Contractor
Depot Maintenance	Government	⟵	⟷	⟶	Total Contractor
Training System	Government	⟵	⟷	⟶	Total Contractor
Facilities, Manpower Other Gov't	Government	⟵	⟷	⟶	Total Contractor

Figure 5.2 *- Business Approach Options*

Business model attributes need to be qualitatively and quantitatively analyzed to determine the impact (cost, effectiveness, or other) of moving from one category to one of the other four categories. Such analyses will provide the functional alternatives and present the economic and technical arguments for executing alternative solutions over the product lifecycle to achieve the stated customer and business objectives and imperatives.

While it is clear that prognostics influences various attributes (i.e. support equipment, maintenance, supply chain, spares ownership/management, propulsion support, etc); the real challenge is the development of valid and measurable metrics for quantifying the impact of the various prognostic technologies on the individual model attributes and to include these in the business model so that the resultant analyses reflect the true value of prognostics to the proposed business case.

For example, to find out the metric to evaluate the consequence of having "P" in a typical electronic component as Power Supply device, it could be necessary to include statistical/damage/physics of failure accumulation models and sensor-based condition monitoring.
Above-modeled performances (as predictive mean life or failure/damage accumulation) would provide time savings equivalent to monetary savings for each Business Model Element. This will enable to assign the Attributes of each Element to the appropriate category, enhancing the optimization of the available resources.

6. ISSUES

One of the major challenges to the designers of modern PHM systems is the need to develop diagnostic and prognostic methods that are truly capable of handling real-world uncertainties – as the real world is not deterministic. Such real world uncertainties cause havoc with deterministic approaches leading to high false alarm rates, inaccurate predictions, incorrect decisions and an overall PHM system that is not very robust. Some of the issues uncertainty presents to the designer are elaborated on below, including issues associated with various steps in the predictive process, the estimate of current condition, the prediction of time to failure (or time remaining) and the choice of appropriate lead times (how far ahead to predict); and on the choice of an overall prognostic methodology.

The first step in predicting remaining life or time to failure is to know where you are at any time along the continuum of condition from normal to degraded to fail – i.e. the current condition. Uncertainty associated with the estimate of current condition flows through to and compounds the inaccuracies of the predictive process itself and so good prognostics starts with good diagnostic design.

For affordability reasons, most PHM designers cannot rely on a PHM specific sensor based solution but rather have to rely on sensors placed for other functional purposes, such as control. This is especially so when attempting to maximize diagnostic coverage. Consequently, direct measurement of condition is more the exception than the rule in operational PHM systems. Rather the condition has to be inferred from an appropriate choice of sensors and parameters – or features.

Naturally, good features will exhibit wide separation - distance- between the normal and failed values with low uncertainty and bad features will exhibit small separation with high uncertainty. The latter means there is a high likelihood that the distribution of feature values that represent a given healthy state - normal condition - will overlap the distribution of feature values that represent a failed state. Such an overlap means at the same feature value one gets both a probability of a false positive and a false negative. The relative probabilities are determined by the shape of the feature distributions and the location of the threshold value. The implication is that the diagnostic performance of a badly chosen feature is not necessarily improved by tweaking the thresholds but rather by paying closer attention to uncertainty in the design process. Furthermore, any such feature that does not provide good discrimination between the normal and failed condition is unlikely to be a good feature for prognosis. The designer needs to go back and choose a better feature.

Given the future is not fully known or knowable the prediction of remaining time (i.e. time to failure or life remaining) needs to be treated as a probabilistic process where the predicted remaining time is represented by a probability density function. Importantly, the accuracy and uncertainty (precision and confidence) of the predicted remaining time are adversely impacted by the stochastic processes yet to happen. While it is useful to try and quantify such future events and behavior, and determine the most likely scenarios, one cannot know exactly what will happen and so the resultant inaccuracy and uncertainty of the predicted remaining time can only be minimized at best but not completely eliminated.

The use and interpretation of probability density functions throws up the following paradox: the more precise the prediction of remaining time, the less probable this prediction will be correct. This is related to the probability of occurrence being equal to the area under the probability density function between two points in time. The two extremes to this are not very useful to the designer, (1) predicting a precise / exact time by definition has zero probability of being correct; and (2) predicting a time that covers the whole distribution is 100 % probable but also totally useless. For this reason, the designer must balance the choice of precision by choosing appropriate lower and upper bounds to the prediction and by result provide the measure of confidence to the prediction e.g. 95% confidence the predicted remaining time is free of failure. The other behavior of the prediction process is that the initial predictions – very far ahead - have the largest error (inaccuracy) and the greatest uncertainty, and those closer to failure become more accurate and precise. This behavior means the designer must pay close attention to how far ahead failure, life or condition needs to be predicted.

What is good enough for a prognostic estimate in terms of accuracy and precision is more a function of how far ahead one needs to predict – the lead-time interval. This, in turn, is a function of the users' needs and the overall logistic system that the predictions fit into. Simple analysis or judgment calls can provide appropriate lead times for various needs such as: turn around - sortie generation rate; spares ordering - supply chain management; and scheduling the removal of life limited components - fleet life management. More complex is determining whether the lead times so chosen provide the necessary system wide impact – optimal system of system behavior with overall lower cost of ownership. Currently, the designer is short of such tools and like in most PHM cost-benefit studies the designer would benefit from improved models of how PHM interacts with and impacts the total logistic system.

The designer having determined the appropriate lead times for a given, failure mode / LRC can then proceed to develop and use different prognostic methods as the different lead times can tolerate different accuracy and precision. For example, in life management the initial fleet-wide planning can use a-priori estimates based on analysis of the likely design life and use, the higher uncertainty of this long look ahead - 20 years - is acceptable for planning purposes. In operation, the method then turns to tracking individual aircraft usage and updating remaining life estimates for the actual usage. When life gets closer to removal time – other methods capable of discerning the presence of cracks, such as vibration analysis can be introduced to provide the necessary confidence – confirming evidence, that the predicted remaining life will be met without failure. Furthermore, such methods can be then used to manage life (risk) beyond the initial design life. While uncertainty is a limiting feature – one only needs to be as accurate as the situation requires and the correct choice of lead-time intervals helps this. Finally, some self-compensation is evident as it is the short-term predictions that are and need to be more accurate.

The issues of uncertainty mean that no one method will provide a useful, robust and comprehensive prognostic capability for a total aircraft logistic system of system. Rather an integrated approach is needed combining information from the individual detection, diagnostic and prognostic functions at the appropriate component, Line Replaceable Unit, sub-system, system and system of system levels and from both on-board and off-board sources. The actual prognostic approach will also vary depending of how well the system behavior and failure mechanisms are understood and the how well the aircraft sensors, and maintenance and logistic systems capture and use the required information. The approaches will necessarily involve a mixture of physics / model based, rule-based and data driven algorithms embedded in an artificial intelligence reasoning framework to enable the desired automatic decisions to made in the presence of uncertainty.

Performance Based Logistics and the PHM capabilities required to enable its successful implementation presents another set of issues and a new paradigm for sub-system suppliers. Suppliers will not only be required to design for functional performance but also to design for PHM. The requirement for an inherent PHM capability will necessarily drive the sub-system suppliers to better know their equipment, how it operates, its operating environment, and most importantly, how it fails. This rightly is the task of the sub-system supplier but the requirement will also cascade down to the individual component suppliers. Some of the challenges and issues imposed by this new paradigm are as follows.

Most sub-system suppliers do not have design teams with the necessary PHM skill sets and tools. Such skill sets and tools are in short supply and expensive to develop. As well, the validation and test of PHM algorithms can impose significant costs. So suppliers need to develop strategies to engage outside help from PHM specialists and experts and to share the development costs across a variety of industry projects and /or defense platforms.

The new paradigm means greater interdependencies between the sub-system suppliers and the prime system integrator. Here the supplier provides the foundation of the PHM capability through the level of sub-system diagnostics and prognostics; and the prime provides the enhanced diagnostics and prognostics of failures that propagate across several sub-systems. Both levels of PHM are needed to meet the stringent diagnostic and prognostic requirements imposed by Performance Based Logistics. Successful PHM design and development will require close collaboration between sub-system suppliers and prime. Over the life of type, this interdependency means more configuration control on proposed sub-system changes. Whereas, previously components in sub-system designs could be changed if it appeared functionally the same to the system but now how it fails must be taken into account and if different it must be quantified and passed onto the system designer, so the designer can change the higher level reasoning. A successful outcome will be most likely if the sub-system suppliers and primes seek to develop a strategic partnership that can support a platform over its life of type.

Suppliers need to take up these challenges if they want to stay competitive in bids for modern systems where Performance Based Logistics will be the new way of doing business.

The challenges and issues of prognostics have been aggressively addressed for mechanical systems for some time but are only recently being fully explored for electronics systems. In some aspects, these challenges and issues are the same or very similar for both electronic and mechanical system prognostic applications. In other aspects, they are very different and unique for electronic system prognostic applications. This is particularly true and challenging with respect to electronic systems employing many digital devices and components in their designs.

The fact that digital devices and components appear to fail very randomly and do not have adequate pre-failure attributes or characteristics leads many to conclude that prognostic capabilities are not possible. The truth is that prognostic capabilities for electronic systems are currently a "mixed bag" of approaches and techniques and are evolving just as they have for mechanical systems. Many of the more traditional prognostic approaches, when applied to the analog devices and components used in electronic systems will most likely work fine. Most electronic systems use many analog devices and components along with their digital ones so at less some partial prognostic coverage is likely possible. Also just being able to detect some form of system performance degradation, even without an accurate useful life remaining prediction, can at times be very beneficial.

There are several approaches that can be applied to the electronic prognostic challenge. Some of these can be described as follows: The legacy time based approach. Very specific data interrogation techniques using attributes, measurands, or characteristics that lead to the indication of repeatable degradation trends. Environmental monitoring to determine cumulative exposures and replacement of the system before some pre-defined limit is exceeded. A combination of environmental monitoring and modeling that uses cumulative exposure to drive an actual predictive cumulative damage model that will cause replacement of the system when a sufficient amount of life has been consumed. Canary Devices and/or witness samples that are integrated into a specific component, device, or system design and that incorporate failure mechanisms that occur first in the embedded device. These embedded Canary Devices would be non-critical elements of the overall design; and providing early incipient failure warnings before actual system or component failure would provide very prognostic information. There are and will be many other approaches developed but an important point to remember is that effective electronics prognostic will utilize a combination of approaches.

The use of physics of failure modeling approaches for electronic components and devices, like those used for mechanical systems will be a powerful tool in support of electronic prognostic capabilities. It is felt that this is the likely possibility because the root cause of most all electronic device or component failure is actually mechanical – something physically breaks at a sub-component, solder joint, connection, layer, delamination, etc., level. Solder fatigue models are already under development and show promise. If it breaks mechanically then there should be some combination of prognostic approaches that could be used to predict failure.

7. Lessons Learned

As various programs progress to develop comprehensive diagnostic, prognostic and health management (PHM) systems many lessons are learned. Similarly, as specific prognostic directed projects are undertaken additional lessons are accumulated. The following paragraphs represent and discuss some of these accumulated lessons learned to date.

Prognostic capabilities are new, mostly very hard to develop, and often take time to mature, but are doable in many, many cases. Having said this, prognostics are certainly not doable for all cases. There is a need to identify the clearly not doable cases and then not worry about them. Rather, focus your limited resources on those that are both doable and directed at high value systems, subsystems, and components.

It is unlikely that any single platform program can afford all the resources required to support all the work necessary to develop and validate accurate useful life remaining prognostic capabilities on every desired system, subsystem, and component. This is true even for just the doable and high value components desired of a complex platform like a modern fighter aircraft. To mitigate this high expenditure of resources from a single platform; a smart strategy needs to be implemented. This strategy should attempt to share prognostic capability development costs across similarly interested program platforms and across groups of similar component types. Smart use of specifically targeted science and technology efforts, outside of a particular platform program, may also offer additional resources, focus, and needed help.

One has to have good diagnostic capabilities before you can do prognostics. The corollary is also very true and real; if you have developed very capable and comprehensive diagnostic capabilities then it necessarily follows you will attempt to develop prognostics. There are a few exceptions to these statements. For one, system or subsystem overall performance degradation trending can be a very useful prognostic capability by itself.

To achieve prognostic capabilities with accurate useful life remaining predictions, presents the need for multiple types of integrated models. These types of models include: sensor calibration or correlation models; models of accumulated usage up to crack initiation; incipient fault or crack propagation models; statistical and/or probabilistic based models; and others. All these individual supporting models need to be developed for accuracy, validated, and integrated into a global prognostic model for the specific component and system application.

To successfully develop global prognostics models and accurate useful life remaining predictive capabilities requires a development team with expertise in various disciplines. This development teams needs a mixture of material science experts; state awareness sensor experts; several types of modeling and data fusion experts; the specific component design specialists; statistical and probabilistic modeling specialists; specific component of interest design specialists; traditional diagnostic experts; and others. Legacy attempts at prognostics development have often left themselves short of material science expertise to develop the required physics of failure modeling.

To successfully develop global prognostics models and accurate useful life remaining predictive capabilities requires a development team with expertise in various disciplines. This development teams needs a mixture of material science experts; state awareness sensor experts; several types of modeling and data fusion experts; statistical and probabilistic modeling specialists; specific component of interest design specialists; traditional diagnostic experts; and others. Legacy attempts at prognostics development have often left themselves short of material science expertise to develop the required physics of failure modeling.

It takes significant data, experience, and maturation time to develop prognostic and accurate useful life remaining prediction capabilities. This is true in some cases more than others; since there are always "low hanging fruit" examples where degradation always occurs along easy to understand trend lines. This statement is particularly true for accurate useful life remaining prediction capabilities and for specific components where physics of failure models are not well understood or where root cause failure mechanisms are very random. The point to remember is sophisticated prognostics capabilities require lots of data to develop and time to mature; plan your program to include enough maturation time.

Often straightforward system or subsystem performance degradation trending, without any accurate useful life remaining predictions, can be a very useful form of prognostics. This represents a prognostic capability in the simplest form but can be extremely useful to the fleet operator. Coupled with case based reasoning or before specific component physics of failure models are available would be good applications for this type of prognostic trending.

Good Failure Modes and Effects Analysis (FMECA) won't give us prognostics by themselves but as with the development of good diagnostics capabilities; they are a necessary starting point for many reasons. Thorough and up to date FMECA's identify failure modes and related symptoms; prioritize high occurrence failure modes; and critical failure path components. This is very necessary to help identify high value components for prognostic coverage. FMECA's can also be very useful for "flushing out" interrelationships for dependent elements or components within a single system or across interconnected subsystems.

Subsystem component seeded fault testing is extremely useful but can be prohibitively expensive. Plan for a few and use them wisely. Back up this strategy with a resourced plan to capture performance degradation AND incipient fault to failure data whenever you can in a piggy backed environment. Examples would include planning to capture this invaluable data during all possible system, subsystem, and component developmental testing, qualification testing, environmental tests, final acceptance tests, etc.

The Maturation of comprehensive prognostic capabilities takes time, lots of data, and significant analysis resources. It also takes dedicated management commitment, a well thought out maturation strategy and coordinated plan, and adequate funding resources at many levels of the platform development team. For more complex systems, for many components where failure mechanics are less understood, and for more "detailed" prognostic capabilities with accurate useful life remaining prediction capabilities, the maturation time will likely exceed well past the original platform development program.

When the platform program funding cuts come, and they always will, prognostic capabilities will often be the first to be scrutinized for cost reduction. Prepare for this and develop a well thought out strategy to counter these threats and to well articulate the specific benefit justifications.

8. SUMMARY AND CONCLUSIONS

This paper explored the background, benefit impacts, and current playing field for predictive prognostics; highlighted some specific design challenges and issues; discussed the various degrees of prognostic capabilities; and drew heavily on lessons learned from previous and current prognostic development efforts. The fleet maintainer needs and benefit impacts of real predictive prognostics, the "Big P" is obvious, real, and great. There are several types of prognostic capabilities of varying levels of sophistication; some based on simple trending techniques to those using multiple integrated modeling approaches. The prognostic definitions used for these different approaches may also vary and are evolving with the technology.

Real predictive prognostic capabilities are just one element among many interrelated and complementary functions in the playing field of PHM. The challenges involved with developing and implementing prognostic capabilities are many; and this paper focused very much on those impacting and enabling benefits associated with new performance based logistic support concepts, global sustainment strategies, and new business practices. Similarly, the technical issues involved with developing real predictive prognostic capabilities are large and many. This paper focused on some of the issues involving uncertainty.

In conclusion it can be said that real predictive prognostics, the "Big P", is hard, challenging, and often involving very complex and difficult technical issues; but with wise development management and the appropriate resources, is doable. Real predictive prognostics represents an essential combination of available, developing, and maturing capabilities critical to an evolving technology discipline called PHM. The challenges, issues, and the lessons learned associated with predictive prognostics are many and will certainly continue to grow, as this technology discipline matures and is more widely applied. Future papers will continue to discuss and document these new found and additional challenges, issues, and lessons learned. We will still be chasing the "Big P" for some time to come and on many new applications. So stay tuned for several continuing versions of this paper.

9. REFERENCES

[1] Hess A. and Fila L. "The Joint Strike Fighter (JSF) PHM Concept: Potential Impact on Aging Aircraft Problems", paper # 403, 2001 IEEE Conference, March 2001.

[2] Hess A. and Fila L. "Prognostics, from the Need to Reality – from the Fleet Users and PHM System Designer / Developers Perspectives", paper #116, 2002 IEEE Conference, March 2002.

[3] Hess Andrew, Capt. Calvello Giulio, Dabney Thomas, Firth Peter: "PHM the Key Enabler for the Joint Strike Fighter (JSF) Autonomic Logistics Support Concept" 58th Meeting of the MFPT Society, 2004.

AUTHOR BIOGRAPHIES

Andy Hess is a graduate of both the University of Virginia and the U. S. Navy Test Pilot School.

Andy is world renowned for his work in fixed and rotary wing health monitoring and is recognized as the father of naval aviation propulsion diagnostics. Beginning with the A-7E engine health-monitoring program of the early 70's, Andy has been a leading advocate for health monitoring in the Navy and has been instrumental in the development of every Navy aircraft application since the A-7E through the F/A-18 and V-22. His efforts have largely led to the successful transition of a development program Helicopter Integrated Diagnostic System into production (IMD HUMS) for all H-60, H-53, and AH-1 aircraft. More recently, Andy has been a strong advocate of prognostic capabilities and has been involved in many efforts advancing the development of these predictive capabilities for a sundry of systems and their component elements. Currently, Andy is hard at work leading the development and integration the Prognostic and Health Management (PHM) system for the Joint Strike Fighter program and is a frequent participant in the international technical community.

Major Giulio Calvello is an Italian Air Force Corps of Engineers Officer, graduated in 1994 of both the Italian Air Force Academy and the University of Naples "Federico II" in Aeronautical Engineering.

He is Professional Engineering (PE) certified as well as a Non Destructive Inspections and Corrosion Analysis Inspector. He served for several years in the F104 Depot Maintenance Center in Grosseto as Mechanics and Avionics Supervisor.

Assigned to the Italian Air Force Logistic Command as Deputy of the National Supply Authority for the Tornado IDS, he was selected to serve as co-chairman in the Financial Sub Group in the F3 Tornado ADV Leasing Management between Italian and UK Government. He is currently serving as Autonomic Logistics leader for the Prognostic and Health Management (PHM) in the Joint Strike Fighter Program Office.

Dr Peter Frith is a principal research scientist from the Australian Defence Science and Technology Organization (DSTO). He has over twenty years experience in the development of advanced diagnostic and health management technologies for military propulsion systems and until recently was Head of Engine Performance at DSTO. He is currently attached to the JSF Program Office as a Collaborative Program Personnel in the Air System PHM Team. His prime focus areas are requirements, verification, prognostics and propulsion.

MATERIAL SIMULATION-BASED ELECTRONIC DEVICE PROGNOSIS

Loren Nasser, Robert Tryon, Animesh Dey

VEXTEC Corporation
750 Old Hickory Blvd, Bldg 2, Suite 270, Brentwood, TN 37027
615-372-0299
lnasser@vextec.com; rtryon@vextec.com; adey@vextec.com

Abstract: Electronic failures can result in immediate system shutdown with no advanced fault or warning signal. As a device is operated thermal and/or mechanical loads are induced across the package. These loads are translated from the global package level to the localized interconnect level. VEXTEC's patent pending material-based simulation software accounts for real world interconnect variability and predicts probably of failure as a function of device operating time. This NAVAIR sponsored SBIR project demonstrated feasibility for using conventional, off-the-shelf sensing, to predict solder degradation due to thermal cycling as a means to prognosticate electronic device reliability.

Keywords: electronics; fatigue; interconnect; modeling; package; prognosis; solder; thermal;

1. Introduction:

It is simply not practical to directly sense the degradation of electronic components. These damage states are usually structural and, due to their size, structural response signatures are not monitored on electronic components. For example, it would be both difficult and expensive to directly sense the cracking of a single emitter wire bond on a circuit board comprised of thousands of emitter wires. Yet, the failure of a single emitter wire can cause the failure of the device.

Conventionally the reliability of electronic devices is assessed using empirically-based models. Design of experiments is a commonly used tool in which the experimental conditions are systematically varied and a mathematical relationship is "fit" to the data that represents the influence of the conditions to the time or cycles to failure. However, one problem is the fact that there is so much variation in the time or cycles to failure that device life can only be conveyed in the form of a statistical average, such as mean time to failure (MTTF) or the mean time between failure (MTBF). These metrics are appropriate in the context of manufactured fleet lot reliability but they lack the fidelity for accurate representation of individual fielded device reliability.

Electronic device reliability prognosis requires that the analysis consider actual field operating conditions for individual devices. These field conditions will be diverse sequences of loads and environments. Theoretically, empirical methods can be used to assess highly reliable systems if a statistically significant number of experiments were performed for every possible sequence of field conditions. As electronic components become smaller and they are used in more diverse environments with diverse sequencing, development of highly accurate empirically-based models is not practical.

VEXTEC has been developing a materials-based, large rotating component prognostic framework under DARPA, Air Force, and Navy program funding. At the foundation of this framework is VPS-MICRO™, which is fatigue crack initiation and total life prediction software. In early 2004, VEXTEC used VPS-MICRO™, under a series of blind validation tests facilitated by Pratt & Whitney, to successfully predict fatigue life for 5 alloys of Ti-6Al-4V material. Statistical analysis showed that the VPS-MICRO™ predictions were within 95% confidence bounds of the P&W test database.

This paper documents the work carried out under a Navy project to demonstrate the feasibility for using a similar prognosis framework for prediction of electronic system failure. Of specific concern to JSF readiness is the reliability of electronic power supplies. Power supply failure can result in immediate JSF system shutdown with no advanced fault or warning signals. According to JSF supplier sources, power supply systems are – compared to other electronic systems – highly susceptible to failure due to the high voltage and current conditions in which they routinely operate. Unlike mechanical systems, these electronic systems do not actively display conventional fault signals prior to failure.

Electronics, specifically power supplies, are reliant on the integrity of interconnects or solder bonds. The reliability of interconnects is a concern because it is widely expressed that fracture failures in solder joints accounts for up to 70% of failures in electronic components. The work under this project thoroughly documents the fact that interconnect or solder degradation and failure is attributed to thermomechanical fatigue mechanisms. Since VEXTEC has proven its fatigue life prediction methodology for large structural components, this project demonstrated a similar approach to predict small-scale interconnect reliability.

2. Solder Fatigue:

The most commonly used solder fatigue models are based on Coffin-Manson plastic strain equations developed for solder interconnects. Due to the low melting temperatures, most electronic devices operate at temperatures above solder creep thresholds. Since Coffin-Manson models do not explicitly address creep strains, specific solder models have been developed to incorporate creep (Knecht and Fox, 1991; Syed, 1997). All of the fatigue models in use today assume large scale similitude i.e., the structure (weld footprint or wire diameter) is large compared to the crack, the crack is large compared to the crack tip plastic zone and the crack tip plastic zone is large compared to the microstructure. These assumptions are fundamentally incorrect for today's small scale devices.

Energy-based solder fatigue models have been developed to account for the large crack size compared to the weld geometry (Liang et al., 1997) and Morris and Reynolds (1997) point out the importance of the size of the material microstructure compared to the size of the damage. Emitter bond wires are usually 300 to 500 microns in diameter and the metalization layer to which they are bonded have a thickness of 3 to 5 microns. With polycrystalline grain sizes as large as 150 microns, microstructural similitude cannot be assumed for crack initiation and growth. Figure 1 shows the relative size of the grains compared to the overall solder contact area.

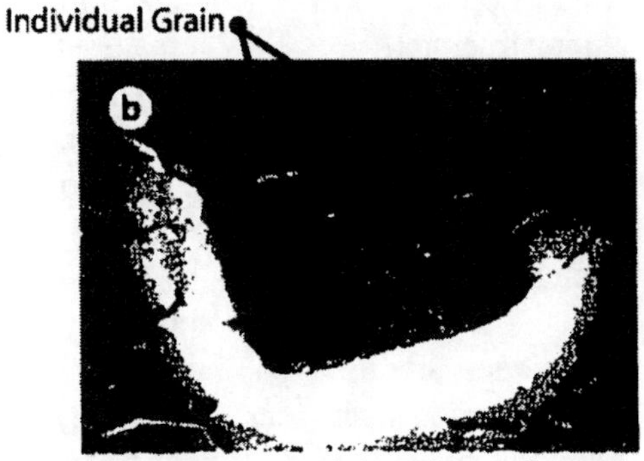

Figure 1. Few material grains make up a typical package interconnect.

None of the previously described fatigue models account for the large scatter in the solder weld properties. The nature of the packaged components and variability within the assembly process creates large variations in the solder welds for even the strictest manufacturing tolerances. The very small size of the welds causes variation of the weld footprint from weld to weld. In addition, the microstructural development of the weld is greatly controlled by rate of cooling from the melt stage. Some of the components within a package are specifically designed with heat sinks for thermal management during operation. These heat sinks are activated during the assembly process by the heat of the welding although they may not have been specifically designed for this purpose. This heat transfer action causes uneven cooling of the welds from component to component and uneven cooling of the different emitter bonds within a single component. Conventionally used electronic device reliability prediction, largely empirically derived, do not account for this real world variability. Thus, variation in the geometry and material properties of the weld must be considered in order to prognosticate reliability accurately.

Industry research indicates that thermo-mechanical fatigue damage mechanisms in Sn/Pb solder as time dependent (creep) shear stress induced cracks initiate and propagate in the solder near the intermetallic interface. The fatigue response of the solder changes due to instability (coarsening) of the microstructure. Although, both intergranular and transgranular fracture surfaces are observed, depending on loading and temperature, transgranular appears to predominate.

VPS MICRO™ is a probabilistic fatigue prediction capability that explicitly models fatigue damage (dislocations, slip band, small crack and long cracks) and the damage interaction with material microstructural features (crystallographic planes, phase boundaries, grain boundaries, etc.). VPS MICRO™ uses Monte Carlo techniques to simulate damage accumulation on a cycle by cycle basis. Monte Carlo reliability techniques are used so that model parameters such as loading, temperature and microstructure can simulate real world conditions by varying randomly with time. The VEXTEC overall electronics package prognosis approach is depicted in Figure 2.

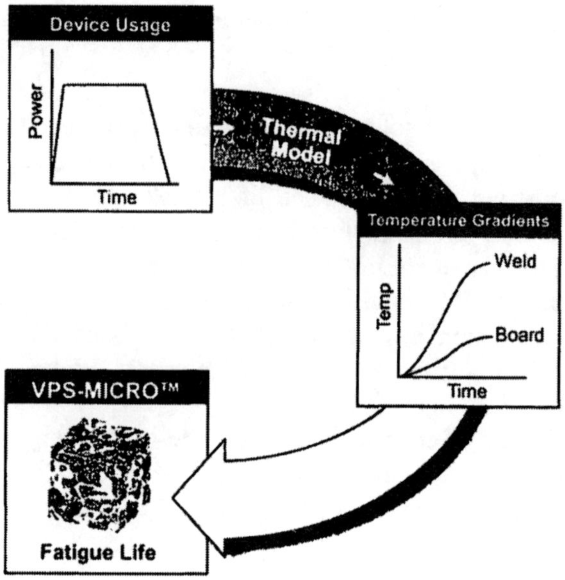

Figure 2. VEXTEC overall prognosis approach.

3. Electronic Prognosis Issues:

The developed prognosis model had to incorporate mission sequencing (i.e.: turning the device on/off and creating thermal swings), geometric design parameters (i.e.: variation in footprint of the solder welds), and material parameters (i.e.: intermetallic interface zone thickness variation).

Electronic device operating principles require that the design parameters be accurately coupled. In order to assess the physical changes in variables at the electronic system level, appropriate system reliability modeling techniques must be applied. For example, temperature gradients are caused by the complex heat transfer characteristics within the device. A chip consists of material layers of various makeup and thicknesses. Some layers radiate heat were others absorb it. Still further, neighboring chips and devices influence the thermal characteristics of the chip as well. Computational analysis techniques such as finite element analysis (FEA) are often used to assess the complex thermal environment during design. Such an environment can be simulated as a temporal thermal mission with start-up and shutdown transients along with steady state conditions. Electronics industry research indicates that the highest stresses imposed within the device are the result of temperature gradients created during on/off transients. The thermal stresses can also be combined with the purely mechanically induced stresses (such as vibration) within the FEA to predict the overall device stresses. The resulting stress prediction becomes the basis for VPS-MICRO™ cyclic life prediction.

The electronics industry routinely combines physics-based thermal or vibration models with empirically-based fatigue models to make a rough order of magnitude prediction of device life. However, the traditionally used semi-empirical approach lacks the required fidelity to be useful in the prognosis of thermo-mechanical fatigue for individual components. Lau and Pao (1997) point out some of the issues for using the current semi-empirical MTTF or MTBF approach. These issues, along with VEXTEC's methodology for overcoming these, are as follows:

Geometry: Dimensional variation of components within electronic packages can be of several orders of magnitude, e.g. a 0.08 x 0.1 mm solder bump versus a 150 x 150 mm Printed Wiring Board (PWB). This necessitates breaking down the numerical analysis into multi-scale (local and global) analyses. Also there is a difference in the actual and simplified geometries of the component or interconnect. For example, the actual geometry of a solder joint depends on a number of parameters such as solder-paste volume, reflow temperature and flux, which can result in a wide range of different shapes. Oversimplification may eliminate geometric discontinuities or defects, which may be intimately associated with a significant part of fatigue life. To address geometry issues, VEXTEC has developed multi-scale finite element models to relate the global stresses and strains caused by the thermal mismatch between the electrical device and the circuit board to the local stresses and strains in the intermetallic solder layers. These methods allow for statistical variations in the complex geometry at the global, local and any intermediate level.

Meshing: Electronic packages are bi-material and tri-material bonded edges and interfaces. Stress concentrations cause interfacial damage or cracks to initiate. Capturing detailed stress-strain distributions with reasonable meshes near these concentrations is difficult. Also, many materials used in packaging are of thin layer shapes, such as thin and thick films. Thus, aspect ratio becomes a problem and a large number of elements are required to accurately model the behavior of these thin layers. The multi-scale finite element models used by VEXTEC addresses meshing issues by incorporating a representative volume element (RVE) approach in which different size scales are modeled with separate RVE's. These RVEs are properly linked using "nested models". This approach allows for small regions of highly concentrated stresses and strains to be modeled with separate RVE's and then linked to the next higher size scale.

Material Properties: Properties of internal solder layers, such as intermetallic regions between the solder and silicon substraight, are not well known. Most properties are temperature and time dependent. The mechanical behavior of a solder joint, for example, depends on strain rate, temperature, cooling rate, loading history and microstructure. Mechanical properties measured using relatively large scale and bulk specimens generally are not the same as those for actual solder interconnects. To address material property issues, VPS MICRO™, a physics-based fatigue analysis approach models the interaction of the fatigue damage with the material microstructure. The material microstructure is explicitly modeled at the crystalline level; no bulk assumptions are made.

4. Prognosis Approach:

Figure 3 presents an overview of VEXTEC's Phase I feasibility demonstration. Temperature cycling produces stresses at the global component level that are translated to local stresses at the material microstructural level. VEXTEC modified the VPS-MICRO™ fatigue analysis approach for use with electronic materials. VPS-MICRO™ was used to predict weld probability of failure as a function of system usage. Further, the approach used to predict the response of a board with many welds and associated failure interrelationships.

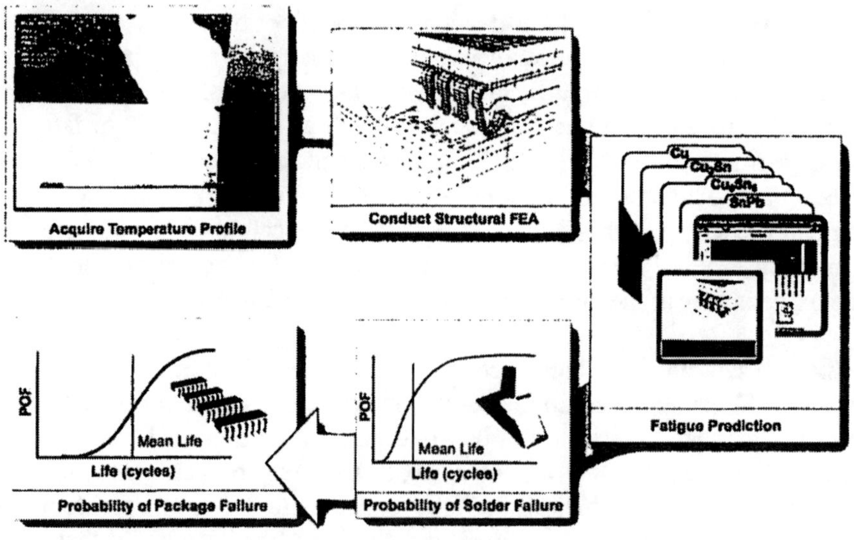

Figure 3. Prognosis demonstration approach.

4.1 Global Modeling:

Conventional FEA analysis is incorporated within the VEXTEC analysis approach. SRI (Menlo Park, CA) performed a thermo-mechanical stress analysis for a J-leaded device for several different packages with different material layering combinations. Under this project an Alloy 42 lead is soldered to a FR-4 board with a copper metalization layer. For the feasibility demonstration described later in this paper, only the corner joint was simulated since it had the highest stresses and can be considered a representative example.

4.2 Local Modeling:

Fatigue failure is a localized, material-driven process. An extensive amount of research has been conducted as to how and why cracks initiate and grow within solder connections. This knowledge was adopted within the developed local simulation modeling approach. High stresses are translated from global loadings to the local material. In particular these stresses exist at the interface between the copper lead and the solder. These localities are of compounded significance due to the existence of a complex microstructure of intermetallic layers between the copper and the solder. When molten Sn-Pb solder contacts the lead, intermetallic compounds (IMC) are formed between the solder and the lead. To model the complex stress state of the microstructure at the copper/solder interface, two dimensional ANSYS finite element models were created under this project by VEXTEC for the copper/intermetallic/solder region. Because the thickness of the intermetallic layers change with time, a series of finite element models were created for various Cu_3Sn and Cu_6Sn_5 thicknesses to simulate this real world variability. Figure 4 shows the example FEA model for 4 microns of Cu_3Sn and 7 microns of Cu_6Sn_5 used for demonstration purposes. The bottom layer shown in the model is copper.

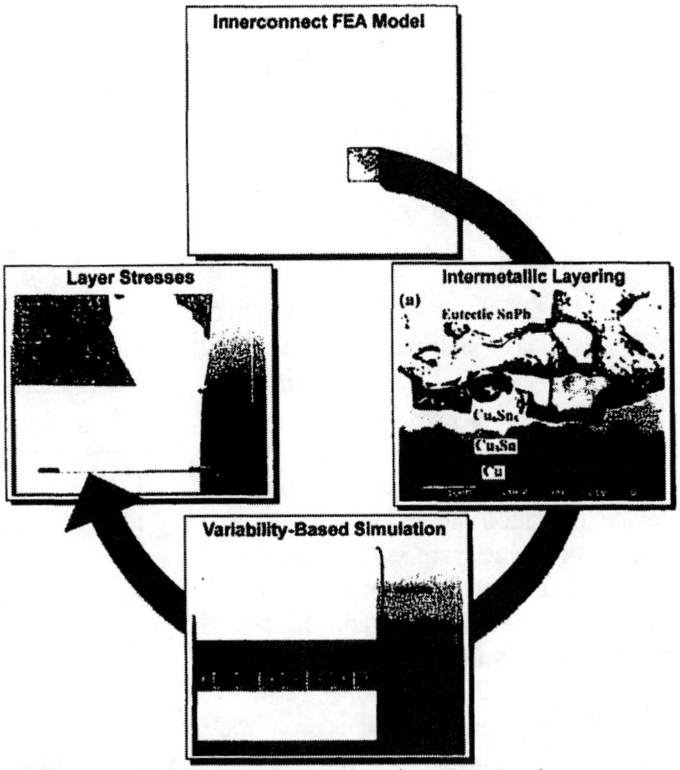

Figure 4: Global stress translated to local microstructure through simulation of intermetallic region.

4.3 System Reliability.

Response surface techniques are used to combine the global and local stress analysis with the microstructural growth kinetics to create a simple mathematical equation of the complex relationship between temperature, dwell time and the stress at the solder joints.

$$\sigma = \frac{T}{256}\left(c_1 + c_2 d_1 + c_3 d_2 + c_4 d_1^2 + c_5 d_2^2 + c_6 d_1 d_2\right)$$

where:

$$d_n = \left(D_n e^{\left(-Q_n/kT\right)} t\right)^{\frac{1}{2}} \tag{1}$$

$n = 1$ for Cu6Sn5

$n = 2$ for Cu3Sn

Where σ is the stress in the solder joint, T is the temperature, t is the cumulative time at temperature, d_n are the thicknesses of the intermetallic layers, c_i are the least squares fit coefficients and D_n, Q_n and k are material properties.

5. Electronic Prognosis

Demonstration:

For feasibility demonstration purposes, an individual solder connection was simulated as being thermally cycled between room temperature and 256 F over 15 minute dwell periods. During actual device operation these temperature inputs will come from conventional thermocouple sensors mounted on the device being analyzed. If a thermocouple can not be mounted on or near the device then conventionally sensed voltages or currents can be translated to board temperature through currently available thermal modeling approaches. Either way, this prognosis approach builds upon state of awareness capabilities in existence today.

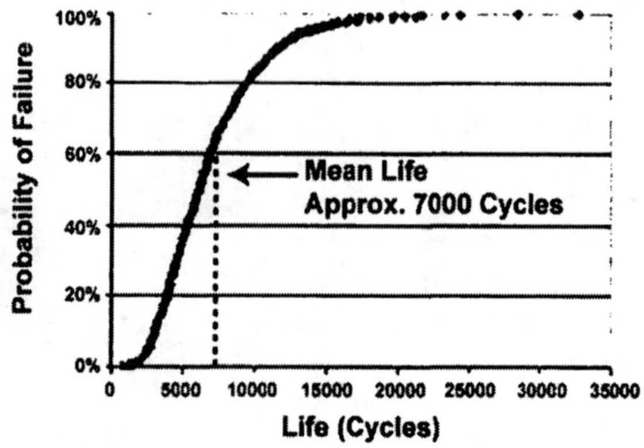

Figure 5: Thermal cycling induced failure prognosis for an individual solder joint.

Based on the industry identified driving mechanisms that cause solder materials to degrade and initiate cracks, VPS-MICRO™ fatigue software was used to predict solder joint life. This feasibility demonstration found that a single, commonly used, simulated solder bond was predicted to have a mean life of approximately 7000 cycles (Figure 5).

5.1 Device with Many Interconnects:

Electronic devices are typically attached to circuit boards or housings through multiple solder connections. To prevent against excessive heating, electronic OEMs use finite element modeling to predict the stresses imposed on device interconnects during design. This Phase I showed how these global stresses can be translated to the material microstructural level, and that these stresses vary for each interconnect location due to the realistic simulation of a thermal gradient across the device. VPS-MICRO™ software was used to prognosis the life expectation difference for an interior solder location compared to an edge solder location. Using a standard industry system reliability approach, it was demonstrated that a thermally cycled device with 28 individual solder connections (7 per side, 4 edge joints) was predicted to have a mean life of 2554 cycles.

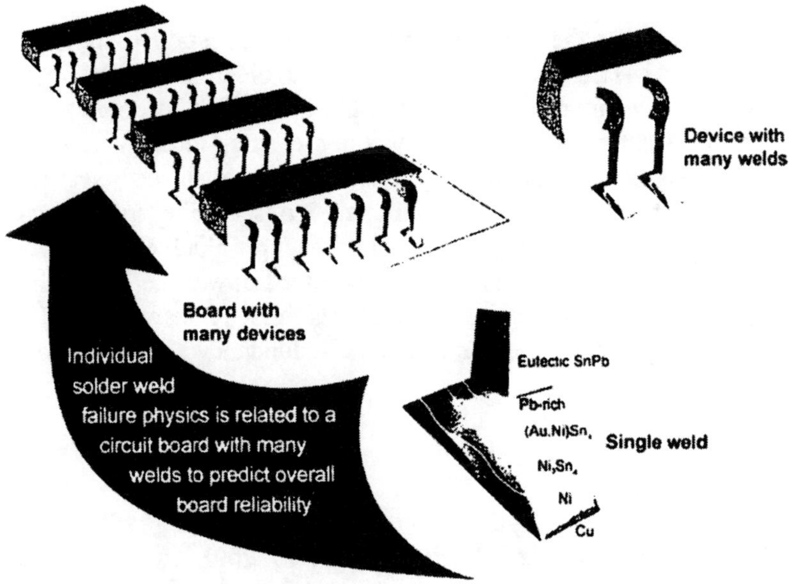

Figure 6. Package reliability predicted through a systems approach.

5.2 Electronic Package with Many Devices:

An electronic system, such as a power supply, will be comprised of many individual devices. In order to demonstrate package prognosis proof of concept, a circuit board with 20 individual devices (each with 28 solder joints) was thermally cycled through the previously described simulation process (Figure 6). Just as expected in actual operation, a thermal gradient was simulated across the board causing the thermally induced stresses to vary from device to device. This demonstration predicted that the board, with 20 devices, will have a mean life of 1283 cycles which was lower than the life predicted for any single device considered alone.

6. Summary:

This SBIR Phase I has successfully demonstrated how VPS-MICRO^{TM} fatigue simulation can be used for electronics systems prognosis. Electronics, specifically power supplies, are reliant on the integrity of interconnects. Industry reliability data indicates that fracture failures in solder joints accounts for up to 70% of failures in electronic digital devices. Electronics industry specific thermal models along with generalized structural finite element software can be used with VPS-MICRO^{TM} software to predict solder fatigue response.

Phase I used simulated, conventionally sensed thermal cycling data, from room temperature to 256 F, as input to the VEXTEC prognosis approach. It was demonstrated that individual interconnect life as well as that for a device or chip with many interconnects could be prognosed. Thereafter life prognosis was demonstrated for a circuit board with many devices that also considered the realistic simulation of a thermal gradient applied across the board.

This paper identifies that the inherent weaknesses of currently used empirically-based approaches such as MTTF or MTBF have been avoided in the VEXTEC prognosis methodology. One day this prognosis approach will be used to predict board or device degradation prior to failure. Such capability is non-existent today within aircraft systems (as an example). Today operational performance is assured through systems redundancy and through the implementation of extremely conservative maintenance practices. Redundancy is inefficient, adds weight and takes up valuable real estate within the aircraft. Conservative maintenance results in increased downtime and higher operational costs due to excessive inspections, tear-downs and replacements. This prognosis technology will mitigate the need for electronic redundancy and allow for maintenance for cause decisions.

References:

[1] Tryon, R.G., "Onboard, Prognostic Micro-structural Reliability Tool for Mechanical Systems," DARPA Contract DAAH01-01-C-R127, 2001

[2] Knecht, S., Fox, L. (1991) "Integrated matrix creep: application to accelerated testing and lifetime predictions, "Solder joint reliability theory and applications," Van Nostrand Reinhold, New York.

[3] Liang, J., Gollhart, N., Lee, P. S., Heinrich, S., Schroeder, S (1997) "An integrated fatigue life prediction methodology for optimum design and reliability assessment of solder inter-connectors," Advances in Electronic Packaging, Vol 2.

[4] Morris, J. W., Reynolds, H. L. (1997) "The influence of Microstructure on the mechanics of eutectic solders," Advances in Electronic Packaging, Vol 2.

[5] Lau, J.H. and Pao, Y.H. (1997). Solder Joint Reliability of BGA, CSP, Flip Chip and Fine Pitch SMT Assemblies, McGraw Hill, NY.

Biography:

Loren Nasser is COO responsible for day to day operations at VEXTEC Corporation. He has been actively involved in the development of reliability prediction software within the aerospace, electronic and automotive marketplace. He has an M.S. in engineering science from the University of Tennessee and B.S. in mechanical engineering from Rose-Hulman Institute of Technology.

Robert Tryon is co-founder and CTO at VEXTEC Corporation. He has lead the development of the patent-pending "VPS-MICRO" software. He served as principal investigator on VEXTEC's DARPA and NAVAIR prognosis technology development projects and has served in a similar role on numerous other projects. He received a Ph.D. in structural engineering from Vanderbilt University and an M.S. and B.S. in mechanical engineering from Rose Hulman Institute of Technology.

Animesh Dey is co-founder and CPDO at VEXTEC Corporation. He has been involved in the development of reliability software for the automotive industry and has served in a similar role on numerous other projects. He received a Ph.D. and M.S. in structural engineering from Vanderbilt University and B.S. from the Indian Institute of Technology.

PHM/PROGNOSTICS FOR DC TO DC POWER CONVERTERS

Douglas L. Goodman
Ridgetop Group, Inc.
6595 N Oracle Road, Suite 153B
Tucson, Arizona 85704
Doug@Ridgetop-Group.com

Abstract: An easily-adopted method of adding Prognostics/Health Management (PHM) to electronic systems is described. As an example, a PWM-type switching power supply was chosen for analysis in this paper, due to the ubiquitous nature of these Power Converters in Mil/Aero systems, and the criticality of their reliable performance.

Switch mode power supplies (SMPS) provide significant advantages to users in terms of conversion efficiency, weight, and size. According to Industry Analysts, over $7.7 Billion of SMPS Power Systems were produced worldwide in 2002. However, SMPS's have unique reliability issues when used within a system due to size constraints, heat dissipation and varying load conditions.

The internal components of a SMPS, including the Power Switch (MOSFET or IGBT), catch diode, capacitors, and magnetics are under repetitive stress conditions that contribute to elevated failure rates. Often the SMPS is considered the most vulnerable and weakest part of a system. Having an early warning of an impending failure, (through PHM), would be very useful information and critical to mission success.

The efficacy of applying PHM to a commercial, off-the-shelf power converter will be explored, and a proposed foundation for future extensions will be covered.

Key Words: Converters; MOSFET; PHM; power systems; prognostics; reliability;

Prognostics/Health Management (PHM) has been described as a key element in the new Joint Strike Fighter (JSF) to reduce the overall life cycle cost of the aircraft.[1] PHM provides the means to predict an impending failure before it occurs, and thus can mitigate potentially catastrophic failure modes, improve the logistics support framework and increase the operational readiness of the JSF. With a focus on the Electronics portion of the problem, Ridgetop has developed a methodology for Electronic Prognostics that provides deterministic techniques of failure prediction, and thus a foundation for PHM.

A common element for most EW and Mil/Aero systems are high efficiency power supplies. The Navy has determined a significant problem in power supplies with premature failures so this was chosen as the test case for testing the efficacy of PHM.

Also, by initially focusing on such a ubiquitous component, the maximum commercial leverage could be realized from this work.

Ridgetop selected a high efficiency, off-the-shelf 50 watt, DC to DC converter from a commercial power supply manufacturer, C & D Technologies. This supply was chosen due to the topology's being well understood and widely-used, and with the understanding that the techniques would be easily extensible to other power supply topologies and wattage sizes.

TABLE I - DC to DC converter characteristics:

Characteristic	Description
Switch Mode Power Supply (SMPS) Topology	Forward Converter Point of Load
DC Input Voltage Range	Up to 48 volts
DC Output Voltage	5 volts
Current Consumption	10 Amps
Efficiency	Up to 92%
Model Name	WPA-50
Manufacturer	C & D Technologies

Electronic Prognostics Methodology:

For electronic prognostics, there are three key factors that must be understood:

- Circuit topology
- Component Failure Rate ranking
- Physics of Failure Models (precursors to failure)

This information is needed so that a suitable collection design can be developed and implemented for the power supply. The collection of data is, in turn, processed and a single composite result is determined. The general process is implemented as shown in figure 1.

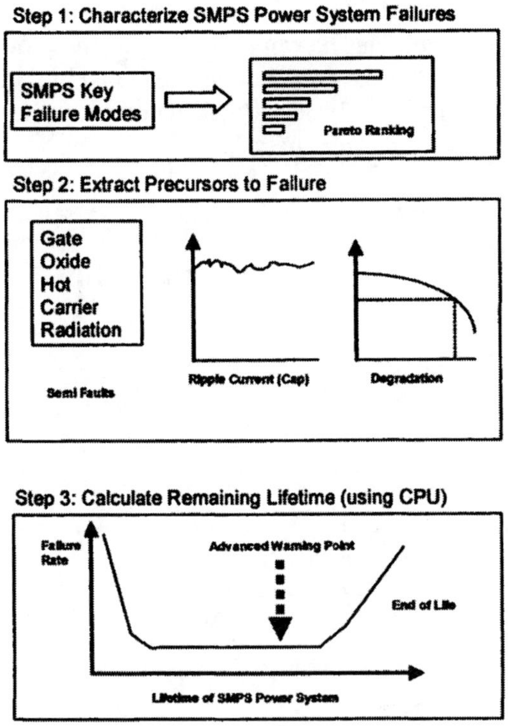

Figure 1 – Electronic PHM Methodology

The WPA50 is a forward switching converter with up to 92% conversion efficiency. A simplified schematic for a forward converter is shown in figure 2.

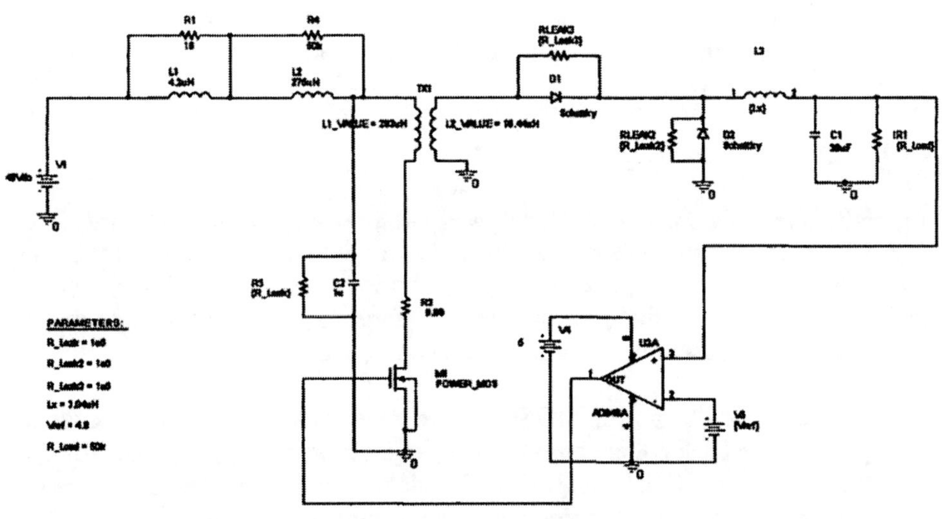

Figure 2 – Simplified Forward Converter Topology

The first step in providing a prognostic solution is to establish a Pareto Chart ranking of the components by their failure rates. C & D Technologies has an extensive database of component data, and this was used for the analysis. For the components contained in the power supply, a ranking was developed, as shown in figure 3; this allows the developer to visually examine the most problematic components for further analysis.

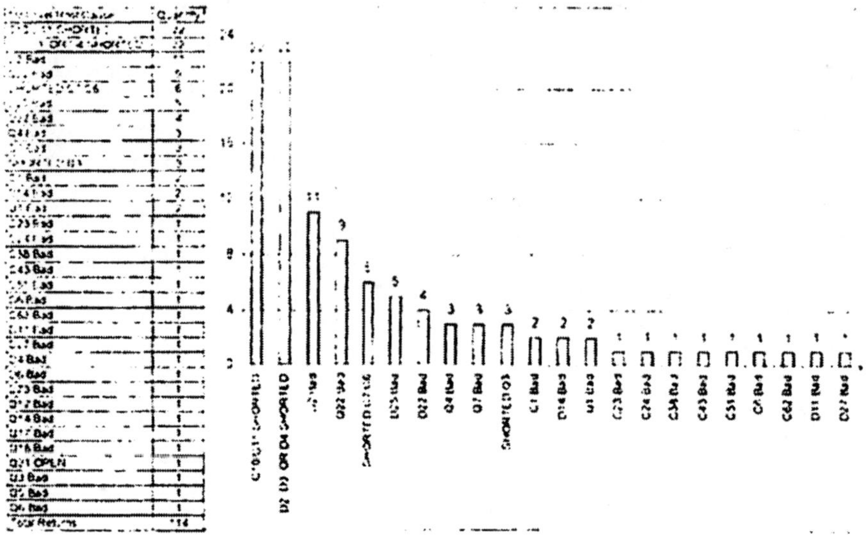

Figure 3 – Pareto Ranking of Parts. With a Pareto, the highest failure rate components are on the left side of the bar chart and descend from there.

Ridgetop examined the power supply topology, along with the components having the highest failure rates. Based on this analysis, there was an examination of the physics of failure associated with each device. The reasoning is that if precursors associated with failure can be extracted, they can be processed into a meaningful prognostic for the power supply.

Physics-of-Failure Model Development:

For the highest failure rate components, the highest failure rates were found on the power MOSFET's, capacitors, opto-isolator and PWM control IC. These failure rates vary based on service environment (e.g. temperature, radiation) and with shrinking process geometries, can have lifetimes that are less than 10 years. [2] Several novel prognostic design approaches are proposed for the four critical components of the power supply.

Due to the importance of the power MOSFET two independent prognostic extraction solutions can be considered for this component depending on the application. The first alternative consists of a sacrificial MOSFET. This device will be placed on the power supply so that it will be subjected to the same environmental conditions as the other components on the board. The operating conditions of the sacrificial MOSFET will result in slightly higher stress conditions so that it will fail prior to the power MOSFETs used in the supply. The increased failure rate will be calibrated, and thus the failure of the sacrificial MOSFET will be an accurate predictor of an impending failure in the power supply's power MOSFETs. The other prognostic solution for the MOSFETs involves monitoring the degradation of device parameters. Since physics of failure models suggest that the R_{DS} (ON) value shows the greatest rate of degradation, the second prognostic alternative for the MOSFET predicts a failure in the device by monitoring this value. Prognostic solutions are also proposed for the other three components. These involve known current monitoring techniques that are proven predictors of failure in such devices. In order to devise and verify the prognostics designs the power supply was simulated using an electronic circuit simulator and simulations were verified using lab measurements of the supply under different operating conditions.

For each device, there was a unique signature associated with degradation that leads to device failure. These models were developed and installed into Ridgetop's device library.

An example for the Opto-Isolator is shown in figure 4.

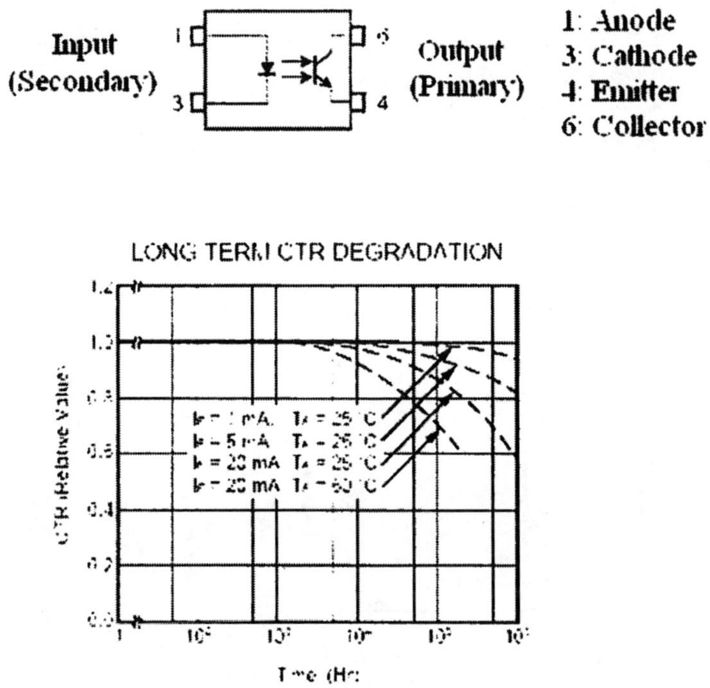

Figure 4 – Physics of Failure Model. For the opto-isolator, there is a defined degradation in current transfer ratio (CTR) over time that leads to eventual failure. This degradation can be monitored in-situ. Other devices exhibit other kinds of degradation that can be measured as precursors to failure.

Prognostic Sensor Array Attachment:

As a design objective, the prognostic sensor array was to be co-located with the host DC to DC Converter, with minimal impact to its performance, and operate in a "fail-safe" manner. The test points for this array are defined by access to the individual components that are to be observed.

To reach that objective, Ridgetop separated the test point sensor problem into three categories:

- In-situ observation of current or voltage is possible during normal operation.
- In-situ observation of current or voltage only possible if the component is disconnected from load (for example, as part of a power-up sequence)
- In-situ observation is not possible at all due to circuit topology considerations

The first two categories can be addressed through the use of a special current measurement hybrid design that attaches to key points within the circuit. This is shown in figure 5.

If no suitable means of direct observation is possible, as in the last category, then there is a fall-back to the statistical models that must be utilized.

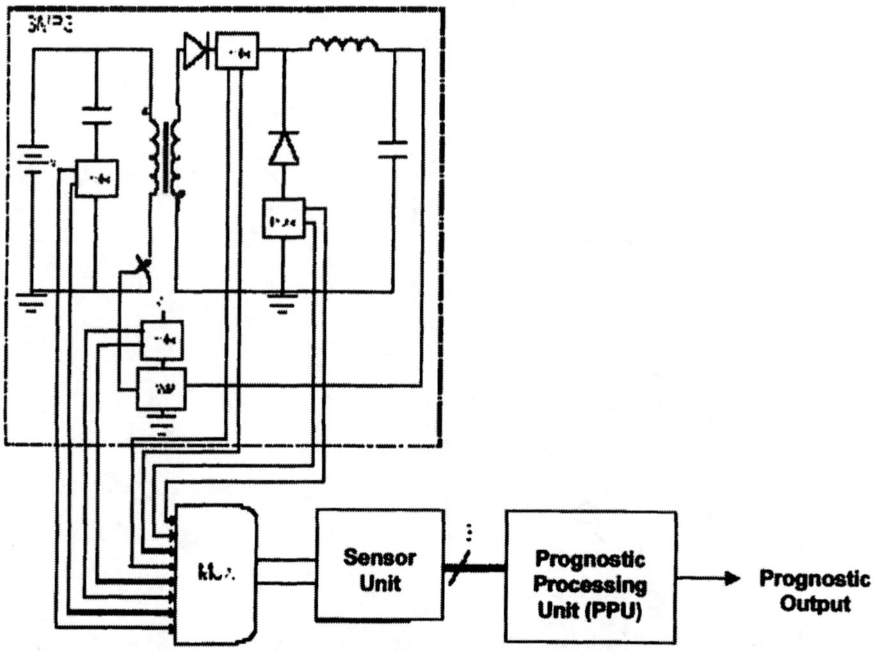

Figure 5 – DC to DC Converter (simplified) with Probes and Sensor Unit

Prognostic Processing Unit (PPU):

It is necessary to collect the distributed prognostic data from the SMPS, and develop a composite calculation on the health of the system based on this data. The calculation is initially simple; the components that have exceeded the prognostic distance will weighted by their importance (for example, bypass capacitors that fail open have a low weight, while Power MOSFETs have a larger weighting), and trigger an alert at the system level to take mitigating action to avoid a catastrophic failure. The calculation will be done by a microcomputer attached to the I2C bus. (Figure 6)

Hardware Interface:

The I2C bus is a commercially-supported, 2-wire bus that is used by C & D Technologies (and others) in systems for power supply monitoring applications. [3] The Master on the bus always sends the clock signal and data is transmitted between master and slave at speeds of 100 KHz, 400 KHz or 3.4 MHz. One of the chief advantages of this bus is that the devices that are attached to the bus are addressed completely by software and new functions can simply be clipped on the existing bus. (A JTAG-type data bus could also be utilized in systems not having the I2C).

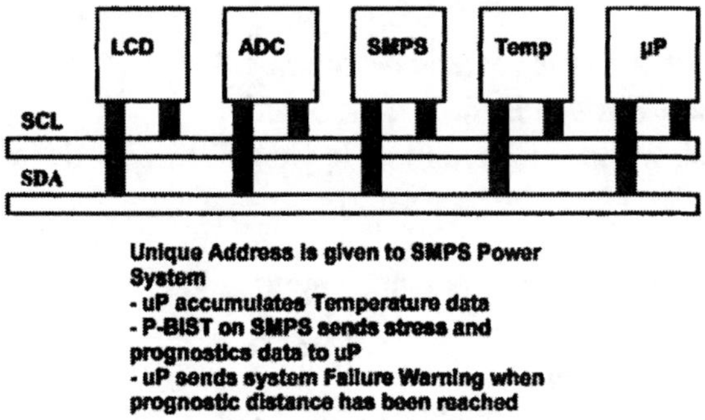

Figure 6 - I2C Bus Configuration with BIST/Prognostics from the SMPS

Using the 8051-type CPU operating as a Master on the bus, the overall stress factor can be calculated, by integrated voltage and temperature over time with the 8 bit processor provided by means of the I2C bus. A composite BIST/Prognostics model can then be built and a means of transmitting the prognostic warning to the system. Initially, Ridgetop has developed a weighted average model to process the data. Further work has been proposed that extends the PPU capabilities for other prognostic detection algorithms that are appropriate for electronic systems. This will help provide more accuracy and reduce the false alarm rates.

The work described has been done using a simple forward switching converter, but the elements in this converter are also present in other power converter topologies, including a boost converter, sepic, cuk, resonant and other forms of high efficiency power conversion systems. Ridgetop will be extending its work into these other areas over time.

Summary:

Ridgetop has completed the first phase of a multi-phase PHM implementation on a high efficiency DC to DC converter. The results from this analysis indicate that certain failure rate components can be directly observed for precursors to failure. This kind of Electronics PHM implementation could aid in providing a higher amount of operational readiness to electronic power systems for the JSF Program. Further work will be conducted by Ridgetop on larger sample sets under HALT Testing regimes at C & D Technologies and further algorithm development.

The benefits to the Fleet in adding Electronics Prognostics/Health Management are as follows:

- Helps avoid catastrophic failures, preventing loss of life
- Addresses most troublesome and pervasive electronic subsystem – the power supply
- Increases the operational readiness for fleet assets
- Supports maintenance strategies tuned to operations and not strictly to elapsed time
- Provisioning and spares pipelines can be optimized with modules that have better lifetime estimates.
- Roots out the latent defects that otherwise escape detection.
- Augments information available through Built-in Test functions
- Redundancies may be reduced, with associated module weight, if the efficacy of Prognostics is established.

While the focus of this work is for the JSF, the potential applications for Electronic Prognostics are very large, and the solutions developed with this work will find wide applicability in other DoD System requirements. These include systems in aircraft, helicopters, ships and land-based systems. To the largest extent possible, Ridgetop has sought to make its Electronic Prognostics as modular as possible, so that the solution would (conceptually) snap-on with minimal rework of the host power conversion system.

As part of the Phase III follow-on, Ridgetop proposes to work closely with the larger JSF contractors such as Lockheed Martin and Northrop Grumman to identify candidate power systems for upgrades with Prognostics.

Acknowledgements:

The authors would like to thank Andrew Hess and Tom Dabney from the JSF Program office for their support of this work under SBIR No. N68335-04-C-0088.

In addition, thanks go to Steve Wood and Don Gerstle of C & D Technologies and Jim Houston of Lockheed Martin Corporation for their support and encouragement.

References:

[1] Hess, Andrew "The Joint Strike Fighter (JSF) Prognostics and Health Management" NDIA Conference, 2001

[2] Goodman, Douglas L. "Prognostic Methodology for Deep Submicron Semiconductor Failure Modes", *IEEE Transactions on Components and Packaging Technologies*, Volume 24, Number 1, March 2001, pp. 109-111

[3] Flynn, James, "Understanding and Using the I 2 C Bus", *Embedded Engineering*, 11/97

[4] Simpson W.R., Sheppard J.W., "Analysis of false alarms during system design", Aerospace and Electronics Conference, 1992. NAECON 1992., Proceedings of the IEEE 1992 National , 18-22 May 1992 Pages:657 - 660 vol.2

[5] Morel H., Gamal S.H., Chante J.P., "State variable modeling of the power pin diode using an explicit approximation of semiconductor device equations: a novel approach", Power Electronics, IEEE Transactions on, Volume: 9, Issue: 1, Jan. 1994 Pages: 112 – 120

[6] Sorensen H.W., "Kalman Filtering techniques", Advances in control systems, Vol.3, Academic Press

[7] Maybeck Peter, "Stochastic Models, Estimation, and Control", Vol. 1, Academic Press, Inc

[8] Zaugg D.A., Samuel A.A., Waagen D.E., Schmitt, H.A., "A combined particle/Kalman filter for improved tracking of beam aspect targets", Statistical Signal Processing, 2003 IEEE Workshop on Statastics, 28 Sept.-1 Oct. 2003 Pages:535 – 538

[9] Dinh-Tuan Pham, Dabia K., Musso C., "A kalman-particle kernel filter and its application to terrain navigation", Information Fusion, 2003. Proceedings of the Sixth International Conference of ACM, Volume: 2, July 8-11, 2003 Pages:1172 - 1179

CONDITION MONITORING AND MAINTENANCE TECHNOLOGY

Chair: Mr. Timothy Wechsler
Chesapeake Reliability Services, LLC

CONDITION BASED MAINTENANCE: A SUPPORT PILLAR TO RELIABILITY CENTRED MAINTENANCE

Lindsay K Galpin[1], Fraser D Fleming[1], and Martin Briddon, MBE[2]

[1]WSA/MLSIPT/MLSCG3, Cedar 1b, Abbey Wood Bristol BS34 8JH
[2]James Fisher Mimic Ltd, The Greenheys Centre, Manchester M15 6JJ

ABSTRACT

Since the early 1980's the Royal Navy has systematically revised its maintenance policy from one of preventative maintenance carried out at prescribed calendar periods to maintenance procedures that, where applicable, relate to the condition of individual equipment. This has been achieved through the introduction of modern data collection equipment and suitable diagnostic software. The development of this approach, now recognized as Condition Based Maintenance (CBM), has enabled a Reliability Centred Maintenance (RCM) strategy to be established. However, the road to RCM has not been a smooth one and this paper seeks to enlighten the reader on the problems that have been overcome, the problems that still exist, and the challenges that lie ahead.

KEYWORDS:

Condition Based Maintenance, Reliability Centred Maintenance, Condition Monitoring, Maintenance Information, Preventative Maintenance, Diagnostics, Prognostics, Data Collection, Trend Analysis

BACKGROUND:

In the middle 80's Condition Monitoring (CM) was identified as a means for reducing costs by giving the engineer the conditional data whereby he could use sound judgement to defer specific intrusive maintenance, with the added possibility of an increase in availability. The Royal Navy (RN) made the decision to embrace CM and revised its maintenance policy to one which incorporated CBM whenever possible. This policy revision was the easy bit; finding a solution to the problems of deciding what data to collect, and subsequently the means for collecting and trending the data, proved more difficult.

What was really lacking was a comprehensive discipline for recording trends other than by paper and log book. By the late 80's the emerging power of the desktop PC was seen as a possible solution to the data collection and trending problems and some simple objectives were set for a potential system. These objectives were:

- It was to be able to be used for the collection and analysis of hourly process parameters and monthly vibration levels on all rotating marine equipment on board RN warships.
- Data was to be clearly presented and easily analysed: it had to be operable by Senior and Junior Engineer Rates.
- It had to have reporting facilities available to the Ship's Engineer Officers.

A review of the CM systems available at that time concluded that the only software system which had the capability of providing both a Watchkeeping and Vibration Monitoring capability was the Maintenance Information on Micro Computers (MIMIC) System. It should be remembered that at this time, the Information Technology (IT) we today take for granted, was in its infancy.

The initial MIMIC system allowed for the following functions:

- Hourly Watchkeeping rounds using data collectors.
- Hourly data log sheets and trends.
- Automatic alert at data upload for parameters in warning or alarm.
- Long term trending.
- End-of-Watch reports.
- Start-Stop Log.

These functions allowed the collection of all process parameters that should, over time, indicate a deteriorating condition leading to machine failure. To collect the data, hand held collection devices were employed. For Watchkeeping a Psion II based instrument was initially developed and for vibration data collection a state of the art data collector analyser was introduced [1, 2, 3]. The introduction of this vibration data collection method gave a greatly enhanced analysis and diagnostic capability by providing: 4-level alerts with multi-level alarm reporting, expert set-up using fault parameters to simplify diagnosis (as shown in Figure 1), fault reporting and trend graphs, and vibration spectral analysis.

From these data collection capabilities it can be seen that MIMIC enhanced the existing paper based system without removing any of the skills of the trained vibration analyst. By using modern hand held vibration data collectors and down loading the information into MIMIC, all rotating equipment had an established baseline set and with regular updates, trending could be achieved. MIMIC also allowed for the trending of the different parts of the spectrum. These trends could also contain relevant process parameters to assist with fault diagnosis.

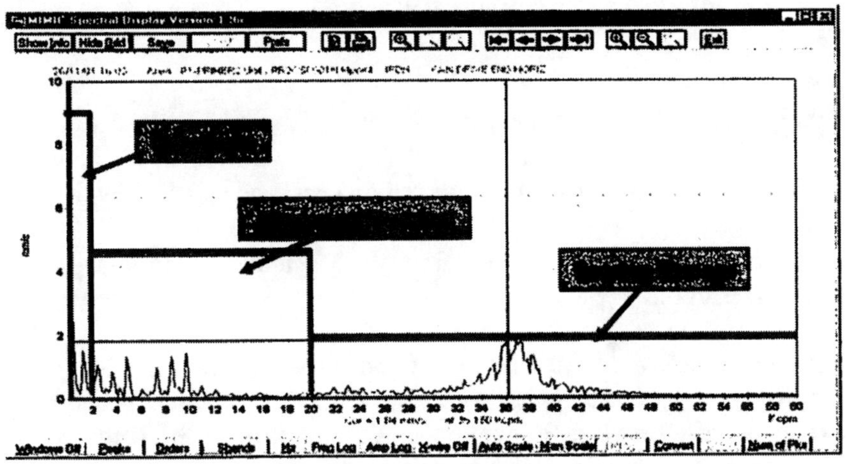

Figure 1. Spectral data with individual fault parameters set.

So having decided what was required, the practical implementation was to prove the most challenging. It was soon discovered that:

- Software works well in an industrial environment, but does not necessarily work well in a marine environment.
- Hand held data collectors would fail if continually dropped.
- Hand held data collectors would fail if immersed in bilge water.
- Software would fail if operators turned off the computers.
- Computers were new onboard and users could load additional software that adversely affected the operation of MIMIC.
- Each data collector was used by many.

The lack of good quality historical data was to be the most vexing problem, as the maximum life of most data collected before the introduction of MIMIC was six months. Therefore, when initially setting data base parameter warning and alarm levels, there was very little Availability, Reliability, Maintainability (ARM) data available to assist in this task. Initially the warning and alarm limits were set using good old fashioned engineering judgement and reviewing similar equipment on ships of the same class. Some of the solutions to the implementation problems were:

- Ruggedising data collectors which entailed encasing them in protective hard cases.

- Reducing the impact of collector failure by provision of transferable Ram and Data Pacs.

- Configuring PC's so as to prevent additional software being loaded.

- A continual hand holding supportive training package supplied by the contractor.

- A provision of first level support from the Machinery Vibration Analysis Teams at the waterfront.

The MIMIC system and associated data collectors were subjected to a lengthy sea going trial, for whilst those ashore thought they knew how ships went about their business, the ships were to prove otherwise. By continuous development of the software, Mod State Zero for the software was declared in November 1994. Perhaps a better way of expressing this software state is; this state of software is what should have been asked for if it had been known what was wanted when the contract was first placed.

Like all simple concepts once proven, one finds the user community requesting additional functionality; hence a continual development programme was maintained where the DOS-based software eventually reached Version 18. Although a Windows variant of MIMIC, was initially available for the VM module in 1996, a conscious decision was made to continue with the stable DOS version and let industry find most of the bugs in the Windows version.

THE INTRODUCTION OF RCM:

At this time the RN embarked on a further maintenance initiative that attempted to produce a cohesive solution to the whole maintenance management chain. This was initially to be known as project Upkeep before changing to Phoenix and ultimately to Unit Maintenance Management System (UMMS). This total maintenance solution was to be based on the philosophy of RCM, the software for which was under development for trials in Hunt Class Vessels. As a part of this whole solution MIMIC was mandated as the data collection tool for all engineering data. It was therefore decided to install the Windows version of MIMIC, MIMIC 2001, on Her Majesties Ship Atherstone, a Mine Counter Measure Vessel , in advance of the RCM RN software. This was undertaken in November 1998. This class of vessel was chosen for the RCM trial because it was small enough to be manageable, but large enough to provide a representative platform to validate the RCM process. An added incentive to the fitting of MIMIC 2001 to all applicable RN vessels by the end of 1999 was to ensure Y2K compliance. This software has been the basis from which MIMIC has continued to be developed to meet the ever increasing needs of the RN.

The initial MIMIC systems consisted of three desktop stations and a server networked together and isolated from any other IT system. This allowed an independence from outside interference, but was not to last. Due to the proliferation of IT systems onboard RN vessels, an IT strategy was implemented to combine all systems into one integrated solution. This removed the independence and brought not only a need to co-exist with other software, but also removed some of the security measures that were in place to prevent inadvertent operating system changes. With the introduction firstly of the RCM RN software and then the total UMMS package, there became a need to develop a standard machinery type template from the data held on ships databases so as to provide a class standard database with warning and alarm levels based on historical data.

This development of machinery templates and the production of a class database led to ease of application and only required minor amendments due to the specific machinery fit of each ship. The added bonus of class generated warning and alarm levels allowed an assessment to be made upon completion of a base line survey for vibration parameters. This allowed the operators to judge the equipment mechanical health against a class average rather than setting the alarm levels relative to the base line levels taken.

INTERFACE DEVELOPMENT:

In addition to the parameters already within the MIMIC database, the RCM studies derived a requirement for parameters that were available from within a disparate set of machinery control and surveillance systems, but were themselves not stored or trended. This led to a development path for system interfaces between ships legacy systems and MIMIC and then MIMIC to UMMS. The first interface was to be for the Type 23 Frigate where MIMIC needed to interface to the Machinery Control Acquisition Surveillance System (MCASS) working with the MCASS system manufacturer.

The original remit was to take an hourly download of the required parameters into MIMIC for trending and passing to the RCM functional software when an alarm condition was met. This proved difficult as the MCASS software was already 20 years old and the hardware had a limited computing capability. To achieve the required output, an interface card (PC104) was developed and inserted into the MCASS system communications rack. The interface when tested worked as designed, capturing the state of a measured channel once an hour. It did not, however, capture any changes in the channel state during the hour, thereby missing any out-of-tolerance readings for recording in the RCM software for maintenance and reliability purposes. The additional requirement to capture events as they happened had therefore to be incorporated.

After the rework and during the implementation phase it was discovered that the MCASS system did not function as thought. This again led to the failure to capture running hours and process data critical to the gas turbines. It was also discovered that the system failed to re-synchronise after any interruption to the data transfer although this was an original design requirement of the interface. A further stage of development has now been embarked upon to overcome these unforeseen problems.

The MIMIC to UMMS interface was required in order to pass the required parameter data in support of the RCM function of UMMS. Ultimately the interface had to be developed in conjunction with the UMMS software supplier. This required all parameters that changed to a warning or alarm state to be displayed to ship's staff for validation prior to sentencing to the UMMS software. This in turn required the alignment of both (MIMIC and UMMS) databases to allow the electronic linking of each parameter to ensure the relevant machine received the correct maintenance. It also provided a rejection facility within each part of the interface whereby an alarmed parameter can be deleted and not sentenced if required.

This interface was designed for the requirements of the T23 Class Frigate and was not generic across all classes of ships. As MIMIC is used in various different manners across the fleet, the use of the interface on all ships causes operational problems. Data produced in support of day-to-day running of Watchkeeping and the provision of a power propulsion trial report facility all appear on the UMMS interface screen when in warning or alarm although they are not an RCM requirement. This did confuse operators and certainly made the system harder to manage. Hence a further remedial package has had to be put in place to adapt the interface to allow the filtering of parameters not linked to UMMS and to align the databases.

FURTHER DEVELOPMENTS:

On Hunt and Single Role MCMV Class ships MIMIC provides the Machinery Trials Assessment Unit (MTAU) with the ability to produce the results of Propulsion Power Trials (PPT-on form S346) whilst still onboard the ship by means of trial specific data tours. This allows a quick review of trial results, thus reducing the time required of the ship to be at full power and an added benefit of a fuel saving. The trial results can also be compared to previous trial results as a means of determining performance degradation. Individual ships can carry out these trials independent of the MTAU as a means of self assessment prior to any tasking or deployment allowing the maintenance package to be tailored to the ships immediate programme.

Whilst not a requirement for the UMMS application, it was asked if Mimic could monitor and trend the torque being transmitted through propeller shafts as there was concern that a class gearbox was being over torqued and thus reducing its designed life. As condition monitoring of large marine gearboxes is notoriously difficult, the measurement of the torque output was seen as a parameter that could be reliably and accurately measured and trended.

The initial interface to the ships control system was a simple one: it was linked to the signal conditioning unit which processes the output signal from the torsion meter. This was chosen to ensure a pre-processed raw signal. This informal requirement soon grew as other factors were considered, these being:

- Single or twin engine drive.
- Manoeuvring or direct drive.
- Throttle position.
- Torque limits per configuration.

From a simple request, a complex CM requirement developed: trend plotting capabilities which accommodated for torque comparisons with engine configuration, clutch position, throttle position and ship direction. To further complicate the system, the rate of data collection varies with torque levels whereby a reading is taken once a minute until the levels reach 90% of the torque limit where it then changes to once a second. This allows for graphical trending of torque excursions and cumulative totals of over torque to be calculated. Examples of these trend plotting capabilities are shown in Figures 2 – 5.

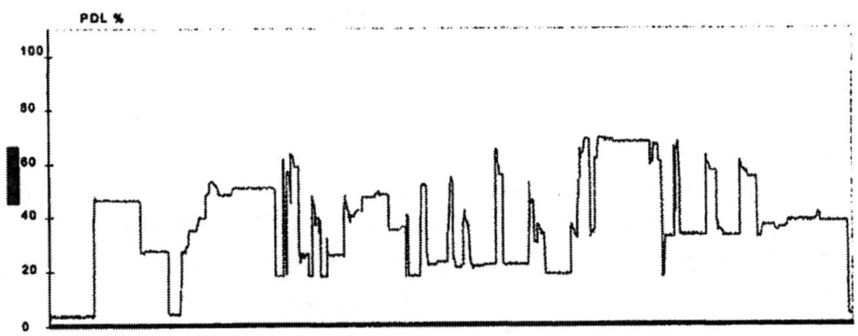

Figure 2. Power Demand Lever % Demand.

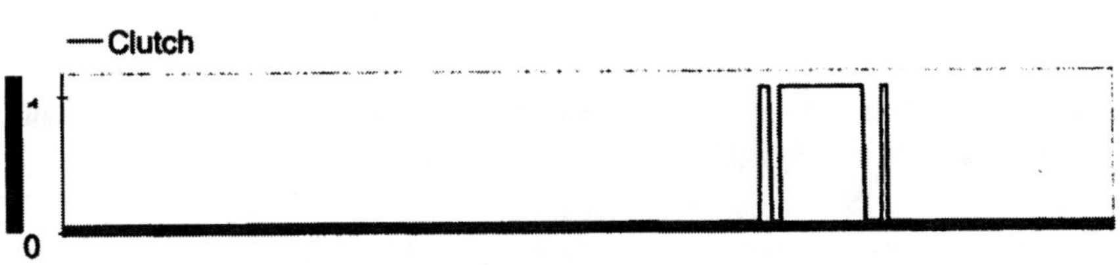

Figure 3. Clutch Command.

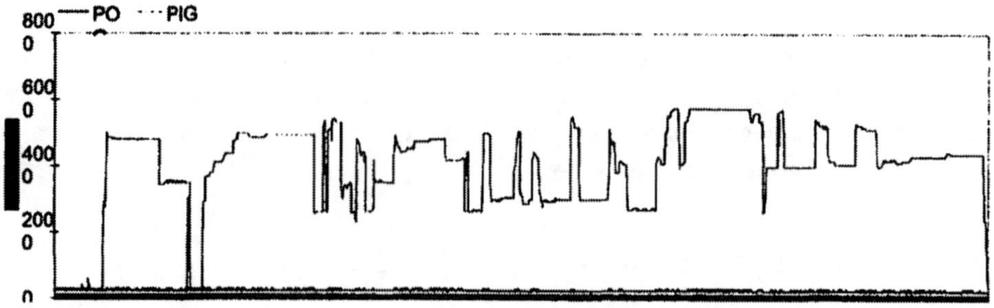

Figure 4. Gas Turbine LP Speed (RPM).

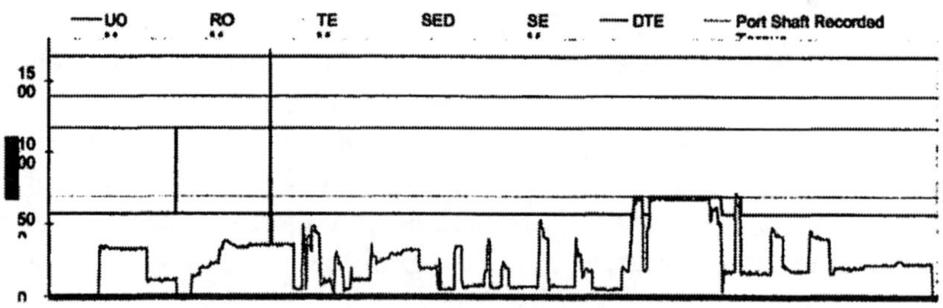

Figure 5. Recorded Torque in kNM.

It can be seen that CM (via MIMIC) has had to overcome many challenges throughout its time in use with the RN. Whilst the system is now stable, the procedures familiar to its operators and the philosophy deeply engrained into naval life, MIMIC needs to remain adaptable. The move from planned maintenance to RCM, along with elements of maintenance planning moving ashore, (through the UMMS software system) has changed the way maintenance is carried out; and, therefore its costs are managed. MIMIC has to recognise its new role and meet the challenges this presents.

The use of CM data and MIMIC's role in the new maintenance structure and the future has yet to be fully utilised, however, a number of indicators are apparent. These include:

- Traditional use of CM data, with the production of alarms and warnings regarding machinery health in support of the RCM requirements, must continue.

- The collection, storage and presentation of performance assessment data to ships' staff will enable mission profiles to be predicted.

- Electronic linking to machinery control systems and other data collection systems to provide an integrated and real time data collection service will enhance data quality.

- The use of MIMIC's data collection devices to assist in low level maintenance tasks, where the recording of parameters and/or simple questions (such as "is level correct"), need to be developed to reduce data input requirements.

- The use of modern web-based and .net software systems for ease of operation and for data transfer to other systems need to be developed.

- The use of predictive trending, by trend graph extrapolation, will aid fleet deployment planning.

- The use of CM data to aid asset defect root cause analysis for RCM development needs to be developed.

- The production of prognosis and diagnostic aids are essential.

- The use of permanently mounted transducers to ease the burden on reduced numbers of operators and to ensure data accuracy needs to be considered where cost effective.

- The incorporation of wireless technology to transfer data should be considered.

- The integration of various CM devices, such as on-line oil analysis and exhaust emission controls, should be pursued.

- Development of reporting tools to facilitate simple use should continue.

- The use of CM data to satisfy classification society rules will become a requirement.

LESSONS LEARNT :

Whilst automatic data collection from fixed sensors can provide quality data, the majority of data required for RCM is still manually collected and the one area for continued improvement must be in the quality of this data. With maintenance being generated from within the RCM software, bad or corrupt data could soon negate the savings predicted by the implementation of RCM based maintenance.

At the outset of a CM strategy, all stakeholders must be included within the overall implementation; failure to do this will lead to failure of the corporate plan. Each stakeholder has different views on what they want from the software and may lead you away from the overall target and compromise the time line of the project if it is not corporately managed.

Understanding the customer and the operator is a key element. Failure to do this may well lead to the collection of inadequate data due to the lack of ownership of the equipment.

Training is expensive but ignorance is more so. This is the most vexing of problems exacerbated by the constant changing of ships staff. A fully trained crew will be dispersed within an 18 month period so a comprehensive training system is essential.

Keep to the standard terms and layout of the paper forms that are to be replaced this will bring the most important stakeholder the end user into the process. If the end user decides not to participate or offers passive resistance the overall project will fail or at best be delayed and costs will rise as the battle to win the hearts and minds takes place.

CONCLUSION:

Having been onboard RN ships for over a decade, MIMIC has been accepted as a composite data collection and analysis tool for CM. It is now well placed to play a pivotal role in the move from a calendar based maintenance management system using CM to a full RCM based maintenance management system.

ACKNOWLEDGMENT:

The authors would like to acknowledge the invaluable help provided by Miss T L Calton, Prognostics and Health Management Centre of Excellence Lead, Sandia National Laboratories, in the proof reading of this paper, thus ensuring that it could be understood by others besides the authors.

REFERENCES:

1. Galpin, L. K. and Fowlie J. "Robust data collectors for the Royal Navy." In *Comadem 2000 Proceedings*, Houston, Texas December 2000.

2. Gosden, S. R. and Galpin, L. K. "Breaking the chains of preventative maintenance – the Royal Navy experience." In *Comadem 1998 Proceedings*, Launceston, Tasmania December 1998.

3. Fowlie, J. and Galpin, L. K. "Data collection for maintenance." In *Technology Showcase 2000 Proceedings*, Mobile, Alabama April, 2000.

NON-STATIONARY CONDITION MONITORING WITH THE AEWATT TOOLBOX

Niels Henrik Pontoppidan, Jan Larsen and Sigurdur Sigurdsson

Informatics and Mathematical Modeling, Technical University of Denmark
Richard Petersens Plads 321, 2800 Lyngby, Denmark
Email: {nhp, jl, siggi}@imm.dtu.dk

Abstract: We are developing a specialized toolbox for non-stationary condition monitoring of large 2-stroke diesel engines based on acoustic emission measurements. The main contribution of this toolbox has so far been the utilization of adaptive linear models such as Principal and Independent Component Analysis, as combined modeling and feature reduction methods. These models describe the, say 1024 or 2048, acoustic emission samples per engine revolution, i.e. data are in the crank angle domain. In this framework we have applied unsupervised learning using only one feature – the log-likelihood of an example given the trained linear model. The setup is semi unsupervised, as model parameters are learnt from normal condition data only, thus the system is not directly capable of error identification. However, it should be noticed that the adaptive linear models allow for some diagnosis based on the angular location of residual energy. Also, the framework can be extended, for instance by post modeling of repeated faults. Furthermore, we have investigated the problem of non-stationary condition monitoring when operational changes induce angular timing changes in the observed signals. Our contribution, the inversion of those angular timing changes called "event alignment", has allowed for condition monitoring across operation load settings, successfully enabling a single model to be used with realistic data under varying operational conditions.

Key Words: Component analysis; Condition monitoring; Event alignment; Non-stationarity; Unsupervised learning

Introduction: We are working on condition monitoring with acoustic emission measurements from large 2-stroke diesel engines used for ship propulsion and power generation. The acoustic emission allows for non-intrusive monitoring as the sensors can be attached on the outside of the cylinder. The AEWATT toolbox is developed for the detection of increased friction between piston and liner, a problem that eventually lead to a severe fault: Scuffing. In recent publications we have suggested and analyzed a collection of algorithms capable of non-stationary condition monitoring. In [9] and [10] we outlined the use of adaptive linear models for stationary condition monitoring and in [11] and [12] we added the event alignment that adds the non-stationarity to the system. The data is non-stationary as the engine control optimizes performance by advancing and delaying events, e.g. prolonging fuel injection time when the load increases. In this paper we apply the event alignment algorithm to experimental data, not used for the development of the system, and show that we obtain the same performance using event alignment as if we had handled each load setting with an independent model. One experiment was conducted by MAN B&W Diesel A/S on their test bed engine in Copenhagen – experimental data used for development of the toolbox was acquired on this engine as well. Additionally we have obtained normal condition data from an in service engine used for power generation by Public Power Corporation on Kos island, Greece.

Data setup and pre-processing: Although it is not directly a part of the AEWATT toolbox we will describe the data acquisition setup as design choices in the toolbox are based on the properties of the acquisition. The engine is equipped with acoustic emission sensors (ultrasonic, 100 kHz – 1 MHz). Further the engine is equipped with tachometer that allows for sampling in the crank angular domain with a resolution of 1024 / 2048 samples per revolution (ppr) depending on the actual system (several systems have been used). Also, the sample-rate is considerably lowered from 2MHz to 20 kHz, by use of analogue root mean squaring, thus the data becomes non-negative.

The crank angle sampling is performed indirectly using two 20 kHz signals containing the *top dead center* and *crank* pulses. The flanks in these two signals indicate when a new cycle begins and when a new crank sample should be calculated. With a running speed of 1-2 Hz and 2048 ppr, the new sample rate does not exceed 4 kHz thus the conversion from time to crank angular domain is also a downsampling. The toolbox default is to re-calculate the RMS; taking the square root of the mean of the squared values between two crank pulses. When the conversion has taken place, each engine cycle is represented as vector of length D with non-negative elements. Stacking N such cycles gives an observation matrix **X** that is later used for training of the linear models (**x**: Dx1, **X**: DxN).

$$\mathbf{X} = [\mathbf{x}_1 \quad \mathbf{x}_2 \quad \dots \quad \mathbf{x}_N] \qquad (1)$$

As the experiments are very expensive to conduct, we are faced with limited data, so in order to test our models and hypotheses, data resampling is utilized, meaning that **X** does not have to be constructed from N consecutive cycles. The data obtained from the test bed engine in Copenhagen has different known load settings – contrary to the data acquired on the Kos engine where such control information is not available. The engine at Kos was monitored by acquiring a few cycles every hour for 9 hours – and in this context we regard those data to be acquired under a stationary normal condition.

Non-stationary data alignment: After data preprocessing we have transformed the data into the crank angle domain, where each pattern is a single engine cycle, showing different events occurring. These patterns have usually high dimensionality, e.g. the AE RMS signals used for the experiments have 1024 and 2048 dimensions, depending on the angle encoder resolution. Under different running conditions with engine load changes, these patterns become highly non-stationary as both the timing and amplitude of different events changes dramatically. This makes it impossible to directly compare the patterns in the crank angle domain. Thus, the patterns need to be alignment prior to feature extraction and detection. Figure 1 illustrates the event time changes during an injection period of a diesel engine. AE RMS signals at 25%, 60% and 90% are shown. The points indicate the landmarks, indicating the time positions that should be aligned. Note that the individual events are in the same order, regardless of load.

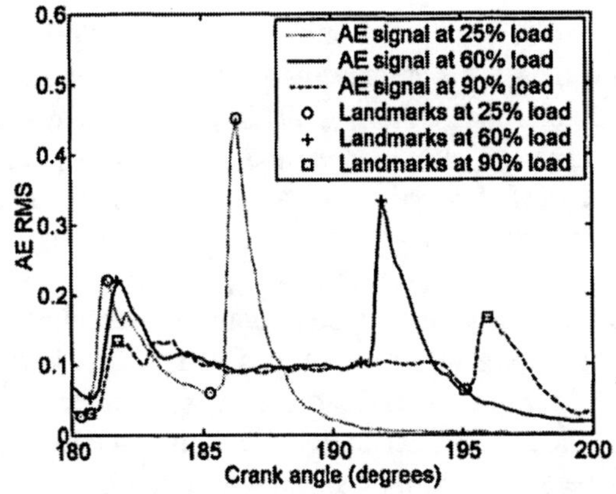

Figure 1: The AE RMS signals during the injection period with different load setting. The markers are the landmarks, indicating the time positions that should be aligned.

Automatically aligning the events of the patterns by means of, e.g., dynamic time warping, have shown poor results, as the patterns are very complex. Instead, we have relied on manually constructing landmarks from the data and used spline-interpolation to align the events [11][12]. All patterns are aligned to a selected reference patterns. Currently, the AEWATT toolbox implements first (piece-wise linear) and third order (cubic) spline-interpolation for event alignment. The left panel in Figure 2 illustrates the event alignment of a single AE RMS pattern using the piece-wise linear alignment.

On top of the event alignment, an amplitude alignment should take place. A scaling of the amplitude of the data has been shown to work well. The scaling is the ratio between the reference pattern and the average pattern at a constant load. The reference pattern is the average of the AE RMS patterns at 25% load. The right panel in Figure 2 shows the results after amplitude and event alignment.

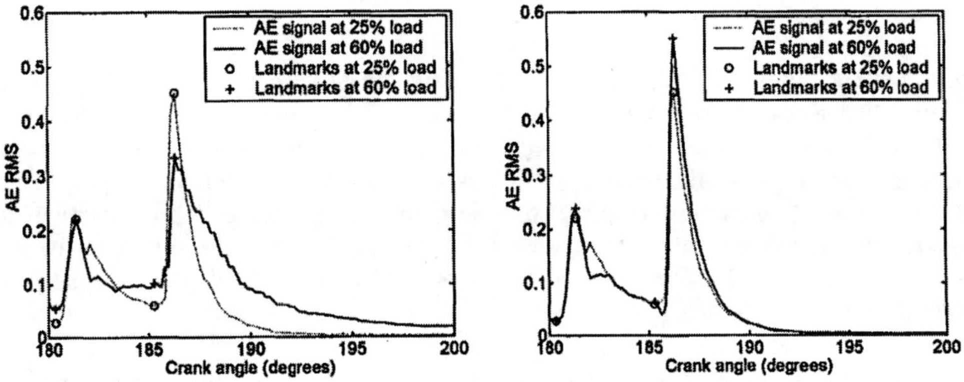

Figure 2: The left panel shows the AE signals at 25% and 60% load from Figure 1 after event aligning the signal at 60% load with the signal at 25% load, using a piece-wise linear spline-interpolation. The right panel shows the results after event and amplitude alignment.

Feature extraction: Feature extraction aims at extracting relevant information from the measured/preprocessed data. This is extremely important when the size of the measured data is large. The patterns of the AE data considered here have 1024 and 2048 dimensions and condition monitoring of such large and complex signals is very difficult. By extracting the relevant information, the dimension may be reduced by orders of magnitude or even to a single feature.

Simple single feature can easily be extracted from the patterns, e.g., empirical average or maximum value. The problem with these types of features is that they do not take into account more general changes in the patterns, e.g., changes that do not influence the average or maximum value. For instance, a fault that would cause a time shift in the patterns will not be detected with these simple features. It is important that feature extraction methods detect such changes, to be able to cover a wider range of engine faults.

An important property of a feature extraction method is to be able to learn the difference between normal and faulty conditions using only normal patterns. It is extremely time consuming and expensive to induce all possible faults in an engine to obtain faulty measurements. Feature extraction methods that only learn from normal data may be considered as *semi unsupervised learning* models and have great advantage compared to models applying *supervised learning*.

In the AEWATT toolbox we have focused on feature extraction based on linear transformation or components analysis of the measured patterns, using only normal patterns. We assume that each pattern $\mathbf{x}$ with size $D \times 1$ is generated from the noise model $\mathbf{x} = \mathbf{A}\mathbf{s} + \varepsilon$, where $\mathbf{s}$ is a $K \times 1$ source signal, $\mathbf{A}$ is a $D \times K$ mixing matrix and ε is a $D \times 1$ additive noise variable. Further, we assume that the dimension of the source signal is much lower than the measured data $\mathbf{x}$, i.e. $K \ll D$. A common way of extracting features is to estimate the source signals, thus reducing the dimensionality from D to K dimensions.

There exist a number of methods for estimating the system $\mathbf{x} = \mathbf{A}\mathbf{s} + \varepsilon$. Here we will mention two data adaptive methods in the AEWATT toolbox, *principal components analysis* (PCA) and *independent components analysis* (ICA). Both models can be formulated in a probabilistic way, which opens for the possibility to apply a very effective way of extracting features, by using the so-called *likelihood* function. The negative log-likelihood values for patterns that are similar to the training set patterns, which are normal, will have lower values compared to patterns that are different, e.g. faulty patterns.

The main goal of PCA is to retain as much variance of the original data as possible. Moreover, the columns of $\mathbf{A}$ are constrained to be orthogonal. This may be done by applying singular value decomposition, given by $\mathbf{X} = \mathbf{U}\mathbf{D}\mathbf{V}^T$, where $\mathbf{X}$ is a $D \times N$ matrix of N measured patterns, $\mathbf{U}$ is a $D \times N$ orthonormal matrix, $\mathbf{V}$ is a $N \times N$ orthonormal matrix and $\mathbf{D}$ is an $N \times N$ diagonal matrix of singular values, where the elements, in ascending order, corresponding to the standard deviation of the data. The matrix $\mathbf{A}$ is then estimated by retaining the first K columns of $\mathbf{U}$. Previously, the PCA has been modeled in a probabilistic framework [4], by assuming $\mathbf{x}$ to be a multivariate Gaussian variable with ε as isotropic Gaussian noise. The left panel of Figure 3 illustrates the use of PCA for feature extraction with two dimensional toy patterns.

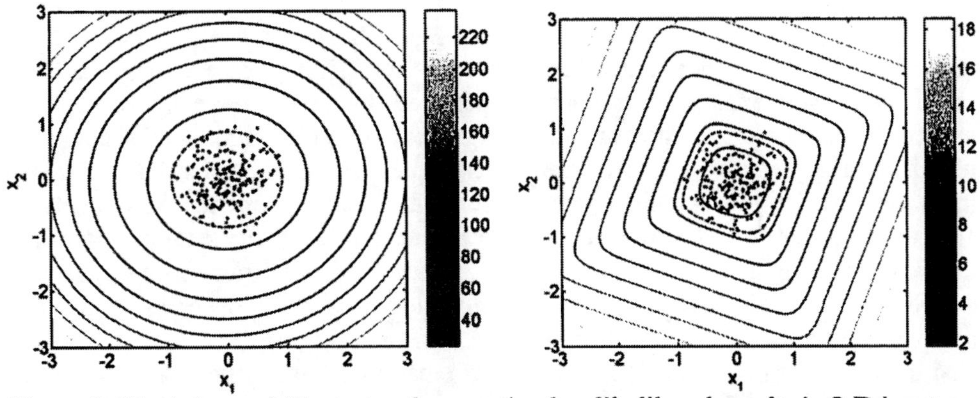

Figure 3: The left panel illustrates the negative log-likelihood results in 2-D input space for the PCA and the right panel for the ICA. The dots are the measured normal patterns, the solid lines are contours of the negative log-likelihood surface and the dashed line is a threshold found by computing the 5% fractal of the normal patterns. Measured patterns that lie outside the area marked with the dashed lines would be detected as faulty. Note the difference between the negative log-likelihood contours of these two methods. The PCA is optimal if the patterns are Gaussian distributed, while the ICA is better at arbitrary distributions.

The ICA method has recently gained popularity in data analysis. The method assumes that the source signals $\mathbf{s}$ are statistically independent. ICA was introduced as information maximization [1] and separation [8], and has recently been extended in a Bayesian framework [5] that allows for specification of certain prior assumptions. For instance, the AE RMS signals are positive and may be considered as positive addition of positive sources, which constrains the elements of $\mathbf{A}$ to be non-negative. This is possible to obtain with the Mean Field ICA algorithm developed at DTU [5][7], which is incorporated into the AEWATT toolbox. As with the PCA, the likelihood function of the ICA acts as a feature extraction for the data. The right panel of Figure 3 illustrates the use of ICA for feature extraction with two dimensional toy patterns.

For both PCA and ICA we need to estimate the number of components K. With too few components the underlying structure of the data is not captured, while too many components will result in a noisy estimate. An optimal number of components keeps the most important components, while disregarding the less important and noisy components. Selecting the optimal number of components is not trivial and many different methods exist. In the toolbox we have focused on the Bayesian Information criterion [4] and well established partitioning schemes, e.g. cross-validation [3], [13].

Note that it is possible to consider both the PCA and ICA as dimension reduction methods. By estimating the source signals $\mathbf{s}$, the dimension of the data is reduced from D to K, thus obtaining K features. Further modeling may then be done with, e.g., density models like the Gaussian mixture model, which is also included in the toolbox. Although we have not experienced increased performance applying this scheme to AE signals, it may be beneficial for other applications.

Classification: We have adopted a semi unsupervised, single feature, fault detection based on single cycle examples. Its semi unsupervised since parameters are learned from normal condition data only. The combined feature extraction and modeling schemes are trained using a subset of the known normal examples. From another subset of known normal examples we build the histogram and cumulated density function for our selected feature: the *negative log-likelihood* (NLL). Thus no known faulty examples are used during model training and threshold estimation. As we expect that normal examples have lower NLL-value than faulty, so the classification boundary becomes a maximal acceptable NLL-value, e.g. a rejection level. A very similar approach was successfully applied to (truly) unsupervised classification of emails by Szymkowiak et al. [13]. We enforce a *tight* boundary by aiming for a rejection rate of normal examples in a test set of say 5%. Unfortunately, this also corresponds to a (design) false alarm rate of 5%. Figure 4 display two NLL-feature time series obtained with data from the Kos engine. For this example 4 of 70 examples in the training set and 5 of 78 in the test set were rejected. The cumulated density functions in Figure 5 show that it is impossible to select a threshold that separates the training and test set; this is good as both sets contain normal condition data. This also shows how the rejection rate is converted to a false alarm rate. One well known way of reducing the false alarm rate is through multiple independent classifiers combined with majority vote system.

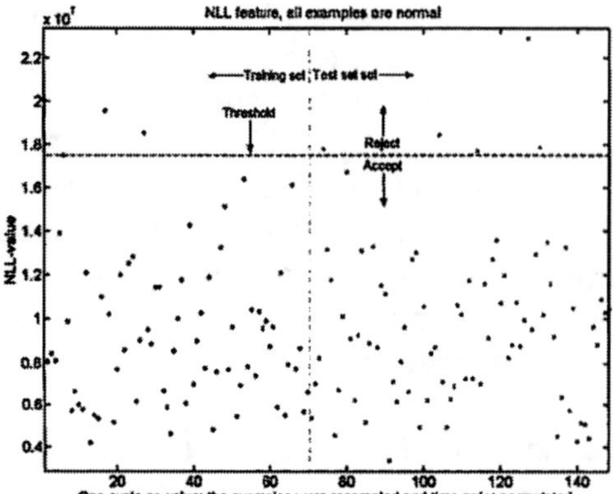

Figure 4: Negative Log-likelihood for training and test set of known normal examples (2 components MFICA)

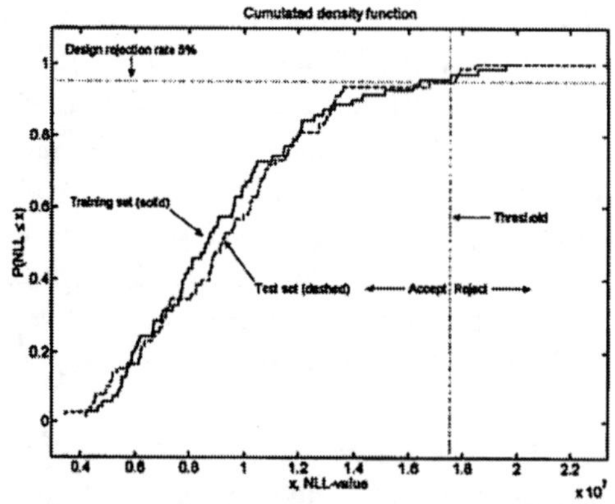

Figure 5: Empirical cumulated density function of NLL values in Figure 4. The two CDF's are very similar.

<u>Binomial hypothesis test</u>: A 5% false alarm rate is way too high for condition monitoring of large diesel engines. Even with a 0% rejection level on an independent validation set, we could still encounter false alarms since the model and threshold is learned from only a

subset of data. The intuitive way to deal with the false alarms is to monitor the rejection rate and only react when the rate of alarms gets high enough. This corresponds to *binomial hypothesis testing* [1]. We can use a binomial hypothesis test to address the inherent false alarms, as well as reduce the false alarm rate by considering a number of successive engine cycles as a whole. We treat the normal/faulty classification of each engine cycle as an independent binomial experiment (normal=0, faulty=1). If the observed binomial sequence can be accepted, with a given confidence level, as being drawn from a binomial process with hit rate equal to the set rejection rate, the classification of the set as a whole is *normal.*

From an engine producing power at Kos we have acquired a few cycles every hour under assumed normal conditions. PCA and MFICA models were trained on a subset of examples (not in time order) and the target rejection rate was 5%. The critical value for the binomial hypothesis test was calculated with confidence level of 0.01 on 78 normal examples to be 10 [2]. 80 such resampled data sets (with 78 examples) where accepted as being normal regardless of model and number of components. The binomial cumulated density function for B(5%,78) is shown in Figure 6. To the right in Figure 7 two binomial sequences generated from the test bed engine at MAN B&W in Copenhagen show that the rejection rate under faulty conditions is much larger than under normal conditions. Actually, they were close to 1 and therefore would this sequence as a whole be classified as faulty.

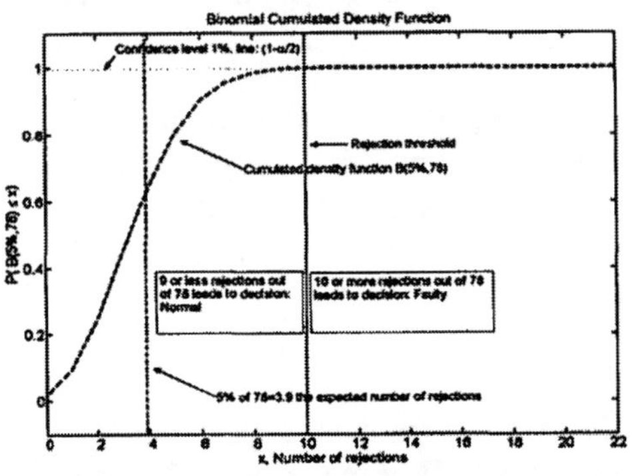

Figure 6: Binomial hypothesis testing with Kos engine data. Target rejection rate 5%, confidence level 1% on 78 examples give new threshold 10 (10/78=13%)

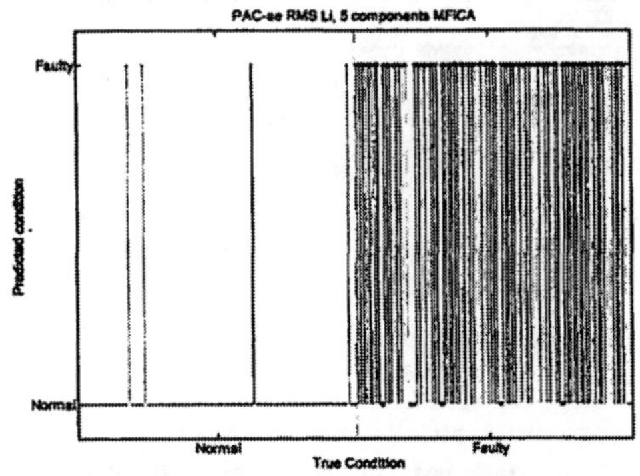

Figure 7: Binomial sequence using MFICA with 5 components on data from Copenhagen test bed. The number of rejected examples rises at the condition change (NB examples not in time order).

Alignment validation: The event alignment is validated using the area under the receiver operation curve (AUC) performance metric. By using the AUC it is possible to evaluate the quality of the feature extraction without making any assumptions on the classification system. The following section will look into the performance of the classification. The AUC may be considered as the probability that the feature value of a faulty pattern is higher then the value of a normal pattern. This gives the highest AUC value as 1, indicating that the two classes may be completely separated using correct threshold for classification. A feature that does not discriminate between classes has the

AUC value of 0.5, indicating that the feature values for the classes are completely overlapping and give random results.

The data used to validate the event alignment is composed of AE RMS signals where scuffing is induced at three different loads, obtained from the cylinder liner of the electronically controlled 2-stroke engine at MAN B&W Research Copenhagen. The loads are at 25%, 60% and 90%. Due to load difference, the data is highly non-stationary. The data is preprocessed by converting the AE RMS signal to angle domain, giving patterns having 2048 dimensions. We considered three methods for feature extraction; the empirical average (MEAN), PCA and MFICA. Note that MEAN is very good feature for detecting scuffing, while has shown poor performance at detecting other faults. We apply 4 different preprocessing schemes; (1) weighted average of models trained on data at each load, (2) models trained on all data without alignment, (3) models trained on all data with event alignment, (4) models trained on all data with event and amplitude alignment. Note that preprocessing (1) corresponds to a stationary condition at each load. The results are shown in Table 1.

	(1) Average of stationary	(2) No event alignment	(3) Event alignment	(4) Event/amplitude alignment
MEAN	0.977	0.709	0.714	0.980
PCA	0.966	0.894	0.946	0.957
MFICA	0.954	0.899	0.940	0.947

Table 1: The AUC for three methods; empirical average (MEAN), PCA and MFICA, using different data preprocessing; (1) weighted average of models trained on data at each load, (2) trained on all data without alignment, (3) trained on all data with event alignment, (4) trained on all data with event and amplitude alignment. The results show that it is possible to obtain the similar performance using event/amplitude alignment as training models at each load.

The results show that by applying event/amplitude alignment it is possible to obtain similar performance as training models at each load. Without event alignment the performance of all methods decreases significantly. Using only event alignment improves the performance of PCA and MFICA dramatically, obtaining almost the same performance as stationary modeling. This shows that event alignment is the most important alignment for the advanced methods. On the other hand, event and amplitude alignment is necessary to improve the performance of simple methods on non-stationary data, while the improvement is marginal for PCA and ICA.

Summary: We have demonstrated some key components for non-stationary condition monitoring with the AEWATT toolbox and showed how these components can be utilized to decrease the number of false alarms significantly. Especially, the results obtained with event alignment are promising, as we cannot learn PCA and MFICA parameters for all possible load settings. Furthermore, we are currently investigating interpolation between known load models within the event alignment framework.

Acknowledgements: This work is supported by EU Competitive and Sustainable Growth Programme GRD2-2001-50014 – acronym AE-WATT (http://isp.imm.dtu.dk/aewatt). The experimental data was provided by AEWATT project partners: MAN B&W Diesel A/S (Denmark), Heriot-Watt University (Scotland, UK.), Envirocoustics (Greece) and Public Power Generation (Greece).

References:

[1] Bell, A. and Sejnowski, T., *An information-maximisation approach to blind separation and blind deconvolution*, Neural Computation, Volume 7, number 6, pp. 1129-1159, 1995

[2] Conradsen, K., *En introduktion til statistik*, IMM-DTU, 6. edition, 1995

[3] Efron, B. and Tibshirani, R.J., *An Introduction to the Bootstrap*, Chapman & Hall, Monographs on Statistics and Applied Probability, 1993

[4] Hansen, L.K. and Larsen, J., *Unsupervised Learning and Generalization*, Proceedings of the 1996 IEEE International Conference on Neural Networks, vol.1, pp. 25-30, 1996

[5] Hansen, L.K., Larsen, J., and Kolenda, T., *Blind detection of indepen-dent dynamic components*, In Proceedings of ICASSP'2001, vol. 5, pp. 3197-3200, 2001

[6] Højen-Sørensen, P.A.d.F.R. and Winther, O. and Hansen, L.K., *Mean Field Approaches to Independent Component Analysis*, Neural Computation, 2001

[7] Kolenda, T., Sigurdsson, S., Winther, O., Hansen, L.K. and Larsen, J., *DTU:Toolbox*, Internet: http://mole.imm.dtu.dk/toolbox/, IMM-DTU, 2002

[8] Molgedey, L. and Schuster, H.G., *Separation of a mixture of independent signals using time delayed correlations*, Physical Review Letters, 1994

[9] Pontoppidan, N.H., Larsen, J., Fog, T orben, *Independent component analysis for detection of condition changes in large diesels*, COMADEM 2003, no. 16, pp. 493-502, 2003

[10] Pontoppidan, N.H. and Larsen, J., *Unsupervised Condition Change Detection In Large Diesel Engines*, 2003 IEEE Workshop on Neural Networks for Signal Processing, pp. 565-574, 2003

[11] Pontoppidan, N.H. and Douglas, R., *Event alignment, warping between running speeds*, COMADEM 2004, no. 17, pp. 621-628, 2004

[12] Pontoppidan, N.H. and Larsen, J., *Non-stationary condition monitoring through event alignment*, IEEE Workshop on Machine Learning for Signal Processing, pp. 499-508, 2004

[13] Ripley, B.D., *Pattern Recognition and Neural Networks*, Cambridge University Press, Cambridge, 1996

[14] Szymkowiak, A. and Larsen, J. and Hansen, L.K., *Hierarchical Clustering for Datamining*, Proceedings of KES-2001 Fifth International Conference on Knowledge-Based Intelligent Information Engineering Systems & Allied Technologies, pp 261-265, 2001

GAS TURBINE TRANSIENT SYSTEM EMPIRICAL MODELING FOR INCIPIENT FAULT DETECTION AND DIAGNOSIS USING ECHO STATE NETWORKS

Sunil Menon

Honeywell ES&S Laboratories
3660 Technology Drive, Minneapolis, MN 55418
sunil.menon@honeywell.com

Abstract: We present a method of fault detection and diagnosis in turbine engines using a type of temporal neural network called the Echo State network (ESN). Temporal neural networks allow us to represent the complete engine operating range by complementing the first-principle models, which are usually restricted to takeoff and cruise phases. This complete coverage leads to more accurate anomaly detection and fault diagnosis systems because faults that are manifest only in particular phases can be detected. The time series sensor data from the engine is collected during particular aircraft flight phases such as startup, takeoff, cruise, and shutdown.

We use the echo state network to develop an incipient fault detection and diagnosis system. Echo state networks have several advantages over conventional types of temporal neural networks, including accuracy and ease of training. We demonstrate the efficacy of using the echo state networks to focus on flight phases that are difficult to model. In addition, using actual propulsion engine transient flight data, we compare the accuracy of the ESN with other methods of dynamic system modeling such as the autoregressive with exogenous (ARX) inputs, neural network ARX models, and the Elman recurrent neural network. The ESN is found to outperform other types of dynamic system models.

Key Words: Turbine engine fault detection, diagnosis, empirical dynamical models, echo state network, neural networks.

INTRODUCTION:

Accurate turbine engine fault detection and diagnosis (FDD) is vitally important to reducing airline operating costs and improving safety. Detecting incipient (slowly occurring) faults in turbine engines can significantly reduce costs related to in-flight shut-downs, delays and cancellations, unscheduled engine removals, and take-off aborts. More effective preventive maintenance programs depend on accurate incipient fault detection.

Turbine engine FDD involves the early detection of incipient as well as sudden faults. Detection is followed by diagnosis of a sensor, actuator, or system component fault. Many approaches to FDD are discussed in the literature. The most common approaches to FDD are model-based methods. Model-based FDD methods use a physical or empirical model of the system to generate residuals (difference between the model outputs and actual system measurements). These residuals are then passed through a decision logic system to determine a diagnosis [1]. Since the accuracy of the model is of utmost importance in model-based FDD methods, a significant amount of effort is expended to ensure that the system model is accurate enough in the operating region of interest. Inaccuracies in the system model can negatively influence the FDD method false alarm rate and missed alarm rates.

Model accuracy is particularly important when the FDD method is developed for incipient faults, because model uncertainties can mask the effects of incipient faults, making the FDD method effective only when the fault is sufficiently progressed.

Most engine FDD is performed using measurement data from engines during steady-state conditions. A major reason for using steady-state data is that system transients can confuse FDD methods, resulting in reporting of incorrect results. A more robust approach to developing FDD methods that explicitly accounts for transient data is required. Furthermore, most turbine engine FDD methods are developed with engine performance models that are validated only under steady-state conditions or with actual engine data at steady-state conditions.

Developing FDD methods that operate on system data during transient as well as steady-state operation has several important advantages. First, during system transient conditions, certain system faults have a distinct signature that is not normally discernible during steady-state conditions. Second, the effect of feedback control action is less dominant during transient conditions than during steady-state conditions. Because feedback control suppresses the effect of sensor and system faults, faults are more evident during transient conditions. Finally, incipient faults in certain engine components are manifest only during transient conditions such as startup and shutdown (e.g., starter and igniter system faults).

In this paper, we present an approach to engine FDD using temporal neural networks. Temporal neural networks are nonlinear, dynamic model structures that can be used to develop models of complex, dynamic systems. Feedforward neural networks are static nonlinear mappings that model dynamic systems by using time-delayed variables as inputs [2]. These models are known as the Nonlinear AutoRegressive models with eXogenous inputs (NARX). The NARX method of using a static map to approximate a dynamic system is fairly accurate in many applications; however, temporal—or recurrent—neural networks are inherently dynamic mappings by virtue of their internal states. Therefore, empirical model structures based on recurrent neural networks have the potential to be very accurate models of complex, dynamic systems [3, 4].

The main drawback to developing recurrent neural networks is the complexity of training methods for these networks. Most of the current methods of training recurrent neural networks are based on computational methods of obtaining the gradient of the error function (based on difference in the actual output and network prediction) [5]. In our work, turbine engine models for FDD are developed using a different type of recurrent neural network known as the echo state network (ESN) [6, 7]. ESNs are much easier to train than conventional, error gradient-based recurrent neural networks. Instead, ESN training methods culminate in a pseudoinverse computation which can be greatly optimized using standard methods.

We develop the turbine engine FDD method for the engine in the takeoff operating phase. Data is collected from a midsize commercial propulsion turbine engine corresponding to the flight phases of interest. The data collected corresponds to a healthy engine and is used to develop and test the FDD method based on ESN engine models.

The following sections describe the engine FDD method, the ESN architecture and training method, the turbine engine application, and the engine flight operating regions of interest. The results of the developed engine FDD method are discussed followed by a summary and conclusions that can be drawn from this work.

MODEL-BASED ENGINE FAULT DETECTION AND DIAGNOSIS:

Figure 1 illustrates our model-based FDD method. The term $\hat{\mathbf{x}}(t+1|t)$ denotes the prediction of the state **x** at time instant *(t+1)* given the inputs and past predictions at time instant *t*. The output prediction term $\hat{\mathbf{y}}(t+1|t)$ is similarly defined.

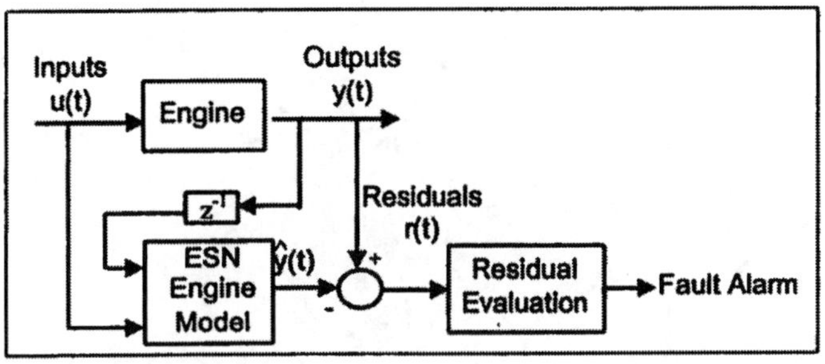

Figure 1. Engine Fault Detection and Diagnosis using ESN Engine Models

The difference between the output prediction of the engine model and the actual engine output measurement is calculated. The output differences are referred to as the *residuals*, **r***(t)*. Ideally, the residual is zero when the engine is operating normally and not zero when a fault has occurred. This characteristic of the residual is used to determine whether a fault has occurred. The residual evaluation block in Figure 1 processes the residual, determines when a fault occurs, and performs the fault classification. We do not develop the residual evaluation block in this work even though this is an important step in the FDD system; refer to [1] for further details on this step.

It is evident that the sensitivity and accuracy of a model-based FDD method depends strongly on the fidelity of the system model used to generate the output predictions; consequently, much effort is expended in developing these models. While a significant number of model-based FDD methods reported in the literature are based on linear system models, most systems—such as the turbine engine—are complex and nonlinear in the operating regions of interest, making nonlinear, dynamic models more suited for FDD methods on these systems.

Several promising nonlinear, dynamic model structures are discussed in the literature: structures based on spline functions [8], polynomial-based NARX models [9], and NARX models based on recurrent neural networks [10]. Nonlinear, dynamic system models based on recurrent neural networks have been shown to be very accurate; however, the common training methods for recurrent neural networks such as back-propagation through time (BPTT) [11], real time recurrent learning (RTRL) [12] and extended Kalman Filter (EKF) approach [3] (mostly based on minimizing the output error gradient) are computationally intense and often slow to converge [5]. The ESN is a model structure that, while retaining the advantages of recurrent neural networks in terms of its nonlinear, dynamic system modeling capabilities, has a much less computationally intensive training algorithm than the recurrent neural network methods discussed above. In the next section, we present the ESN and discuss its development.

ECHO STATE NETWORKS:

ESNs were first developed by Jaeger [6] to improve training for recurrent neural networks. An ESN consists of input nodes, output nodes, and a large network of nodes forming the *dynamic reservoir* (DR). Figure 2 is a diagram of the ESN architecture.

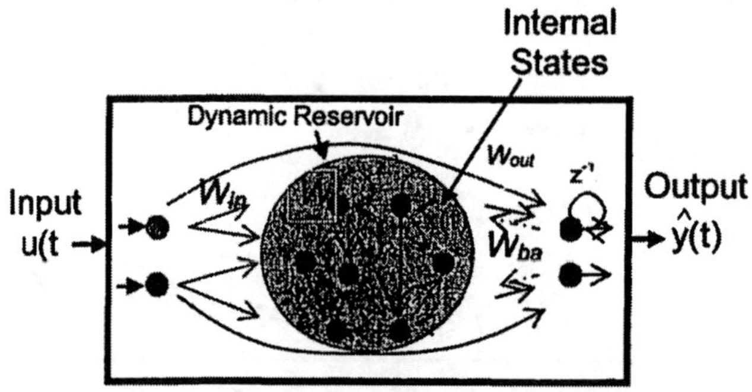

Figure 2. Echo State Network Architecture

Consider a network with K inputs, N internal nodes and L outputs. The network inputs are represented by $\mathbf{u}(t)=[u_1(t), \ldots, u_K(t)]$; the internal nodes are represented by the vector $\mathbf{x}(t)=[x_1(t) \ldots x_N(t)]$, and the network output is represented by $\mathbf{y}(t)=[y_1(t) \ldots y_L(t)]$, where t represents the time step. The real value connection weights are randomly generated and are stored in an $N \times K$ weight matrix $\mathbf{W}_{\mathbf{in}}$ for the inputs, an $N \times N$ weight matrix $\mathbf{W}$ for the internal connections, an $L \times (K+N+L)$ weight matrix $\mathbf{W}_{\mathbf{out}}$ for the connections to the outputs, and an N x L matrix $\mathbf{W}_{\mathbf{back}}$ for the connections that feedback from the outputs to the internal units. We use the same notation for the ESN parameters as in [6]. Note that ESNs allow direct connection from inputs to the outputs. The size of the internal layer should be sufficient to store the dynamics of the problem. Thus, if the system involves complex dynamics, then the number of internal nodes should be relatively high. The "echo state" property is a condition specific to the ESN; it is defined by the values of the weight matrices. The echo state property is the current network state uniquely determined by the past input and teacher-forced outputs. For more details on the properties of echo states, refer to [6].

ENGINE APPLICATION:

In this section, we describe the specific turbine engine application for FDD. The engine is a two-spool, high-bypass turbofan gas turbine engine. The engine system can be divided into two main components based on the operating pressure: low pressure and high pressure. The low pressure section uses a gear reduction unit to drive a single-stage fan off the two-stage, low-pressure turbine. The high-pressure system consists of a double compressor and a gearbox, both driven by the high-pressure turbine.

The engine measurements that are available as inputs to the engine models are: total air temperature (deg. F, *TAT*), altitude (feet, *Alt*), Mach number (no units, *Mach*), engine oil temperature (deg. F., T_{Oil}), and fuel flow (lbs/hr, *Wf*). The outputs of the engine models are: low-pressure spool speed (percentage of full speed, *N1*), high-pressure spool speed (percentage of full speed, *N2*), and the exhaust gas temperature (deg. C, *EGT*).

Broadly, a typical aircraft flight has the following distinct stages: *startup*—when the engine is brought to a certain idle speed from rest, *takeoff*—when the aircraft is taken from the ground to a specific altitude, *climb*—when the aircraft is steadily taken from a lower to higher altitude after the takeoff phase, *cruise*—when the aircraft is maintained at a constant speed and altitude, *approach*—when the aircraft descends and lands, *spin down*—when the engine is shutdown from idle condition. Additional minor stages include the aircraft taxiing before takeoff and after landing. The engine is designed to operate with maximum efficiency around certain operating points—usually around cruise conditions, which corresponds to quasi-steady state operation.

Our objective in this work is to develop engine models that capture the dynamics of the engine during transient conditions. The effects of certain incipient faults, for example, in the fuel delivery system, will be more pronounced during transient operation [13]. Therefore, we selected the startup, takeoff, and spin down flight stages as the stages from which engine data will be collected to develop the engine model.

The transient data for engine startup has a special significance when it comes to engine FDD. The details of the engine startup profile are given in [13]. The types of incipient faults that can be determined during the startup phase relate to igniter wear, start system degradation, fuel system faults, and power section faults. The engine takeoff phase is useful in identifying fuel system faults and power section faults. The engine spin down phase is particularly useful in identifying turbine bearing degradation. In this work, however, only the takeoff flight phase data is used to develop transient engine models. The development of these ESN engine models are described in the next section.

EMPIRICAL DYNAMIC MODEL DEVELOPMENT

The empirical dynamic engine models that are developed in this work are based on ESNs, linear ARX, neural network ARX (NNARX), and Elman network (a type of recurrent neural network) structures. We compare the output predictions from all these models.

The ESN models are trained using the data from a healthy engine during the takeoff flight phase. The takeoff flight phase begins when the aircraft is on the ground and the engines are running at idle speed. The takeoff phase is initiated when the pilot moves the power lever angle (PLA) up. The PLA change, in turn, signals the fuel pump to increase the fuel flow. This sets in motion the engine controls to increase engine low pressure fan speed (*N1*), increasing the engine thrust to enable the aircraft to takeoff from the ground and reach a specific altitude.

Data corresponding to the engine takeoff phase is selected from an engine's complete flight data set by using a selection logic that is based on the PLA movement and maximum EGT. Each engine takeoff takes about 25 seconds. In this case, data is sampled at 4 Hz, so 100 data points are used for each takeoff. One week's takeoff data is used to represent the engine in a healthy condition. The complete data set is divided into: the training data set, test data set, and validation data set. The training data set comprises 15 separate takeoffs (1500 data points) and is used to develop the ESN weights (parameters). The test data set comprises 6 takeoffs (600 data points) and is used to help prevent overtraining the ESN network. The validation data set comprises 5 takeoffs (500 data points) and is used to assess the performance of the final ESN network.

Three separate, nominal ESN models are developed: one each to predict *N1, N2*, and *EGT* engine variables. This type of model development improves the ESN model fidelity—separate models can predict all three engine variables better than a single model. The inputs to each of the engine models are the same: altitude (*Alt*), total air temperature (*TAT*), pilot lever angle (*PLA*), and fuel flow (*Wf*). The ESN engine model is shown in Figure 3.

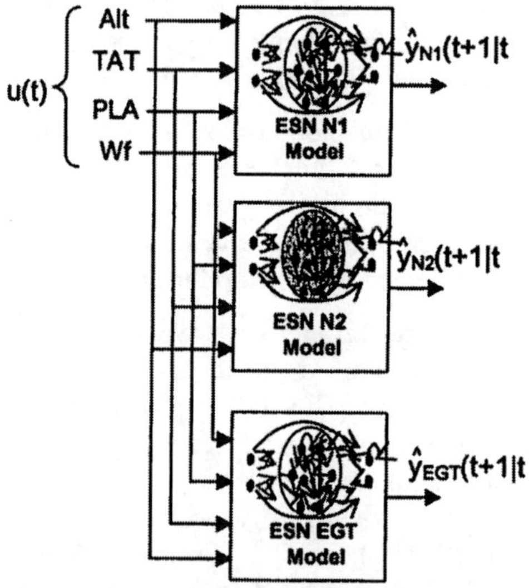

Figure 3. ESN Engine Takeoff Prediction Models

The main ESN model parameters to be set during training are the spectral radius, α, the percentage connectivity of the DR weights, γ, and the weight scales. The weight scales are multiplicative values on the weights $\mathbf{W}_{back}$ and $\mathbf{W}_{in}$ and take values between 0 and 1. Several numerical experiments were conducted by developing ESN models with different parameter values to find those values that gave the best generalization performance. In general, the best results were for values of α between 0.8 and 0.98, γ between 0.7 and 1.0, and weight scales of 0.1 for $\mathbf{W}_{in}$ and 0 for $\mathbf{W}_{back}$. The ESN parameter values for the takeoff models were: α was set to 0.98 and γ was set to 1. Further details of the training results are given in Menon [14]

The ARX model, NNARX, and Elman network-based dynamic engine models are developed with the same engine data used to develop the ESN models. Table I shows the final model parameters for all the dynamic models developed.

Table I. Empirical Dynamic Model Parameters. *ny* and *nu* correspond to the number of output and input time delayed feedback terms, *nh* are the number of hidden nodes.

	N1	N2	EGT
ARX [ny nu1 nu2 nu3 nu4]	[0 1 1 3 4]	[0 1 1 2 4]	[0 1 4 4 4]
Neural Network ARX nh [ny nu1 nu2 nu3 nu4]	9, [0 1 2 4 4]	11, [0 1 2 2 4]	9, [0 1 2 3 4]
Elman Network [nh]	3	3	7
Echo State Network [n, alpha, percent connectivity]	[100 0.84 90]	[100 0.84 100]	[100 0.84 90]

RESULTS AND DISCUSSION

The engine takeoff ESN and other dynamic model predictors were developed (trained) and then evaluated against the validation data set. The validation data set is data that was not used in the training and testing process discussed in the previous section.

Table II: Empirical Dynamical Engine Model Performance Results

	N1		N2		EGT	
	NRMSE (%)	Correlation Coefficient	NRMSE (%)	Correlation Coefficient	NRMSE (%)	Correlation Coefficient
Naive Predictor	23.2	0.93	7.4	0.93	3.5	0.95
ARX	25	0.89	9.9	0.84	4	0.92
Neural Network ARX	8.1	0.97	5	0.94	2.3	0.97
Elman Recurrent Neural Network	7.8	0.99	3.1	0.98	2.9	0.95
Echo State Network	3.8	0.99	3.3	0.98	1.9	0.98

The performance of the dynamic engine model results is shown in Table II. In Table II, the naïve predictor (where $\hat{y}(t+1|t) = y(t)$) results are also shown for comparison. The normalized root mean square error (NRMSE) is defined in Eq. (1):

$$NRMSE(\%) = \sqrt{\frac{1}{N}\sum_{t=1}^{N}\frac{\left(y(t)-\hat{y}(t|t-1)\right)^2}{y(t)^2}} \times 100 \qquad (1)$$

From Table II, the nonlinear dynamical models (NNARX, Elman and ESN) performed the best compared to the linear ARX model. Further, it is seen that two recurrent model structures, the ESN and the Elman network, outperform the other dynamic models. The

ESN outperforms the Elman network for the N1 and EGT outputs. The ESN and Elman model performance on the N2 output are very close to each other.

From the engine model results, we see that the nonlinear dynamical empirical models are most accurate for modeling the engine takeoff behavior. The ESN engine model performance is superior to the ARX models, both linear and nonlinear. This implies that the ESN (and to a large extent the Elman network as well) captures additional nonlinear dynamics which the linear ARX and NNARX model structures do not.

SUMMARY AND CONCLUSIONS:

This paper describes a model-based method of FDD on a propulsion gas turbine engine using a nonlinear, dynamic model structure—the ESN. The ESN is a temporal neural network with several advantages over currently used recurrent neural networks—especially in ease of training. The ESN-based engine model performance is compared with the performance of conventional dynamical models, linear and nonlinear. These other models are ARX, NNARX, and the Elman network. The takeoff flight phase is chosen as the operating regime in which to develop the empirical dynamical predictors. Transient data is taken over several flights from a field operated propulsion engine. This data, representing the nominal engine condition, is used in the development of the ESN model predictors. Separate model predictors were developed for each of the engine outputs, *N1, N2,* and *EGT.* Each of the prediction models used the same input measurements.

The results of the ESN-based FDD methods indicate that the ESN is very promising as a nonlinear modeling tool. The ESN significantly out-performed the ARX and NNARX engine models. In addition, the ESN out-performed the Elman network which is another recurrent model. Therefore, FDD methods for transient data based on ESN models have great potential due to their accuracy and relative ease of development. Further work in ESN training methods and architecture for FDD is currently underway and the results will be reported in future papers.

REFERENCES

[1] Chen, J. and Patton, R. J., *Robust Model-based Fault Diagnosis for Dynamic Systems*, Boston: Kluwer Academic Publishers, 1999.

[2] Narendra, K. S., and Parthasarathy, K., "Identification and control of dynamical systems using neural networks," *IEEE Transactions on Neural Networks*, **1**, 1990.

[3] Feldkamp, L. A., and Puskorius, G. V., "A Signal Processing Framework based on Dynamic Neural Networks with Applications to Problems in Adaptation, Filtering, and Classification," *Proceedings of the IEEE*, **86**, 11, 1998.

[4] Parlos, A. G., Chong, K. T., Atiya, A. F., "Application of the Recurrent Multilayer Perceptron in Modeling Complex Process Dynamics," *IEEE Transactions on Neural Networks —Special Issue on Dynamic Recurrent Neural Networks: Theory and Applications*, 5, 1994.

[5] Atiya, A. F., Parlos, A. G., "New Results on Recurrent Network Training: unifying the algorithms and accelerating convergence," *IEEE Transactions on Neural Networks*, 11, 2000.

[6] Jaeger, H., "The Echo State Approach to Analyzing and Training Recurrent Neural Networks," GMD Report 148, GMD – German National Research Institute for Computer Science,2001. http://www.gmd.de/People/Herbert.Jaeger/Publications.html
[7] Jaeger, H., "Short Term Memory in Echo State Networks," GMD Report 152, GMD—German National Research Institute for Computer Science, 2002.
[8] Friedman, J. H., "Multivariate Adaptive Regression Splines," *The Annals of Statistics*, 19, 1991.
[9] Leontaritis, I. J., and Billings, S. A., "Input-output Parametric Models for Nonlinear Systems. Part 1: Deterministic nonlinear systems," International Journal of Control, 41, 1985.
[10] Parlos, A. G., Rais, O. T., Atiya, A. F., "Multi-step-ahead Prediction in Complex Systems using Dynamic Recurrent Neural Networks," Neural Networks, 13, 2000.
[11] Werbos, P., "Backpropagation through time: what it does and how to do it," *Proceedings of the IEEE*, 78, 1990.
[12] Williams, R., and Peng, J., "Gradient-based Learning Algorithms for Recurrent Networks and their Computational Complexity," in *Backpropagation: Theory, Architecture and Applications*, Hillsdale, NJ: Lawrence Erlbaum, 1992.
[13] Menon, S., Uluyol, Ö., Kim, K., and Nwadiogbu, E., "Incipient Fault Detection and Diagnosis in Turbine Engines using Hidden Markov Models," *Proceedings of the ASME Turbo Expo 2003—Atlanta, GA*, 2003.
[14] Menon, S., Uluyol, Ö., Kim, K., and Gupta, D.., "Turbine Engine Modeling using Temporal Neural Networks for Incipient Fault Detection and Diagnosis," *Proceedings of the ASME Turbo Expo 2004—Vienna, Austria*, 2004.

UTILIZING AIRBORNE/STRUCTURE BORNE ULTRASOUND FOR PREDICTIVE MAINTENANCE INSPECTIONS

Mark A. Goodman, VP Engineering

UE Systems, Inc.
14 Hayes Street
Elmsford, NY 10523
Tel: 914-592-1220
Fax: 914-347-2181
mkgoodman@att.net

Abstract: As equipment has become more complex and demands for determining actual and potential sources of failure have become more acute, the need for portable, field effective diagnostic instrumentation has become evident. In addition to standard technologies such as vibration and infrared, airborne/structure borne ultrasound is being recognized as an effective tool for predictive diagnostics and condition analysis. Instruments based on this technology are used for a multitude of inspection activities ranging from mechanical condition to detection and analysis of electrical emissions and leakage. This paper will introduce the technology, its instrumentation and applications. The portability of the instrumentation and how it is used to identify stages of bearing failure, locate leaks and determine types of electrical faults such as partial discharge, arcing, tracking and corona in operating equipment will be described.

Key Words: Ultrasound; Leak detection; Mechanical inspection; Electrical inspection; Predictive maintenance; Condition monitoring; Energy conservation; bearings; lubrication

BACKGROUND AND INTRODUCTION

Current economic conditions have produced a compelling need to find areas of cost reduction, which has led to the "downsizing" of many departments. This has left many managers trying to do more with less. In order to maintain an efficiently running operation, and improve on worker productivity there has been a trend to revise and upgrade Proactive Maintenance programs. Along with this trend has come a need for effective technologies and instruments that enable personnel to become more productive in terms of equipment inspection and problem solving. Opportunities in predictive maintenance and condition monitoring have led to improvements in production, reduced maintenance costs, reduced energy consumption, improved efficient use of personnel and increased profitability.

One of the main reasons for the advances in condition monitoring programs has been the development and use of technologies such as oil analysis, vibration, infrared and ultrasound. This paper will examine one of these technologies, airborne/structure borne ultrasound. This technology can be considered an integrating technology since it can be used with infrared and vibration inspections as well as stand alone to perform a multiplicity of inspection activities. Instruments based on this technology can monitor a wide range of plant operations and yet are simple enough to be used with minimal training for basic, effective inspection routines.

Overview of the technology:
Light weight and portable, ultrasonic translators are often used to inspect a wide variety of equipment. Some typical applications include:

> Bearing inspection, testing gears/gearboxes, pumps, motors, steam trap inspection, valve testing, detection/trending of cavitation, compressor valve analysis; leak detection in pressure and vacuum systems such as, boilers, heat exchangers, condensers, chillers, tanks, pipes, hatches, hydraulic systems, compressed air audits, specialty gas systems and underground leaks; testing for arcing, tracking, corona in electrical apparatus

What makes airborne ultrasound so effective? All operating equipment and most leakage problems produce a broad range of sound. The high frequency ultrasonic components of these sounds are extremely short wave in nature. A short wave signal tends to be fairly directional. It is therefore easy to isolate these signals from background plant and operating equipment noises and to detect their exact location. In addition, as changes begin to occur in mechanical equipment, the subtle, directional nature of ultrasound allows these potential warning signals to be detected early, before actual failure.

Airborne ultrasound instruments, often referred to as "ultrasonic translators", provide information three ways: qualitative with the ability to "hear" ultrasounds through a noise-isolating headphone, quantitative via intensity (dB) readings on a meter or display panel and analytical with the use of spectral analysis software to review recorded sound samples.

Although the ability to gauge intensity and view sonic patterns is important, it is equally important to be able to "hear" the ultrasounds produced by various equipment. That is precisely what makes these instruments so popular. They allow inspectors to confirm a diagnosis on the spot by being able to clearly discriminate among various equipment sounds. This is accomplished in most ultrasonic instruments by an electronic process called "heterodyning" that accurately translates the ultrasounds sensed by the instrument into the audible range where users can hear and recognize them through headphones.

The high frequency, short wave characteristic of ultrasound enables users to accurately pinpoint the location of a leak or of a particular mechanical sound.

The basic advantages of ultrasound and ultrasonic instruments are:

1. They are directional and can be easily located
2. They provide early warning of impending mechanical failure
3. Instruments can be used in loud, noisy environments
4. They support and enhance other PDM technologies or can stand on their own in a maintenance program

INSTRUMENTATION

Airborne Ultrasound translators are relatively simple to use. They consist of a basic hand held unit with headphones, a meter, a sensitivity adjustment, and (most often) interchangeable modules that are used in either a scanning mode or a contact mode. Some instruments have the ability to adjust the frequency response from between 20 to 100 kHz. These instruments may be analog or digital. Digital instruments indicate intensities as decibels. Digital instruments generally have on-board data logging with data management software to provide trending information and create alarm groups for equipment needing special attention. Some of the digital instruments also have on-board sound recording, which enables users to grab sound samples and review them on spectral analysis software.

APPLICATIONS

Generically, applications for ultrasonic translators fall under three basic categories: mechanical inspection, leak detection and electrical inspection.

MECHANICAL INSPECTION

Mechanical equipment produces a "normal" sound signature while operating effectively. As components begin to fail a change in the original sonic signature occurs. This change can be noted as a shift in intensity on a display panel and/or as a qualitative sound change that can be heard through headphones and recorded for further analysis. An ultrasonic translator may be connected to a vibration analyzer or the sound samples may be reviewed through spectral analysis software on a PC.

According to NASA research[1], ultrasonic monitoring of bearings provides the earliest warning of bearing failure. They noted that an increase in amplitude of a monitored ultrasonic frequency of 12dB over baseline would indicate the initial (incipient) stages of bearing failure. This change is detected long before it is indicated by changes in vibration or temperature.

Other opportunities for ultrasonic mechanical inspection include: cavitation in pumps, compressor valve leakage, faulty gears, excessive rubbing, poor connections, to name a few.

Should a vibration program already exist for bearing analysis, it can be enhanced by adding an ultrasonic bearing monitoring program.

Ultrasound translators can be used to aid a diagnosis. The high frequency, short wave characteristic of ultrasound allows the signal to be isolated so that a user can hear and determine if, in fact, a bearing has been correctly diagnosed as failed.

At times there can be signals generated by equipment or components interconnected with a particular bearing that might mask a condition or make a definitive diagnosis difficult. By adjusting the sensitivity, the frequency and listening to the sound, often the actual sound source can be identified. The ability to hear what is going on can prove very important.

Ultrasound detectors work well on slow speed bearings. In some extreme cases, just being able to hear some movement of a bearing through a well-greased casing could provide information about potential failure. The sound might not have enough energy to stimulate classic vibration accelerometers, but will be heard via ultrasonic translators, especially those with frequency tuning. In addition, most slow speed bearings take a long time to analyze with typical vibration equipment, by connecting the ultrasonic translator to the vibration analyzer, the resolution improves and the time to grab a sample drops dramatically.

The following demonstrates another application in which ultrasound can be used with vibration analysis.

The following time waveforms were taken on identical gearboxes in a tubing manufacturing facility. There are 6 gearboxes coupled together each driving tooling that bends steel to form tubing. An inspector was out collecting data with a vibration analyzer and an Ultraprobe ultrasonic translator when he came across these gearboxes. The vibration time waveform was taken using an accelerometer. When listening to the gearboxes with the Ultraprobe, most of them had a "hissing" type sound. One gearbox had a "ticking" sound to it. A time waveform was captured using the accelerometer and then using the Ultraprobe ultrasonic translator. These were then compared to the gearbox right beside it. The following time waveforms show the results:

Gearbox A

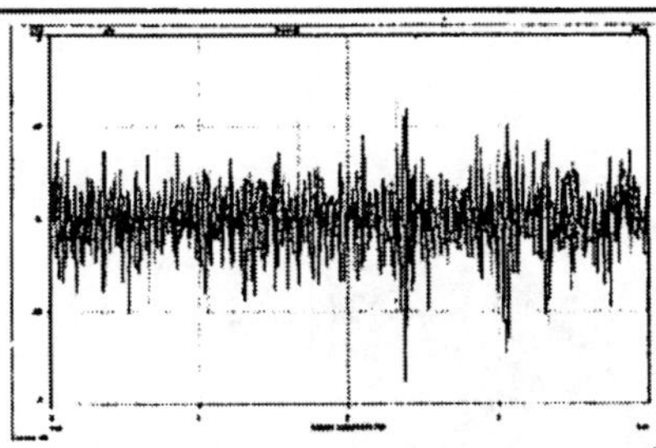

Gearbox B

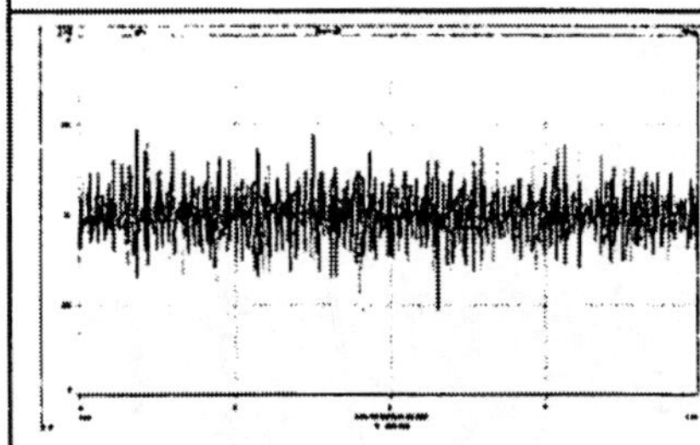

By comparing the two time waveforms, you can see that there is a slight difference in the two gearboxes. Now when you look at the following time waveforms that were captured with ultrasonics the problem becomes more defined.

Gearbox A

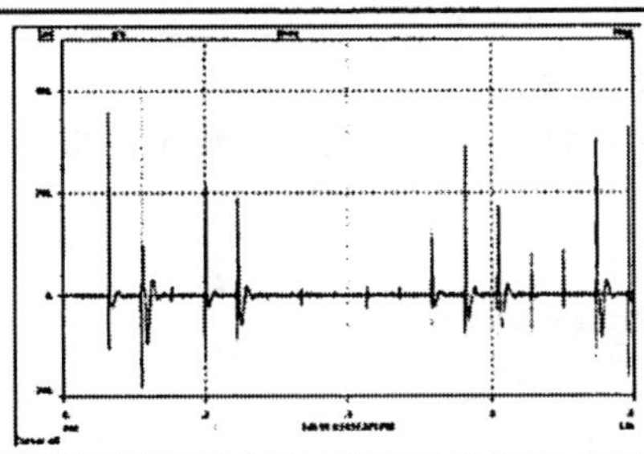

Gearbox B

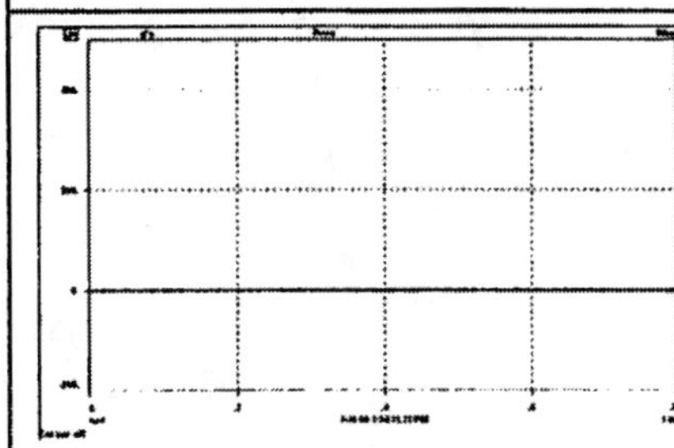

The ultrasonic time waveform shows the broken tooth or teeth in gearbox A. By the use of ultrasonics this problem was caught in an earlier stage.

Spectral Analysis Software

With the introduction of spectral analysis software, a similar type of FFT diagnosis can be performed on a standard Personal Computer as long as the PC has a sound card. These programs not only provide the spectral and time series views of sound but enable users to hear the sound samples simultaneously as they are viewing them on the PC monitor. Typically a sound sample is recorded using a mini-disc or tape recorder. Some ultrasonic instruments have on-board sound recording, which can be downloaded to a PC via a Compact Flash Card. When played back in real time, the acoustic properties can be analyzed. Based on a known "good" or "normal" condition, an anomaly can be quickly determined.

The following is an example of a good bearing. If one were to imagine the sound a good, lubricated bearing makes, the sound quality would be a smooth, "rushing" sound. Here is a sample captured on Spectral Analysis software:

Good bearing Spectra view

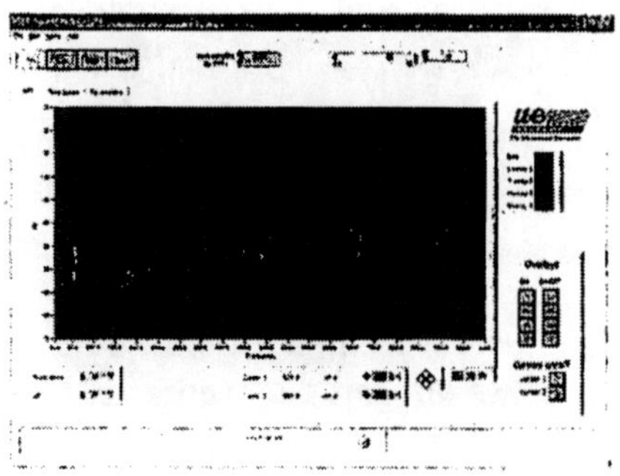

Good bearing Time Series view

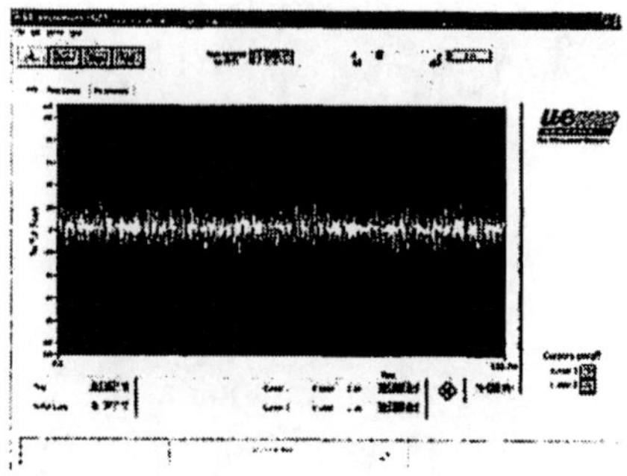

Note there are no distinguishing characteristics. You are observing an example of "white noise". Now let's view a bad bearing. There has been a shift from the expected smooth, rushing sound produced by a good bearing. A fault has occurred in this high-speed bearing. The difference is noted two ways, a shift in the sound quality from a nice, recognizable rushing sound to a jarring buzzing noise. Observe the change in acoustic quality in the spectra view below. Note the fault frequency has been captured as 90 Hz harmonics.

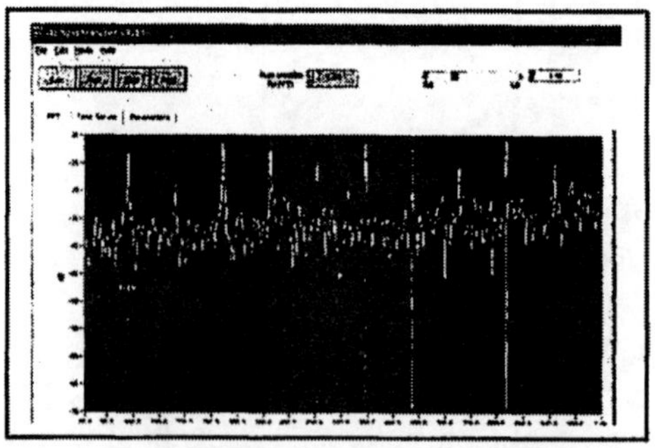

The advantage of ultrasound inspection, aside from providing a finer resolution for viewing the sound image, is inspections can go beyond what is possible to do with typical vibration or infrared scans.

Reciprocating compressor valves are very noisy and produce a lot of extraneous vibration. By isolating the sound with the advantage of the short-wave nature of ultrasound, it is possible to listen to and view the sounds of these noisy valves in real time and to determine a leaking valve. How would you imagine the sound of a "good" compressor valve? As it opens and closes there will be a definite, pronounced clicking. What about a bad valve?

Here are two samples of compressor valves viewed in the time series. Which is the bad valve, the one on the left or the one on the right?

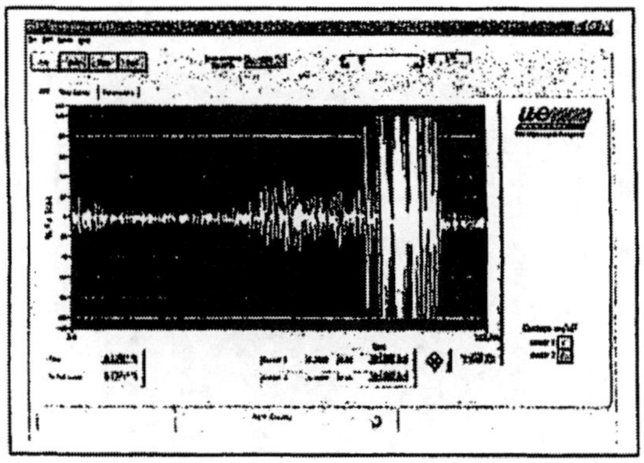 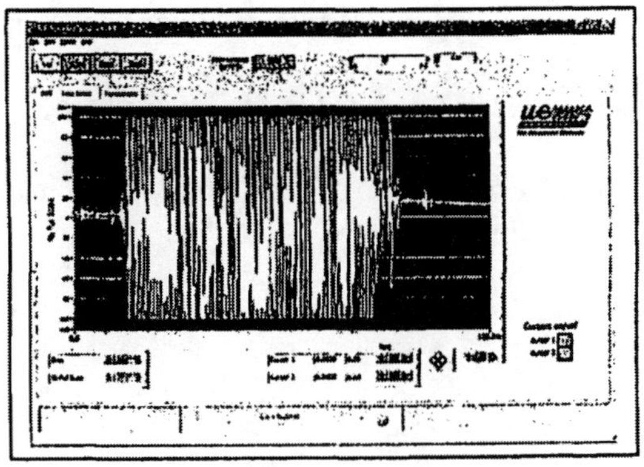

The sample on the left has a definite closed period followed by a distinct "open" period. The view on the right shows a very long "open" period. Therefore the leaking valve is the sample on the right. If one were to listen to these samples as they were viewed, the conclusion would be very obvious.

Electrical Emissions

Below are some examples of typical electrical emissions. Again, the advantage of being able to hear while the sound sample is viewed can help an inspector understand the type and possible the severity of the condition.

Some preliminary experimentation has demonstrated that the main harmonic of an electrical emission (60 Hz in the USA, 50 Hz elsewhere) will be most prevalent in corona. As the condition becomes more severe, there will be fewer and fewer 60 Hz harmonics observed. As an example, arcing has very few 60-cycle components. Mechanical looseness will demonstrate harmonics other than 60 Hz with little to no frequency content between peaks.

When possible, listen to the sound characteristics as they are played in the Spectra View and note that the harmonics of 60 Hz diminish as the condition becomes more severe. The frequency content will move up and down in relation to the ionization produced by the particular emission (arcing, tracking or corona).

This is a view of Corona. Note the steady display of 60 Hz harmonics with a lot of frequency content between the spikes.

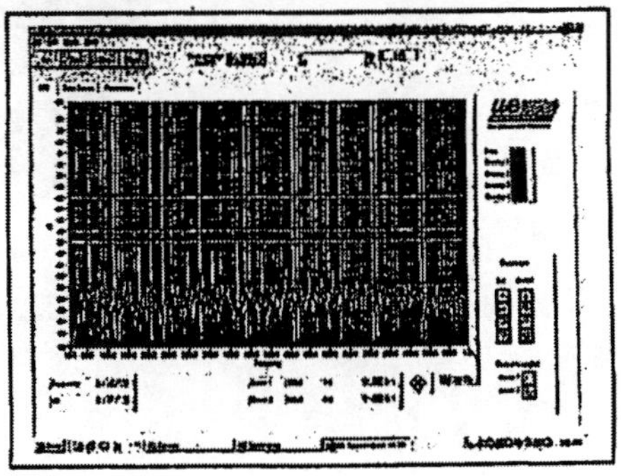

If one were to listen to corona as the sound sample was being played, it would have a distinct "buzzing" and crackling noise.

Tracking, sometimes referred to as "destructive corona" has a different acoustic property. This type of emission has a rapid build up and discharge called "baby-arcing". Below are sound samples viewed in the spectral view.

Note the harmonics of 60 Hz have diminished in this tracking example.

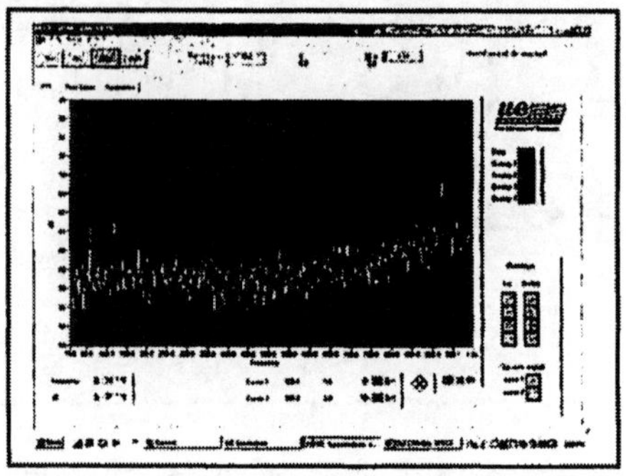

Here is another view of the tracking sound as seen in the "time series". If you were to imagine the "popping" sounds, they can be seen as the long, vertical lines in this view.

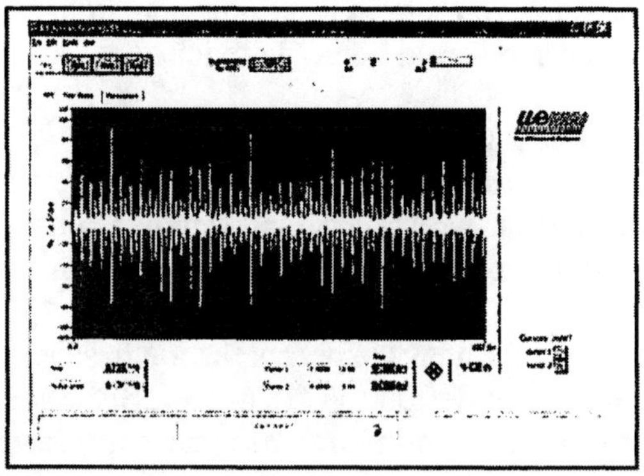

What do you think arcing would sound/look like? Here is an example.

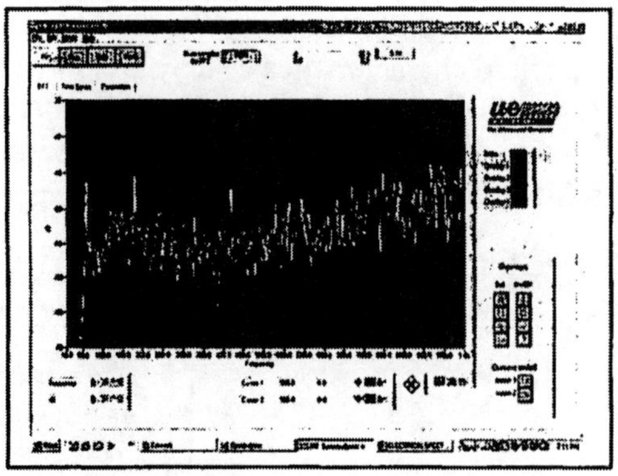

Note the irregular starts and stops. You'll also notice very few harmonics of 60 Hz are present. Arcing will contain almost violent starts and stops. The time series view below might help complete the sound image.

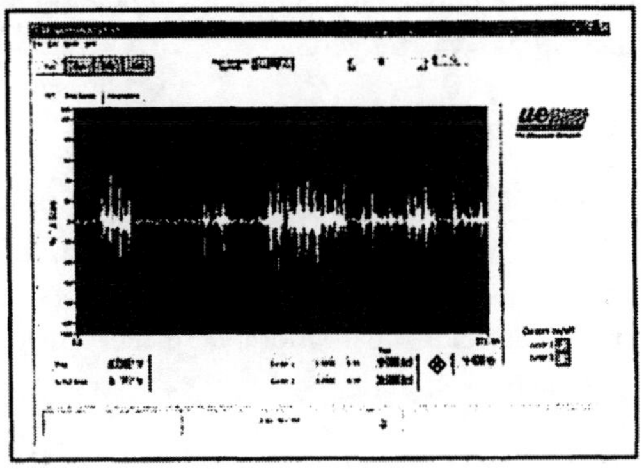

Lubrication

As the concept of "Predictive" lubrication versus time-based ("Preventive") lubrication has emerged, there are times when it is useful to use spectral analysis combined with sound. Instead of lubricating bearings on a routine, "time-based" schedule, inspectors can routinely test bearings and identify those that need lubrication, leaving the others alone. In this manner, lubrication technicians can be taught how to effectively apply just enough lubricant to prevent over lubrication.

When a bearing has exceeded a baseline by 8 dB with no change in acoustic quality, the bearing should be lubricated. While applying lubrication, the technician should stop when the sound level has dropped to the pre-determined baseline level.

A way to demonstrate this process is to view a sound image while noting changes in amplitude and listening to acoustic properties in real time.

Below is spectral view of a sound sample taken while a bearing was being lubricated. The top, red overlay is the non-lubricated state, note how the sound level drops as the lubricant is applied.

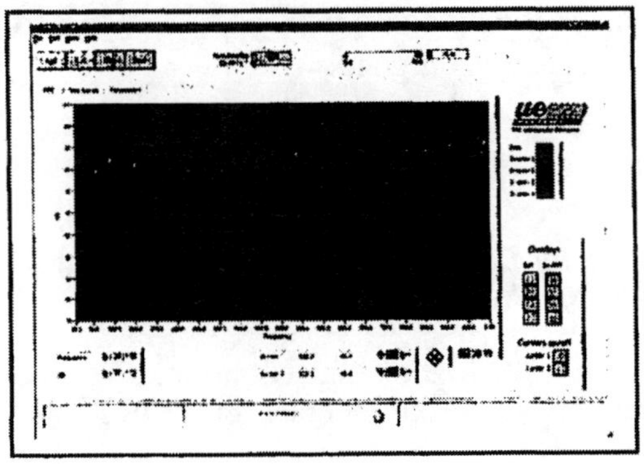

As the proper lubrication level is reached, the decibel level drops by 8 dB with no change in the sound quality. There are no distinctive harmonics or spikes, only a representation of the "white noise" sound quality we would expect to hear if we were to listen as this sound sample was played.

Leak Detection

The category of leak detection covers a wide area of plant operations. It can be looked on as a way of keeping a system running more efficiently. Some plants include it as part of an energy conservation program while others refer to it as fugitive emissions. No matter what, leaks can cost money, effect product quality and wreak havoc with the environment.

Leakage can occur in liquid or gas systems. The greatest advantage of ultrasonic detection is that it can be used in a variety of leak situations since it is sound sensitive and not "gas" specific.

The reason ultrasound is so versatile is that it detects the sound of a leak. When a fluid (liquid or gas) leaks, it moves from the high-pressure side of a leak through the leak site to the low-pressure side where it expands rapidly and produces a turbulent flow. This turbulence has strong ultrasonic components. The intensity of the ultrasonic signal falls off rapidly from the source. For this reason the exact spot of a leak can be located. This can apply to pressure leaks and negative pressure (vacuum) leaks, leaks in valves and in steam traps.

Conclusion

Airborne ultrasound instruments are becoming an important part of Predictive Maintenance, Fugitive Emissions and Energy Conservation programs. Their versatility, ease of use and portability enable managers to effectively plan and implement inspection procedures. By locating leaks, detecting high voltage electrical emissions and sensing early warning of mechanical failure, these instruments contribute to cost reductions, improved system efficiencies and reduced downtime. For optimum effectiveness, it is recommended that all major technologies, infrared, vibration and ultrasound, be used as part of a comprehensive inspection program.

REFERENCES

1. B72-10494, " A System for Early Warning of Bearing Failure" NASA Tech Brief, Marshall Space Flight Center (August 1972)

2. Thatcher, B. et al, "Using Vibration and Ultrasound Technologies as an Integrated Inspection Tool" Application Note from Enteract: Entek IRD.

GENERAL-PURPOSE REAL-TIME MONITORING OF MACHINE SOUNDS

Stephen V. Rice
The University of Mississippi
Dept. of Computer and Information Science
P.O. Box 1848
University, MS 38677 USA
rice@cs.olemiss.edu

Stephen M. Bailey
Comparisonics Corporation
P.O. Box 1960
Grass Valley, CA 95945 USA
sbailey@comparisonics.com

Abstract: The "Comparisonics" sound-matching algorithm computes signatures to characterize audio data and compares the signatures to measure the similarity of sounds. This algorithm is general purpose: it can compare any sounds and can monitor sounds from any machine. After a baseline recording is characterized, signatures are derived from the current sound and compared in real time with the baseline signatures. Similarity scores are computed continuously, reflecting the similarity of the current sound to the baseline, and an alert can be given if the similarity falls below a threshold. In addition, the current sound can be compared with known "error" sounds to identify specific faults. This algorithm can operate in a wireless sensor network and be deployed in a factory or on a ship to monitor an unlimited variety of machinery.

Key Words: Acoustic emissions; audio comparison; condition monitoring; wireless sensor networks.

Introduction: A human listener accustomed to the "normal" sound of a machine readily detects changes in the sound. It is well known that the sounds made by machines can indicate the health of the machinery. A change in sound may portend trouble and warrant investigation by skilled maintenance personnel. However, a change may go unnoticed by human listeners due to unfamiliarity with the usual sound, or because the change is gradual, sporadic, or masked by the din of neighboring machinery. If the machine is remote and unsupervised, there is no one to listen.

There is a great need for computers to listen to machinery and many millions of machines could be usefully monitored; however, a robust low-cost solution has been unavailable. Let us consider some existing approaches:

- *amplitude monitoring* – a simple device detects whether the sound is louder or softer than expected; it is a low-cost, general-purpose solution but it does not perform any frequency analysis; it detects changes in the quantity, but not the quality, of sound

- *machine-dependent pattern recognition* – a custom pattern-recognition system is developed to monitor the sounds from a particular machine; accurate monitoring may be achieved for this machine but the cost to develop the system is high

- *machine-class-dependent pattern recognition* – characteristics of a particular class of machines (for example, rotating machinery) are exploited to monitor machines belonging to the class; the cost is high but not as high as developing a custom pattern-recognition system for each machine in the class

The current situation, as we see it, is a choice between low-cost amplitude monitoring and high-cost pattern recognition. The semi-automated solution in which trained humans analyze FFT output belongs to the latter category.

We have developed a solution that might be considered "machine-independent pattern recognition." Its cornerstone is a general-purpose "sound-matching" algorithm that compares any sounds and measures their similarity. This algorithm can listen to and compare sounds from any type of machine. After characterizing a baseline recording of a machine, it compares the current sound with the baseline to detect changes in real time.

The algorithm is ideally deployed in a wireless sensor network, where each wireless node is equipped with a microphone. The wireless nodes can be easily placed by personnel throughout a factory or ship, near machinery to be monitored. Although the algorithm can compare vibration data, we avoid the use of accelerometers, which are more expensive and more difficult to install than microphones. After some initial configuration, the system begins real-time monitoring of sounds.

By employing a general-purpose sound-comparison algorithm and inexpensive, easy-to-install wireless nodes with microphones, the goal of robust, low-cost sound monitoring can be achieved.

Sound Matching: The Comparisonics sound-matching algorithm was developed by S. V. Rice in 1997. The initial goal was to develop a method for "content-based" retrieval of sound effects. Previously, sound-effect collections could be searched only by entering a text description for each sound and then performing a keyword search of the text descriptions. In Rice's "sounds-like search," a user can specify any example sound and the system automatically retrieves perceptually similar sounds. This unique search capability has been incorporated into FindSounds.com, the first Web search engine for sound effects [1,2].

The sound-matching algorithm processes the uncompressed sequence of sample values in any digital audio recording, provided the duration of the recording is at least ten milliseconds and the sample rate is at least 8 kHz. The algorithm characterizes the sounds in the recording by a "signature," which is a 16-byte quantity that encodes a vector of perceptual features. Given any two signatures, the algorithm returns a score ranging from 0 (least similar) to 100 (most similar) describing the similarity of the recordings from which the signatures are derived. In assigning a score, the algorithm emulates the human perception of sound similarity: the higher the score, the more a human listener will perceive the two recordings to be alike. Nearly all of the perceptual features are characterizations of frequency content and are extracted from the time domain by a proprietary transform. A nonlinear distance measure in the multi-dimensional feature space is used to compute similarity scores.

The sound-matching algorithm was designed to produce a meaningful measure of sound similarity for all audible sounds. It is general purpose, not trained to a particular class of sounds. For sound-effects retrieval, it is impractical to develop a custom algorithm for each type of sound effect: rain, sirens, elephants, etc. Likewise, we cannot afford to build a custom algorithm for each machine.

The efficiency of the Comparisonics algorithm is notable. On a modern personal computer, the time required to compute a signature is less than one percent of the duration of the recording; thus, signatures can easily be computed in real time. The time required to compare two signatures and compute a similarity score is less than 0.5 microseconds. A signature occupies only 16 bytes for efficient manipulation and storage. (Contrast this with the storage required for FFT output, which is measured in kilobytes.)

One signature can be computed for a long recording and represents the "average" of the sounds in the recording. For machinery monitoring, it is desirable to divide the recording into consecutive 100-millisecond intervals and compute one signature to characterize each interval. This sequence of signatures can be compared with another sequence of signatures if the temporal ordering of sounds is to be matched. However, in our standard implementation, we ignore the temporal sequence.

Let $S = \{s_1, s_2, \ldots, s_n\}$ be an unordered set of n signatures derived from one or more baseline recordings of a machine. We apply a clustering algorithm to find a set of signatures $C = \{c_1, c_2, \ldots, c_m\} \subseteq S$ representing the distinct sounds in the recording, where each c_i is the exemplar of a cluster of perceptually similar sounds. In this way, the set C characterizes the natural variations of the "normal" sound of the machine. These variations may occur in one operating mode or be produced by different modes (for example, the cycles of a washing machine).

To monitor the sound of the machine in real time, a signature s is computed for the current 100-millisecond interval and the similarity of s and c_i is computed for each $i = 1,2,\ldots,m$. The maximum similarity score represents the best match between the current sound and the baseline recording. If the maximum score falls below the value of a threshold parameter τ, then an alert can be raised immediately, or the alert might be issued only after the maximum score has fallen below the threshold for multiple time intervals. In Fig. 1, the maximum score is plotted versus time for a recording of a compressor. The score declines as the sound of the compressor deviates from the baseline recording.

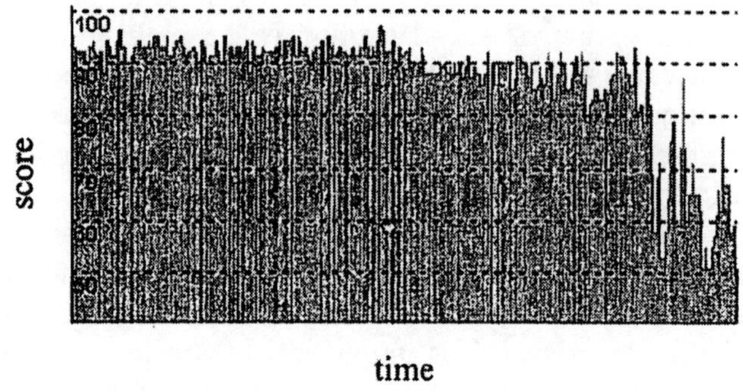

Fig. 1. Monitoring the Sound of a Compressor

The choice of τ can be determined manually through experimentation or automatically by comparing the signatures derived from one baseline recording with the exemplars of another baseline recording. In the latter approach, τ is chosen to be less than the observed scores.

The algorithm currently does not factor in the relative "strengths" of the clusters. That is, some clusters represent more sounds in the baseline recording than others. However, this can be considered to determine whether the current sound is rare or common in the baseline recording.

The algorithm can characterize not only baseline recordings but also recordings of known "error" sounds. If the current sound deviates from the baseline, it can be compared with the signatures of faults in an attempt to diagnose the problem.

Testing: The robustness of the Comparisonics sound-matching algorithm has been demonstrated for a wide variety of sounds in its application to sound-effects retrieval. FindSounds.com is utilized by more than 150,000 users per month.

A collection of recordings of 600 different machines has been assembled from sound-effects compilations for testing the machinery monitoring application. The sound-matching algorithm was not modified or specially trained for this test, yet it reliably matches the sounds in this collection. The diversity of this collection demonstrates the general-purpose nature of the algorithm. Below are some of the sound sources in this collection:

- acetylene torch
- air conditioner
- anesthesia ventilator
- arc welder
- band saw
- boiler
- cement mixer
- centrifuge
- clothes dryer
- clothes washer
- compressor
- conveyor belt
- drill press
- fan
- furnace
- gears
- generator
- grinder
- lathe
- microwave oven
- milling machine
- motor
- oxygen mask
- packing machine
- photocopier
- pile driver
- planer
- printing press
- pulse oximeter
- pump
- respirator
- stamping machine
- steel cutter
- table saw
- threshing machine
- transformer
- turbine engine
- winch

In addition, the test collection includes sounds from vehicles: airplane, bus, car, elevator, ferry, helicopter, motorcycle, rollercoaster, ship, subway, tank, tractor, train, and truck.

Proof of concept has been demonstrated using personal computers. However, it is impractical to place a PC next to each machine to monitor. The next step is to port the sound-matching algorithm to a wireless sensor platform and conduct tests of the algorithm on this platform. A project is underway to implement and test the algorithm on Crossbow Technology's MICA Mote platform [3].

Conclusion: Recent issues of the *Communications of the ACM* and *IEEE Computer* are devoted to the burgeoning field of wireless sensor networks [4,5]. Technologies for monitoring are rapidly evolving. For machinery monitoring, comparing audio and vibration signals is a challenge. A general-purpose sound-matching algorithm avoids the costs and complexities of custom development, and deployed on a wireless platform, provides a cost-effective solution for monitoring the sounds of an unlimited variety of machinery.

References:

1. S. V. Rice and S. M. Bailey, "Searching for Sounds: A Demonstration of FindSounds.com and FindSounds Palette," *Proceedings of the International Computer Music Conference*, Coral Gables, Florida, November 2004, pp. 215-218.
2. http://www.FindSounds.com
3. http://www.xbow.com
4. *Communications of the ACM*, vol. 47, no. 6, June 2004.
5. *IEEE Computer*, vol. 37, no. 8, August 2004.

MONITORING ELECTRICAL PROFILES IN ROTATING MACHINERY SHAFTS WARNS OF DEVELOPING PROBLEMS

Paul I. Nippes, P.E.
Magnetic Products and Services, Inc.
2135 Highway 35, Holmdel, NJ 07733, USA
www.gaussbusters.com

Abstract: A unique and independent unit condition monitoring technology for rotating machinery, called Shaft Condition Monitoring (SCM), provides early warning of developing problems by monitoring and analyzing shaft voltage and grounding current profiles. The process begins with capturing and trending of critical information inherent in raw signals derived from reliable shaft-riding brushes. Through patented analysis of this critical information, problems in rotating machinery are detected at their inception, long before they are apparent on traditional instruments. With this data in hand, a proactive plan of monitoring and maintenance is facilitated, possibly averting the extreme costs of failure and forced shutdown.

Key Words: Alarm systems; compressors; electric machines; generators; monitoring; motors; pumps; rotating machines; shaft currents; shaft voltage; signal analysis; turbines.

Introduction: Monitoring of shaft grounding currents and shaft voltages of rotating machinery, called Shaft Condition Monitoring (SCM), is the culmination of early measures employed to remedy the effects of stray currents on shaft bearings. SCM is also extended to identification of machine problems, including an influx of new problems attendant to variable frequency and PWM power sources.

Technology: Shaft condition monitoring gives a broad and often specific picture of a train's health. Shaft-riding brushes must have low resistance contact to the shaft to obtain accurate measurement of the shaft's electrical condition. These provide critical data on the magnitude and frequency of shaft grounding current and/or voltage[1]. In all trains, the source of shaft currents can be either electrostatic charge build-up or voltage generation from residual magnetism. Trains having rotating electrical machines provide many additional voltage sources, such as unbalance or asymmetry in magnetic fields [2]-[14], shorted turns in the rotor and/or stator winding, shorted rotor/stator core laminations, shorted bearing, seal and coupling insulation, uneven air gap, and/or harmonics and transients from power electronics, power supplies or loads.

A number of papers have been written on the mechanisms of stray shaft current generation [2]-[4], [7]-[16]. The location, direction and power delivery of stray current sources are difficult to anticipate and track. It is possible, however, to measure and monitor shaft grounding currents and voltages. Measured currents and voltages provide early warning signs of the potential for damage[1]-[2]. Shaft grounding current peaks range from 20 mA to 20 A on healthy rotating machines rated above 75 kW (100 Hp), up to the largest, 1500 MW. Voltages on the electrical machine shaft end, which is opposite from the grounding brush location, are customarily insulated from ground to prevent flow of damaging shaft currents. These shaft voltages can be very high, reaching hundreds of volts. Significant deviations from normal grounding currents and shaft voltages are usually indicators of problem development. In some well-maintained trains with proper shaft grounding and monitoring, measured peak grounding currents are in the low milli-ampere range. On some trains, rubs occurred resulting in severe current damage and unusually high magnetism. Electrical machines have the greatest potential for generating shaft currents and voltages. The technology of monitoring shaft grounding currents and voltages in rotating shafts is based upon the following factors (proven field cases are marked with an asterisk):

A. Shaft or rotor rubs onto casing or stationary members:
   1. Shaft voltage decreases at rub occurrence [16].*
   2. Shaft grounding current increases, compounding magnetism in the rotating and stationary members. *
   3. Shaft voltage can be generated by a rub, depending upon the type of materials that come into contact.
B. Static charge build-up on the rotating member due to:
   1. High liquid or gas velocity.
   2. Liquid or gas flow through filters.
   3. Turbine saturated steam. *
   4. Turbine dry steam under certain conditions. *
   5. Fogging or wet compression in gas turbines. *
C. Voltage generation from residual magnetism from:
   1. Magnetic particle inspection. *
   2. Welding with improper ground return. *
   3. Improper demagnetizing. *
   4. Magnetic bases. *
   5. Lightning strike nearby. *
   6. Plant electrical faults or ground currents.
   7. Compounding of residual magnetism [13]. *
D. Electromagnetic asymmetries in electrical machines:
   1. Loss of bearing or seal insulation [12].*
   2. Loss of coupling insulation [12].*
   3. Core lamination shorting. *
   4. Unequal gap between the stator and rotor.
   5. Rotor winding turn shorting. *
   6. Rotor winding ground faulting. *
   7. Stator winding loss of a parallel circuit. *
   8. Stator winding transposition or turn shorting.
   9. Stator winding faults near wye connection.
   10. Induction motor broken rotor bars.

E. Voltage harmonics and/or transients in electrical machines:
   1. Variable frequency drives [9]. *
   2. Pulse-width modulated supplies [7]-[8]. *
   2. Excitation system harmonics [15]. *
   4. Unbalanced phase loads on generators.
   5. Rectifier or SCR loads or power supplies [10].
   6. Power system transients, unbalance or electrical discharge.

Components Needed: Quality shaft riding brushes are essential for both shaft grounding and shaft voltage sensing[1]-[2]. Experience has demonstrated that conventional carbon and babbitt brushes are unreliable for shaft grounding duty, especially in the presence of oil and dirt. Even in a clean atmosphere, carbon brush current density should be maintained above 7 A/cm2 (40 amperes per square inch). Otherwise, performance is impaired by high resistance glazing of the shaft contact surface and by chipping away of the brushes. The most reliable brushes have proved to be those that have metal contacting the shaft. Even wth these, cleanliness of the brush and its contact to the shaft are necessary.

Reliable shaft grounding protects bearings against damage from shaft currents and provides a means to measure shaft-grounding currents. The brush must be electrically insulated and connected to the brush shaft-riding element(s) only. With quality grounding, accurate measurements of the grounding current and shaft voltage are obtained. Voltage and current signals are then conditioned and compressed, maintaining critical signal aspects needed to provide early warning of impending unit problems. The monitor should be mounted as close as possible to the current shunt or tapped resistor in the brush grounding cable to avoid interference from extraneous electromagnetic signals.

Typical Configuration:

A. Turbine Generator Trains (Figure 1): Dual grounding brushes are commonly employed on large and critical machines and are located between the turbine and the generator on large turbine generator sets. Small voltage sensing brushes are usually located at the generator outboard, or excitation end of the shaft, and on one of the turbine shafts.

B. Mechanical and Electro-Mechanical Sets (Figure 2): Where the train has electrically-conductive couplings, two brush locations are usually adequate for proper train shaft grounding and voltage sensing.

Additional brushes for shaft grounding and voltage sensing may be necessary on trains where elements are isolated by electrically-insulated couplings and/or train shaft oil films, such as in gear-type couplings and speed-changing gears. Each such electrically-isolated shaft should have its own brushes for shaft voltage sensing and shaft grounding, along with its own voltage-current monitor.

DUAL GROUNDING BRUSHES, DUAL VCM'S IN ONE BOX, DUAL VOLTAGE SENSING

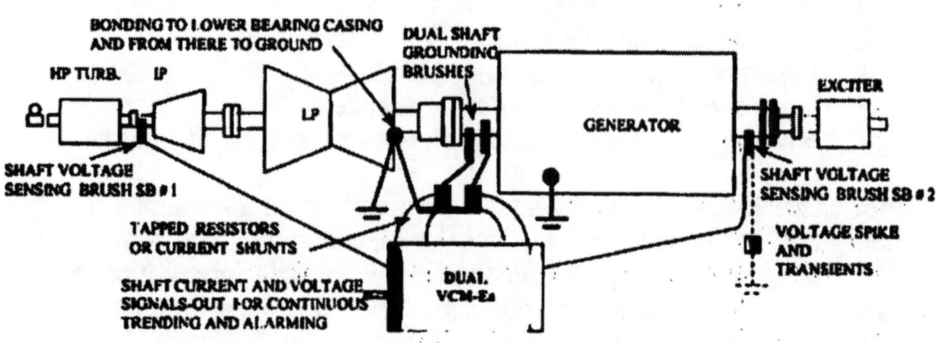

Figure 1 – Turbine Generator Shaft Grounding Current and Voltage Sensing

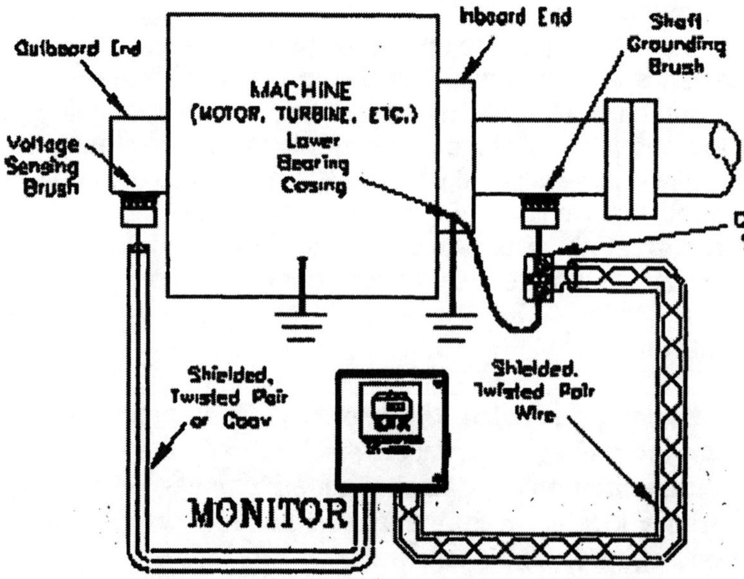

Figure 2 - Sensing Grounding Current and Voltage on an Electrically-Isolated Shaft

Analysis, Detection and Pinpointing: Changes in the shaft electrical profile are the first indicators of a developing problem. If this does not point to a specific problem, conventional instruments and monitoring devices should be trended very carefully, narrowing the possibilities to specific problems. For example, identification of an indicated problem and its further development can usually be obtained by investigating and/or tracking changes of the following:

In all rotating shaft machines:

1. Vibration of shaft, bearing, frame and/or core [10].
2. Temperature of bearings.
3. Oil particle analysis.
4. Audible levels and frequencies.
5. Shaft torque.
6. Electromagnetic signal indicators.

In electrical machines and trains:

1. Temperature of armature and/or field windings of electrical machines.
2. Electrical current and/or voltage harmonic analysis.
3. Stator winding partial discharge level and changes.
4. Stator winding turn and/or circuit short or open circuits.
5. Field winding ground fault indications.
6. Current side band analysis on induction machines.
7. Gap flux-probe symmetry on synchronous machines.
8. Field AC impedance changes on synchronous machines.
9. Field current for a given load and power factor on synchronous machines.
10. Cooling gas particle monitor indications.
11. Sparking of brushes on commutator and/or collector rings.
12. Diode, fuse or other component failure in excitation system.
13. Electrical system transients, unbalance, discharges.

Abnormal trends of shaft condition monitor signals should encourage operators to watch carefully for correlated changes in the above factors. Even gradual changes could be overlooked as a syndrome worsens without the benefit of early warning provided by effective shaft sensing and monitoring. By narrowing the possible problems, effort can then be focused on effective troubleshooting, leading to orderly planning and/or execution of maintenance or shutdowns, as appropriate.

SCM Technology Case Histories:

A. A Stator Coil Failure Due to Core Iron Shorting on a 750 MW Turbine-Generator: In July 1995, VCM shaft grounding current readings were recorded on a strip chart. The author inspected this chart and found that the unit operated less than half time. Of six periods of operation, 10 to 20 hours, five showed normal low level current of less than three amperes, followed by bursts of excessive currents. Figure 3 shows the period of highest currents exceeding the scale maximum of ten amperes. The shaft grounding current then settled down to either a normal or a slightly higher continuous level. The author concluded that either shorted stator laminations or shorted rotor winding turns alternately blew open or shorted together. No other possible operation or instrument condition would explain the trace abnormalities. In year 2000, the unit was pulled from service because of a stator coil failure due to excessive heat caused by shorted core end lamination packets (see Figure 4). The 1995 prediction, after five years, is considered reasonable since this generator was on peaking duty, operating only for short periods during the 1995-2000 interval.

B. Electrostatic Charges from Wet Steam [1]: The shaft grounding peak current on a large turbine-generator changed from 3 amperes to 6 amperes, coincident with a drop in steam temperature entering the turbine, from 521°C (970°F) to 510°C (950°F). The correlation was duly noted and when the steam temperature was restored to 521°C (970°F), the shaft grounding current consequently returned to 3 amperes. This is attributed to wet steam due to lowered steam temperature [12], which is undesirable since it causes turbine blade erosion and rotor electrostatic charging, possibly damaging bearings.

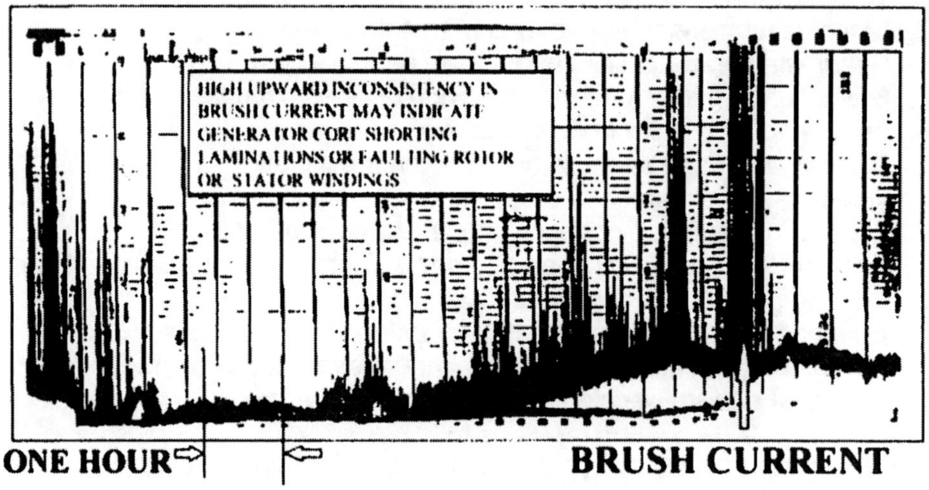

Figure 3 - 750 MW Generator VCM Output Indicating a Unit or System Recurring Condition

Figure 4 – Shorted Core End Laminations

C. Shaft Residual Magnetism [2],[4]-[6]***:*** Trains sometimes experience current attack on their bearings, seals and couplings from currents generated by residual magnetism. This occurs even on purely mechanical trains. Thrust, anti-friction and/or journal bearings are affected and can exhibit increased temperature and vibration, shaft axial or vertical displacement, and in some instances, unit crashes. Figure 5 shows electrical current damage from residual magnetism in the turbine on a turbine-compressor train. Excessive magnetism was measured in the turbine casing and rotor.

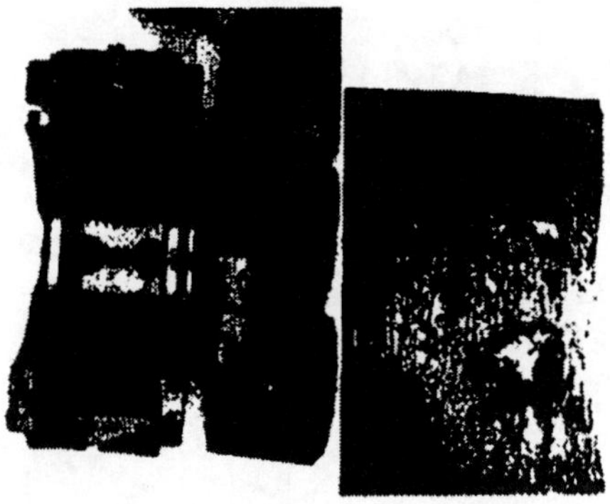

Figure 5 - Results of Electrical Current Damage from Residual Magnetism

Down-cycle demagnetizing was effective in reducing residual magnetism, the source for stray voltage generation. A steel mill furnace turbine-driven blower had failure of its outboard bearing.

On a blower the peak voltage reduced from 1.8 volts to 0.5 volts when the turbine shaft grounding brush was lowered onto the shaft, restoring shaft grounding.

D. Excitation Circuit Failure: A large utility generator employing static excitation had severe current damage to its bearings even though the bearings were properly insulated. The cause was traced to a high spikes and harmonics generated by a failed diode in the excitation circuitry passing current through the capacitance of the bearing insulation. With the installation of reliable shaft grounding and a voltage current monitoring system calibrated to warn of excitation diode failure, the operators are now alerted at the onset of such a problem.

E. Rotor Excitation Circuit Ground Fault: A large nuclear station generator experienced ground fault indications starting in January 1, 1999. The fault indication would clear when field current was reduced; however, there was a limit below which further reduction of the field current would be permitted. Since it was important to locate the cause, an MPS monitoring system (VCM) and an event recorder were installed to determine what was occurring. During the next ground fault indication, the VCM recorded peak shaft grounding current increase from 3 to 17 amperes and the event recorder showed complete offset of the field voltage (see Figure 6). Generator rotor dismantling in March 1999 revealed a torn shim in one of the rotor winding excitation leads coming into contact with the shaft (see Figure 7). Correcting this and the other DC lead, which was developing a similar problem, probably prevented possible disaster. The VCM provided completely independent confirmation of a valid alarm, proving that the ground fault alarm was due to a rotor winding fault and not an instrument or excitation system failure or error.

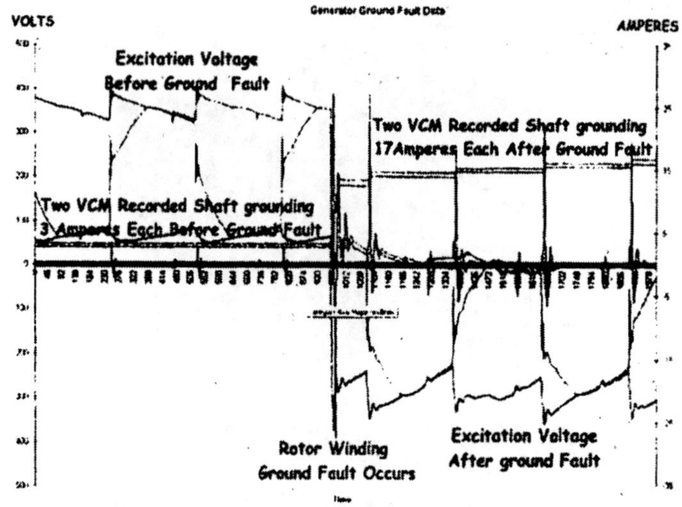

Figure 6 - Rotor Winding Ground Fault Occurrence

Figure 7 – Ground Fault at Torn Copper Leaf

F. Loss of Bearing Insulation: Numerous cases have occurred where the bearing insulation on a train has been bridged or defeated. Damage was averted by properly insulating the bearings and by monitoring the shaft voltage. It is observed that shorting of the insulated bearing caused a voltage drop by at least one half, with a proportional rise in grounding brush current.

G. Shaft Rubs: During shop testing of a motor, its shaft voltage dropped in value by one half. It was also noted that the oscilloscope trace of this voltage became half wave rectified, rather than sinusoidal. Disassembly of the motor revealed that a rub had developed. When the rub was cleared, the full wave character of the shaft voltage was restored, half wave discharge having occurred at the initial shaft rub contact.

Problems Encountered With VCM Installations: The first VCM's were sold in 1989 and since that time more than 600 have been sold. Very few problems with VCM's or their installation have been reported. Users have been very forthcoming with performance information and have been very cooperative in ensuring proper VCM installation and performance. Listed below are VCM problems that have been encountered requiring correction to ensure technology reliability:

A. Installation Issues:
   1. Poor performing shaft riding brushes.
   2. Shaft-riding brush insulation missing or shorted.
   3. Improper routing and/or connection of shaft grounding brush cable.
   4. Improper VCM installation or wiring connections.
B. External Factors:
   1. Shaft or cable pickup of interfering electromagnetic signals.
   2. Plant ground grid has voltage or is unreliable.
   3. Signal interference from stray plant voltage and/or current spikes.

Summary and Conclusions: Shaft condition monitoring is an ever-expanding technology based upon in-the-field performance of rotating machinery. It is designed to warn of developing problems in rotating machinery and to pinpoint specific problems, either directly or through confirming and predictive capabilities of normal plant system monitors and instruments. Warning of impending damage is provided in advance of normal plant alarms and/or monitors that indicate only after damage has already occurred. As demonstrated in this paper, considerable data has already been collected over the past 15 years and more is being gathered, leading to increased accuracy in the interpretation of shaft voltage trends [1]-[2]. The economic potential in savings is enormous when taking into account the losses from forced shutdowns and the replacement of parts due to failure, which may have been averted if the electrical source had been properly diagnosed and treated in its embryonic stages. Furthering the economic appeal of this treatment is the fact that installation does not require disassembly of the machine and VCM's can be installed on operating machines with insulated shaft-riding brushes, precluding unwanted and unnecessary shutdowns. As even more advances in this innovative technology take place, applications to monitoring for specific causes of shaft current damage will be tailored to the needs of users worldwide. The trend is clearly toward the implementation of shaft condition monitoring as an industry standard in the science of preventing machine damage before or as it starts.

REFERENCES:

Papers Presented at Conferences (Unpublished):

[1] P. Nippes, D. David and A. Peniazev, "Monitoring of Shaft Voltages and Grounding Current Brushes," presented at the EPRI Motor and Generator Predictive Maintenance & Refurbishment Conference, Orlando, FL, Nov. 29, 1995.

[2] P. Leonard, P. Nippes, J. Sohre, B. Sandner and L. Sutherland, "Panel on Electromagnetic Shaft Current Control," presented at the 10th Turbomachinery Symposium, Texas A&M University, Houston, TX, Dec. 1981.

[3] P. Nippes and J. Sohre, "Electromagnetic Shaft Currents and Demagnetization of Rotors of Turbines and Compressors," presented at the 7th Turbomachinery Symposium, Texas A&M University, Houston, TX, Dec. 1978.

Periodicals:

[4] P. I. Nippes, "Magnetism and Stray Currents in Rotating Machinery," Journal of Engineering for Gas Turbines and Power, Transactions of ASME, vol. 118, pp. 225-228, Jan. 1996.

[5] J. S. Sohre, "Shaft currents Can Destroy Turbomachinery," Power, pp. 96-100, Mar. 1980.

[6] J. S. Sohre, "Are Magnetic Currents Destroying Your Machinery?" Hydrocarbon Processing, pp. 207-212, Apr. 1979.

[7] D. Busse, J. Erdman, R. Kerkman, D. Schlegel and G. Skibinski, "The Effects of PWM Voltage Source Inverters on the Mechanical Performance of Rolling Bearings," IEEE Transactions on Industry Applications, vol. 33, no. 2, pp. 567-576, Mar/Apr 1997.

[8] D. Busse, J. Erdman, R. Kerkman, D. Schlegel and G. Skibinski, "System Electrical Parameters and their Effects on Bearing Currents," IEEE Transactions on Industry Applications, vol. 33, no. 2, Mar/Apr 1997.

[9] S. T. Chen, A. Lipo and D. Fitzgerald, "Modeling of Motor Bearing Currents," IEEE Transactions on Industry Applications," vol. 32, no. 6, Nov/Dec 1996.

[10] R. W. Jones and D. E. Seaver, "Investigation and Results of Eddy Currents on DC Motor Bearings," pp. 145-150, 1990.

[11] R. Von Kreutzer and R. Schubert-Auch, Ismaning "Gefugemwandlung in Lagern durch Stromubergang," Der Maschinenschaden, 60, heft 3, pg. 132, 1987.

[12] L. T. Rosenberg, "Eccentricity, Vibration, and Shaft Currents in Turbine Generators," AIEE Transactions, pp. 38-41, Apr. 1955.

[13] V. Schier, "Selbsterregte Unipolare Gleischstrome in Maschinenwallen," Elektrotechische Zeitschrift, Nov. 1965. A translation of this article was prepared by the Associated Electrical Industries, Ltd., Power Group, Research Laboratory, Trafford Park, Manchester 17, UK, Translation #3925, "Self Excited Homopolar Direct Currents in the Shafts of Machines."

[14] H. O. von Seinsch, "Lagerstrome bei Drehstrom-Induktionsmaschinen, Ursachen und Methoden iher Unterdrukung," Conti Elektro, pp. 43-51, Jan. 1969.

[15] C. Ammann, K Reichert, R. Joho, Z. Posedel, "Shaft Voltages in Generators with Static Excitation Systems-Problems and Solutions", IEEE Transactions on Energy Conversion, Vpl.3, No. 2, June 1988.

[16] Costello, Michael J., "Shaft Voltages in Rotating Machinery", IEEE Transactions on Industry Applications", Vol. 29, No. 2, Mar.-Apr. 1993, pp 419-426.

Patents:

[17] Paul I. Nippes, "Shaft Voltage Current Monitoring System for Early Warning and Problem Detection," U. S. Patent 6 460 013 B1, Oct. 1, 2002.

SENSORS TECHNOLOGY

Chair: Mr. Henry R. Hegner
Engineering Consultant

PREDICTING REMAINING USEFUL LIFE IN ROLLING ELEMENT BEARINGS

Thomas J. Walter and Hyungdae Lee
Foster-Miller
431 New Karner Road
Albany, NY 12205
twalter@fosmiltech.com and hlee@fosmiltech.com

ABSTRACT: This paper relates the progress of an ongoing project to develop a practical sensor system to assess current conditions and predict remaining useful life in rolling element bearings. A methodology is presented that includes: integration of a wireless smart sensor platform to extract condition indicators from bearing vibration and acoustic emission signals; a pattern recognition technique based on a neural network and accelerated life testing; and a bearing life equation to provide the types of advance information needed. In-house and field tests are being performed to verify the feasibility of the method.

Key Words: Condition indicator; neural network; pattern recognition; remaining useful life; rolling element bearing

INTRODUCTION: To assess current bearing condition and predict remaining useful life (RUL) in rolling element bearings, we have developed a methodology that combines a wireless smart sensor platform, a bearing life equation, bearing life prediction based on accelerated life testing, and a pattern recognition technique. The following sections review the available techniques to predict RUL, present the overall concept of the approach implemented at Foster-Miller, and summarize the experimental work accomplished to date and conclusions drawn.

TECHNIQUE REVIEW: The techniques available to predict bearing RUL can be widely categorized in two ways: statistical methods and condition-based maintenance.

Bearing Life Estimation Based on Statistical Methods. Lundburg and Palmgren [1, 2] developed life prediction methods for ball and roller bearings using the Weilbull distribution function [3].

$$L_{10} = \left(\frac{C}{P}\right)^p \text{ and } L_{10h} = \frac{1000000}{60n}\left(\frac{C}{P}\right)^p \quad (1)$$

where $L10$ and $L10h$ are the bearing life in millions of cycles and operating hours respectively; C is the dynamic load capacity of the bearing or the theoretical load that will produce a life of one million cycles with a 90 percent probability of survival; P is the applied equivalent radial load derived from the combination of radial and thrust loads and their associated coefficients; p is the load-life exponent, 3 for ball bearings and 10/3 for roller bearings; and n is shaft rotational speed.

As bearing design and materials in use improved, other researchers [4] have proposed some modifications of the equation based on: a material fatigue limit and discrete finite elements; a Weilbull-based life theory using discrete finite elements that accounts for deviations in the accepted material Hertz stress-life relation; and a fatigue limit and debris contamination.

Condition-Based Maintenance. The objectives of condition-based maintenance are to reduce total maintenance costs, avoid unplanned equipment outages and even catastrophic failures, and reduce the number of the costly scheduled maintenance activities based on bearing service life. Key factors for condition-based maintenance consist of defect detection, severity assessment, and life prediction. For bearing defect detection, widely used methods include: the high frequency resonance technique (HFRT) using the envelope technique and its spectrum; short time energy and wavelet transform using the sliding window technique; band-weighted demodulation; auto regressive models; time-frequency distribution; and statistical parameters such as peak and root mean square (RMS) or their normalized values, Crest factor, Kurtosis, shape factor, and impulse factor [5, 6].

Barkov and Barkova [7] proposed a modulation factor of a random vibration signal to determine the severity levels of defected bearing. Researchers at Georgia Tech [8] proposed a diagnostic method based on the hypothesis testing procedure, which permits defects to be assessed under bearing conditions that have not been previously assessed. They also proposed an analytical model that integrates a dynamic model and finite element model to relate measurable vibration signals to the geometrical structure of tapered roller bearings, the rotating speed of the system, and the defect width of the outer race.

For bearing RUL prediction, researchers have also suggested the following techniques: deterministic defect propagation model and adaptive prognostics; stochastic modeling of defect propagation; stiffness-based life-signal prognostic model based on vibration response analysis and damage mechanics; bearing life prediction based on accelerated life testing; and expert estimation that represents an array of diagnostic symptoms for each type of defect and the threshold values for defects of different severity [8].

APPROACH: Figure 1 shows the overall concept of the approach we are implementing to predict RUL in rolling element bearings. It consists of bearing condition indices calculation and trending from vibration and acoustic emission signals; training a pattern recognition technique based on a neural network; and implementing the trained network for RUL prediction from current condition indices.

Bearing Condition Indices. Many signal processing techniques can be used to calculate condition indices. However, the following techniques were chosen because they provide good trends of bearing condition. Peak and Root-Mean-Square (RMS) directly reflect the energy level of the vibration signals, and since the localized bearing defects result in structural vibrations, these two parameters are generally used to indicate the presence and

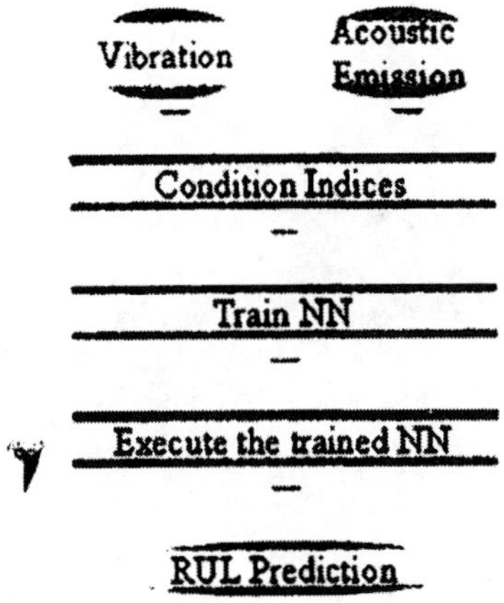

Figure 1. Overall Concept of Foster-Miller Approach to Bearing RUL Prediction

severity of the localized bearing defects. Crest factor and Kurtosis are less dependent on the vibration level but are sensitive to the spikiness of the vibration signals. As such, they can provide early indication of significant changes in the vibration signals. However, as the localized defects spread around the bearing races and balls, the vibration signals become more random and the values can fall back to the lower levels. Shape factor and Impulse factor are the normalized RMS and peak to averaged value of the absolute signal, and M6A is the sixth normalized statistical moment. The aforementioned three parameters have similar properties with Crest Factor and Kurtosis [9]. The ratios of peak to mean in the regions that involve the spectral energies at bearing defect frequencies and their surroundings are calculated using the envelope of a vibration signal and a spectrum of the enveloped signal in sequence as shown in Figure 2 and can be used to locate defects and determine the severity levels.

Bearing condition indices can be affected by operating conditions, applied load, speed, temperature, and lubrication. Therefore, to apply condition indices extracted from the accelerated life tests for a machine that operates with normal speeds and loads, one must find the correlations between the test-rig-generated indices and the machine's operating profile and then adjust them to fit this profile. Figure 3 shows an example of mapping the condition indices acquired at variable load conditions to those at our target accelerated load, 5,000 lb. To obtain compensation data, a defect-free bearing was run in the test rig. Data were taken at various loads and the condition indices for each load were calculated. Then a neural network was applied to approximate a function representing the compensation data to express the relationships between condition indices and the load points. For this process, a generalized regression neural network [10] was used. The condition indices at the target and measured loads were estimated using this function. Subsequently, ratios of these indices were used to obtain the compensated indices at the target load. The first and second graphs in Figure 3 show the measured load variations around 2,500 to 5,000 lb and their corresponding RMS values, respectively. The third curve shows the result with compensation. For the compensated RMS, it is evident that the effect of load variations on the RMS values has been smoothed. This prepares the indices for use in the following analysis.

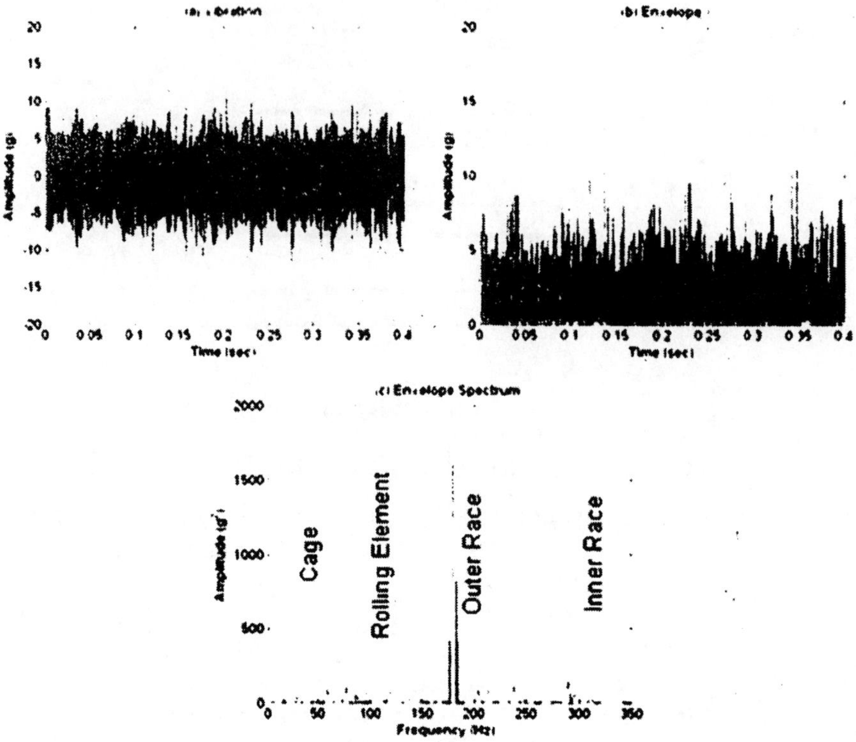

Figure 2. Time Waveform of a Bearing Vibration Signal, its Envelope, and a Spectrum of the Enveloped Signal

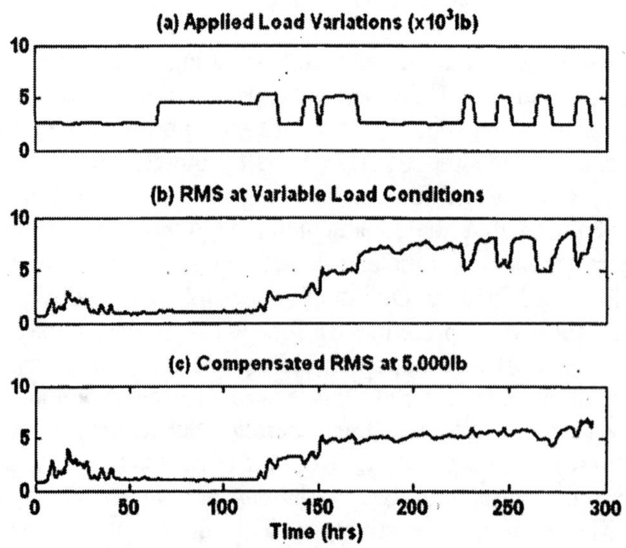

Figure 3. Compensating Condition Indices to account for Variation in Operating Conditions

Feature Extraction and Selection. In most cases, pattern recognition methods benefit greatly by inputting only a few features, with those features principally containing all of the relevant change information. This enables a reduction in the feature space generated by the method. It is desirable to make a feature space 2-dimensional because it can be easily perceived and analyzed. A goal therefore is to extract two parameters from the aforementioned bearing condition indices. At present, we have chosen Kurtosis as one of the features because of its insensitivity to operating conditions (as shown in Figure 4). The second feature we are evaluating is the Fusion Factor [6], the logarithmic value of a feature that combines RMS, Crest factor, Kurtosis, impulse factor, and the RMS value of the undamaged bearing data. The resulting vector defining the feature space is given by the 2×1 matrix of equation 2.

$$\left[\begin{matrix} Kurtosis \\ log\left(\frac{RMS}{RMS_0}\cdot\frac{CrestFactor}{Kurtosis}+ImpulseFactor\right)\end{matrix}\right] \qquad (2)$$

Bearing Severity Assessment. A feed-forward neural network (FFN) [11] is employed to assess the levels of bearing severity. To train the network, we used the trends of compensated condition indices with severity levels categorized in five steps: (1) "wear-in", immediately after bearing installation; (2) "normal" operating conditions; (3) "incipient defect", as defects initiate; (4) "medium defect", as defects propagate; (5) "severe defect", as the bearing gets close to the end of its service life. From the condition indices measured, one can assess bearing health using the trained NN.

Bearing Defects Localization. A spectral energy detection technique is applied to vibration data and calculates spectral energies at bearing defect frequencies, and the energies are compared with threshold values. When the energy levels exceed the threshold values, it can be stated that the bearing has a defect. We assess the spectral energy at frequencies corresponding to the inner race, outer race, rolling elements, and cage frequencies. As a general rule of thumb, we set the threshold within each band at 5 times the equivalent energy level in a defect-free bearing.

RUL Prediction. Bearing RUL prediction is performed whenever a defect is detected or the severity level becomes an incipient defect as described above. A FFN is chosen for the RUL prediction and trained according to the following steps: (1) input the trends of compensated condition indices with accumulated run time from accelerated run-to-fail tests; (2) determine the network structures and hidden neurons at each layer and their transfer functions and number of iterations, and (3) train the network and obtain weight vectors. From the measured indices, one can estimate an accumulated run time for the real bearing using the trained NN. By assuming that the ratios of predicted end of life for the accelerated test and the lifetime of a machine's bearing (which is calculated from the bearing life equation (equation 1) are proportional, one can estimate the remaining life for the real bearing.

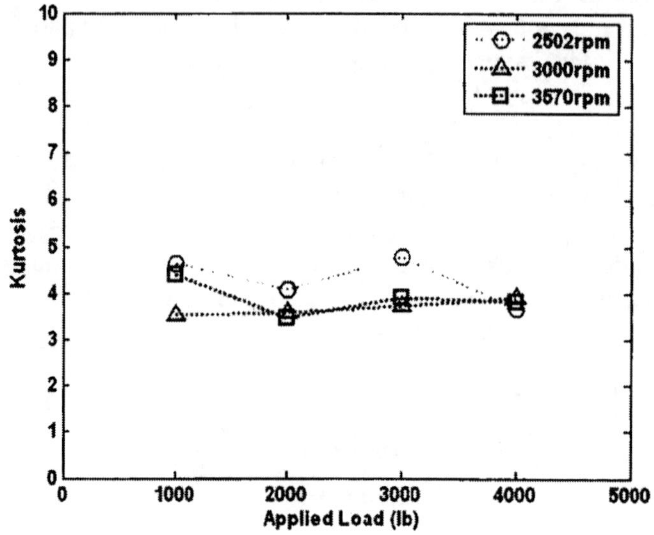

Figure 4. Kurtosis Uniformity for Variations in Rotating Speed and Applied Load

EXPERIMENTAL STUDY: To assess the feasibility of this approach, in-house and field tests are in progress. Initial run-to-fail test data were collected from oil and grease lube bearings on an in-house test rig and used to extract the bearing condition indices and threshold values for bearing defect localization and to train the neural networks for severity assessment and RUL predictions. The threshold values and the trained networks were then applied to subsequent in-house and field tests.

In-House Bearing Test. The bearing test rig was designed to test two bearings simultaneously, at the free end (FE) and coupling end (CE) of the driven shaft. For detailed descriptions of the in-house bearing test rig, test bearing, and data acquisition system, refer to the reference 9.

Bearing Condition Indices. Figure 5 shows the condition indices extracted from the FE bearing with grease lube and no shields and modified to fit the operating conditions in a target bearing. The FE bearing failed after 295 hours operation with defects on the outer race, lubrication deficiencies, and indications of the inner ring spinning on the shaft. It is apparent in the trend of RMS that the FE bearing exhibited typical wear-in and normal patterns categorized in previous sections before defects initiated around 120 hours. The bearing continued to show significant changes in RMS and fusion factor as the bearing progressed towards failure. These types of changes provide important information for trending bearing condition and predicting RUL.

RUL Prediction. As described above, RUL prediction starts automatically as defects emerge. Figure 6 shows the main window of the LabVIEW™ graphical user interface (GUI) used with the in-house test rig. The GUI relates the data acquisition, condition indices calculation, severity assessment, defect localization, bearing RUL prediction, and displays the results. All of these functions are implemented in sequence.

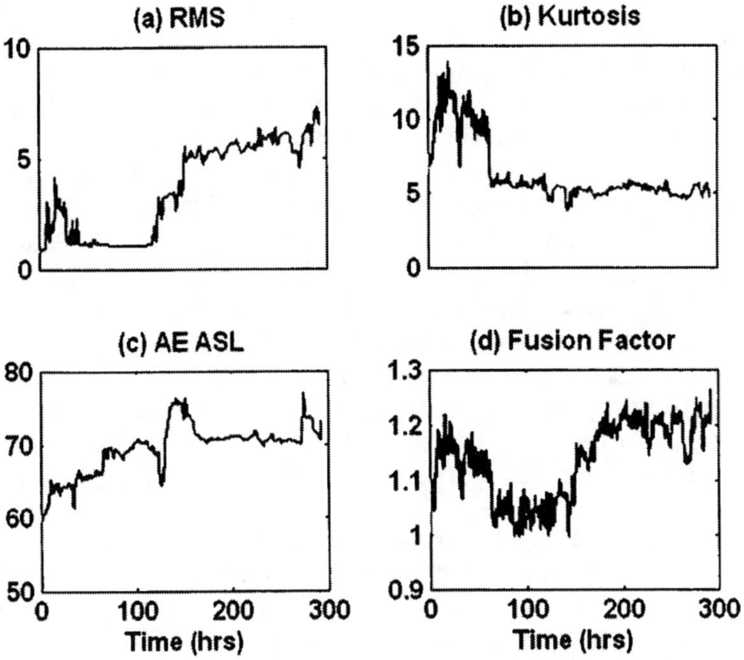

Figure 5. Condition Indices at FE Bearing with Grease Lube and No Shields (Foster-Miller bearing test 4)

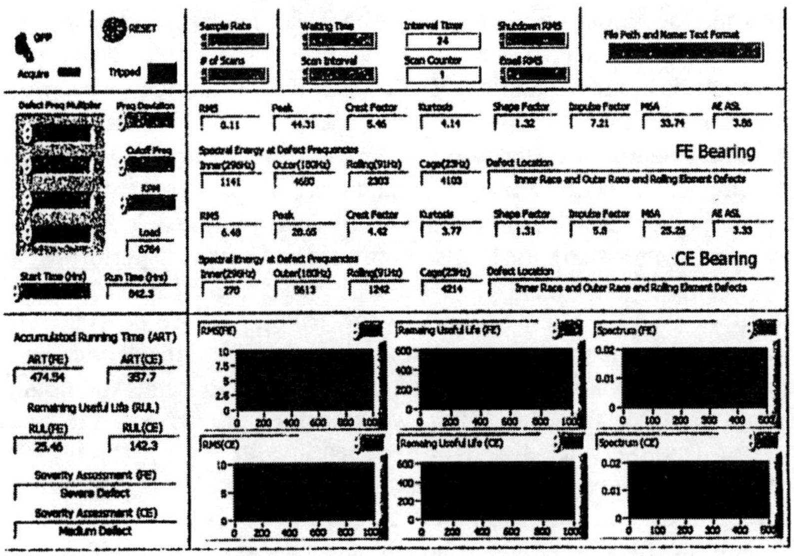

Figure 6. LabVIEW™ Graphical User Interface Main Window (Foster-Miller bearing test 5)

As a first step, the program acquires 20,000 samples of vibration and acoustic emission average signals at sample rates of 50 kHz each. These data are stored. The GUI then executes external MATLAB® scripts, which include signal processing techniques written in MATLAB®, to calculate the condition indices from the raw signal. Spectral energies are obtained from the enveloped vibration signal and its spectrum as described above. The spectral energies are compared with pre-defined threshold values to locate the bearing defects noted earlier. Subsequently, a neural network (also written in MATLAB® and trained using the data from an earlier test) is applied (through the external MATLAB® scripts) for on-line severity assessment and RUL prediction. Results are displayed in the GUI window with numbers, text strings, and trend plots. As the lower center portion of Figure 6 shows, the RUL decreased gradually as the RMS, peak, and spectral energies increased. Several other parameters did not change significantly. When the test rig automatically shut down due to high vibration, the bearings were visually checked and found to be out of lubricant. In this case, it can be concluded that the lubrication loss is a principal factor in determining the service life of the bearing.

Field Test: A cargo supply and crane ship in the MARAD ready reserve fleet, the SS Cornhusker State (T-CAS 6), was selected as a demonstration platform. From available candidates, a small seawater circulating pump and a service air compressor were selected for installation of an engineering prototype version of our system.

Prototype System. As shown in Figure 7, the prototype system consists of an acoustic emission ASL (average signal level) sensor and signal conditioner; an accelerometer; a sensor platform for data acquisition and initial analysis; and an ICP® power supply and wireless transmission to the base station that controls the sensor platform which trends bearing condition, predicts RUL, and displays the results. The water pump and air compressor motor have grease-lubricated bearings. For oil-lubricated bearings (in the air compressor) a particle monitor in the oil drain line is used. The system was installed in August 2004 and began 24/7 observation of the monitored bearings. While the ship is in port, the system can be accessed remotely via a cell phone (data only) and static internet IP address. Throughout the ship's time in port (it deployed in early January 2005 as this document was being prepared), the system hardware and software performed as planned.

RUL Prediction. We trained a neural network using in-house bearing run-to-fail test data and deployed it to the prototype system. Initial data from the service air compressor motor indicated a reasonable RUL prediction (several thousands of hours), and this coincided with the fact that the motor was recently rebuilt and the bearings were relatively new. The motor has the same size bearings as those we ran in the in-house bearing test rig, which provided a good base of data for neural network training. For the seawater pump, the system initially predicted very low RUL values (hundreds of hours or less). Although this bearing exhibits fault frequencies associated with a worn bearing, the bearing has more life remaining than the system predicts. We believe the discrepancy results from the differences in bearing size between the pump (smaller bearings) and the larger rig bearings that form the basis of our NN training experience. Seasonal changes of the pump discharge pressure (for example, the ship's air conditioning was secured for in-port winter conditions, and refrigeration needs are less) can also affect the bearing condition indicators and cause variation in RUL prediction.

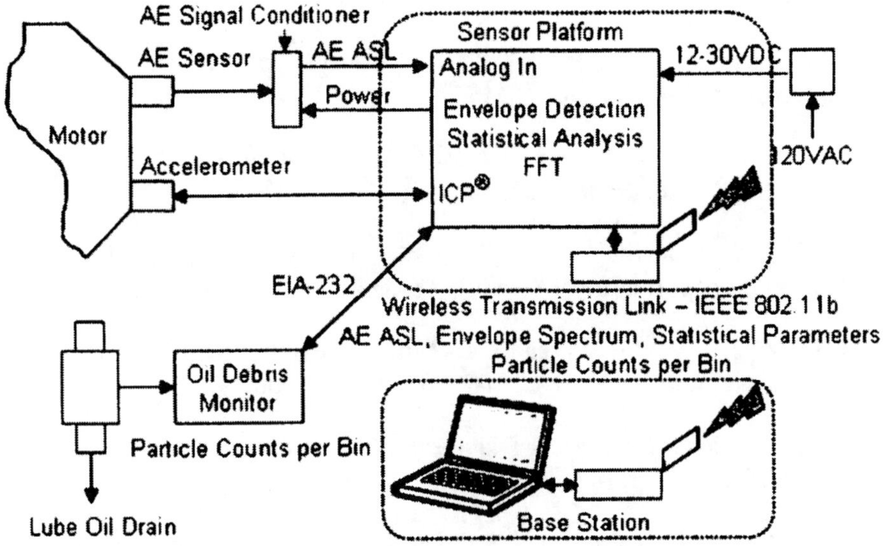

Figure 7. Prototype System for Field Test

CONCLUSIONS: Our work to date holds promise for the feasibility of integrating a pattern recognition technique based on a neural network and condition indicators extracted from bearing vibration and acoustic emission signals for RUL prediction of rolling element bearings. The accelerated run-to-fail test data derives reasonable RUL prediction in the same size of bearings but not in bearings of significantly different size. The wireless sensor platform successfully performs low-level data acquisition functions, extracts specified condition indices using onboard software, and wirelessly transmits the indices over feasible distances in a practical shipboard environment for prognostic evaluation.

FUTURE WORK: Our in-house test rig continues to operate 24/7. We have made preparations to test a bearing closer to the size of the seawater pump bearing and will initiate this test at the next bearing change-out. We anticipate that developing rig data for generic classes of bearing will be sufficient for most bearing types, thus avoiding the infeasible task of rig testing all bearing models. We are also looking further into optimal sets of condition indices for improved RUL prediction accuracy and a generic system that is robust in the variations of operation conditions and combines a rule-based knowledge representation and a neural network implementation.

ACKNOWLEDGEMENTS: This work has been conducted under a Phase II Small Business Innovative Research (SBIR) contract N00178-03-C-3057. Mr. Kisan Pandit is the NAVSEA Technical Point of Contact. Principal subcontractors include Imes Inc. (Norwell, MA) and GasTOPS Ltd (Ottawa ON). The authors are especially appreciative of the support and assistance rendered by the crew of the SS Cornhusker State.

REFERENCES

1. Lundberg, G. and A. Palmgren. "Dynamic Capacity of Rolling Bearings." Acta Polytechnica Mech. Eng. Series, 1, 3, 1947.

2. Lundberg, G. and A. Palmgren. "Dynamic Capacity of Roller Bearings." Acta Polytechnica Mech. Eng. Series, 2, 3, 1952.

3. Weibull, W. "A Statistical Theory of the Strength of Materials." Ingeniors Vetenskaps in Proc. of the Royal Swedish Academy of Eng., 151, 1939.

4. Zaretsky, E.V., J.V. Poplawski, and C.R. Miller. "Rolling Bearing Life Prediction-Past, Present, and Future." NASA/TM-2000-210529, 2000.

5. Li, C. J. and K. McKee. "Bearing Diagnostics." Encyclopedia of Vibration, Vol. 1, Academic Press, 2001, pp. 143-152.

6. Krishnappa, G. and M. Donovan. "Railway Bearing Diagnostics: Laboratory Data Analysis." TP 13472E, National Research Council Canada, Vancouver, BC, 1999.

7. Barkov, A.V. and N.A. Barkova. "Condition Assessment and Life Prediction of Rolling Element Bearings." Sound and Vibration, June and September, 1995.

8. Liang S., T. Kurfess, and S. Danyluk. "Dynamic Metrology as a Bearing Wear Diagnostic." Georgia Tech, www.me.gatech.edu/diagnostics/c3.pdf.

9. Walter, T.J. and H. Lee. "Development of a Smart Wireless Sensor for Predicting Bearing Remaining Useful Life." The 58th Meeting of the Society for Machinery Failure Prevention Technology, Virginia Beach, Virginia, April 25-30, 2004.

10. Wasserman, P.D. "Advanced Methods in Neural Computing." New York: Van Nostrand Reinhold, pp. 155-61, 1993.

11. Hagan, M. T., and M. Menhaj. "Training Feedforward Networks with the Marquardt Algorithm." IEEE Transactions on Neural Networks, 5:6, pp. 989-993, 1994.

REMOTE CONDITION MONITORING USING OPEN SYSTEM WIRELESS TECHNOLOGY

Jeffrey M. Rybak

Oceana Sensor Technologies, Inc.
1632 Corporate Landing Parkway
Virginia Beach, VA. 23454
757-426-3678
jrybak@oceanasensor.com

Abstract: Condition monitoring involves the utilization of sensors and data acquisition equipment to assess the quality of a given process by determining the health of its components. Machinery condition monitoring, which affords one the ability to implement condition-based maintenance (as opposed to time-based maintenance or run-to-failure) assesses the health of critical machinery components, in order to reduce catastrophic process downtime and extend the useful life of the machinery. The advent of robust, open system, spread spectrum radio technology has brought about the ability to reduce unplanned maintenance by continuously monitoring the condition of critical machinery and processes remotely and affordably via LAN or PC direct interface. This paper will look at the value of continuous condition monitoring through remote data access, and the means in which that data can be cost effectively transmitted wirelessly without compromising its integrity. It will provide information on available wireless system technologies, address the suitability of each for various applications and briefly describe product solutions that are commercially available today.

Key Words: Data Acquisition; Prognostics; Remote monitoring; Sensors; Wireless integration

Rotating machinery is a vital part of virtually every manufacturing process. Motors, gearboxes, pumps and compressors are relied upon to operate efficiently in order to maintain a steady stream of production at maximum throughput. It's this dependency upon the machinery used in critical operations that has prompted the evolution of machinery health monitoring. The lifespan of every piece of machinery is limited by the speed at which it is operating, the load to which it is subjected to, the quality of it's components, assembly and installation, the environment in which it is placed and the level at which it's maintained. While all of these factors are extremely important, it's the attention that's placed upon the latter that will dictate how efficiently the machinery performs. It's the incorporation of intelligent maintenance practices, which predict impending failure of critical machinery components like bearings, which prevent costly downtime and excessive overhead due to inventory and costly capital expenditures.

Performing maintenance on a piece of machinery relative to its actual condition as opposed to how many hours it has been in service is referred to as Condition-Based

Maintenance or CBM. This practice, which requires accurate machinery health monitoring techniques, allows machinery to operate to its maximum useful life without being subjected to premature component replacement or unplanned downtime. Machinery health monitoring utilizes sensors to monitor for one or more of many "symptoms" associated with the degradation of rotating machinery. Some of these are temperature, vibration, oil condition, acoustic emission, pressure and motor current. These parameters are typically gathered and analyzed by specialized "predictive maintenance" systems implemented by the end-user within the factory environment. The data are often gathered with portable data collection equipment by a technician who must physically walk from machine to machine. Vibration signals are obtained using a hand-held data collector and then compared to a vibration signature or baseline associated with that specific data collection point. Symptoms of impending bearing or gear failure due to pitting, fretting, spalling, hairline cracks, corrosion and improper load often show up as low level signals across a broad range of frequencies. These indications can then be used to order replacement components in advance without holding excess inventory and to schedule corrective maintenance during off-peak production periods.

A permanent and continuous monitoring system offers several advantages over a periodic route-based collection system. The first is that, by mounting sensors directly on the machinery in a fixed location, consistent and accurate results are provided for valid trending, analysis and decision support. Most importantly, alarm levels can be set so operators can be made aware of potential problems at early stages of development before a condition becomes critical. While data is more readily available in a continuous monitoring / remote data access system, which better assures that the data will actually be processed and utilized, traditional cabling is very expensive, making this solution cost prohibitive to most companies interested in monitoring a large number of points. Cabling unnecessarily tethers equipment to fixed locations, thus reducing flexibility in equipment placement and reorganization. Cabling can also be very expensive to install and maintain in terms of both material and labor costs. New runs, moves, or upgrades easily disrupt operations while cable is accommodated, and re-positioning or upgrading equipment can necessitate completely new runs. Moreover, as the distance between equipment and control or monitoring devices increases, cable run length maximums are quickly exceeded. Cable installation accounts for roughly 2/3 of the total cost of obtaining a channel of data in an industrial environment. At a modest $40/ft., a typical 2-channel vibration monitoring system would cost over $13,000 just to run 100 meters of cable for remote alarm notification, analysis and trending.

Distance and cost limitations associated with wired links surface quickly on sprawling factory floors and in large industrial settings, and running cable to new or relocated equipment can interrupt production. These hardwiring drawbacks have led many to seek a longer range and more flexible alternative in wireless networks. "Wireless Ethernet," for example, the general descriptor applied to wireless links within an Ethernet network, is any over-the-air connection between Ethernet network nodes or devices. As is often the case with new technology applications, there is a wide range of wireless network implementations on the market, and there is no single wireless standard. Wireless Network solutions typically fall into one of two classes of over-the-air protocols: those based on open system standards (like IEEE 802.11 for WLANs and Bluetooth for WPANs) and those based on proprietary protocols designed specifically for a given application. The big difference between the two is cost. The ubiquitous nature of open system solutions lends themselves to economies of scale, offering the consumer a more cost effective solution. Open system wireless data acquisition has greatly enabled the continuous monitoring of critical assets. Wireless systems are now available that reduce that $13+k 2-channel installation cost that we spoke of earlier, to under $2,000 for a savings of over 80%. Wireless network deployment issues can be best understood if placed into the following three service classifications:

- Wireless Personal Area Networking (WPAN)
- Wireless Local Area Networking (WLAN)
- Wireless Wide Area Networking (WWAN)

Wireless wide area networks (WWANs) use various devices—telephone lines, satellite dishes, and radio waves—to service an area broader than that which can be covered by WLANs and WPANs, although typically with lower bandwidth (referring to the amount of data that can be transmitted in a fixed period of time, typically measured in "bps"). WWANs are generally publicly shared data networks designed to provide coverage in metropolitan areas and along traffic corridors. WWANs are owned by a service provider or carrier, wherein data rates are low and charges are based upon usage. Specialized applications are characteristically designed around short, burst messaging. Examples of WWAN technologies are CDPD, ARDIS, GSM and GPRS, and these WWAN technologies are often categorized into "generations" 1G, 2G, 3G, and 4G. While a WWAN system may be practical for mobile phones, pagers and even open field sensor data acquisition, they are generally poor choices for industrial applications due to low bandwidth, spotty coverage and the expense of service contracts. While satellite based systems can be viewed as an augmentation to WWAN services, the cost value of service is often limited to specific applications, most notably those applications that benefit from one-direction broadcasting of content or communication to locations on the globe not serviced by other means. Bi-directional communication using satellites to support high-speed and/or pervasive connectivity is not a very practical approach, so here; the major cellular systems are listed being most representative of WWAN technology.

- **GSM** (Global System for Mobile Communications) variations are used in Europe, Asia, and North America and operate in 900, 1800, and 1900 MHz bands with a typical maximum data rate of 14.4 Kbps.
- **GPRS** (General Packet Radio Service) is a digital mobile phone technology enhancement to GSM, providing data rates to 150+ Kbps. GSM and GPRS are considered rivals to CDMA technologies.
- **CDPD** (Cellular Digital Packet Data) is a data transmission technology developed for use on the 800 to 900 MHz cellular phone frequencies, to transmit data in packets at rates up to 19.2 Kbps. Low cost but slow.

- **CDMA** (Code Division Multiple Access) is a transmission technology that accommodates multiple signals in the same channel ("multiplexing"). It uses direct sequence spread spectrum technology to vary the transmission frequency according to a defined code pattern.
- **1xRTT** (1x Radio Transmission Technology) also known as CDMA2000 increases data transmission rates over existing CDMA networks. Provides 144 Kbps of data and voice.

As stated earlier, wireless wide area networks are not the best choice for continuous monitoring of industrial machinery and processes due to bandwidth limitations, spotty coverage and the monthly cost of service contracts.

The term wireless networking refers to technology that enables two or more computers to communicate using standard network protocols, but without network cabling. Strictly speaking, any technology that does this could be called wireless networking. The current buzzword however generally refers to wireless LANs. This technology, fuelled by the emergence of cross-vendor industry standards such as IEEE 802.11, has produced a number of affordable wireless solutions that are growing in popularity with business and schools as well as sophisticated applications where network wiring is impossible, such as in warehousing or point-of-sale handheld equipment. WLAN systems are designed to supplement and in some cases replace traditional wired-based Local Area Networks. The predominant standards-based WLAN technology being deployed in the United States is based on the IEEE 802.11b standard.

In a basic IEEE 802.3 Ethernet LAN, Category 5 (Cat5) cable connects LAN stations to a hub. In a wireless LAN, Cat5 cable is replaced by a radio channel, connecting stations to wireless Access Points (APs). Each wireless station -- laptop, desktop, or server -- has a radio Network Interface Card (NIC). APs are essentially hubs, outfitted with a radio transceiver, Ethernet uplink, and 802.1d bridging software. Wireless stations transmit to an AP over a shared channel, carved out of the unlicensed 2.4 GHz band. There are two types of wireless networks: Ad-hoc or Infrastructure. An ad-hoc, or peer-to-peer wireless network consists of a number of computers each equipped with a wireless networking interface card. Each computer can communicate directly with all of the other wireless enabled computers. They can share files and printers this way, but may not be able to access wired LAN resources, unless one of the computers acts as a bridge to the wired LAN using special software. (This is called "bridging"). A wireless network can also use an access point, or base station, for an Infrastructure configuration. In this type of network the access point acts like a hub, providing connectivity for the wireless computers. It can connect (or "bridge") the wireless LAN to a wired LAN backbone, allowing wireless computer access to LAN resources, such as file servers or existing Internet Connectivity.

There are two types of access points:

- Dedicated hardware access points (HAP) offer comprehensive support of most wireless features.
- Software Access Points, which run on a computer equipped with a wireless network interface card as used in an ad-hoc or peer-to-peer wireless network.

Approved in 1997, the original IEEE 802.11 standard uses the 2.4 GHz band to provide shared bandwidth at a maximum rate of 1 to 2 Mbps. In 1999, the IEEE approved the 802.11b High Rate (Wi-Fi) amendment, increasing the rate to 11 Mbps. Together, these standards specify WLAN Physical (PHY), Media Access Control (MAC), and Logical Link Control (LLC) layers for fixed wireless broadband access and WLANs.

802.11 is an IEEE standard for wireless local area networks (WLANs) that covers the wireless LAN media access control (MAC) and physical layer specification. 802.11b and 802.11a are extensions of this standard, also referred to as WIFI (Wireless Fidelity), which is an interoperability certification.
802.11b is a well-accepted standard for WLANs optimized for the unlicensed 2.4 GHz band, with speeds up to 11 Mbps when using DSSS.
802.11a is a standard that improves upon 802.11b with support for speeds up to 54 Mbps in the less-crowded 5 GHz band by using OFDM (Orthogonal Frequency Division Multiplexing, which splits a high speed signal into a number of low-speed signals transmitted in parallel, thus more efficiently using bandwidth, but decreasing wireless range.
802.11g is a WLAN standard comparable to 802.11a (uses OFDM for speeds up to 54 Mbps) but operates in the 2.4 GHz spectrum.

WPAN systems have evolved from "cord" replacement technologies. Some examples are; cordless communication between your keyboard and computer and cordless communication between your Personal Digital Assistant (PDA) and your computer. Because of their initial function focus, WPAN wireless implementations to date have been low-powered and offer limited coverage range. The most hyped of all WPAN wireless technologies today is called "Bluetooth®," which got its unusual name in honor of Harald Bluetooth, king of Denmark in the mid-tenth century. *Bluetooth* is a product of the telecommunications and computer industry "*Bluetooth* SIG" and is rapidly gaining wide acceptance throughout the industry. *Bluetooth* is a telecommunications industry specification that describes how mobile phones, computers, and personal digital assistants (PDAs) can be easily interconnected using a short-range wireless connection. Using this technology, users of cellular phones, pagers, and personal digital assistants can buy a three-in-one phone that can double as a portable phone at home or in the office, get quickly synchronized with information in a desktop or notebook computer, initiate the sending or receiving of a fax, initiate a print-out, and, in general, have all mobile and fixed computer devices be totally coordinated. *Bluetooth* requires that a low-cost transceiver chip be included in each device. The transceiver transmits and receives in the

2.45 GHz frequency band that is available globally (with some variation of bandwidth in different countries). In addition to data, up to three voice channels are available, and each device has a unique 48-bit address from the IEEE 802.15 standard. Connections can be point-to-point or multipoint and can lock out other devices selectively, preventing needless interference or unauthorized access to information. The maximum range is typically 10 meters for battery-powered devices, as transmission distance is directly linked to power consumption, and data can be exchanged at a rate of 1 megabit per second (up to 2 Mbps in the second generation of the technology). A frequency-hopping scheme allows devices to communicate even in areas with a great deal of electromagnetic interference, and built-in encryption and verification are provided for security. There are three classes of *Bluetooth* radio: Class 1 - 100 meters, Class 2 - 15 meters, Class 3 - 10 meters. The lowest power radio within the network defines the maximum transmission distance allowed.

WLAN and WPAN systems, i.e. IEEE802.11 (otherwise referred to as WIFI or wireless Ethernet) and IEEE802.15.1 (otherwise known as *Bluetooth*), both utilize spread spectrum technologies for battling interference and noise, but there are differences in the manner that these two systems attack the problem. *Bluetooth* uses a frequency hopping method, while WIFI uses a direct sequencing method. The proper choice of direct sequence or frequency hopping as a spread spectrum technique depends on the actual environment in which the system will be deployed. If there are narrowband interferers of moderate level, then a DSSS system that will completely reject them may be designable. Should there be any large interfering signals, then a DSSS link may completely fail while FHSS is likely to continue operating, even though the interference is not completely rejected.

Direct Sequence (DSSS): designed originally by two vendors to increase the available speed on the wireless network. Divides the available 83.5 MHz spectrum (in most countries) into 3 wide-band 22 MHz channels. Uses an 11 bit spreading code to reduce the possible interference on signals in each wide-band channel. (Fig.1a)

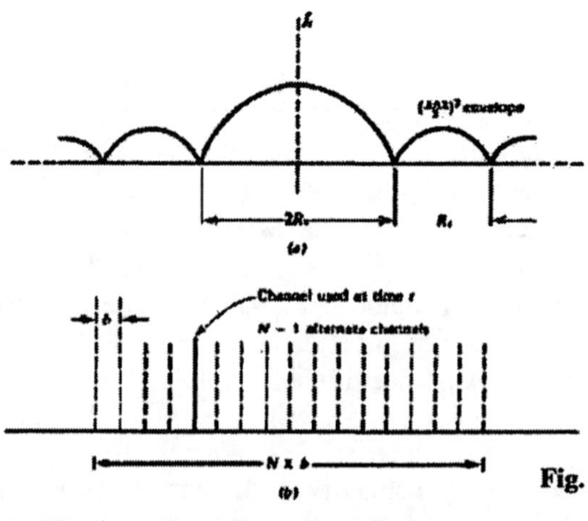

Fig. 1

Frequency Hopping (FHSS): derived from military radio technology where it was designed to be inherently secure and reliable under adverse battle conditions. Divides the available 83.5 MHz spectrum (in most countries) into 79 (or 75) discrete 1 MHz channels (the 4.5 MHz left over provides 'guard bands' at either end of the spectrum), the Radio then hops around these 1 MHz channels in a 'pseudo-random' sequence, using a minimum of 75 frequencies every 30 seconds and using any single frequency for a maximum of 400 milliseconds. (Fig 1b)

Wireless communications obviously provide potential security issues, as an intruder does not need physical access to the traditional wired network in order to gain access to data communications. However, 802.11 wireless communications cannot be received --much less decoded-- by simple scanners, short wave receivers etc. This has led to the common misconception that wireless communications cannot be eavesdropped at all, however, eavesdropping is possible using highly specialized equipment. While it should be noted that traditional Virtual Private Networking (VPN) techniques would work over wireless networks in the same way as traditional wired networks, additional security measures are rapidly evolving to reduce the threat of eavesdropping and data theft.
Bluetooth security utilizes 128bit encryption, authentication and authorization schemes. Flaws in security found on mobile phones have undergone software modifications to improve the level of protection from hackers. 802.11 wireless communications initially addressed security in much the same way via a function called WEP (Wired Equivalent Privacy), a form of encryption, which provides privacy comparable to that of a traditional wired network. If the wireless network had information that should be secure, then WEP should be used, ensuring the data is protected at traditional wired network levels. In addition to 128-bit encryption, the WEP key also provided authentication when a new client accessed the network. (PROBLEM) The key management structure had two likely outcomes: When WEP was used, keys were rarely updated, leaving the network vulnerable to break-in, and WEP itself was rarely used. Needless to say, neither outcome was desirable from a security standpoint, so the 802.11 committee had formed its "i" subcommittee to explore the problems with and correct the WEP protocol. WPA2 is the second generation of WPA security; providing enterprise and consumer WiFi users with a high level of assurance that only authorized users can access their wireless networks. WPA2 is based on the final IEEE 802.11i amendment to the 802.11 standard and is now eligible for FIPS 140-2 compliance.

So, which system is best suited for continuous monitoring of sensor data in an industrial environment? Bandwidth rich WIFI, utilizing direct sequence spread spectrum, which has some limitations in noisy, interference-ridden environments, or bandwidth sufficient *Bluetooth*, utilizing frequency hopping spread spectrum technology, which is very robust and noise immune? In reality, they both have earned there places within a corporate / industrial setting. Many existing infrastructures are suitable for wireless Ethernet interface to sensor data, wherein the environment may also not be too hostile for WIFI's direct sequencing spread spectrum technology limitations. Where the environment is more appropriate for a noise immune frequency hopping spread spectrum system, *Bluetooth* could be used in conjunction with WIFI. While *Bluetooth* and Wireless LAN were earlier labeled as competing technologies, manufacturers have discovered over time that this is not necessarily the case. Some have even gone so far as to develop products that feature both technologies, such as wireless access points. There is now widespread market acceptance of both *Bluetooth* and WLAN, which has lead to a greater incidence of coexistence between the two, most commonly in computer network environments. Coexistence in the unlicensed 2.4 GHz band however, comes with a price. Unlicensed means that competing, or complementary, technologies are free to operate in this frequency band, which has in turn given rise to interference that impinges on the quality of communication. To most users, deterioration of quality may be more apparent in voice centric applications than in data centric applications. For example, one is more likely to be aware of poor sound quality while using a *Bluetooth* headset than of the extent to which data packets must be retransmitted between one's notebook PC and a network access point. The *Bluetooth* industry, through the Bluetooth SIG, has responded by taking measures to reduce interference in environments where multiple wireless technologies coexist. Version 1.2 of the *Bluetooth* Specification, adopted in 2003, includes Adaptive Frequency Hopping, a technique proven to be an effective remedy to the problem of interference in WLAN and similar environments. Also referred to as AFH, this technique can be implemented through various methods, each with its own inherent set of advantages and drawbacks. Ericsson, a leader in the field of *Bluetooth* wireless technology, uses a method well suited for its broad based Bluetooth design solution sold as intellectual property (IP). Ericsson's implementation of AFH is further enhanced through the use of other standard and proprietary techniques, providing excellent audio quality for voice centric applications in the presence of multiple wireless technologies. Note that there are proprietary systems out there that battle noise and interference problems with proprietary radio solutions, and while these systems may solve an immediate application requirement, proprietary solutions will always be more costly than those adopting open system standards. Vendors ensure considerable repeat business by locking customers into their proprietary systems, which don't allow for interface to cost competitive open system standard third party hardware and application software.

As for commercially available wireless data acquisition systems, there are several that incorporate proprietary technology, yet very few that utilize open system standards. Here, however, is a wireless sensing system offered by Oceana Sensor (Fig. 2) that is suitable for industrial applications, employing both open system *Bluetooth* and WIFI technology. The system is based upon the ICHM® 20/20 programmable, wireless data acquisition platform and processing module with on-board diagnostics. The ICHM®'s Digital signal processor executes complex algorithms, Fourier analysis, digital filtering, band level comparisons, and advanced math calculations and is capable of functioning autonomously and communicating reliably in

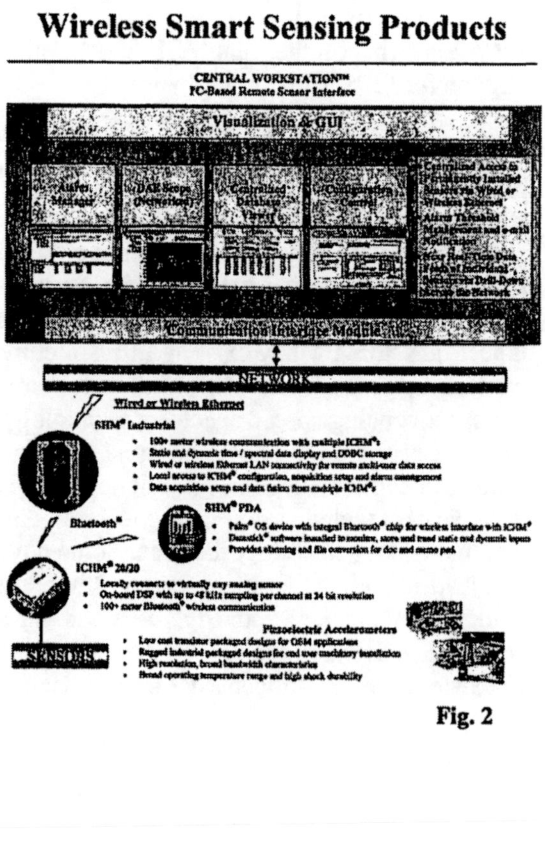

Fig. 2

industrial environments through the use of *Bluetooth* spread spectrum technology. Each module is equipped with as many as 2 dynamic channels (48kHz sampling per channel at 24 bit resolution) and 4 static channels (DC to 3Hz at 12 bit resolution), for local interface to a variety of analog sensors. The ICHM® 20/20 is available with a Class 1, 20dBm (100 meter range) *Bluetooth* radio system or serial output to interface with other wireless systems, including WIFI. An integral WIFI based ICHM is also currently under development.

The ICHM® 20/20 (Fig.3) offers remote process and condition monitoring without the enormous expense of running cable and conduit. It is built on open system architecture for lower installation and life cycle costs, and provides localized health monitoring at the component level. The ICHM® 20/20 is a low power, noise immune device that incorporates Frequency Hopping Spread Spectrum (FHSS) technology and transmits data continuously or only when needed. ICHM®s can communicate directly to *Bluetooth* equipped PDAs and PCs, including the System Health Monitor (SHM®) Industrial PC, for 100+ meter *Bluetooth* interface to multiple ICHMs®

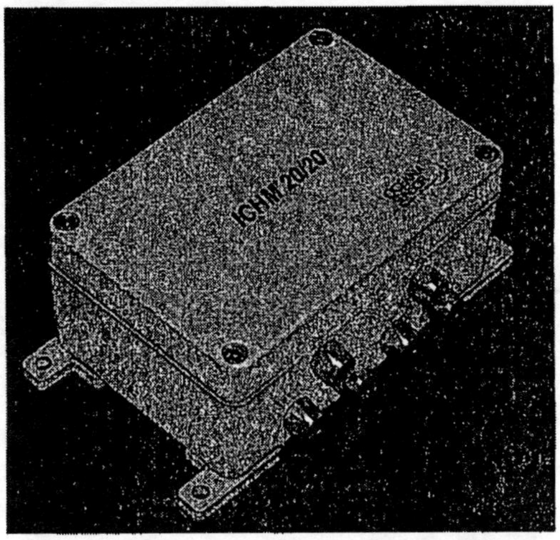

Fig. 3

The SHM® Industrial is a PC-based LAN access point for wired or wireless Ethernet connection to a network gateway that contains logic software for periodically acquiring data from multiple ICHM®s. It's ODBC compliant database provides necessary data for local monitoring or LAN based access with Central Workstation™ and third party software systems. While *Bluetooth* offers a robust yet cost effective wireless interface between a multitude of sensors and the SHM® Industrial, wireless Ethernet can be used for even longer-range communication between stored data and the user. Central Workstation™ provides LAN based access to data and features inherent to the SHM® Industrial, including alarm monitoring and management, waveform analysis, and a sensor setup function, which allows independent configuration of all static and dynamic inputs. Alarms are observed as a Red/Yellow/Green status indication, and individual alarm histories can be obtained for historical trending and reporting purposes. Time stamped data is stored in MS Access ODBC format for viewing at any time. Issues that continue to drive technology forward are:

- Size
    - Self-contained, miniature wireless sensors
- Power
    - Power scavenging features for extracting power from the environment

- OEM Integration
    - Incorporation of wireless, remote monitoring systems into machinery at the point of manufacture
- Open System Architecture
    - Inter-operability amongst system components and data handling applications

As for the future of wireless networking standards:

- 802.11e is being developed to help wireless LANs handle interference--by moving away from it--and to provide better support for those big streaming multimedia files by using error-correction and better bandwidth management.

- The IEEE P802.15.3 High Rate (HR) Task Group (TG3) for Wireless Personal Area Networks (WPANs) is chartered to draft and publish a new standard for high-rate (20Mbit/s or greater) WPANs. Besides a high data rate, the new standard will provide for low power, low cost solutions addressing the needs of portable consumer digital imaging and multimedia applications.

- Zigbee is a standard for wireless personal area networking designed to be simpler and cheaper than other WPANs such as *Bluetooth*, and is aimed at applications with low data rates and low power consumption. It operates in the unlicensed 2.4 GHz, 915 MHz and 868 MHz bands. The data rate is 20 kbps per channel and the transmission range is between 10 meters and 75 meters.

- WiMAX: a marketing name bestowed on a new technology standard that adheres to a certain derivation of the IEEE 802.16 standard, provides wireless broadband Internet connections at speeds similar to Wi-Fi but over distances of up to 30 miles from a central tower.

Once again, open system architecture is at the core of universal acceptance and the economies of scale that offer cost effective industrial wireless solutions.

References:

[1] http://www.bluetooth.com

[2] http://www.ieee.org

[3] Erik Strom, Tony Ottosson, Arne Svensson (2002), *An Introduction to Spread Spectrum Systems.*

[4] F.L. Lewis (2004), *Wireless Sensor Networks.*

[5] Earl McCune (2000), *DSSS vs. FHSS narrowband interference performance issues*

THE USE OF ACOUSTIC EMISSION FOR ESTIMATION OF BEARING DEFECT SIZE

D. Mba

School of Engineering, Cranfield University, Bedfordshire. MK43 0AL, UK
E-mail : d.mba@cranfield.ac.uk

Abstract: Vibration monitoring of rolling element bearings is possibly the most established diagnostic technique for rotating machinery. The application of Acoustic Emission (AE) for bearing diagnosis is gaining ground as a complementary diagnostic tool, however, limitations in the successful application of the AE technique have been partly due to the difficulty in processing, interpreting and classifying the acquired data. Furthermore, the extent of bearing damage has eluded the AE diagnostician. The investigation reported in this paper was centered on the application of the Acoustic Emission technique for identifying the size of a defect on a radially loaded bearing. An experimental test-rig was designed such that defects of varying sizes could be seeded onto the outer race of a test bearing. It is concluded that AE can provide an indication of the defect size, thus allowing the user to monitor the rate of degradation on the bearing.

Keywords: Acoustic Emission, bearing defect, condition monitoring, defect size, vibration analysis.

Introduction

Acoustic emissions (AE) are defined as transient elastic waves generated from a rapid release of strain energy caused by a deformation or damage within or on the surface of a material [1]. In this particular investigation, AE's are defined as the transient elastic waves generated by the interaction of two surfaces in relative motion. The interaction of surface asperities and impingement of the bearing rollers over the seeded defect on the outer race will generate AE. Due to the high frequency content of the AE signatures typical mechanical noise (less than 20kHz) is eliminated. There have been numerous investigations reported on applying AE to bearing defect diagnosis [2,3,4,5,6,7,8,9,10,11]. This paper investigates the relationship between the defect size and AE activity.

Experimental Test Rig and Test Bearing

The bearing test rig employed for this study had an operational speed range of 10 to 4000rpm with a maximum load capability of 16kN via a hydraulic ram. The test bearing employed was a Cooper split type roller bearing (01B40MEX). The split type bearing was selected as it allowed defects to be seeded onto the races, furthermore, assembly and disassembly of the bearing was accomplished with minimum disruption to the test sequence, see figure 1.

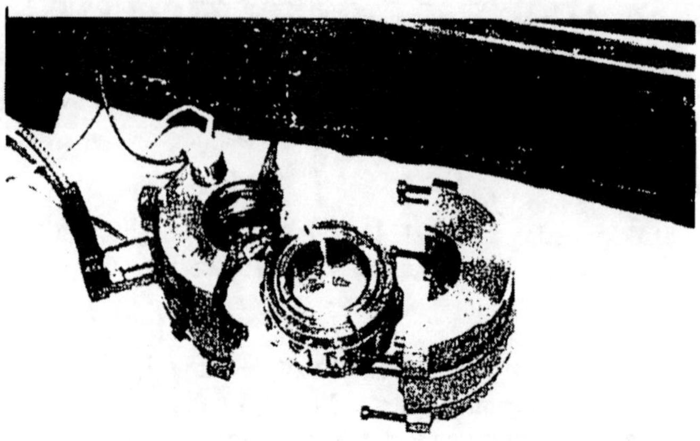

Figure 1 Bearing Test Rig

Data Acquisition System and signal processing

The transducer employed for AE data acquisition was placed directly on the housing of the bearing, see figure 1. A piezoelectric type AE sensor (Physical Acoustic Corporation type WD) with an operating frequency range of 100 kHz – 1000 kHz was employed. This acquisition card provided a sampling rate of up to 10MHz with 16-bit precision giving a dynamic range of more than 85 dB. In addition, anti-aliasing filters (100 KHz to 1.2MHz) were built into the data acquisition card. A total of 256,000 data points were recorded per acquisition (data file) at a sampling rates of between 8MHz and 10MHz, dependent on simulation. Twenty (20) data files were recorded for each simulated case; see experimental procedure. The defects were seeded with an engraving machine.

Experimental procedure

The aim of this programme is to establish a correlation between AE activity with increasing defect size. This was accomplished by controlled incremental defect sizes at a fixed speed. For this particular test programme, two experiments (E1 and E2) were carried out to authenticate observations relating AE to varying defect sizes. Each experiment included seven defects of different lengths and widths, see table 1. A sample defect is shown in figures 2 and 3. In experiment-1 (E1), a point defect (D1) was increased in length in three steps (D2 to D4) and then increased in width in three more steps (D5 to D7). However, in experiment-2 (E2), a point defect was increased in width and then in length interchangeably from D1 to D7. Both experiments were run at 2000 rpm and at a load of 4.4kN. The AE sampling rates for the first and second experiments are 8MHz and 10 MHz respectively. The seeded defects resulted in material protrusions above the mean surface roughness, see figure 2.

Table 1 **Notation for test programme-2 with seeded defect dimensions**

	Experiment 1		Experiment 2	
	Defect Size, (width x length) mm		Defect Size, (width x length) mm	
Defects of Different Sizes	E1-D0	No defect	E2-D0	No defect
	E1-D1	0.85 x 0.85 mm	E2-D1	0.85 x 1.35 mm
	E1-D2	1 x 2.95 mm	E2-D2	2. x 1.35 mm
	E1-D3	1 x 7.12 mm	E2-D3	2 x 4 mm
	E1-D4	1 x 15.83 mm	E2-D4	4 x 4 mm
	E1-D5	3.98 x 15.83 mm	E2-D5	8 x 4 mm
	E1-D6	8.66 x 15.83 mm	E2-D6	13 x 4 mm
	E1-D7	13.6 x 15.83 mm	E2-D7	13x 10 mm

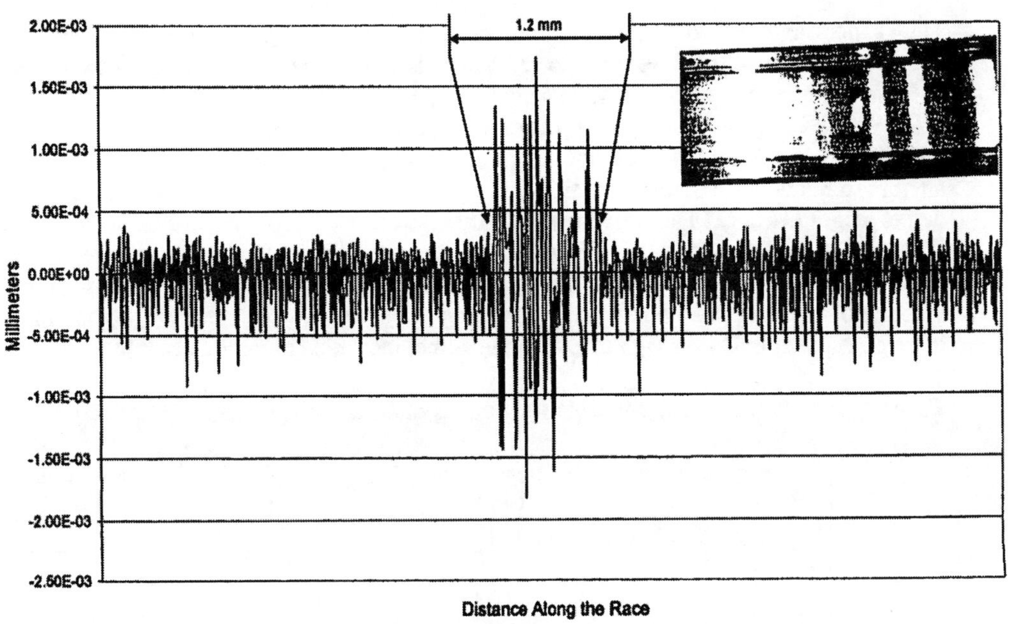

Figure 2 **Surface profile of a line defect condition**

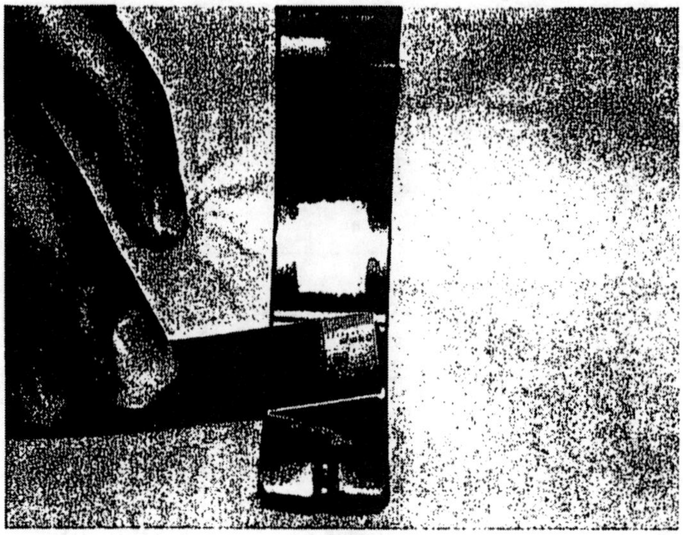

Figure 3 The largest seeded defect, E1-D7, 13.6 x 15.83 mm

Observations of experiment-1 (E1)

This experiment included seven defect conditions; Dl to D4 had a fixed width with increasing length while defects D4 to D7 had a fixed length with increasing width; see table 1. From observations of the AE time waveforms, AE bursts were clearly evident from defects D4 to D7, see figure 4. The x-axis in these figures corresponds to one shaft revolution at 2000rpm. For defects D4 to D7 the width of associated defect AE burst was measured in an attempt to relate the AE burst duration to the defect size. Interestingly, the bursts with equal defect length D4 to D7 had near identical burst durations, see figure 5. For the other defects (Dl to D3), the burst duration could not be separated from underlying background noise, most probably due to the size of the seeded defect. Also it was observed that the ratio of amplitude of the AE burst to underlying operational background noise increased incrementally from D4 through to D7, as the defect increased in width, see table 2.

Table 2 Burst to noise ratio's for defects of fixed defect length (15.8mm)

Defect	Burst duration (second)	Burst amplitude (volt)	Noise amplitude (volt)	Burst to noise ratio
D4	0.0061	0.13	0.09	1.4:1
D5	0.0056	0.18	0.09	2.0:1
D6	0.0056	0.33	0.11	3.0:1
D7	0.0056	0.46	0.10	4.6:1

Two conclusions can be drawn, firstly, increasing the defect width increased the ratio of burst amplitude to operational noise (i.e., the burst signal was increasingly more evident above the operational noise levels, see figure 4). Secondly, it was deduced that increasing the defect length increased the burst duration. To confirm this, a second experiment was performed utilizing the same running conditions but with different defect size combinations. Furthermore, the second test was undertaken to ensure repeatability, and a new bearing of identical type to that used in experiment-1 was employed.

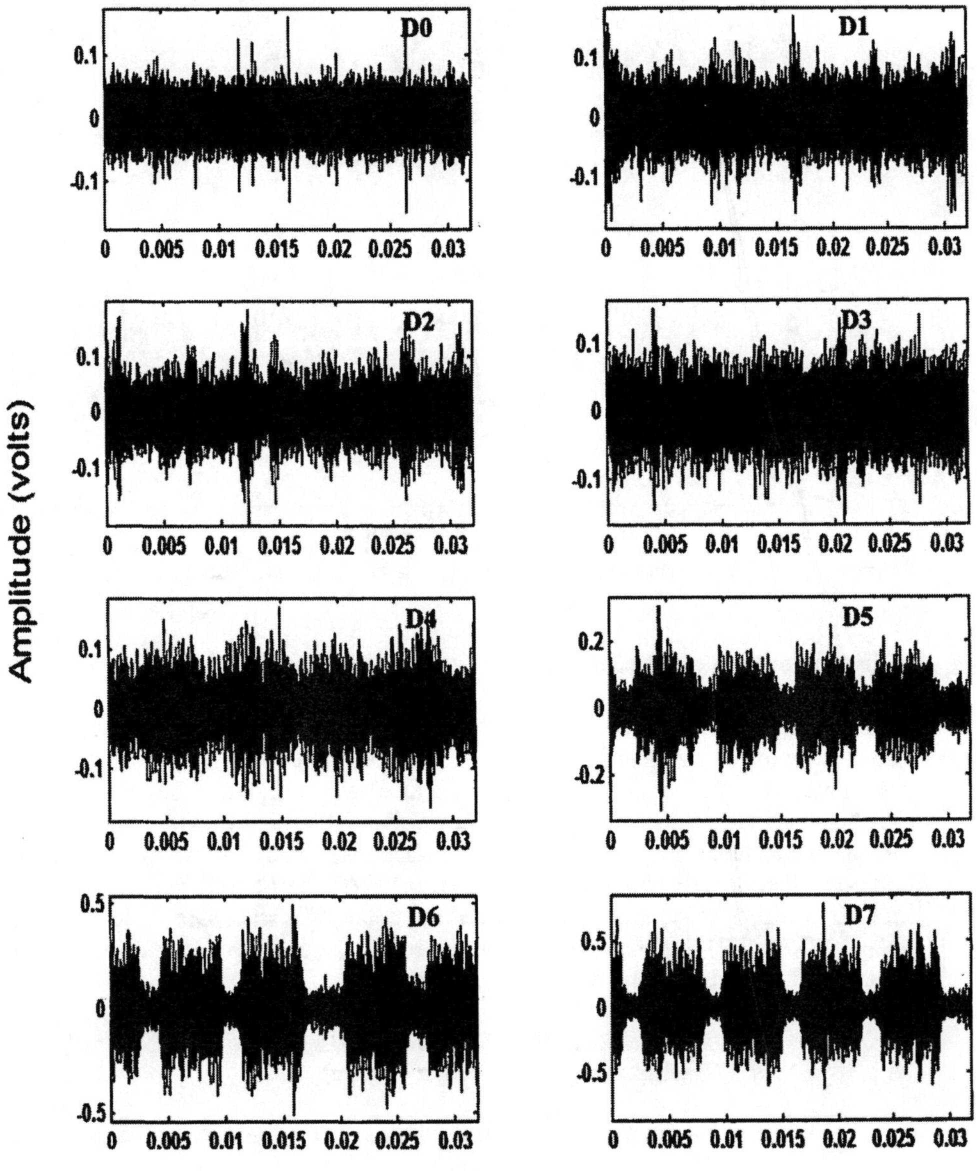

Figure 4 Sample AE time wave forms for defects D0 to D7 (Experiment-1)

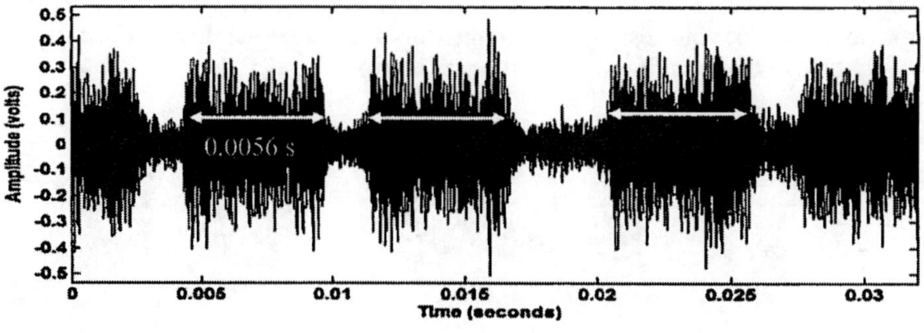

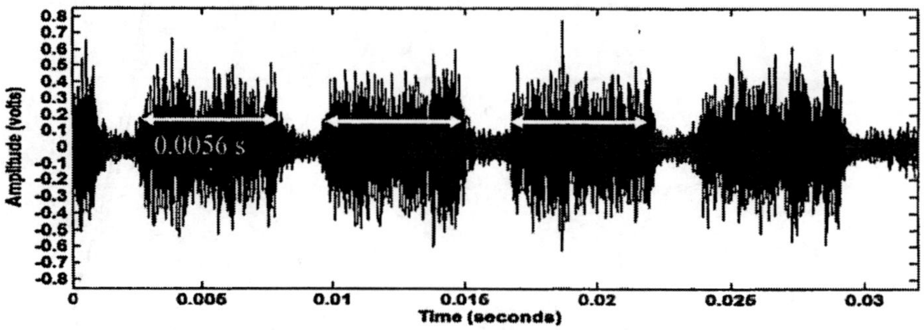

Figure 5 **Burst duration for defect D6 & D7 (Experiment-1)**

Observations of experiment 2 (E2):

The difference between this experiment and that reported in the previous section is two fold. Firstly, the defect size and arrangement of the seeded defect progression was different from experiment-1 and secondly, the AE data captured was sampled at 10 MHz. Observations of the bursts durations for defects D1 to D7, see figure 22, identified that for defects D3 to D7, the burst duration was discernable; these defects had widths of at least 2 mm. It was also observed that the AE bursts for defects D3 to D6 (figure 6) were similar; these defects had the same length of 4 mm. Clearly when the defect was increased in length from 4mm to 10mm (D6 to D7) the burst duration increased dramatically, see figures 6, 7 and 8. Table 3 details information on burst duration, amplitude, etc.

Table 3 **Burst to noise ratio's for defects with a fixed length (4mm) and a defect (D7) with an increased length (10mm)**

Defect	Burst duration (second)	Burst amplitude (volt)	Noise amplitude (volt)	Signal to noise ratio
D3	0.0018	0.17	0.10	1.7:1
D4	0.0018	0.24	0.09	2.7:1
D5	0.0019	0.32	0.10	3.2:1
D6	0.0019	0.43	0.11	3.9:1
D7	0.0036	0.42	0.11	3.8:1

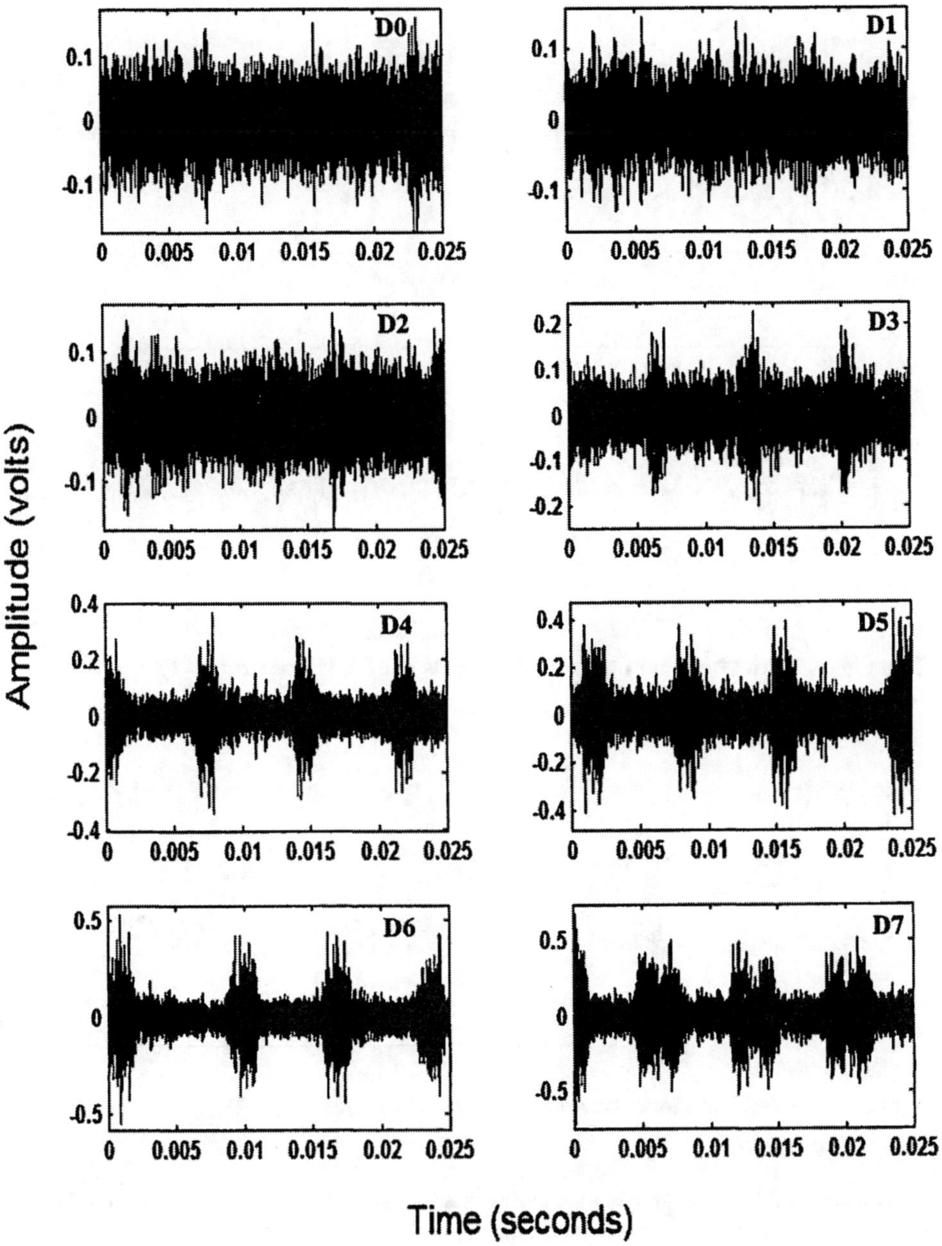

Figure 6 Sample AE time wave-forms for defects D0 to D7 (Experiment-2)

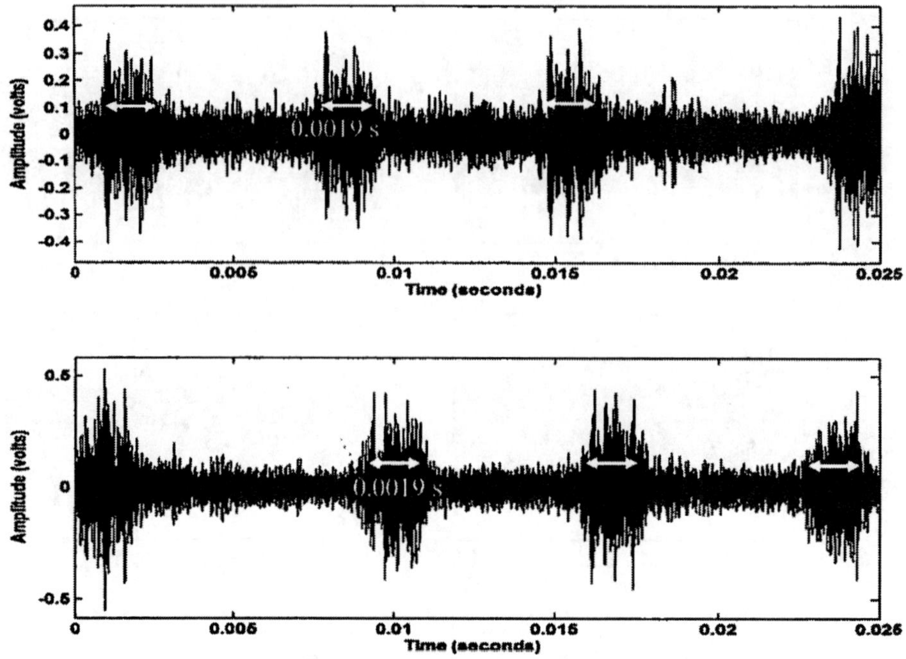

Figure 7 AE waveform bursts for defects D5-D6 (Experiment-2)

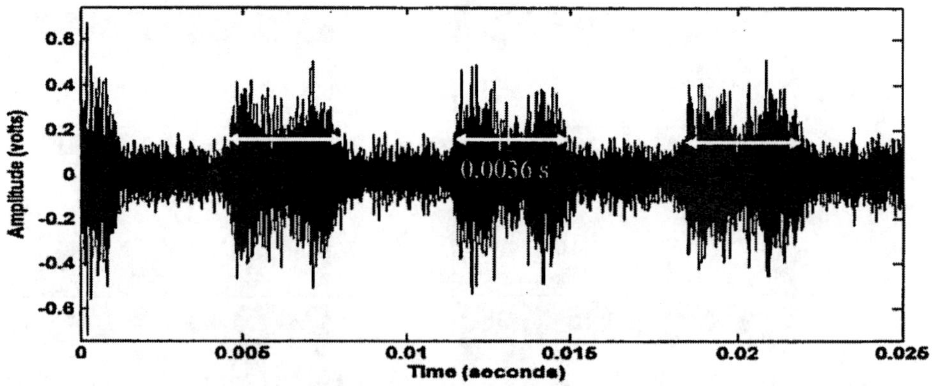

Figure 8 AE waveform burst for defect D7 (Experiment-2)

A summary of AE burst duration for experiments-1 and -2 are detailed in table 4. From table 4 a linear relation between the burst duration and the defect length is noted.

Table 4 Defect length and width vs. burst duration from experiments 1 and 2

Exp.	Defect	Length in mm	Width in mm	Burst Duration in seconds
2	D3 to D6	4	4, 8 and 13	0.00185
2	D7	10	13	0.00360
1	D5 to D7	15.83	4, 9 and 14	0.00564

Discussions and conclusions

The source of AE for seeded defects is attributed to material protrusions above the surface roughness of the outer race. AE transients associated with the defect frequency were observed. The relationship between defect size and AE burst duration is a significant finding. In the longer term, and with further research, this offers opportunities for prognosis. AE burst duration was directly correlated to the seeded defect length (along the race in the direction of the rolling action) whilst the ratio of burst amplitude to the underlying operational noise levels was directly proportional to the seeded defect width.

References

1. Mathews, J. R. *Acoustic emission*, Gordon and Breach Science Publishers Inc., New York. 1983, ISSN 0730-7152.
2. Roger, L. M., The application of vibration analysis and acoustic emission source location to on-line condition monitoring of anti-friction bearings. *Tribology International*, 1979; 51-59.
3. Mba, D., Bannister, R.H., and Findlay, G.E. Condition monitoring of low-speed rotating machinery using stress waves: Part's I and II. Proceedings of the Instn Mech Engr 1999; 213(E): 153-185
4. N. Jamaludin, Dr. D. Mba, Dr. R. H. Bannister Condition monitoring of slow-speed rolling element bearings using stress waves. Journal of Process Mechanical Engineering, I Mech E. Pro. Inst. Mech Eng., 2001, 215(E):, Issue E4, 245-271.
5. Yoshioka T, Fujiwara T. New acoustic emission source locating system for the study of rolling contact fatigue, *Wear*, 81(1), 183-186.
6. Yoshioka T, Fujiwara T. Application of acoustic emission technique to detection of rolling bearing failure, *American Society of Mechanical Engineers*, Production Engineering Division publication PED, 1984, 14, 55-76.
7. Hawman, M. W., Galinaitis, W. S, Acoustic Emission monitoring of rolling element bearings, *Proceedings of the IEEE, Ultrasonics symposium*, 1988, 885-889
8. Holroyd, T.J. and Randall, N., (1993), Use of Acoustic Emission for Machine Condition Monitoring, *British Journal of Non-Destructive Testing*, 1993, 35(2), 75-78.
9. Holroyd, T. Condition monitoring of very slowly rotating machinery using AE techniques. *14th International congress on Condition monitoring and Diagnostic engineering management (COMADEM'2001)*, Manchester, UK, 4-6 September 2001, 29, ISBN 0080440363
10. Bagnoli, S., Capitani, R. and Citti, P. Comparison of accelerometer and acoustic emission signals as diagnostic tools in assessing bearing. *Proceedings of 2nd International Conference on Condition Monitoring*, London, UK, May 1988, 117-125.
11. Morhain, A, Mba, D, Bearing defect diagnosis and acoustic emission Journal of Engineering Tribology, I Mech E, Vol 217, No. 4, Part J, p 257-272, 2003. ISSN 1350-6501.

WIRELESS SENSING NODES POWERED BY PIEZOELECTRIC VIBRATION ENERGY HARVESTING DEVICES

Ken Deng and Clay Shipps
Wilcoxon Research, Inc.
21 Firstfield Road, Gaithersburg, MD 20878
Phone: (301) 216-3017, Fax: (301) 330-8873
kendeng@wilcoxon.com

Abstract: Wireless sensor networks are gaining attention and becoming a favorable solution in a myriad of applications. However, cable supplied power to such a system may negate its advantages. Though batteries can be used, the time required to change the batteries can be costly. It has been long desired that a truly wireless sensor could be powered by the ambient energy in its environment. Light, heat, and kinetic energy are the most likely ambient energy sources. Of these energy types, only vibration energy is ubiquitous and has a sufficiently energy level in the places where the wireless sensors will be mounted. In this work, we have successfully developed a high efficiency, piezoelectric vibration-energy harvesting device. In addition, this energy-harvesting device is integrated with a low power, wireless-sensing system to provide a fully self-sustained wireless sensing node.

It is well known that the highest strain energy level of a mechanical structure can be reached when it's resonating. Therefore, the harvesting device is designed to work around its resonance. Furthermore, in order to accommodate different vibration signals, a 2DOF piezoelectric, harvester with wider peak range is devised. A matched harvesting circuit is designed to efficiently accumulate and store the electric energy generated by the piezoelectric element. The conversion efficiency of our harvester has reached an unprecedented level of over 50%.

Key Words: Energy scavenge, power transducer, self-powered, wireless sensing

Introduction: As the advances in wireless and microelectromechanical systems (MEMS) technology, the wireless sensing systems have been penetrating a variety of application fields and growing rapidly. Different wireless transmitting techniques have been developed and each focuses on a specific application. Table I summarizes the three most popular wireless standards in today's market.

Most wireless devices carry their own power supply cable due to high power consumption, which makes the 'wireless' system less attractive. One trend is to reduce the power consumption level during transmitting and receiving period [1]; the emerging ZigBee technology is the result of this effort. As the power requirement is lowered, batteries can power the wireless sensing systems. This is a step forward to achieve a fully autonomous sensing node, however, batteries still present problems. Since batteries have limited lifespan, they still require maintenance. Most wireless sensors are installed in remote locations or on the moving objects (where conventional, cable connected sensors are hard to be installed). So, replacing batteries becomes difficult and costly. Ideally, it is desired that a wireless sensor can scavenge energy from its surroundings and support itself. If this becomes true, the power unit of the wireless sensor will be constantly replenished from the environment and the lifespan of the unit could be infinite. In recent years, tremendous research and development efforts have been devoted toward this goal [2]. Great progress has been made in the area of energy harvesting circuit design and control schemes that facilitate the electric energy flowing efficiently from the transduction element to the electric storage element. Regarding the energy transducer design, which is the key to converting environment energy into electric energy, no high efficiency harvester has been reported to date. Most published power harvesting devices suffer from low energy conversion efficiency, are not easy to implement, and are at a high cost. In this paper, we present a newly developed piezoelectric vibration power harvester. To our knowledge, it has the highest energy conversion efficiency to date. It is compact in design and can be easily installed on machines or structures. We have demonstrated that under small vibration levels the energy generated by our harvester is sufficient to support a low power wireless accelerometer.

Table I. Wireless Standard Comparison

	ZigBee 802.15.4	Bluetooth 802.15.1	Wi-Fi 802.11b
Application Foucus	Monitoring & Control	Cable Replacement for Portable Devices	Web, Video, Email
System Resource	28 kbytes	250 kbytes	1 Mbytes+
Battery Life (Days)	> 100	1 to 7	0.1 to 5
Nodes per Network	255	8	30
Bandwidth (kbytes/sec)	250	1	11,000+
Range (Meters)	-30	-10*	1-100
Success Metrics	Reliable Low Power Cost Effective	Cost Effective Convenience	Speed Flexibility
Usage Status	Toddler	Teenage	Teenage

*Without Power Amplifier   Source: Helicomm, Inc.

Available Environmental Energy Sources: The most available energy sources in the environment are solar, acoustic, kinetic (vibration, or movement), and thermal energy. Table II lists all the power density levels of these energy sources side by side with the battery power density for comparison purpose. Apparently, among these environmental energy sources, vibration energy has the strikingly highest energy density, and it is comparable to that of a fuel cell battery. However, this energy is in the mechanical domain, and can not be directly used by electronic devices. If one can efficiently convert and store this energy into electrical energy, the application potential of vibration energy harvesting will be enormous. In this study, we attempt to develop a high efficient vibration energy harvesting device as well as the electronic extracting and storage circuit, and the power stored is sufficient to support a wireless sensor.

Table II. Comparison of environmental available energy forms [3]

Energy Source	**Power Density (μW/cm³)**
Solar	**6** (office desk)
Acoustic	**0.96** (100 dB)
Temperature	**15** (10°C gradient)
Vibration	***100****
Battery (Lithium)	**8** (5 year lifetime)
Fuel Cells (methanol)	**112** (5 year lifetime)

*At 0.15g vibration level, and based on the efficiency of our harvester.

Vibration Power Harvesting Device Design and Modeling: Piezoelectric materials are a medium by which kinetic energy can be converted into electrical energy. The properties of the piezoelectric materials allow them to produce electrical energy when a force is applied to the material and vice versa. Based on these properties, this type of material was selected as the conversion material in our vibration energy-harvesting device. To help achieve a high energy transfer efficiency, the piezoelectric energy-harvesting device needs to use the most efficient piezoelectric material available, and in its most efficient excitation mode. Lead Zirconate Titanate [PZT] ceramic is a well-established piezoelectric material that has been used for years in transducer applications. However, a new piezoelectric material, Lead Magnesium Niobate/Lead Titanate [$Pb(Mb_{1/3}Nb_{2/3})O_3$-$PbTiO_3$], known as PMN-PT crystals, has recently emerged as a much more efficient material. Table III shows typical piezoelectric material properties for both PZT-4 and PMN-PT in three different excitation modes. To date, all the piezoelectric energy harvesting devices reported in the literature utilized a flexural bending mode (k_{31}) piezoelectric structure resulting in total conversion efficiencies below 10%. Since the compression mode (k_{33}) or the shear mode (k_{15}) has a much higher piezoelectric coupling factor, we propose using either modes for our vibration energy harvesting device.

Table III. Typical properties of piezoelectric materials

Parameters / Material	**Symbol**	PZT-4	PMN-PT
Relative Dielectric Constant	ε_{33}, ε_{31}	1350	4500
Relative Dielectric Constant	ε_{15}	1300	10000
Dielectric Loss Factor	tan δ	0.3%	0.3%
Compression Piezoelectric Coefficients (pC/N)	d_{33}	300	2000

Flexural Piezoelectric Coefficients (pC/N)	d_{31}	-115	-900
Shear Piezoelectric Coefficients (pC/N)	d_{15}	580	4500
Piezoelectric Coupling Factor	k_{33}	0.58	0.91
Piezoelectric Coupling Factor	k_{31}	-0.34	-0.55
Piezoelectric Coupling Factor	k_{15}	0.71	0.95

It is known that the highest strain energy of a mechanical structure is reached when it's resonating. To take the advantage of amplification effect of resonance, the harvesting device is designed to work at its resonating condition. Implicitly, all the energy in a resonating system comes from its attached substrate. Figure 1 depicts the dynamic model of a single-degree-of-freedom (SDOF) mechanical system. The salient feature of this SDOF harvester design is that the piezoelectric transduction element is integrated with the spring element K. Through this design, the mechanical potential energy is translated in the piezoelectric element that converts it into electrical energy.

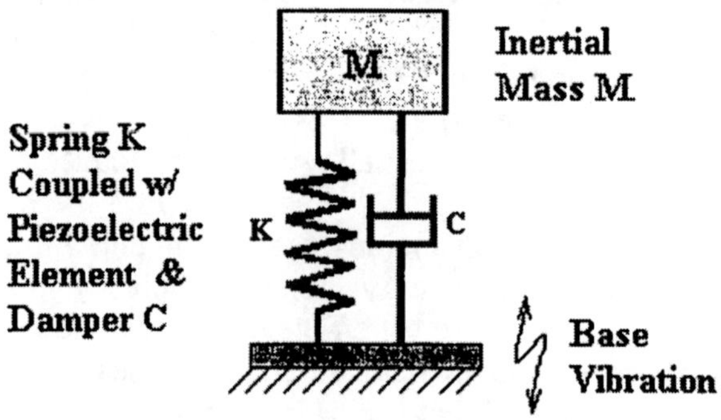

Figure 1. Dynamic model of a SDOF piezoelectric kinetic energy harvester

The total kinetic energy of a SDOF resonating system is expressed in term of its input acceleration A (peak),

$$E_{kinetic} = \frac{M_t \cdot A^2_{vibration}}{16\xi_t^2 \cdot \omega_o} \tag{1}$$

and in terms of its input displacement Y (peak),

$$E_{kinetic} = \frac{M_t \cdot Y^2_{vibration} \cdot \omega_o^3}{16\xi_t^2} \tag{2}$$

where M_t is the total effective inertial mass; ω_o is the vibration angular frequency; ξ_t is the system total damping coefficient [3,4,5].

From equations (1) and (2), it is shown that the total energy level is proportional to the inertial mass M_t, so the energy output can be scaled up or down by adjustment of the mass. For a system having smaller overall damping coefficient, it possesses a higher energy level. To illustrates the energy level of a SDOF system, let's look at a miniature SDOF resonator and assume it has the following parameters:

$M_t = 15\text{gram}, \quad \xi_t = 0.025, \quad A_{vibration} = 0.1\text{g peak}, \quad f_o = 60\text{Hz}$

Then the available kinetic energy is calculated by equation (1), **$E_{kinetic}$ = 3.83 mW**
This is a very high continuous energy level with such a small input vibration level. If one can harvest this energy, it could support a myriad of electronic devices. In fact, the final electric energy output can be simply expressed as,

$$E_{electric} = \eta_e \cdot E_{kinetic} \tag{3}$$

The goal for a designer is to achieve the highest transduction efficiency η_e.

We pointed out earlier that the piezoelectric coupling coefficients of the compression and shear modes are higher than that of the flexural mode. In this project, we intend to integrate the more efficient piezoelectric modes and materials in the harvester design. The first prototype of our harvester is shown in Figure 2. The so-called 'cymbal' composite structure exerts both compression and extensional stresses on the piezoelectric crystal when the base moves up and down; also it acts as the mechanical spring in the SDOF system.

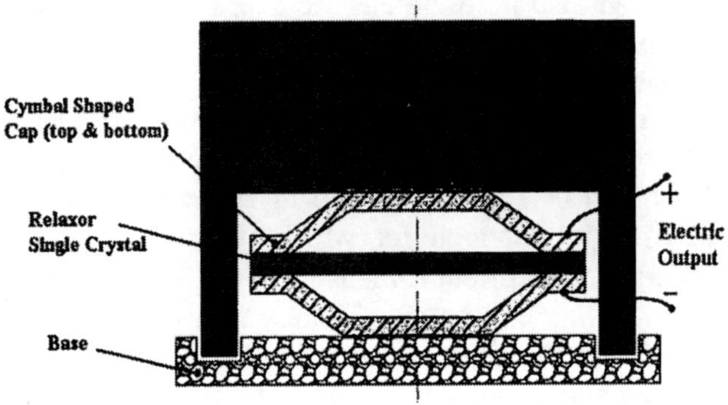

Figure 2. Schematic of Cymbal Harvester Design

The resonance of the cymbal stack can be changed through the variation of the cymbal cap dimensions. The resonance is ideally tuned to match with the vibration peak in the environment where the harvester will be installed. That way, the highest transduction efficiency is achieved.

Harvesting Circuitry and Power Storage: Piezoelectric material is highly capacitive and can be modeled as an active capacitor in an electric circuit model. The reactance of the piezoelectric element hinders the electric energy flowing out to load. Therefore, impedance compensation for the piezoelectric element is necessary to effectively extract the electric signal from the piezoelectric element. In addition, the electric power gathered from the piezoelectric transducer needs to be accumulated to significant level in order to support a connected wireless sensing system. Figure 3 depicts the system layout of the piezoelectric power harvesting system.

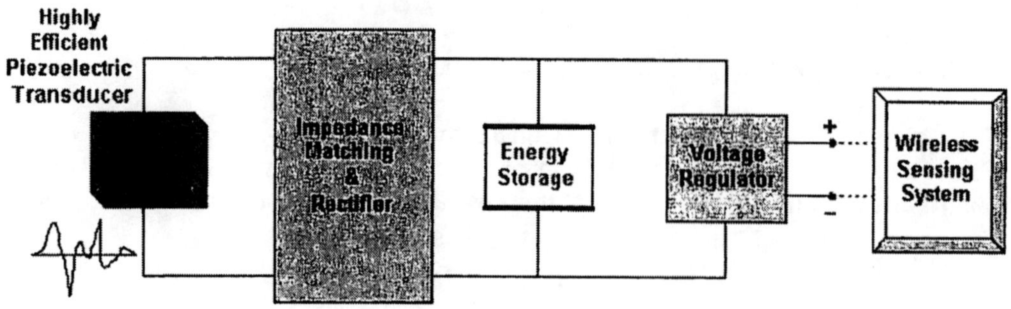

Figure 3. Architecture Layout of a Piezoelectric Power Harvesting System

There are tw

super capacitor or a rechargeable battery. Typically, the energy density of a super capacitor is smaller than 5 Whr/kg and the capacitor is able to hold the accumulated energy for days. Cost of super capacitor is low and its charging circuit is also very simple. Another method of storing electrical energy is the rechargeable battery. A battery can have an energy density between one and two orders of magnitude higher than a super capacitor. A battery can also hold the electric energy (lower self-leakage) for a much longer time than a capacitor does, usually for months. However, traditional battery has a limited number of charge-discharge cycles compared to a capacitor which can be charged and discharged indefinitely. Table 4 summarizes the performance characteristics of different existing battery types. The state-of-the-art battery is the thin-film solid state lithium battery. As shown in Table IV, this battery has the highest energy density and highest rechargeable cycles (enough for whole life service of an energy-harvesting device), it can be charged at a current level as low as 10's of μA, and it has a low profile (thin plate). The cost of this cutting edge battery is acceptable at 50 cents per square centimeter in high volumes. Based on all these factors, the thin film lithium battery seems ideally suited for energy harvesting storage applications.

Table IV. Characteristics of Current Rechargeable Batteries [8,9]

Type	**Ni-Cd**	**Ni-MH**	**Pb-Acid**	**Li-Ion**	**Li-Polymer**	**Thin-Film Solid State Lithium**
Energy Density (Whr/kg)	45-80	60-120	30-50	110-160	100-130	~200
Cycle Life	500	500	250	1000	500	> 60,000

The circuit model used in our prototype vibration harvester is shown in Figure 4. Because the power consumption of the wireless accelerometer used in this study is quite low (about 60mW for 20 seconds during transmitting period), a fully charged 0.2 farad super capacitor is able to provide for its power requirements. The charging and rectifying circuit for capacitors is shown in Figure 4. It is worthy to point out that the two single diode rectifiers in Figure 4 not only have a lower threshold voltage than the full bridge rectifier, but also doubles the voltage of the final output. The inductor L compensates the capacitive impedance of the piezoelectric transducer. The inductor L can also be replaced by an electronic version inductor, known as gyrator [6].

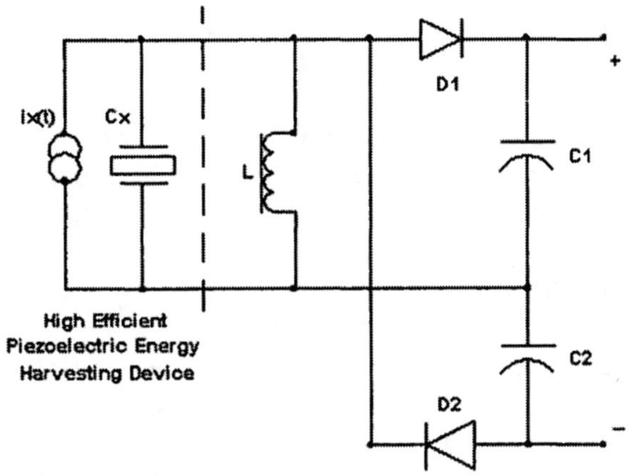

Figure 4. Schematic of Power Harvesting Circuit

Measurement and Conversion Efficiency: The frequency response of the prototype is first measured. Figure 5 presents the cymbal stack's response in the frequency domain at two extreme electrical loading conditions: open circuit and short circuit. Clearly, there is a resonant peak around 3600 Hz for the closed circuit load, and the low quality factor of the resonant peak indicates that a significant amount of mechanical energy is turned into electric energy (loss).

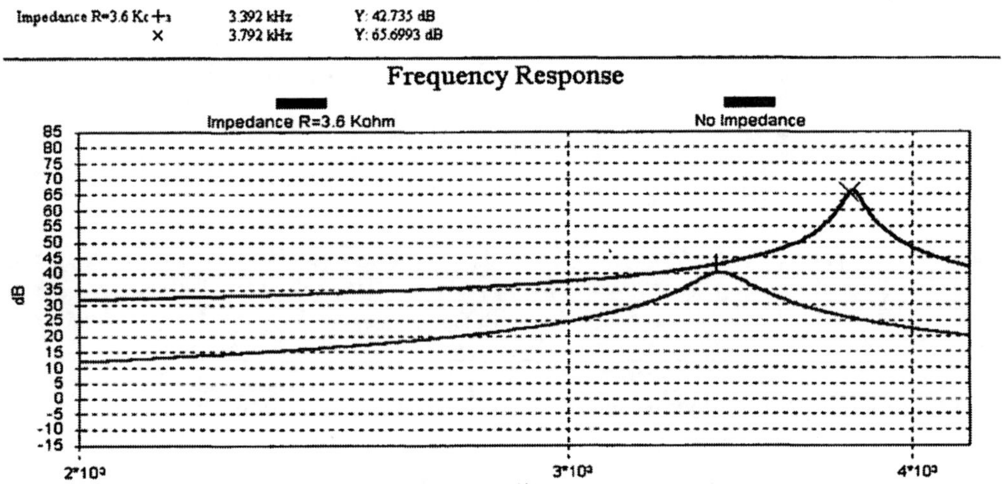

Figure 5. Frequency Response of the Cymbal Vibration Harvester

The effective transduction coefficient, k_{eff}, of a piezoelectric element can be calculated easily from the open circuit resonance f_a and close circuit resonance f_r using the following formula [7].

$$k_{eff} = \sqrt{1 - \left(\frac{f_r}{f_a}\right)^2} \tag{4}$$

If we substitute the measured resonance frequencies, in Figure 5, into the above equation, we find this cymbal prototype has achieved a coupling efficiency, k_{eff}, of **53%**. To our knowledge, this is the highest conversion efficiency achieved to date for a piezoelectric vibration harvester. We have demonstrated that our piezoelectric transducer coupled with our effective harvesting circuit is able to continuously light up an LED (~1 mW power consumption) or continuously power a thermometer under a vibration condition of 400Hz band white noise centered around 3.8 kHz and at vibration level **0.07g**.

To quantify the harvester's output, its electric output vs. vibration input level is measured. Figure 6 gives the measurement results. The high output electric power again validates the above obtained harvester conversion efficiency.

In most industrial environment, the vibration level is higher than 100mg. Based on our high efficiency, a continuous power output around 1mW can be easily obtained. If a wireless sensing system consumes 100mW for 20 seconds to transmit a measured data block, our harvester would need an accumulation period of 30 minutes to provide enough power for the operation. Polling a data from a RF sensor for a half-hour period is usually more than enough to meet the condition based health monitoring requirements.

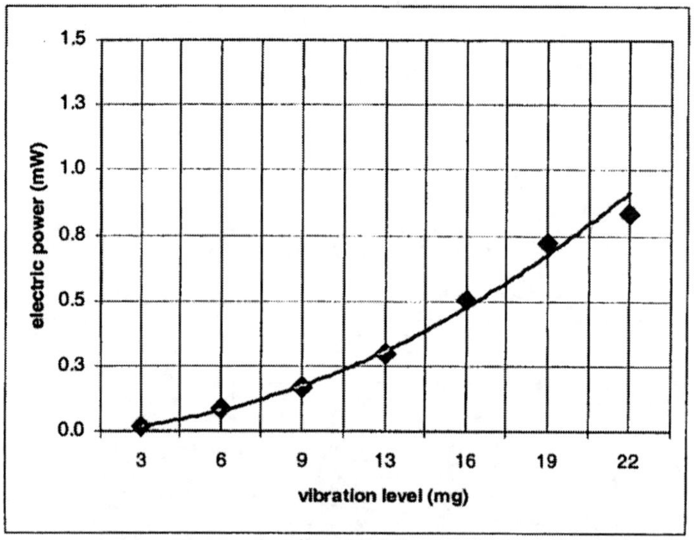

Figure 6. Electric Power Output of the Prototype vs. Vibration Levels
Load impedance: 3.6 kohm; Vibration signal: 3.2-3.6 band limited white noise

Next Generation Vibration Harvest Design: For many applications, the vibration energy is spread in the frequency domain or the major peak may vary in some extent. In these situations, it is ideal to have a broader frequency band of high conversion efficiency for vibration harvesting device so that a single device can cover a broad range of vibration conditions. Theoretically, a multiple degree of freedom dynamic system can be designed to have a wide band peak. Considering the cost and complexity of such system, a 2DOF piezoelectric resonator is more practical. A new version of vibration harvester is therefore proposed based on this concept. Figure 7 depicts the dynamic model of a 2DOF piezoelectric system [6].

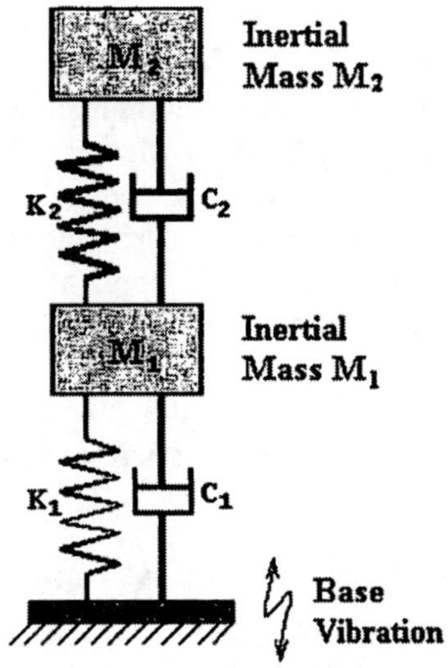

Figure 7. Mechanical Model of a Two-degree-of-freedom Piezoelectric Transducer

The theoretical simulation results of such a resonator are illustrated in Figure 8. The solid curve is the projected spectral response from the piezoelectric element, which is coupled with the spring element K_2.

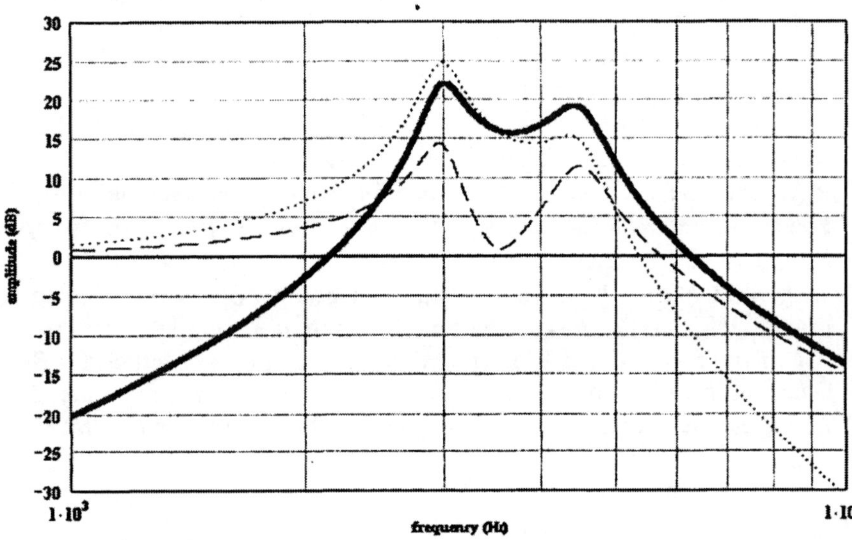

This new design is expected to have the equivalent high conversion efficiency as the cymbal design, but its high conversion efficiency range will be tripled in the frequency domain. Currently, a prototype of this new design is under construction.

Conclusion and Future Work: A compact and highly efficient vibration energy harvester has been devised and built. An unprecedented conversion efficiency of 53% has been achieved through the utilization of a highly efficient mode of the piezoelectric material, the proper transducer design, and an efficient electric energy extracting and storing circuit. The milli-Watt level power output from our vibration harvesting device would open the door for a variety of applications. Specifically, it will enable a fully autonomous, self-powered, wireless sensing system.

Future work will be to build and evaluated a wide-band version of our vibration harvester and to integrate our vibration harvester with a low power, wireless sensing system and evaluate it in an industrial environment.

Acknowledgement: We'd like to thank the Department of Energy (DOE) for sponsoring this project under grant no. DE-FG02-04ER84119. Also, we want to thank Techkor Instrumentation for providing their low power wireless acclerometer system.

References:

1. Christopher McLean, et al, "Wireless Data Gathering System for Condition Based Maintenance", 7th *Annual Reliability Conference*
2. Henry A. Sodano, et al, "A Review of Power Harvesting from Vibration Using Piezoelectric Materials", *The Shock and Vibration Digest*, 36, pp197-205, 2004
3. Shad Roundy, "Energy Scavenging for Very Low Power Wireless Sensor Nodes", Research Presentation, UC Berkeley, 2001
4. Paul D. Mitcheson, et al, "Architectures for Vibration-Driven Micropower Generators", *J. of MEMS*, Vol. 13, No.3, pp.429-440, June 2004
5. P Glynne-Jones, et al, "The Modelling of a Piezoelectric Vibration Powered Generator for Microsystems", *Transducers'01 and Eurosensors XV*, pp.46-49, June 2001
6. Ken Deng, et al, "Piezoelectric Vibration Energy Harvesting Device and Method", *Patent Pending*, Wilcoxon Research, Inc.
7. K.D. Rolt, et al, "Finite Element Modulus Substitution Method for Sonar Transducer Effective Coupling Coefficient", *Proceedings of the Third International Workshop on Transducers for Sonics and Ultrasonics,* Lancster, PA, 1993
8. http://www.batteryuniversity.com
9. Technical notes of Infinite Power Solutions

HYBRID CONDITION BASED MONITORING

Steve Wichtendahl and Ron Denton
Wilcoxon Research, Inc
21 Firstfield Rd, Gaithersburg, MD 20878
stevew@wilcoxon.com

Abstract: Vibration Monitoring is a critical component of any sound Predictive Maintenance (PdM) Strategy. Traditionally, data from the vibration sensors is recorded and analyzed using hand-held, walk-around vibration analyzers, or via expensive and elaborate online analysis systems. For an extremely large operations and/or very expensive equipment, these approaches represent cost effective PdM and have repeatedly shown to save money. In many applications, however, neither of these approaches is cost effective. In recent years we have seen an increased use of 4-20 mA Loop Powered sensors in the market. These sensors have proven extremely valuable, since they can be tied directly into the existing PLC/DCS Networks and do not require the skilled analyst or the expensive systems. However, these sensors only provide trending information to identify that a potential problem exists, and do not provide the detailed information necessary to determine what the cause of the problem. A hybrid approach to vibration monitoring provides the best of both worlds to establish a PdM strategy that is not only extremely cost effective but also simple to deploy.

Background: In the early days, equipment maintenance was conducted only when equipment actually failed. The work was more "fix it" than maintenance. Shortly thereafter, came the recognition that performing regular maintenance and refurbishment tasks on equipment could keep equipment operating longer between failures. This became known, variously, as Periodic Maintenance, Calendar Based Maintenance or Preventive Maintenance (PM). The goal was to have most of the equipment be able to operate most of the time until the next scheduled maintenance outage. This approach provided more control over the maintenance schedule; however, the system was still susceptible to failure between maintenance cycles.

With the development of the Fast Fourier Transform (FFT) mathematical method in 1964, portable spectrum analysis equipment became available that could be used with special transducers to measure machinery vibration. Early research showed that the mechanical condition of a machine could be deduced while it was still in operation. While this early FFT equipment was called portable, "transportable" might have been a better word to describe it. Early users often installed it in vehicles such as vans or Suburbans to move it to equipment and ran long cables to the equipment to make measurements. These efforts laid the groundwork for Predictive Maintenance (PdM) of machines.

When portable FFT-based data collectors became available in the 1980's the use of vibration as tool for diagnosing machinery faults underwent explosive growth. Tens of thousands of data collectors have been sold worldwide for vibration diagnosis. Vibration analysis technology has found its way into a variety of industries to include the petrochemical, paper, and power industries. Instead of waiting for a machine to fail before working on it, or performing maintenance on a machine regardless of its condition (PM), the idea of performing maintenance on equipment only when it indicates impending faults – Predictive Maintenance (PdM) – took hold. The idea of using PdM to perform maintenance on machines only when they exhibit signs of mechanical failure has become known as Condition Based Maintenance (CBM). This evolutionary process of machinery maintenance has allowed the maintenance operation to become more "proactive" than "reactive" in their maintenance tasking. The result is that the maintenance department can operate at lower costs because staffing and tasking can be better planned when the process of maintenance is under control.

Current CBM Technology: Today there are a variety of technologies for assessing machinery health for a complete CBM strategy. Because of the value of the information provided by, vibration analysis is one of the most prevalent methods for a variety of equipment and the primary method in rotational equipment.

One approach to data collection is the off-line walk-around approach. Portable data collection has many advantages and is the most cost effective in some instances, however it has two major drawbacks: it requires a skilled vibration analysts to interpret the data and, without continuous monitoring, problems in between rounds could be costly. In an attempt to overcome the later problem, industry increased the number of points measured and the frequency of the measurements. However, this is a very labor-intensive approach and in recent years as we move to a global economy, all manufacturing plants have undergone tremendous efforts to reduce staff in cost cutting measures. Hence emerges the Online Vibration Analysis system as a second major approach to CBM today.

On-line systems are now widely used for monitoring vibration with piezoelectric accelerometers as the vibration probes and computer-based specialized data acquisition systems for gathering, storing, and archiving FFT data. These FFT-based data systems exist apart from the plant process control computer system because the data files are so massive and specialized. It still requires trained vibration analysts to view, interpret, and analyze the data collected. The system installation costs are high also because of the capital cost of the hardware as well as the installation labor.

The main advantage of an on-line dynamic vibration monitoring system is that is acquires data continuously. Most of the on-line systems use some kind of acquisition system architecture that involves a single input channel that is multiplexed across many vibration sensors. This results in a scan rate that varies according to the system scheduler. However, most systems will acquire a full set of time and spectrum data for each monitored point about once every three to six hours. The result is four to eight sets of data each day, meaning there are 1,500 to 3,000 data sets acquired for each point every year. This is extremely detailed data! However, because of modern computing storage and power, this no longer represents a serious problem.

Another advantage to an on-line dynamic vibration monitoring system is that there is no labor cost to acquiring the data and minimal labor cost for identifying machine faults.

The system acquires data 24 hours a day. The resulting data is "sorted" based algorithms established by the skilled vibration analyst. Problems are "flagged" by the system and identified on the screen of the control computer.

The continuous monitoring advantage of an on-line dynamic vibration monitoring system enables mechanical fault detection at the earliest detectable time. When plants either need or want to be able to detect machinery faults at the earliest possible moment, on-line dynamic vibration monitoring systems will be the system of choice.

The disadvantages of on-line dynamic vibration monitoring systems are that they are the more costly CBM systems to implement and maintain. The cost of acquiring an expensive elaborate online system, maintaining a full time vibration analyst, and installing a wired network to get the signals from the sensor to the analysis system is significant. Furthermore, the software and hardware that make up the system typically require an extensive maintenance contract as well. Many plants are downsizing and aggressively becoming more cost competitive in the global marketplace, and so many plant managers have resisted the move to elaborate online vibration monitoring systems or avoided implementing a CBM system altogether.

The Critical Component – The Vibration Analyst: Whether your CBM operation requires a walk-around analysis system, a dynamic on-line system or both, a full time vibration analyst is a critical to the support of the CBM effort.

It takes considerable time to become a skilled vibration analyst. Many years of training and experience are necessary to properly apply the techniques of vibration analysis to the wide variety of machines and equipment installed in the plants today. The maintenance department also invests considerable resources in time and training to equip the mechanics and technicians that perform the maintenance on the plant's equipment. In some cases the personnel performing the vibration analysis are the same ones who perform the equipment maintenance. They are not the highly skilled Vibration Analysts, but able to perform the vibration analysis necessary to maintain their particular machines. The skills and equipment necessary to maintain an "in-house" motor repair facility serve as an example. Furthermore, well-trained, experienced Vibration Analysts are a dying breed.

New Age Economics: Implementing and maintaining the technologies necessary to have a totally "in-house" PdM program is a very costly investment. As most companies try to improve their cost effectiveness, outsourcing of services has become a way to achieve cost reduction targets.

Today, the Vibration Analysts can either be on the plant staff or can be an external contracted analyst. Many plants have moderately skilled analysts, but contract the services of vibration consultants when serious problems develop with critical and expensive plant machines. Other plants will contract the entire vibration analysis program to consulting firms and take action based on the report of the consulting analyst.

The size of the plant and the requirement for rapid data analysis will help determine the right mix of in-house versus contracted operations. It will also determine the level of in-house expertise that would be needed to retain some measure of rapid analysis even when most of the data acquisition and analysis is contracted. If there is absolutely no in-house source of independent analysis, the plant is totally at the mercy of the contractor for direction as to equipment condition assessment. The question is what level of internal expertise to maintain so as to be able to effectively use the contractor personnel and data.

Cost Effective Options are Available for CBM Systems: While these CBM systems were driving the evolution of modern plant maintenance programs, real-time plant control computer systems were evolving as well to better monitor and control all plant processes. Over a period of many years, control systems in plants have evolved from the air-powered systems of the past to modern computer control systems.

These systems have evolved from two directions. Older local equipment relay ladder controls were replaced by small Programmable Logic Control (PLC) systems. Meanwhile, the total plant air operated system was being replaced by an integrated, computerized Distributed Control System (DCS) to maintain the operation of the plant equipment and coordinate the flow of the production process. As a result of time and increases in computing power, these two on-line control techniques have gradually merged into a variable-scale Process Control System (PCS).

Although only a small percentage of plants have an installed cabled network for on-line vibration analysis system, most plants have an existing PCS network. Being able to access these existing networks would produce a cost-effective alternative to dynamic on-line vibration analysis systems.

4-20 mA Vibration Data: To interface to existing PCS networks, the vibration data must be converted to a standard PCS format. One of the most common formats employed today is a 4-20 mA current loop signal. The current loop signal standard was established due to its noise immunity performance and the fact that it can be transmitted over very long distances without loosing its integrity. The lowest signal level was standardized at 4 mA to ensure that non-operating conditions were differentiated at lower current levels below 4 mA. And the top end of the scale was standardized on 20 mA to provide enough range to address almost any process.

The 4-20 mA current loop is a low bandwidth system that can not support the transfer of the dynamic vibration data necessary for FFT Analysis. Therefore, the PCS network can never be employed as a direct replacement to the dynamic on-line vibration analysis system. Alternatively, many maintenance engineers have recognized, in recent years, that using a 4-20 mA network to transfer less dynamic information about equipment vibration is a very low cost way to trend the system's general condition. This results in a more cost-effective approach to equipment maintenance than traditional PdM.

Today vibration transducers can incorporate the signal processing to produce a conditioned vibration signal that can be transferred on a standard 4-20mA PCS network.

These devices can be directly powered from the plant DCS or PLC system and yield an overall 4-20 mA signal proportional to the machine vibration level.

Vibration transmitters are also available that utilize the dynamic accelerometers traditionally used by portable data collectors and on-line FFT-based vibration systems. These modules are DIN-rail mounted and convert the dynamic signal to a 4-20 mA loop-powered signal.

Modern plant DCS/PLC systems can accept the 4-20 mA signal as input, and interpreting the overall vibration signal is simple. The signal is a calculation related to the overall energy of the vibration, and ISO 10816 is an industry standard that provides guidelines to determining the machine condition based on the overall vibration level[1]. If the user knows the manufacturers recommended maximum vibration level, that can be used to establish limits to the allowable vibration. Otherwise, ISO 10816 can be used for guidance.

Vibration Velocity in/sec. peak (mm/sec. r.m.s.)	Class I < 20 HP	Class II 20 to 100 HP	Class III > 100 HP typ. rigid rotor	Class IV > 100 HP typ. flexible rotor
2.5 (45)	D	D	D	D
1.6 (28)	D	D	D	D
1.0 (18)	D	D	D	C
0.63 (11.2)	D	D	C	C
0.4 (7.1)	D	C	C	B
0.25 (4.5)	C	C	B	B
0.16 (2.8)	C	B	B	A
0.1 (1.8)	B	B	A	A
0.063 (1.12)	B	A	A	A
0.04 (0.71)	A	A	A	A
0.025 (0.45)	A	A	A	A
0.006 (0.112)	A	A	A	A

The advantage of using 4-20 mA based vibration data is that there is no complex analysis to perform. The data is the overall vibration of the machinery. While the overall vibration level does not change significantly in the early stages of machinery mechanical fault development, it will change significantly for most machine faults during the last three months prior to failure. The overall vibration level does not contain sufficient information to determine the specific machine fault, but it can serve well as a "pointer" to problem machines. The long-term trend data will confirm that a problem is developing as it will show that a true increase in the vibration occurring. An astute mechanical maintenance department can generally find the specific machine fault once alerted to a change in the overall vibration level.

The main disadvantage of 4-20 mA vibration data is that it contains no specific pointers to problems. The overall vibration can usually only point to the machine. Multiple sensors on machinery "sets" such as a motor-pump or motor-fan set can aid in focusing on the component of a machine set. With sufficient sensors or with band-limited signal processing, certain types of problems can be diagnosed. However, these implementations will also increase the monitoring system cost.

Sensors for 4-20 mA Process Loops: The transducers that are available for vibration sensing offer a wide range of monitoring options. They can output 4-20 mA loop current as a velocity of vibration signal, or as a less-useful acceleration signal. The transducers use a piezoelectric acceleration sensing element as the heart of the sensor and include electronic circuitry, powered from the 24 Volt DC power of the process system.

---

[1] For additional assistance see the "Application Guide for PC420-series LPS® Transducers", Wilcoxon Research, Inc.

The internal circuitry allows these sensor to be configured in a variety of ways. They usually offer the output calibrated in true root-mean-square (rms) amplitude or in peak output. The peak output is usually just a derivative of the processed rms signal, equal to 1.414 times the calculated rms as this is the ratio of peak to rms for a sinusoidal waveform. Wilcoxon also offers a 4-20 mA output that is "true peak" and it actually captures the real peak amplitude of the signal being measured. The sensors can be customized with built-in high-pass and low-pass filters to focus on just the range of frequencies of interest to the user, offering trending of specific types of vibration problems.

As the use of these loop-powered sensors has become popular the demand to be able to install and operate them in hazardous classified areas has also increased. 4-20 mA sensors are available as Intrinsically Safe certified units as well as in Explosion-Proof housings. This allows their use in those areas classified under NEC® section 500 as classified hazardous areas.

The real advantage of using the 4-20 mA loop-powered sensors is that they are simple to install and wire. They are usually mounted using a single ¼-28 threaded stud into a hole drilled and tapped into the machine case. The wiring is also very simple. Using shielded, twisted pair wires one conductor brings the 24-Volt power to the sensor, the other carries the 4-20 mA signal back to the DCS/PLC analog 4-20 mA input.

Transmitter Modules for 4-20 mA Process Loops: In some plants, accelerometers have already been installed permanently on machine to aid the vibration data collection process. These are often located on equipment in hard-to-reach areas or for safety reasons. These already existing sensors can also be used for gathering 4-20 mA vibration signals. The transmitters are powered from 24-Volt DC supplies and output a processed signal quite similar to that produced by the 4-20 mA sensors.

These modules offer more power and space for the signal processing than the sensor version. The transmitters can also implement stronger filtering with more filter options than the sensor. Whether the user wants to trend the entire vibration spectrum of signals or focus on a small range within the possible spectrum, the transmitter modules give more options for the user.

In addition to the advantage of being able to use existing installed dynamic accelerometers, the transmitter modules can be tailored for the frequency range of interest to alert specific machine faults. Multiple modules can easily use the same accelerometer to process different frequency bands for a comprehensive fault identification. Using multiple modules also allows developing both acceleration and velocity 4-20 mA loop signals.

One disadvantage of transmitter modules is that they require power to operate. Either AC or DC power must be available. Depending on where they are mounted, this may not be an issue.

A Hybrid Monitoring System: Using vibration trending data from 4-20mA sensors and vibration transmitters is a cost effective alternative to traditional PdM programs. In some applications, i.e., spared equipment and non-essential equipment, 4-20mA trending provides an excellent total cost effective solution. However, in more critical applications where an effective CBM System is required, analysis of the dynamic vibration data is essential.

An ideal alternative would be a hybrid approach that leverages the best of both options, the cost-effective approach of vibration trending and the more complete FFT Dynamic Analysis approach. This is now possible through a hybrid approach that employs dual output devices, either a sensor or Vibration Transmitter, as shown below. The 4-20 mA vibration sensors and the Vibration Transmitters are now available with the 4-20 mA processed output for the DCS/PLC, as well as a "buffered" dynamic signal output available for the Vibration Analyst's portable data collector.

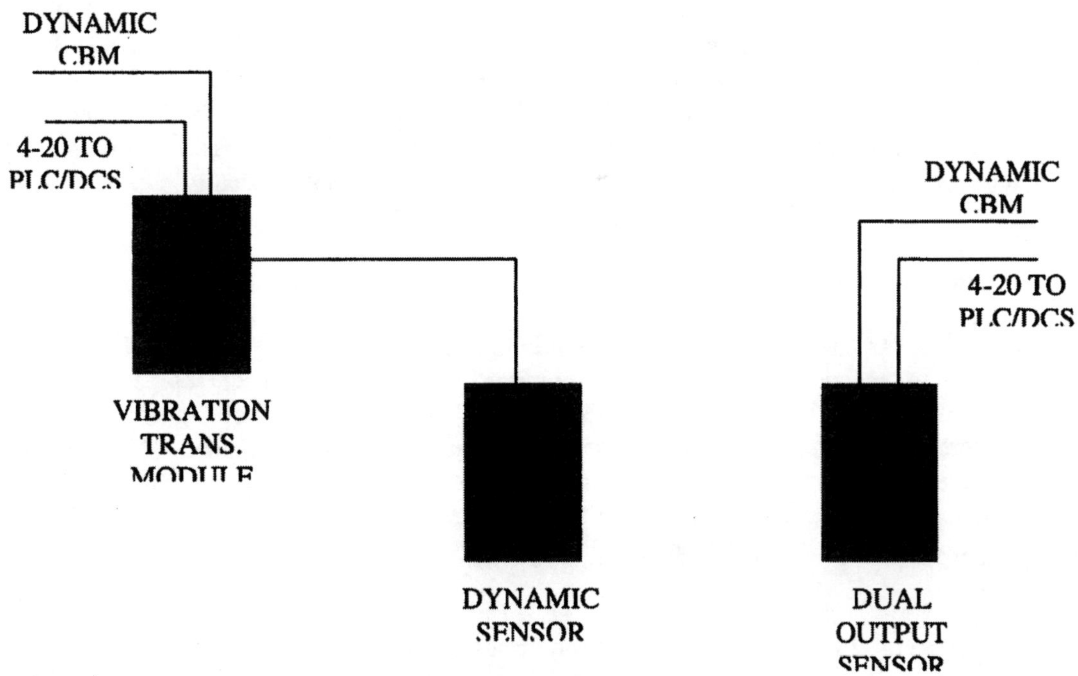

The dynamic buffered output signal allows traditional vibration analyses to be performed. This means the one sensor can perform two monitoring functions. It can trend the overall equipment condition through the DCS/PLC system and the dynamic data is available for use by the Vibration Analysts as a permanently mounted sensor for the further analysis. Besides the buffered dynamic output, some models also have a third option for temperature measurement as well. In those cases the user can have both the buffered dynamic signal and the temperature signal. Wilcoxon offers the triple output option (4-

20, dynamic, temperature) sensor in a low-profile side-exit connector case style. With the addition of the temperature signal, the sensor now can implement temperature trending with vibration. This is a powerful complementary tool for confirming a developing mechanical fault.

The hybrid implementation presents a win-win opportunity for plants. Essentially this provides continuous monitoring of equipment with early warning of pending trouble in sufficient time to employ contract Vibration Analyst to collect and analyze buffered dynamic data on an 'as needed' basis. This truly cost-effective system will allow CBM to expand its market into new industries and equipment for much more cost-effective plant operation.

Even in plants with vibration analysis staff, having an on-line 4-20 mA sensor based monitoring system can aid the vibration program. Machines are often identified that just seem to be "bad actors" and will develop problems more quickly than can be detected on the normal route-based monitoring. Having 4-20 sensors wired to the plant DCS/PLC will maintain vibration monitoring coverage during those intervals between the regular route data collection.

Using 4-20 mA vibration sensors or transmitters with a DCS or PLC system allows plants the ability to have overall vibration data on a very frequent basis. The data can then be trended to allow the system to alarm when trends indicate a problem developing. Wilcoxon Research, Inc. also has an application guide available to assist in interpreting the meaning of the overall levels and setting alarms and automatically editing data to eliminate "wild points" in the data that can cause false alarms.

Having vibration data available on-line for critical machines can allow the maintenance management staff to always be aware of the general vibration condition of those critical machines. As the machines that need monitoring change, the system can be adapted readily to account for the need to monitor other or more machines. The data is straightforward and easy to interpret. The flexibility of 4-20 mA loop signals allows the sensors to be re-positioned where necessary or the system to be expanded. The result is a flexible system that can be used to manage risk regarding machinery health condition monitoring. Managing the risk while minimizing the associated cost of data is a prime objective for modern plant monitoring systems.

Summary: 4-20 mA loop transducers and transmitter offer plants the ability to monitor vibration on a continuous basis. The 4-20 mA loop signal is the de facto standard for plant process control signals. Where plants already have existing process control systems (PLC or DCS) vibration can be added without the need for resident technical staff to assist in data interpretation. While on-line dynamic systems offer the advantage of detailed diagnostic ability, they come with the added cost of requiring an analyst to interpret the data. Off-line dynamic signal monitoring (walk-around data collector) systems also offer detailed diagnostic ability with a low entry cost, but with a very high continuing annual cost.

Modern lean management can take advantage of a hybrid monitoring technique to implement on-line 4-20 mA vibration monitoring of plant machines and employ the services of an experienced Vibration Analyst only when a pending problem is identified. This is an ideal approach for an extremely cost-effective CBM System.

AUTHOR INDEX

Author Index